Phytopharmaceuticals for Brain Health

T0315258

Phytopharmaceuticals for Brain Health

Edited by
Shahnaz Subhan, PhD
Applied Biodiversity Company LLC, Jersey City, New Jersey, USA

Manashi Bagchi, PhD, FACN
Dr. Herbs LLC, Concord, California, USA

CRC Press
Taylor & Francis Group
Boca Raton London New York

CRC Press is an imprint of the
Taylor & Francis Group, an **informa** business

CRC Press
Taylor & Francis Group
6000 Broken Sound Parkway NW, Suite 300
Boca Raton, FL 33487-2742

First issued in paperback 2021

© 2017 by Taylor & Francis Group, LLC
CRC Press is an imprint of Taylor & Francis Group, an Informa business

No claim to original U.S. Government works

ISBN-13: 978-1-03-209632-2 (pbk)
ISBN-13: 978-1-4987-5767-6 (hbk)

Visit the Taylor & Francis Web site at
http://www.taylorandfrancis.com

and the CRC Press Web site at
http://www.crcpress.com

Dedication

Dedicated to my well-respected and beloved parents, Maulana Abdul Subhan and Farida Begum, for always giving me inspiration; and my two beloved young sons, Ehtesham Suhail and Mujtaba Suhail, for helping me by their cute smiles and being patient while I was working on this esteemed book.

Shahnaz Subhan

Dedicated to my beloved husband, Debasis Bagchi, and our only daughter, Dipanjali Bagchi, and my mother, Bakul Bardhan.

Manashi Bagchi

Contents

SECTION I Pathophysiology

SECTION II Phytopharmaceuticals

SECTION III   Molecular Mechanisms

SECTION IV   Autism

Preface

Research studies have demonstrated that appropriate nutrition, in conjunction with mental, social, and physical activities, may have a greater benefit in maintaining or improving brain health. Diet and appropriate nutrition, cognitive activity, social engagement, and regular physical exercise can significantly help in improving brain health with advancing age and potentially reduce the risk of cognitive decline.

Nutraceuticals and functional foods work against neurodegenerative diseases, which are associated with exacerbated oxidative stress in the central nervous system. The fundamental "neurohormesis" principle will also be discussed in this book.

Advancing age can exhibit health-related challenges that may take a toll emotionally, financially, and physically. Furthermore, regular stress and environmental pollution are challenging problems. There is no easy or quick solution. Recently, at the International Conference of Alzheimer's Association in 2014, a two-year clinical trial on older adults at the risk of cognitive impairment demonstrated that a combination of physical activity, proper nutrition, cognitive training, social activities, and management of heart health risk factors slowed cognitive decline.

Research studies with a number of phytopharmaceuticals and medicinal plants demonstrated the efficacy of huperzine A, berry anthocyanins, *trans*-resveratrol, *Bacopa monniera*, *Centella asiatica*, *Curcuma longa*, flavonoids tocotrienols, and palm oil in boosting brain health and physical well-being. A chapter is dedicated to autism treatment with psychotherapy, nutrition and dance movement, a challenging and upsetting problem of the millennium.

Also, consumption of marine fishes and general seafood has been recommended for long-term nutritional intervention to preserve mental health and delay neurodegenerative processes, and to sustain cognitive health in humans. Omega-3 and omega-6 polyunsaturated fatty acids and antioxidants prevent the initiation and progression of many neurological disorders. Several phytochemicals have shown promising results against free-radical-promoted neurodegenerative processes and cognitive impairment.

Overall, this book will bring a classic scenario of neurological problems and their possible amelioration using novel nutraceuticals and functional foods.

About the Editors

Shahnaz Subhan, PhD, is a chief scientific officer at Applied Biodiversity Company, LLC, in Jersey City, New Jersey. She has been working as a researcher and academician with 20 years of experience in plant, medical, and microbial biotechnology. She is also working for the Jersey City Board of Education, New Jersey. Previously, she was an assistant professor at the Amity University in Noida, India, where she taught courses in biotechnology and worked almost 10 years in the same university. She was awarded a Young Scientist Fellowship in India for a government-funded major scientific project by DST. During her tenure in Amity University, she also got a short-term postdoctoral fellowship at The Institute for Molecular Medicine, California. She holds a PhD in botany (plant biotechnology) from the University of Delhi at Delhi, India, and a master's degree in botany with major in microbial molecular genetics and microbial ecology and a BSc in botany (honors) with zoology and chemistry from the University of Delhi at Delhi, India.

Her research areas include the following: plant and medical biotechnology, phytochemistry, applied microbiology, and environmental sustainability. She has published in various peer-reviewed and leading journals, including *Plant Cell Report, Journal of Medical Science, Asian Journal of Chemistry, British Biotechnology, Journal of Food Process Technology, International Journal of Pharma and Bioscience*, CRC Press/Taylor & Francis, etc. She has also presented several research papers in international conferences. She has more than 20 years of experience in research, teaching, and supervising hundreds of undergraduate and postgraduate biotechnology and science students. Under her supervision, PhD thesis was also awarded to a PhD student in 2011.

Dr. Subhan, a mother of two young sons, lives with her family in Jersey City, New Jersey.

Manashi Bagchi, PhD, FACN, earned her PhD degree in chemistry in 1984. Dr. Bagchi is currently the chief scientific officer of Dr. Herbs LLC, Concord, California. Dr. Bagchi is also a consultant for Cepham Research Center, Piscataway, New Jersey, and Purity Products, Plainview, New York. Dr. Bagchi served as associate professor in the Creighton University School of Pharmacy and Health Professions, Omaha, Nebraska, from September 1990 to August 1999. Later, she served as the director of research at InterHealth Nutraceuticals, Benicia, California, from September 1999 to July 2009. Dr. Bagchi is a member of the Study Section and Peer Review Committee of the National Institutes of Health, Bethesda, Maryland. Her research interests include free radicals, human diseases, toxicology, carcinogenesis, anti-ageing and anti-inflammatory pathophysiology, mechanistic aspects of cytoprotection by antioxidants and chemoprotectants, regulatory pathways in obesity and gene expression, diabetes, arthritis, and efficacy and safety of natural botanical products and dietary supplements. She is a member of Society of Toxicology (Reston, Virginia), New York Academy of Sciences (New York, New York) and Institute of Food Technologists (Chicago, Illinois). She is a fellow and currently a board member of the American College of Nutrition (Clearwater, Florida). Dr. Bagchi has 225 papers in peer-reviewed journals and two books, *Genomics, Proteomics and Metabolomics in Nutraceuticals and Functional Foods* and *Bio-Nanotechnology: A Revolution in Food, Biomedical and Health Sciences* from Wiley-Blackwell. She has delivered invited lectures in various national and international scientific conferences, organized workshops, and group discussion sessions. Dr. Bagchi is serving as editorial board member of the *Journal of the American College of Nutrition*, and is a reviewer of many peer-reviewed journals. Dr. Bagchi received funding from various institutions and agencies including the U.S. Air Force Office of Scientific Research, National Institute on Aging, National Institute of Health, Nebraska State Department of Health, and Cancer Society of Nebraska.

Contributors

Cesarettin Alasalvar
Food Institute
TÜBİTAK Marmara Research Center
Gebze-Kocaeli, Turkey

Agarwal Amit
Research & Development Center
Natural Remedies
Bangalore, India

Debasis Bagchi
Cepham Research Center
Concord, California

and

University of Houston College
of Pharmacy
Houston, Texas

Manashi Bagchi
Dr. Herbs LLC
Concord, California

Theeshan Bahorun
ANDI Centre of Excellence for
Biomedical and Biomaterials
Research
CBBR
University of Mauritius
Réduit, Republic of Mauritius

Sabia Bano
Department of Biotechnology
Singhania University
Jhunjhunu, Rajasthan, India

Kartik Baruah
Laboratory of Aquaculture & Artemia
Reference Center
Department of Animal Production
Faculty of Bioscience Engineering
Ghent University
Ghent, Belgium

Bethapudi Bharathi
Research & Development Center
Natural Remedies
Bangalore, India

Peter Bossier
Laboratory of Aquaculture & Artemia
Reference Center
Department of Animal Production
Faculty of Bioscience Engineering
Ghent University
Ghent, Belgium

Leonid Breydo
Department of Molecular Medicine
and Byrd Alzheimer's Institute
Morsani College of Medicine
University of South Florida
Tampa, Florida

Sanjoy Chakraborty
Department of Biological Sciences
New York City College of Technology
The City University of New York
(CUNY)
Brooklyn, New York

**Chinampudur Velusami
Chandrasekaran**
Research & Development Center
Natural Remedies
Bangalore, India

Sui Kiat Chang
Department of Nutrition and Dietetics
School of Health Sciences
International Medical University
Kuala Lumpur, Malaysia

Chien-Fu Fred Chen
Graduate Institute of Life Sciences
National Defense Medical Center
Taipei, Taiwan

Majeedul H. Chowdhury
Department of Biology
Touro College & University System,
   Flatbush Campus
Brooklyn, New York

Soheli A. Chowdhury
Department of Biology, York College
The City University of New York
   (CUNY)
New York City, New York

Mundkinajeddu Deepak
Research & Development Center
Natural Remedies
Bangalore, India

Tom Defoirdt
Laboratory of Aquaculture & Artemia
   Reference Center
Department of Animal Production
Faculty of Bioscience Engineering
Ghent University
Ghent, Belgium

Andrea I. Doseff
Department of Molecular Genetics and
   Department of Physiology and Cell
   Biology
Davis Heart and Lung Research
   Institute
The Ohio State University
Columbus, Ohio

and

Department of Physiology
Michigan State University
East Lansing, Michigan

Bernard W. Downs
VNI Inc
Lederach, Pennsylvania

Shameem Fawdar
ANDI Centre of Excellence for
   Biomedical and Biomaterials
   Research
CBBR
University of Mauritius
Réduit, Republic of Mauritius

Charles Glabe
Department of Molecular Biology and
   Biochemistry
University of California
Irvine, California

Devya Gurung
Department of Biological Sciences
New York City College of Technology
The City University of New York
   (CUNY)
Brooklyn, New York

Noriko Hattori
Nagaragawa Research Center
API Co., Ltd.
Nagara, Gifu, Japan

Syed Ahktar Husain
Department of Biosciences
Jamia Millia Islamia
New Delhi, India

Kenji Ichihara
Nagaragawa Research Center
API Co., Ltd.
Nagara, Gifu, Japan

Abdul Ilah
Faculty of Medical Technology
University of Tobruq
Tobruq, Libya

Faisal Ismail
Faculty of Medical Technology
University of Tobruq
Tobruq, Libya

Shigeru Katayama
Department of Bioscience and
    Biotechnology
Shinshu University
Japan

Emily Kawesa-Bass
Department of Pharmacy Sciences
School of Pharmacy and Health
    Professions
Creighton University
Omaha, Nebraska

Monowar Alam Khalid
Department of Environmental Science
Integral University
Lucknow, U.P., India

Savita Khanna
Department of Surgery
The Ohio State University Wexner
    Medical Center
Columbus, Ohio

I-Hsun Li
School of Pharmacy
National Defense Medical Center
Taipei, Taiwan

and

Department of Pharmacy
Taichung Armed Forces General
    Hospital
Taichung, Taiwan

Eva S.B. Lobbens
Department of Pharmacy
Faculty of Health and Medical Sciences
University of Copenhagen
Copenhagen, Denmark

Mohd Maqbool Lone
Department of Radiation Oncology
SKIMS
Srinagar, Kashmir, India

Odete Mendes
Product Safety Labs
Dayton, New Jersey

Leah Mitchell-Bush
Department of Pharmaceutical Sciences
College of Pharmacy and Health
    Sciences
Texas Southern University
Houston, Texas

Hiroyoshi Moriyama
The Japanese Institute for Health Food
    Standards
Hongo, Tokyo, Japan

Nithyanantham Muruganantham
Research & Development Center
Natural Remedies
Bangalore, India

Rama Nair
Nutriwyo LLC
Wyoming Technology Business Center
Laramie, Wyoming

Sreejayan Nair
University of Wyoming, School
    of Pharmacy
College of Health Sciences
and
Center for Cardiovascular Research
    and Alternative Medicine
University of Wyoming
Laramie, Wyoming

Soichiro Nakamura
Department of Bioscience
    and Biotechnology
Shinshu University
Japan

Yukio Narita
Nagaragawa Research Center
API Co., Ltd.
Nagara, Gifu, Japan

Darakhshanda Neelam
Department of Biosciences
Jamia Millia Islamia
New Delhi, India

Vidushi Shradha Neergheen-Bhujun
Department of Health Sciences
Faculty of Science
and
ANDI Centre of Excellence for
    Biomedical and Biomaterials
    Research
CBBR
University of Mauritius
Réduit, Republic of Mauritius

Ya Fatou Njie-Mbye
Department of Pharmaceutical Sciences
College of Pharmacy and Health
    Sciences
Texas Southern University
Houston, Texas

Parisa Norouzitallab
Laboratory of Aquaculture & Artemia
    Reference Center
and
Laboratory of Immunology and Animal
    Biotechnology
Department of Animal Production
Faculty of Bioscience Engineering
Ghent University
Ghent, Belgium

Sunny E. Ohia
Department of Pharmaceutical Sciences
College of Pharmacy and Health
    Sciences
Texas Southern University
Houston, Texas

Odochi I. Ohia-Nwoko
Department of Psychology
    and Texas Institute for Measurement,
    Evaluation, and Statistics
University of Houston
Houston, Texas

Catherine A. Opere
Department of Pharmacy Sciences
School of Pharmacy and Health
    Professions
Creighton University
Omaha, Nebraska

Arti Parihar
Department of Biological Sciences
Government PG College of Excellence
Ujjain, MP, India

Silia Rafti
10 Stasandrou, Apt 202
Nicosia 1060
Cyprus

Antonios C. Raftis
10 Stasandrou, Apt 202
Nicosia 1060
Cyprus

Chandrasekaran Prasanna Raja
Research & Development Center
Natural Remedies
Bangalore, India

Suhail Rasool
College of Human Sciences
Auburn University
Auburn, Alabama

Jenaye Robinson
Department of Pharmaceutical Sciences
College of Pharmacy and Health
    Sciences
Texas Southern University
Houston, Texas

Zaina Bibi Ruhomally
Department of Health Sciences
Faculty of Science
University of Mauritius
Réduit, Republic of Mauritius

Syed Monowar Alam Shahid
College of Medicine
University of Hail
Hail, Saudi Arabia

Jui-Hu Shih
Department of Pharmacy Practice
Tri-Service General Hospital
and
School of Pharmacy
National Defense Medical Center
Taipei, Taiwan

Hiroshi Shimoda
Oryza Oil & Fat Chemical Co. Ltd.
Ichinomiya, Aichi, Japan

Benjamin A. Stancombe
Department of Pharmacy Sciences
School of Pharmacy and Health
    Professions
Creighton University
Omaha, Nebraska

Richard Stewart
Department of Surgery
The Ohio State University Wexner
    Medical Center
Columbus, Ohio

Shahnaz Subhan
Applied Biodiversity Company LLC
Jersey City, New Jersey

Anand Swaroop
Cepham Research Center
Piscataway, New Jersey

Tanveera Tabasum
Department of Zoology
Srinagar, Kashmir, India

Akhtar uz Zaman
Department of Anatomy
National Medical College
Beniapukur, Kolkata, India

Section I

Pathophysiology

1 A Hidden Etiological Nemesis of Chronic Neurodegenerative Diseases

Bernard W. Downs and Manashi Bagchi

CONTENTS

> To find a solution, you must first be able to (accurately) state the problem.
>
> **Albert Einstein**

1.1 DEFINING THERAPEUTIC PARADIGMS: HISTORICAL OVERVIEW

Neurodegenerative disorders rank among the most disruptive, challenging, and burdensome maladies for which afflicted individuals, their loved ones and medical institutions contend. There are more published reports on the genetic influences predisposing neurodegenerative disease than can be referenced here. A simple literature search will inundate the researcher with a plethora of publications. However, while there is no lack of competent scientific opinion, a complete understanding of the causes and effective treatments of neurodegenerative disorders, such as early onset Alzheimer disease (EOAD), among others, remains elusive. For example, a recent review paper (as of the writing of this chapter in 2016) is focused on exploring new avenues in translational research and therapeutic discoveries in EOAD [1]. The researchers conclude by commenting on the relevance of reinvestigating EOAD patients as a means to explore potential new avenues for translational research and therapeutic discoveries. Translational research can help uncover genetic targets for therapeutic interventions. Although immensely valuable, similar to translational research in cancer, for example, unraveling and understanding molecular genetics continue to be inadequate to effectively eradicate this terrible disease. And, unfortunately, while continued research in this direction could illuminate therapeutic benefits, pursuing a pharmacological premise will not likely reveal a cure, as the disease is not caused by a drug deficiency.

Pharmacological interventions for some neurodegenerative disorders have met with frustrating side effects. For example, in regard to Parkinson disease, dyskinesia is a well-documented pathological development following L-DOPA therapy [2,3]. Other conditions present codisorders, which have met with even more severe results using pharmaceuticals. One such case is a report of a 38-year-old man diagnosed with multiple sclerosis (MS) in 2007. In that year, the man took part in a study examining the comparative effects of fingolimod (0.5 or 1.5 mg daily), interferonβ-1a (Avonex, Biogen Idec, Cambridge, UK), and placebo on MS [4]. The man was given fingolimod, a sphingosine-1-phosphate receptor modulator that sequesters lymphocytes in lymph nodes, preventing them from contributing to an autoimmune reaction. This mechanism of action reduces circulating lymphocytes, which is contraindicated in herpes simplex virus (HSV) infection. Based on evidence to be presented later in this chapter, the authors suggest that herpes viruses are an etiological factor in a large number of MS cases and other chronic neurodegenerative diseases. In fact, suppressing or weakening immune competence and responsivity could amplify HSV activation and the severity of HSV infection.

In the case of the present discussion, over the course of the next seven years, in a progressive and increasing manner, the man's health deteriorated. In April 2014, he was admitted to the emergency department with loss of consciousness, fever, and epileptic seizures. Cranial computed tomography showed a hypodense area in the right temporal lobe. Diagnosis of HSV-1 encephalitis was made by polymerase chain reaction (PCR) analysis of cerebral spinal fluid. Blood count indicated lymphocytopenia; lymphocytes were low. His medical history indicated a previous herpes labialis. Antiviral therapy with intravenous acyclovir was initiated immediately at the day of presentation at the standard dose. Cranial magnetic resonance imaging showed signs of nonhemorrhagic encephalitis in both hemispheres. After 34 days, he was referred to a neurologic rehabilitation center. At discharge, he was alert but showed signs of right dominant tetraparesis (i.e., muscle flaccidity and loss of muscle control in four limbs). He was unable to speak due to a tracheal cannula and received an Expanded Disability Status Scale (EDSS) score of 9.5 (ranging from 0 to 10, with higher scores indicating increasing disability; 10 being death from MS). When he began the study seven years earlier, his EDSS score was 2.5. Follow-up examinations in the next nine months showed clinical worsening and progressive brain atrophy accentuated in the postencephalitic regions. Prior to onset of HSV-1 encephalitis, he had not shown any evidence of immunologically relevant comorbidities during continuous follow-up in the MS outpatient clinic since 2007. This lack of "evidence" regarding the severity of the disease is most likely a "stealth" capability of the herpes virus. The comprehensive study of viral gene structure since the 1990s has revealed that virtually every class of animal virus has incorporated into its genome the machinery to thwart, suppress, neutralize, or evade the mitochondrial "danger alarm system" [5–8]. So, the lack of "obvious" evidence in this case, and others not cited here, is one reason why this chapter is entitled "A Hidden Etiological Nemesis of Chronic Neurodegenerative Diseases." Rather than treating the patient based only on obvious symptoms, the physician should incorporate therapies based on the proposition of existing comorbidities, even when symptomatic signs are not obvious.

The man had not received corticosteroids or other immunomodulatory treatments apart from fingolimod. The only comedications were bupropion (Wellbutrin, GlaxoSmithKline AG, Münchenbuchsee, Switzerland) and, occasionally, methylphenidate (Ritalin, Novartis Pharma AG, Basel, Switzerland). Both bupropion and methylphenidate are dopamine and norepinephrine reuptake inhibitors. As impairments in brain function increased in severity, medications were most likely prescribed to reduce severe symptoms of attention-deficit/hyperactivity disorder (ADHD), depression, anxiety, sleep disturbances, etc. The long-term effect of these drugs and this therapeutic approach are subjects for another chapter at another time. However, interested clinicians and researchers can do a PubMed search of "Downs B" (one of the authors) and/or "Blum K" for extensive research and publications on reward deficiency syndrome (RDS).

It should also be noted from other research reports that fingolimod treatment of MS lowers varicella-zoster virus-specific immunity (VZV; a herpes virus) [8]. This suggests that subclinical VZV reactivation, demonstrated by PCR detection of VZV DNA in the saliva, is higher among MS patients treated with fingolimod compared with healthy controls.

Not to overstate the obvious, impaired immunity is to be avoided when fighting any herpes virus. While following Standard of Care procedures, these therapeutic interventions apparently intensified and accelerated HSV-related pathological progression and neurodegeneration.

1.2   REORIENTATION OPPORTUNITIES

Some very important questions need to be asked to determine whether the collective "we" are heading in the right direction. After so many years of intensive research on so many chronic degenerative diseases, why does the incidence of these diseases continue to rise? Why are so many people afflicted and so many people continue to suffer so much? It seems that we haven't put a noticeable dent in the incidence of neurodegenerative diseases. We have identified the functional impairments in many of the neurodegenerative diseases. However, we have not identified the causes of these impairments. A simple age-old premise has been ignored. If you want to know how to solve a problem, you must first be able to accurately state or define the problem. For the most part, medical technocracy is focused on relieving suffering with pharmacological interventions that reduce obvious symptoms. This approach generally does not effectively identify and/or address the underlying cause or causes. More unfortunate, is that in reality, pharmacologically speaking, in terms of chronic disease, the big money is not in the cure; it is in the "ongoing" treatment.

Another challenge of the pharmacological approach is the "reductionist" paradigm. This paradigm is based on the need to reduce the active substance to a single "active ingredient," reduce the biochemical transaction to a single mechanism of action, reduce the targeted benefit site to some single loci, and reduce the outcome to a single "primary" benefit. For many years, the National Institutes of Health would not award grants for research on multi-ingredient nutraceutical formulas as, they asserted, it would not be possible to determine the "active ingredient" (a pharmacological requirement). As a result of this paradigm, almost all of the nutraceutical research performed since the passing of the Dietary Supplement Health and Education Act in 1994 has been on single ingredients with a well-defined "active molecule." Hence, the default supposition is that the higher the concentration of the active ingredient is, the more beneficial the product will be. This was the perspective that spurred the meteoric rise of mega-dosing therapy of single ingredient nutritional products, i.e., vitamin C, B vitamins, oligomeric proanthocyanidins (OPCs), etc. A natural result of this paradigm was the birth of the evidence-based proprietary ingredient market.

Dietary supplement manufacturers and marketers constructed condition-specific finished product formulas by combining ingredients backed by research on or for specific conditions.

1.3 PARADIGM 'SHIFT' BEGINS

The emergence of "Integrative Medicine" has opened up another dimension of therapeutic opportunities for physicians using natural products. Over the ensuing years since the passage of the Dietary Supplement Health Education Act in 1994, nutraceutical research produced a plethora of products with sufficient scientific validation that health professionals could more confidently apply them in a clinical setting. However, the vast majority of physicians view evidence-based applications through the "lens" of a classically trained reductionist perspective with condition-specific symptomatic assessments. The primary difference from drugs was that the natural "remedies" were less biologically impacting than drugs and that natural products should be devoid of drug-like side effects, even in significantly higher amounts. Natural products and nutraceuticals were just being substituted for drug applications, e.g., glucosamine for joint health, red yeast rice to lower cholesterol, curcumin to reduce inflammation, chromium to promote insulin sensitivity, St. John's Wort for depression, and the list is almost endless. This ingredient/condition-focused approach catapulted the rise of "condition-specific" objectives with natural product formulations.

The holistic principle should be that the "orchestra of nutrition" goes into the body and "plays the symphony of biology," simultaneously and synergistically supporting the structure and function of all the cells, tissues, and organs in the systems of the body.

The experienced physician must have the wisdom of an artful conductor in constructing natural product and lifestyle protocols that address underlying etiological factors and arrest disease pathologies. But most physicians still focus on reducing symptoms to relieve suffering and improve comfort and functionality, which is their logical primary healthcare objective. While providing measurable improvements in symptomatic relief, very often, with this approach, the actual cause still remains elusive and unaddressed. Most physicians are trained to evaluate health status or the severity of illness from a symptomatic assessment and a blood chemistry, although genetic factors are beginning to gain attention for greater diagnostic accuracy. At the very least, a great need to expand the investigation to identify converging etiological cofactors is warranted. Such an investigation would ultimately lead to an imminent paradigm shift in diagnostic and treatment therapies.

1.4 MOST WELL-KNOWN ETIOLOGICAL FACTORS

Numerous papers have been published on the known primary causes of chronic neurodegenerative diseases. Investigators should be cognizant that multiple etiological factors frequently converge to manifest pathological symptoms. In addition to memory and neuromuscular tests and the results of a complete blood count, allergy panels, and other technical analyses, the clinician generally identifies the most obvious symptoms in their diagnostic evaluation. However, once obvious pathologies and symptoms have been identified, the quest for other etiological factors can be overlooked or diminished. Many times, the converging etiological factors are in fact the result of a sequela of etiological events, which is beyond the scope of generally recognized evaluation criteria. The sequence in which these events occur, the severity of each factor, and the underlying chronicity of each factor are important in determining treatment solutions.

For example, inflammation is probably the most noted etiological factor contributing to neurodegenerative pathologies. One of the authors (MB) has published extensively on the role of oxidative stress and damage, inflammation, and antioxidants in various pathologies, including neurological disorders and various stages of neoplastic processes and carcinogenesis including detoxification of carcinogenic metabolites [9–16]. This type of original research into reducing oxidative damage and inflammatory events in brain tissue has been foundational to spurring research by other authors on this topic as well [17].

While oxidative-induced inflammatory mechanisms are a significant etiological factor and occupy a significant amount of attention in the quest for various types of therapeutic interventions in neurodegenerative disorders, another well-known etiologic aspect involves imbalances in neurotransmitter function. Research in this area was the first to disclose a common genetic predisposition to the constellation of conditions categorically termed "reward deficiency syndrome" (RDS) [18]. In subsequent research, the authors further validated that dysfunction of the D2 dopamine receptors leads to aberrant substance seeking behavior with substances such as alcohol, drugs, tobacco, food, and other related behaviors (i.e., pathological gambling, Tourette's syndrome, and attention deficit hyperactivity disorder, etc.). They provide further evidence that variants of the D2 dopamine receptor gene are important common genetic determinants of RDS [19].

Addressing dopamine resistance/insufficiency is a primary target in disorders such as attention deficit disorder (ADD), attention deficit hyperactivity disorder (ADHD), addictions, obsessions, compulsivity, impulsivity, anxiety, posttraumatic stress disorder, Parkinson's disease (PD), and many more. These types of conditions fall under the RDS rubric, a subject of much research and on which one of the authors (BD) and others have also published [2,3,19–22]. Fortunately, nutraceutical interventions are available to optimize gene expression, rebalance the brain, and enhance brain reward, especially in people with a genetic predisposition to excessive reward seeking behaviors.

1.5   NEW INSIGHTS INTO NEURODEGENERATIVE DISEASE (HIDDEN NEMESIS)

Another area that merits intense etiological research pertains to microbial and viral factors, specifically investigating the role of the herpes virus in neurodegenerative disease pathologies. This area of investigation will potentially reveal not only the viral causes and exacerbating viral factors but also effective natural solutions that have the ability to improve the quality of life for people suffering from chronic neurodegenerative disorders. Revelations in this area will add an important dimension of holistic therapeutic interventions, especially for nutraceutical-based "systems biology" modalities.

Numerous authors have reported on the herpetic etiology of various neurodegenerative diseases (i.e., Alzheimer's disease [AD], MS [and acquired demyelinating syndrome], amyotrophic lateral sclerosis [ALS], PD, encephalitis, etc.) [23–33].

As already mentioned, you must first be able to accurately state or define the problem before you can effectively solve the problem. Unless interventions for neurodegenerative diseases include an antiviral therapeutic strategy that also boosts innate immune competence, they will not completely address or reverse the pathological process. Regardless of whether the intervention is pharmacological or natural, limited or one-dimensional therapeutic interventions can at best only attempt to manage symptoms and reduce suffering during the irrevocable progressive decline characteristic of the disease. Keep in mind that deteriorating health is the macrosystem manifestation of an increase in anaerobic metabolism, an increase in anaerobic infections (various microbes), and weakened immune competence. In addition to providing systemic nutritional support, inclusion of an antiviral strategy is mandatory.

The big question is how to arrest herpetic viral progression implicated in chronic neurodegenerative disorders.

HSV infection results in lifelong infection, which can be asymptomatic or present with recurrent lesions [6,32–34]. The herpes virus resides as a long-term underlying presence. Moreover, a growing body of evidence points to chronic bacterial and viral coinfections as potential etiological factors in neurodegenerative diseases, including AD, PD, and ALS. The chronic activation of inflammatory processes and host immune responses cause chronic damage resulting in alterations of neuronal function and viability. Viral and microbial agents have been reported to produce molecular hallmarks of neurodegeneration, such as the production and deposit of misfolded proteins, oxidative stress, deficient autophagic processes, synaptopathies, and neuronal death. Chronic cycles of pathogen replication within the central nervous system (CNS) overburden already weak immune competence, alter neuronal function, and produce premature apoptosis and cell death [5]. The authors' notion is that while microbial infections appear as coinfectious agents with herpes, such microbial species initiate further weakening of an already fragile immune system, enabling opportunistic activation and more severe infection of the herpes virus. Such a scenario is also believed to be common for Lyme disease, for example [35]. Excessive stress apparently increased immune suppression, activating and intensifying both types of infections. Stress from any source exerts overburdening effects on an already fragile or challenged immune system.

1.6   NUTRACEUTICAL INTERVENTIONS

Reducing the severity and/or incidence of active HSV infection with a combination of nutraceutical products holds significant potential. Pharmacological perspectives that influence natural approaches for the most part confine therapeutic approaches to single-ingredient, single-mechanistic tactics. Therefore, the synergistic effect of multifaceted multifactorial strategies is highly recommended. The proposed strategic objectives should include the following:

1. Inhibiting the herpes virus from binding to a cell receptor site
2. Boosting immune competence
3. Reducing factors that weaken immune competence and promote herpetic activation

1.7   HERPETIC THERAPEUTIC NUTRACEUTICALS (HERPECEUTICS)

L-Lysine (an essential amino acid)
Botanicals providing specific phytosaccharides (more on this later)
Para amino benzoic acid (PABA)
Geopropolis (from stingless bees [*Scaptotrigona postica*])
Multinutrient Complex (vitamins, minerals, phytonutrients, phospholipids)

There are numerous other ingredients that could be included. However, what is being presented in this chapter is a foundational baseline. The clinician should use this information as a starting point and diligently investigate other products to expand treatment options.

1.7.1   L-LYSINE

In order to replicate, the herpes virus requires the amino acid L-arginine, another amino acid common in foods and necessary to human life. Lysine is thought to interfere with the absorption of arginine in the intestine and inhibit viral severity. Moreover, the antiviral cell danger response is strongly regulated by the posttranslational state of lysines on histones and immune effector proteins like the double-strand

RNA binding protein, known as RIG1 (retinoic acid inducible gene 1), and the mitochondrial antiviral sensor. Lysine ubiquitination is a necessary prestep for oligomerization of RIG1, required for efficient binding to the mitochondrial antiviral sensor and interferon induction [36].

Although L-lysine efficacy as an antiherpetic agent is the subject of some controversy, the majority of results from numerous studies (in vitro and in vivo) on a range of population sizes and over various time frames confirm dose-dependent inhibitory effects of L-lysine (with and without low-arginine diets) against the incidence, recurrence, duration, and/or severity of HSV infection. Dosages range from 500 mg OID to 1000 mg TID [37–42]. A scant few studies report the failure of L-lysine to exert antiherpetic effects [43]. Dosage levels in this research were low, population size was small, and duration of the study period was short. When dosage levels are more robust, population sizes and duration of the study period are shown to be less influential. The conclusion is that L-lysine, in a dose-dependent manner, has been shown to exert beneficial antiherpetic effects. The potential therapeutic benefits of L-lysine should be synergistically amplified when it is combined with other products and therapeutic strategies.

1.7.2 PHYTOSACCHARIDES

Another therapeutic nutraceutical approach with significant potential is to inhibit the herpes virus from binding to a cell receptor site. If the herpes virus is unable to bind to a cell receptor, its infectious potential remains impotent. Various phytosaccharides have been shown to prevent binding of HSV to receptors, thereby reducing severity and duration of HSV infection.

The mannose receptor (MR) was shown to be an important receptor for the nonspecific recognition of enveloped viruses by dendritic cells (DCs) and the subsequent stimulation of interferon alpha (IFN-α) production by herpes viruses binding to that site. The MR binds several monosaccharides, including fucose, N-acetylglucosamine, and mannose, with high affinity [44,45]. Six different monosaccharides were shown to reduce the frequency of HSV-induced interferon-alpha-producing cells in a dose-dependent manner. The research demonstrated that most sugars had inhibitory effects at high concentrations (0.50 mM). However, fucose had the greatest inhibitory effect, followed by N-acetylgalactosamine (Gal-N-Ac) and N-acetylglucosamine (Glc-N-Ac). While the MR has been identified as an important receptor target for phytosaccharides, no receptor with specificity for both Glc-N-Ac/Man and Gal-N-Ac has yet been identified. These results suggest either a novel receptor with both mannose and galactose specificities or that more than one receptor with different specificities is involved in the stimulation of IFN-α synthesis by HSV and are therefore specific targets for competitive monosaccharide binding [46]. Thus, the MR probably serves as a critical link between innate and adaptive immunity to viruses, especially given the role of the MR in antigen (Ag) capture by DC and the importance of IFN-α in shaping immunity [46]. Therefore, supplying various phytosaccharides can evidently preferentially occupy the MR and block HSV binding. A number of monosaccharides, including fucose, mannose, galactose, N-acetylglucosamine, and N-acetylgalactosamine, exerted a strong inhibitory activity against HSV-1 and -2, with no cytotoxicity [47].

Various botanicals from fucoidan species of brown seaweed to pine cone extracts to aloe vera and others can provide a range of the phytosaccharides mentioned that have been shown to inhibit HSV binding. Caution should be exercised with fucoidan seaweed to ensure that it is free of heavy metals and other contaminants common to the marine environment. In the case of aloe vera, the inner gel is saccharide rich [48]. If the aloe vera being used is not fresh, the nutraceutical product being purchased should use a water or CO_2 extraction as many standard chemical solvent extraction methods break glycosidic bonds and eliminate or significantly reduce the presence of valuable monosaccharides. The same is true for pine cone extracts. Many ancient peoples made pine cone tea as a remedy for various maladies.

One type of aloe vera gel (99% H_2O) exhibited a significant inhibitory effect, one hour after a Vero cell line was infected with HSV-1, of 0.2%–5% on viral growth. The gel could be a useful topical treatment for oral HSV-1 infections without any noticeable toxicity [49].

While this chapter is examining therapeutic nutraceutical strategies to reduce chronic neurodegenerative diseases, owing to the nutritional value of aloe in this instance, certainly, other collateral beneficial effects could be expected. To demonstrate the broader benefit of aloe, we present evidence of its in vitro and in vivo effects on bone health. A quote from the study abstract is presented.

"In an animal study, mandibular right incisors of male Sprague–Dawley rats were extracted and an acemannan treated sponge was placed in the socket. After 1, 2, and 4 weeks, the mandibles were dissected. Bone formation was evaluated by dual energy X-ray absorptiometry and histopathological examination. The in vitro results revealed acemannan significantly increased bone marrow stromal cells (BMSC) proliferation, VEGF, BMP-2, alkaline phosphatase activity, bone sialoprotein and osteopontin expression, and mineralization. In-vivo results showed acemannan-treated groups had higher bone mineral density and faster bone healing compared with untreated controls. A substantial ingrowth of bone trabeculae was observed in acemannan-treated groups. These data suggest acemannan could function as a bioactive molecule inducing bone formation by stimulating BMSCs proliferation, differentiation into osteoblasts, and extracellular matrix synthesis. Acemannan could be a candidate natural biomaterial for bone regeneration" [50]. This data is presented as just one example to confirm the premise that these nutraceuticals are not pharmaceuticals or other type of medications. They are exerting multisystem benefits that are entirely nutritional-type therapeutic effects.

The authors suggest that researchers in the field of glycobiology expand legitimate scientific efforts to investigate and better define the inhibitory interaction of various saccharides against viral and bacterial species underlying many chronic degenerative diseases, especially the herpes virus. But, researchers should maintain a perspective of the structural and functional value of the nutraceuticals and their synergistic benefits when combined with other nutraceuticals that can "play the symphony of biology."

It is also important to note that with the advent and advancing technologies of agribusiness and food treatment/processing practices, societal evidence indicates that our food supply is apparently falling short of supplying the requisite nutrition to sufficiently support optimal health. The collective "we" are not curtailing the escalating juggernaut of chronic degenerative diseases. And, the existing healthcare system is and will continue to be woefully inadequate to keep pace with the increasing incidence of chronic degenerative diseases, especially of the neurological disorders.

1.7.3   GEOPROPOLIS (PROPOLIS FROM BRAZILIAN STINGLESS BEES *S. POSTICA*)

This is one of the first studies investigating the potential effects of geopropolis against HSV-1. In cell culture, the results showed that hydromethanolic extracts of geopropolis (HMG) from stingless bees significantly reduced the number of copies of HSV-1 genomic DNA in the supernatant and in the lysate cell. All concentrations tested against HSV-1 through pre-, post-, and virucidal treatment were found to be effective in inhibiting HSV-1 viral replication. Quantification of viral DNA from herpes virus showed reduction of about 98% in all conditions and concentrations tested of the HMG extract. This indicates that geopropolis inhibited the events in early infection, such as viral binding and viral entry into cells as well as the viral replication [51]. To reiterate, while the scientific community readily uses pharmacological nomenclature, perspectives, and explanations, the effects demonstrated are not pharmaceutical-type actions. These effects have been attributed to various bioflavonoids and glyconutrient molecules within the geopropolis. But this approach would be more accurately presented as a whole food extract.

1.7.4   PABA

PABA demonstrated virucidal effects in a culture of the cell-free virus-containing material. It reduced the death rate of laboratory mice infected with experimental herpetic encephalitis (via intraperitoneal contamination) by an average of 40%. PABA increased the mean life-span of the same species of mice with experimental herpetic encephalitis, also significantly decreasing the virus titer in the mouse brain. In the same study, PABA exhibited a significant ability to potentiate the antiherpetic action of acyclovir (Zovirax, acycloguanosine) in the infected cultures when acyclovir was used in inactive concentrations [52].

While individual studies on single ingredients are by no means a conclusive therapeutic edict or a reductionist recommendation, they add another dimension of therapeutic options to the arsenal of nutraceutical products (already discussed) that reduce herpetic infections that exacerbate chronic neurodegenerative diseases. A new hope for an improved quality of life is presented.

1.7.5   MULTINUTRIENT COMPLEX

In a *JAMA* paper published in 2002, researchers stated that "suboptimal intake of some vitamins, above levels causing classic vitamin deficiency, is a risk factor for chronic diseases common in the general population, especially the elderly." They also stated that "it appears prudent for all adults to take vitamin supplements" [53]. While this is a well-intentioned recommendation, the significant exponential increase in the consumption of dietary supplements over the last 30 years has not put a dent in the incidence of chronic degenerative diseases. Moreover, the number one health malady in the United States is digestive problems, and they are increasing. The implication from these factors is that consumed food and supplements are achieving suboptimal absorption at best. Health is progressively eroding on a societal scale. The informed consumer should seek out multivitamin/mineral supplements that provide nutrition that gets in, i.e., gets absorbed into the body's tissues. Dietary supplements are not all the same. Consumers should ask dietary supplement companies to supply research published in peer-reviewed scientific literature to validate the effectiveness of their products.

Other nutraceuticals having some antiherpetic validation are

–Monolaurin
–Zinc

(The reader is encouraged to investigate the benefits of these ingredients to augment others mentioned in this chapter.)

1.8   CONCLUSION

A preponderance of evidence presented in this chapter confirms that chronic viral infections, specifically various herpes viruses, are important but overlooked etiological factors in chronic neurodegenerative diseases. Clinicians should not wait for the emergence of obvious symptoms to add nutraceutical products to

1. Bolster immune and metabolic competence
2. Inhibit the herpes virus from binding to a cell receptor site
3. Reduce factors that weaken immune competence and promote herpetic activation

We have revealed a little recognized etiological nemesis (i.e., herpes viruses) that demands therapeutic inclusion to more competently understand and effectively address chronic neurodegenerative diseases. There are nutraceutical products that have been shown to be effective in reducing the incidence, severity, duration, and recurrence of the herpes virus. Nutraceuticals and therapeutic strategies mentioned in this chapter are intended to convey valuable information regarding the biological building materials (nutritional/nutraceutical products) that are either unavailable or inadequate from standard American dietary practices and require supplementation. These nutraceutical products should be used in concert with other cofactors mentioned to bolster protection against the scourges of chronic neurodegenerative diseases.

REFERENCES

1. Cacace R, Sleegers K, Van Broeckhoven C. Molecular genetics of early-onset Alzheimer disease revisited. *Alzheimers Dement.* March 23, 2016. Epub ahead of print.
2. Iravani MM, McCreary AC, Jenner P. Striatal plasticity in Parkinson's disease and L-dopa induced dyskinesia. *Parkinsonism Relat Disord.* 18 Suppl 1:S123–5 (2012).
3. Rylander-Ottosson D, Lane E. Striatal plasticity in L-DOPA- and graft-induced dyskinesia; the common link? *Front Cell Neurosci.* 10(16):1–13 (2016).

4. Pfender N, Jelcic I, Linnebank M, Schwarz U, Martin R. Reactivation of herpes virus under fingolimod: A case of severe herpes simplex encephalitis. *Neurology.* 84:2377–8 (2015).

5. Corcoran JA, Saffran HA, Duguay BA, Smiley JR. Herpes simplex virus UL12.5 targets mitochondria through a mitochondrial localization sequence proximal to the N terminus. *J Virol.* 83:2601–10 (2009).

6. Ohta A, Nishiyama Y. Mitochondria and viruses. *Mitochondrion.* 11:1–12 (2011).

7. Scott I. The role of mitochondria in the mammalian antiviral defense system. *Mitochondrion.* 10:316–20 (2010).

8. Ricklin ME, Lorscheider J, Waschbisch A et al. T-cell response against varicella-zoster virus in fingolimod-treated MS patients. *Neurology.* 81:174–81 (2013).

9. Bagchi D, Swaroop A, Preuss HG, Bagchi M. Free radical scavenging, antioxidant and cancer chemoprevention by grape seed proanthocyanidin: An overview. *Mutat Res.* 768:69–73 (2014).

10. Zafra-Stone S, Yasmin T, Bagchi M et al. Berry anthocyanins as novel antioxidants in human health and disease prevention. *Mol Nutr Food Res.* 51(6):675–83 (2007).

11. Bagchi D, Bagchi M, Stohs S et al. Cellular protection with proanthocyanidins derived from grape seeds. *Ann N Y Acad Sci.* 957:260–70 (2002).

12. Bagchi D, Ray SD, Patel D, Bagchi M. Protection against drug- and chemical-induced multiorgan toxicity by a novel IH636 grape seed proanthocyanidin extract. *Drugs Exp Clin Res.* 27(1):3–15 (2001).

13. Bagchi D, Bagchi M, Stohs SJ et al. Free radicals and grape seed proanthocyanidin extract: Importance in human health and disease prevention. *Toxicology.* 148(2–3):187–97 (2000).

14. Bagchi D, Bagchi M, Balmoori J et al. Induction of oxidative stress and DNA damage by chronic administration of naphthalene to rats. *Res Commun Mol Pathol Pharmacol.* 101(3):249–57 (1998).

15. Bagchi D, Garg A, Krohn RL et al. Protective effects of grape seed proanthocyanidins and selected antioxidants against TPA-induced hepatic and brain lipid peroxidation and DNA fragmentation, and peritoneal macrophage activation in mice. *Gen Pharmacol.* 30(5):771–6 (1998).

16. Bagchi D, Hassoun EA, Bagchi M, Stohs SJ. Protective effects of free radical scavengers and antioxidants against smokeless tobacco extract (STE)-induced oxidative stress in macrophage J774A.1 cell cultures. *Arch Environ Contam Toxicol.* 29(3):424–8 (1995).

17. Porat Y, Abramowitz A, Gazit E. Inhibition of amyloid fibril formation by polyphenols: Structural similarity and aromatic interactions as a common inhibition mechanism. *Chem Biol Drug Des.* 67:27–37 (2006).

18. Blum K, Noble EP, Sheridan PJ et al. Allelic association of human dopamine D2 receptor gene in alcoholism. *JAMA.* 263(15):2055–60 (1990).

19. Blum K, Sheridan PJ, Wood RC et al. The D2 dopamine receptor gene as a determinant of reward deficiency syndrome. *J R Soc Med.* 89(7):396–400 (1996).

20. Blum K, Oscar-Berman M, Stuller E et al. Neurogenetics and nutrigenomics of neuro-nutrient therapy for reward deficiency syndrome (RDS): Clinical ramifications as a function of molecular neurobiological mechanisms. *Addict Res Ther.* 3(5):139–82 (2012).

21. Bowirrat A, Chen TJ, Blum K et al. Neuro-psychopharmacogenetics and neurological antecedents of posttraumatic stress disorder: Unlocking the mysteries of resilience and vulnerability. *Curr Neuropharmacol.* 8(4):335–58 (2010).

22. Downs B, Oscar-Berman M, Blum K et al. Have we hatched the addiction egg: Reward deficiency syndrome solution system. *J Genet Syndr Gene Ther.* 4(4):139–42 (2013).

23. Wozniak MA, Mee AP, Itzhaki RF. Herpes simplex virus type 1 DNA is located within Alzheimer's disease amyloid plaques. *J Pathol.* 217(1):131–8 (2009).

24. McNamara J, Murray TA. Connections between herpes simplex virus type 1 and Alzheimer's disease pathogenesis. *Curr Alzheimer Res.* 13(9):996–1005 (2016).

25. Makhani N, Banwell B, Tellier R et al. Viral exposures and MS outcome in a prospective cohort of children with acquired demyelination. *Mult Scler.* 22(3):385–8 (2016).

26. Zivadinov R, Nasuelli D, Tommasi MA et al. Positivity of cytomegalovirus antibodies predicts a better clinical and radiological outcome in multiple sclerosis patients. *Neurol Res.* 28(3):262–9 (2006).

27. Limongi D, Baldelli S. Redox imbalance and viral infections in neurodegenerative diseases. *Oxid Med Cell Longevity.* Volume 2016, March; Article ID6547248, 13 pages.

28. Volpi A. Epstein-Barr virus and human herpesvirus type 8 infections of the central nervous system. *Herpes.* Suppl 2:120A–127A (2004).

29. Irkeç C, Ustaçelebi S, Ozalp K, Ozdemir C, Idrisoğlu HA. The viral etiology of amyotrophic lateral sclerosis. *Mikrobiyol Bul.* 23(2):102–9 (1989).

30. Ferri-De-Barros JE, Moreira M. Amyotrophic lateral sclerosis and herpes virus. Report of an unusual case: A cause or casual association? *Arq Neuropsiquiatr.* 56(2):307–11 (1998).

31. Caggiu E, Paulus K, Arru G, Piredda R, Sechi GP, Sechi LA. Humoral cross reactivity between α-synuclein and herpes simplex-1 epitope in Parkinson's disease, a triggering role in the disease? *J Neuroimmunol.* 15;291:110–4 (2016).

32. DeChiara G, Marcocci ME, Sgarbanti R et al. Infectious agents and neurodegeneration. *Mol Neurobiol.* 46:614–38 (2012).

33. Yamada S, Kameyama T, Nagaya S, Hashizume Y, Yoshida M. Relapsing herpes simplex encephalitis: Pathological confirmation of viral reactivation. *J Neurol Neurosurg Psychiatry.* 74:262–4 (2003).

34. Azwa A, Barton SE. Aspects of herpes simplex virus: A clinical review. *J Fam Plann Reprod Health Care.* 35(4):237–42 (2009).

35. Gylfe A, Wahlgren M, Fahlén L, Bergström S. Activation of latent *Lyme borreliosis* concurrent with a herpes simplex virus type 1 infection. *Scand J Infect Dis.* 34(12):922–4 (2002).

36. Jiang X, Kinch LN, Brautigam CA, Chen X, Du F, Grishin NV, Chen ZJ. Ubiquitin-induced oligomerization of the RNA sensors RIG-I and MDA5 activates antiviral innate immune response. *Immunity.* 36:959–73 (2012).

37. Griffith RS, Norins AL, Kagan C. A multicentered study of lysine therapy in herpes simplex infection. *Dermatologica.* 156(5):257–67 (1978).

38. Miller CS, Foulke CN. Use of lysine in treating recurrent oral herpes simplex infections. *Gen Dent.* 32(6):490–3 (1984).

39. Milman N, Scheibel J, Jessen O. Lysine prophylaxis in recurrent herpes simplex labialis: A double-blind, controlled crossover study. *Acta Derm Venereol.* 60(1):85–7 (1980).

40. McCune MA, Perry HO, Muller SA, O'Fallon WM. Treatment of recurrent herpes simplex infections with L-lysine monohydrochloride. *Cutis.* 34(4):366–73 (1984).

41. Griffith RS, DeLong DC, Nelson JD. Relation of arginine-lysine antagonism to herpes simplex growth in tissue culture. *Chemotherapy.* 27(3):209–13 (1981).

42. Griffith RS, Walsh DE, Myrmel KH, Thompson RW, Behforooz A. Success of L-lysine therapy in frequently recurrent herpes simplex infection. Treatment and prophylaxis. *Dermatologica.* 175(4):183–90 (1987).

43. DiGiovanna JJ, Blank H. Failure of lysine in frequently recurrent herpes simplex infection. Treatment and prophylaxis. *Arch Dermatol.* 120(1):48–51 (1984).

44. Stahl PD, Schlesinger PH, Rodman JS, Lee YC. Evidence for receptor-mediated binding of glycoproteins, glycoconjugates, and lysosomal glycosidases by alveolar macrophages. *Proc Natl Acad Sci USA.* 75:1399 (1978).

45. Pontow SE, Kery V, Stahl PD. Mannose receptor. *Int Rev Cytol B.* 137:221 (1992).

46. Milone MC, Fitzgerald-Bocarsly P. The mannose receptor mediates induction of IFN-a in peripheral blood dendritic cells by enveloped RNA and DNA viruses. *J Immunol.* 161:2391–9 (1998).

47. Ponce NM, Pujol CA, Damonte EB, Flores ML, Stortz CA. Fucoidans from the brown seaweed *Adenocystis utricularis*: Extraction methods, antiviral activity and structural studies. *Carbohydr Res.* 338(2):153–65 (2003).

48. Olwen M, Grace OM, Dzajic A et al. Monosaccharide analysis of succulent leaf tissue in aloe. *Phytochemistry.* 93:79–87 (2013).

49. Rezazadeh F, Moshaverinia M, Motamedifar M, Alyaseri M. Assessment of anti HSV-1 activity of aloe vera gel extract: An in vitro study. *J Dent Shiraz Univ Med Sci.* 17(1):49–54 (2016).

50. Boonyagul S, Banlunara W, Sangvanich P, Thunyakitpisal P. Effect of acemannan, an extracted polysaccharide from aloe vera, on BMSCs proliferation, differentiation, extracellular matrix synthesis, mineralization, and bone formation in a tooth extraction model. *Odontology.* 102:310–7 (2014).

51. Coelho GR, Mendonça RZ. Antiviral action of hydromethanolic extract of geopropolis from *Scaptotrigona postica* against antiherpes simplex virus (HSV-1). *Evid Based Compl Altern Med.* 2015:1–10 (2015).

52. Akberova SI, Leont'eva NA, Stroeva OG, Galegov GA. Action of para-aminobenzoic acid and its combination with acyclovir in herpetic infection. *Antibiot Khimioter.* 40(10):25–9 (1995).

53. Fletcher RH, Fairfield KM. Vitamins for chronic disease prevention in adults: Clinical applications. *JAMA.* 287(23):3127–9 (2002).

2 Modulation of Brain Function with Special Emphasis to Hypothalamic– Pituitary–Gonadal Axis by Xenohormones, Phytohormones, and Insulin

Sanjoy Chakraborty, Devya Gurung, Soheli A. Chowdhury, and Majeedul H. Chowdhury

CONTENTS

2.1   INTRODUCTION

The hypothalamic–pituitary–gonadal (HPG) axis is a critical part in the control of development, reproduction, and aging in humans. In the HPG axis, the hypothalamus controls the release of two peptide hormones from the anterior pituitary by secreting the hypothalamic tropic hormones—a peptide hormone called gonadotropin-releasing hormone (GnRH), by GnRH-expressing neurons, to the hypothalamic-hypophyseal portal system. In response to GnRH, the anterior pituitary gland produces two tropic hormones—luteinizing hormone (LH) and follicle-stimulating hormone (FSH)—that act on the gonads to produce steroid hormones—estrogen and progesterone by ovary and testosterone by testis. Many factors, such as xenohormones or endocrine disrupting chemicals (EDCs), insulin or insulin-like molecules, and phytochemicals (mainly phytoestrogens), have been reported to exert control on the HPG axis, both directly and indirectly, and thereby modulating the growth, metabolism, and reproduction in animals. In this review, the mechanism of such modulation on the functionality of the HPG axis will be addressed, and current and future research in these areas will be discussed (Table 2.1). A clear delineation of such modulations, by further research, may resolve issues associated with fertility, mental health, and metabolism.

2.2   MODULATION OF HPG AXIS BY EDCs

2.2.1   XENOHORMONES

With the increase in industrialization and human activities, several chemicals are produced either intentionally or as a by-product and wastes. These chemicals are called xenohormones or EDCs. Their structure similarities mimic or antagonize hormonal activities by binding to the hormonal receptor and thereby disrupting the endogenous signaling pathway. They can cause several health defects by interfering with the synthesis, metabolism, binding, or cellular actions of natural hormones, mainly estrogen and thyroid. To date, there are several reports of the effect of EDCs on reproductive health but few on mental health. With the growing awareness of the effects of EDCs on neurological disorders, this review attempts to focus on estrogen-like EDCs with their effect on the brain. Some of the major EDCs are dichlorodiphenyltrichloroethane (DDT), dioxin, polychlorinated biphenyls (PCBs), bisphenol A (BPA), polybrominated biphenyls, and phthalate esters.

2.2.1.1   Mechanism of Action

The HPG axis is a critical part in the control of neuroendocrine function in vertebrates [1]. In the HPG axis, the hypothalamus controls the release of two peptide hormones from the anterior pituitary by secreting the hypothalamic tropic hormones—a peptide hormone called GnRH, by GnRH-expressing neurons, to the hypothalamic-hypophyseal portal system [2]. In response to GnRH, the anterior pituitary gland produces two tropic hormones—LH and FSH—that act on the gonads to produce steroid hormones—estrogen and progesterone by ovary and testosterone by testis. Estrogens play a critical role in the growth, differentiation, and homeostasis in various tissues,

TABLE 2.1

Summary of Effects of Estrogen, Xenoestrogens, Phytoestrogens, and Insulin on Brain Functions

		Type of Action	Effect on Brain	References
Natural estrogen		Via ERα and ERβ	Neuroprotective effects; affects multiple pathways by binding via ERs in the brain	Chakraborty et al. [3] Chakraborty et al. [4] Markou et al. [5]
Xenoestrogens				
1.	DDT, p,p-DDE, o,p-DDE	Estrogen mimicker/ blocker (via ERα and ERβ)	Hippocampal neurogenesis, Incomplete DNA methylation	Bulger et al. [6] Shutoh et al. [7]
2.	Dioxin	Estrogen blocker (via ERα and ERβ) via AhR	Hypothalamus, hippocampus, pituitary, and other parts of CNS	Huang et al. [8] Hutz et al. [9]
3.	BPA	Estrogen blocker (via ERα and ERβ or extracellular signal-regulated kinases [ERK]/nuclear factor kappa [NF-kappa] via ERK/ NK-kappa pathway)	Development of brain	Negri-Cesi [10] Krishnan et al. [11] Lee et al. [12]
4.	PCB	Estrogen mimicker/ blocker (via ERβ)	Neurotoxic to nigrostriatal dopamine system	Salama et al. [13] Caudle et al. [14]
5.	Phthalate esters	Estrogen mimicker (via ERα and ERβ)	Development of brain	Takeuchi et al. [15] Lim et al. [16]
6.	Endosulfan	Estrogen mimicker (via ERα and ERβ)	Neurotransmitter circuit in the frontal cortex	Wilson et al. [17] Wozniak et al. [18]
Phytoestrogens				
1.	Isoflavones	Preferably ERβ	Improve short- and long-term memory and cognitive function in women, poor cognitive performance, and low brain weight in men	Collins-Burow et al. [19] File et al. [20]
2.	Lignans	Specificity unknown	Hippocampus and the cortex	Jeong et al. [21]
Insulin				
1.	Insulin	Via IR and IGF-1 receptor	Neuroprotection, neuromodulation, cognition, and memory	Kleinridders et.al [22] Blazquez et al. [23]

including reproductive tracts (both male and female), mammary glands, bone, brain, and liver [24–27], acting through its two receptors, ERα and ERβ, which belong to the class of nuclear receptor superfamily [28]. The robust expressions of estrogen receptor (ER) (α and β) are mainly found in the hypothalamus and hippocampus of the brain [3,4]. The structural similarities between EDCs and endogenous estrogen facilitate binding to ERs and activate the signaling pathway in the absence of estrogen and modify their function in *in vitro* and *in vivo* model systems [29] by altering the feedback mechanism in central or peripheral tissues. Due to their chemical structure, EDCs may have a more intense effect than endogenous estrogen by the following ways: (i) bind with ERs but may have a higher affinity for ERs than endogenous estrogen [30,31], (ii) resistant to environmental degradation such that they can remain in the environment for longer time, and (iii) bioaccumulate in a variety of cells due to their lipophilic property.

To explain the common molecular mechanism of EDC and phytoestrogen (discussed in detail later) action, it has been suggested that they use both genomic and nongenomic (Figure 2.1) 17 β-estradiol (E2) signal transduction pathways and mimic or block the E2 responses [32–34].

In the *genomic pathway*, the mechanism for E2 action on cells involves the binding to nuclear ERs, which are ligand-activated transcription factors that stimulate the expression of several target genes [35,36]. Many important E2-mediated events occur through this mechanism, involving NRs. There are two types of such NRs for the ligand E2: ERα and ERβ. Since many EDCs have similar structures to E2, the NR ligands, they can directly bind to NRs to function either as an agonist and cause gene expression, or as an antagonist and inhibit the activity of NRs [37].

In *nongenomic pathway*, several estrogen responses are via membrane E2 receptor (mER) and occur via G protein-coupled receptor (GPR30) pathways, which cannot be explained by the aforementioned nuclear mechanism [38]. Membrane-associated versions of ER subtype (α, β), complexed with signaling and scaffolding molecules capable of acting via G proteins, have been identified that may explain in part such events [35,36]. In the nongenomic pathway, estrogen exposure causes rapid activation of kinase signaling cascades. Activation of GPR30 by EDCs leads to rapid downstream cellular signaling. This induces subsequent stimulation of protein kinase activation and phosphorylation, which in turn may affect the transcription of target genes. The resulting changes, by interaction between ERs and GPR30, in gene expression and intracellular signaling, can cause cellular response without regulation, which may produce adverse effects of EDCs [32]. EDC ligands activate nongenomic activity, via one of the G-protein coupled membrane receptors (GPR30), at concentrations at which they do not alter the transcription of estrogen target genes in the genomic pathway involving NR [39].

For other pathways, EDCs, in addition to manipulating NRs and mERs, interrupt E2 hormone signaling pathway in target tissues by interfering its biosynthesis, transport to the target tissues, availability of ligand binding proteins, and hormone catabolism [40,41]. Many EDCs may have actions via, or independent of, classic actions at cognate steroid receptors. EDCs may have effects through numerous other substrates, such as the aryl hydrocarbon receptor (AhR), the peroxisome proliferator-activated receptor (PPAR), and retinoid X receptor, signal transduction pathways, calcium influx, and/or neurotransmitter receptors. Thus, EDCs may have

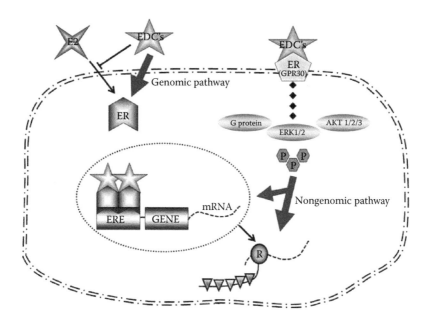

FIGURE 2.1 Potential mechanism(s) of EDC action. In the "genomic pathway" of EDC action, EDCs interfere with estrogen (E2) binding to ERs. EDCs bind to ERs instead of E2 and can thus affect the transcription of target genes in the nucleus by binding to the estrogen response element (ERE) of target genes. The "nongenomic pathway" of EDC action may occur through ER such as GPR30 located in the cytoplasmic membrane. Activation of GPR30 by EDCs leads to rapid downstream cellular signaling. This induces subsequent stimulation of protein kinase activation and phosphorylation, which in turn may affect the transcription of target genes. The resulting changes by interaction between ERs and GPR30 in gene expression and intracellular signaling can cause cellular response without regulation, which may produce adverse effects of EDCs on organs. (From Lee, H.R., Jeung, E.B., Cho, M.H., Kim, T.H., Leung, P.C., Choi, K.C., *J. Cell. Mol. Med.*, 17, 1–11, 2013.)

organizational effects during development, and/or stimulating effects in adulthood, that influence sexually dimorphic, reproductively relevant processes or other functions, by mimicking, antagonizing, or altering steroidal actions [42]. The AhR is a ligand-activated cytoplasmic transcription factor involved in the regulation of biological responses to planar aromatic (aryl) hydrocarbons. This receptor has been shown to regulate xenobiotic-metabolizing enzymes such as cytochrome P450. AhR is a cytosolic transcription factor that is normally inactive, bound to several cochaperones. Upon ligand binding to chemicals such as 2,3,7,8-tetrachlorodibenzo-*p*-dioxin (TCDD), the chaperones dissociate resulting in AhR translocating into the nucleus and dimerizing with AhR nuclear translocator, leading to changes in gene transcription. Enzymes induced by activated AhR, such as CYP1A1, CYP1A2, CYP3A4, and CYP1B1, are involved not only in metabolism of xenobiotics but also in catabolism of steroid hormones. Thus, the induction of these enzymes, upon exposure to xenobiotics, can lead to increased hormone catabolism, reduced availability of endogenous hormones, and, consequently, compromise hormone signaling [37].

2.2.2 XENOHORMONES AND BRAIN

Estrogen is one of the primary hormones involved in brain development. Fetuses and newborns synthesize estradiol locally in their brain. They can also obtain estradiol from their mother and/or their own gonads. In addition, the developing brain expresses high levels of ERs. Estrogen is responsible for sexual dimorphism and differences in males and females and provides trophic and neuroprotective effects which are permanent or irreversible [43]. The developing brain is one of the most vulnerable organs to toxic chemicals. In case of HPG axis, exposure of EDCs (e.g., atrazine) early in development can permanently alter sexual dimorphic circuits in the brain resulting in demasculization in males and defeminization in females [1].

With the increase in the number of xenohormones or EDCs, there is a robust increase in neurological diseases and it is estimated to double every 20 years [44]. Only 30%–40% of the causes seem to be due to genetic factors and the rest are probably due to environmental factors specially EDCs. Even at low levels of exposure, EDCs can cause permanent damage to the developing brain. They are able to cross the placental and blood–brain barrier affecting the neural cell migration, development, proliferation, and differentiation [45]. The effects of EDCs are multifaceted; for example, tetra-cholorodixion (TCDD), phthalates, and BPA can affect the development of the embryonic neural cells [10,16,46], DDT can impair hippocampal neurogenesis and DDT and BPA can cause incomplete DNA methylation in specific genes in the developing brain [7,47,48], and phthalates, PCBs, and heavy metals have been implicated in autism spectrum disorders [49]. In adults, polycholorinated diphenyls have shown to be neurotoxic to the nigrostriatal dopamine system and increase risk for developing Parkinson's disease [14,50]. Fetal, prenatal, and postnatal exposure (developing brain) have shown to have more detrimental consequences than in adults (developed brain) [51,52] when exposed to EDCs. These are critical periods of brain development that can have long-lasting effects on cognitive functions in humans.

On the other hand, plant-derived xenoestrogens, which are known as phytoestrogens, have long been implicated in both beneficial and detrimental effects on health [53].

2.2.3 PHYTOESTROGENS

Phytoestrogens are plant-derived xenoestrogens found in soy and a wide variety of food products; fruits like citrus fruit, cherries, berries, apples, and grapes; and vegetables like celery, capsicum, broccoli, onions, tomatoes, alfalfa, and spinach. In addition, it is also found in red wine, flaxseeds, grains, nuts, chocolate, green tea, soy bean, legumes, and clover. They are naturally occurring plant-derived nonsteroidal compounds that are functionally and structurally similar to steroidal estrogens [54]. Various studies showed that phytoestrogens, unlike xenohormones, have many beneficial effects on humans. They belong to the polyphenolic group of compounds divided into three classes: isoflavones (genistein, diadzein, and biochanin) [55], lignans (enterolactone, enterodiol) [56], and coumestans [57]. Diadzein and genistein are the two most common isoflavones and are found in soy-based food and beverages. Soy has been a staple food for Asian populations for many decades, but over the past 15 years, its use as a food supplement has quadrupled in the Western world because of its health benefits [58–60]). Phytoestrogens have been related to a number of health benefits, including cardiovascular, cognition, prostate health, bone formation promoting properties, and other brain functions [54,61,62].

2.2.3.1   Mechanism of Action

The similarity in structure of both phytoestrogens and estrogen regarding the phenolic ring enables it to bind to ERs, ERα, and ERβ, activating the signaling pathway [63,64]. They can also act as agonists or antagonists, although they exhibit weak binding affinity toward ERα and have a preference for ERβ [19]. The antagonist actions are manifested by blockade of the activity of endogenous ligands or inhibiting the activity of androgen or estrogen metabolizing enzymes, such as aromatase [65]. It can also modulate the signaling cascades related to estrogenic function by preferentially binding to ERs and up-regulating antioxidant defenses [66–68]. The actions of some phytoestrogen like soy isoflavones have been reported to be mediated through the nonclassical mechanism of estrogen [69]. So, the varied actions of phytoestrogens, whether it would have a beneficial or detrimental effect, depend on a number of factors, including tissue type, receptor subtype, dose, and the amount of the endogenous ligand present [65].

2.2.4   Phytoestrogens and Brain

Phytoestrogens have been shown to affect the HPG axis by decreasing FSH and LH in premenopausal women and may increase estradiol in postmenopausal women [70]. This has shown to improve the quality of life in postmenopausal women [71]. Diets rich in soy-phytoestrogens (100 mg isoflavones/day) have been implicated to improve short- and long-term memory and cognitive function in young adults [20]. Women aged 55–74 years showed improved verbal memory (110 mg/day, six months) [72] and women aged 50–65 showed cognitive improvement, which was related to frontal cortical functions (60 mg/day, three months). However, there are studies that have shown no significant association of dietary isoflavone intake and cognitive function in Japanese and Chinese women aged 46–56 years and Dutch women aged 60–75 years [73,74]. Asian males aged 71–91 years on high-soy diet have also been associated with poor cognitive test and low brain weight [75]. On the other hand, other studies have shown isoflavone supplementation (116 mg/day) given to healthy males aged 30–80 years for 12 weeks may enhance cognitive processes [62]. These sexual differences could be due to differences in the expression of oxidative stress, hydrogen peroxide formation, and DNA damage in brain among males and females, which are all estrogen dependent [76,77]. Although the data available are conflicting regarding the beneficial effects of phytoestrogen, it seems that a soy diet in women taken for a short time period may show positive effect on mental health.

2.2.5   Kisspeptin and Brain

On HPG circuitry signaling, the downstream adverse effects of EDCs on gonads are well studied. However, the upstream effect of EDCs on hypothalamus and anterior pituitary was an enigma until kisspeptin protein (formerly known as metastin) was found to play a role in hypogonadotropic hypogonadism in 2003 [78]. This neurohormone is encoded by the metastasis suppressor *KISS1* gene in chromosome 1 in humans and is expressed in a variety of endocrine and gonadal tissue [79,80]. GPR54 is a G-protein-coupled receptor that binds kisspeptins and is widely expressed throughout the brain. Kisspeptin–GPR54 signaling has been implicated in the regulation of pubertal and adulthood GnRH secretion, and mutations or deletions of GPR54 cause hypogonadotropic hypogonadism in humans and mice. GPR54 is necessary for proper male-like development of several sexually dimorphic traits, likely by regulating GnRH-mediated androgen secretion during the critical period in perinatal development [81]. Activation of the kisspeptin receptor is linked to the phospholipase C and inositol triphosphate second messenger pathway regulating multiple ion channels [82]. These findings prompted the research on how kisspeptin is involved during the beginning of puberty, and it led to the discovery that kisspeptin stimulates the neurons of the brain that were involved in the release of GnRH and possibly may have some impact on the release of LH and FSH [78].

In rat and mice models, BPA exposures have been reported to affect hypothalamic kisspeptin fiber density, KiSS-1, and ERα mRNA expression [56,83]. The disruption of the normal functioning of the HPG axis by environmental pollutants like EDCs may lead to an interruption of GnRH, LH, and FSH release and gametogenesis.

2.3 MODULATION OF HPG AXIS BY INSULIN AND INSULIN-LIKE FACTORS

2.3.1 INSULIN AND INSULIN-LIKE FACTORS

Insulin is a key homeostatic factor in the brain, acting through both the insulin receptor (IR) and insulin-like growth factor 1 receptor (IGF-1R) to regulate systemic metabolism and brain function. IRs, as well as IGF-1 receptors, are distributed throughout the brain. Insulin acts on these receptors to modulate peripheral metabolism, including regulation of appetite, reproductive function, body temperature, white fat mass, hepatic glucose output, and response to hypoglycemia. Alterations in insulin action in the brain can contribute to metabolic syndrome and the development of mood disorders and neurodegenerative diseases [22]. The central role of insulin, in addition to its peripheral role, in modulating peripheral metabolism and reproductive function is emerging. Insulin plays a cardinal role in maintaining postprandial normoglycemia. Insulin, proinsulin, and epidermal growth factor (EGF) use high-affinity cell surface receptors, the receptor tyrosine kinases (RTK), for signal transduction mechanism to regulate intracellular activities. In our earlier studies, we found that proinsulin, the prohormone precursor to insulin made in the beta cell of the islets of Langerhans of the pancreas, manifests effects similar to insulin on metabolic pathways, but with much less efficiency. EGF, on the other hand, counteracted the glycogenic effects of insulin in parenchymal hepatocytes. To understand the signal transduction mechanism, in IR binding studies, we noted that proinsulin binds with IRs, albeit less effectively. However, EGF did not share the same subset of RTK receptors, which perhaps explains its counteraction of insulin's glycogenic effect [84–86]. Today, approximately 20 different RTK classes have been identified; EGF receptor family belongs to RTK class I, whereas IR family, to RTK class II. This difference in RTK class explains their differences in signal transduction mechanism. In the following sections, we will review our current knowledge about insulin action in the brain and discuss how this can affect both brain and peripheral metabolism. In particular, we will explore how insulin, and insulin-like molecules, modulates the HPG axis and reproductive health.

2.3.1.1   Mechanism of Action

In its central role, insulin intimately controls the HPG axis. In perfused hypothalamic fragments from female rat, GnRH release was dramatically increased by low concentrations of insulin; this occurs only when glucose was available. Acutely elevated glucose levels alone did not affect GnRH release [87]. Apparently, insulin and/or glucose are metabolic modulators of GnRH secretion and mediate the effects of nutrition on gonadotrophin secretion. Sheep infused with insulin or insulin plus glucose showed an increase in LH pulse frequency, but no increase in FSH concentration [88].

The mechanism of poor reproductive health in type 1 diabetic males is not well understood. Although Sertoli cells of the testis locally secrete insulin [89], this insulin does not appear to be critical for fertility. It is currently unclear whether compromised fertility in diabetes is the result of deficiency of pancreatic insulin, testicular insulin, or both. It was observed in diabetic Akita mice that exogenous pancreatic insulin regenerates testis and restores fertility not through direct interaction with the testis but through restoring the function of the HPG axis and thereby normalizing hormone levels of LH and testosterone. It is noteworthy that exogenous pancreatic plasma insulin cannot pass through blood–testis barrier [90]. Insulin seems to have a significant impact on reproductive endocrine function. It regulates GnRH/LH hormone secretion; targets kisspeptin, agouti-related peptide, and proopiomelanocortin neurons in the brain; and acts as a signal in prenatal programming of adult reproductive function. Insulin resistance in human disease is associated with reproductive dysfunction [91].

Insulin-like factor 3 (INSL3) is expressed in Leydig cells of the testis and theca cells of the ovary. Recent data suggest both paracrine (in the testis and ovary) and endocrine actions of INSL3 in adult animals. INSL3 circulates at high concentrations in serum of adult males and its production is dependent on the differentiation effect of LH. Therefore, INSL3 is increasingly used as a specific marker of Leydig cell differentiation and function [92].

2.3.2   Insulin and Brain

Although extensive work to delineate the peripheral role of insulin in mammalian tissues has been done, only recently has a clear understanding of the central role of insulin in the brain emerged. The IR is expressed in various regions of the brain. Insulin enters the central nervous system (CNS) through the blood–brain barrier by receptor-mediated transport to regulate brain-mediated activity and peripheral insulin action and reproductive endocrinology. On a molecular level, some of the effects of insulin converge with those of the leptin signaling machinery at the point of activation of phosphatidylinositol 3-kinase (PI3K), resulting in the regulation of ATP-dependent potassium channels. These findings indicate that neuronal IR signaling has a direct role in the link between energy homeostasis, reproduction, and the development of neurodegenerative diseases [93].

IRs are widely distributed in the brain, with the highest concentrations in the olfactory bulb, hypothalamus, cerebral cortex, cerebellum, and hippocampus [94,95]. Peripherally circulating insulin crosses the blood–brain barrier in proportion to serum insulin levels via a saturable mechanism [96,97]. It is shown that the chronic lack of IR signaling in the CNS decreases hypothalamic GLUT4 expression, attenuates individual hypothalamic glucose-inhibited neuronal responses to low glucose, impairs hypothalamic neuronal activation in response to hypoglycemia, and reduces the sympathoadrenal response to hypoglycemia by shifting the glycemic level necessary to elicit appropriate sympathoadrenal responses. It is concluded that insulin acts directly in the brain to regulate both glucose sensing in hypothalamic neurons and the counterregulatory response to hypoglycemia. Because insulin-treated diabetic patients have an impaired ability to sense and appropriately respond to insulin-induced hypoglycemia, the mechanism by which insulin regulates CNS glucose sensing needs to be actively investigated as research scientists endeavor to supplant insulin-induced hypoglycemia as the rate-limiting factor in the glycemic management of diabetes [98]. Study on mice with brain-specific IR deficiency, which display obesity and impaired fertility, ultimately assigned brain IR signaling a critical role in the central regulation of fuel metabolism and reproduction [99].

A transcription factor, forkhead box–containing protein O subfamily-1 (FOXO1), that mediates the effect of insulin on the gluconeogenic genes PEPCK and glucose-6-phosphatase catalytic subunit (G6PC) [100] plays a significant role in regulating whole-body energy metabolism in peripheral tissues. In the fasted state, the liver maintains glucose homeostasis, with FOXO1 promoting the expression of hepatic gluconeogenic enzymes. In obese or diabetic states, FOXO1-dependent gene expression promotes hyperglycemia and glucose intolerance. Following feeding, pancreatic beta cells secrete insulin, which promotes the uptake of glucose by peripheral tissues, and can in part suppress gluconeogenic enzyme expression in the liver [101]. Downstream effectors of insulin, such as IR substrate (IRS) proteins and PI3K isoforms, exhibit distinct expression patterns in the CNS partially overlapping with those of the IR [102]. Hormone binding to the IRs leads to rapid autophosphorylation of the receptor, followed by tyrosine phosphorylation of IRS proteins, inducing the activation of downstream pathways such as mitogen-activated protein kinases (MAPKs) and the PI3K cascade [103]. Stimulation of the latter pathway negatively regulates the activity of forkhead-box O transcription factors (FOXO) by promoting its export from the nucleus upon phosphorylation by protein kinase B/AKT [104]. Akt2 is an important signaling molecule in the insulin signaling pathway and is required to induce glucose transport. FOXO1 in neurons in the ventromedial nucleus of the hypothalamus (VMH) also regulates glucose homeostasis. FOXO1 ablation in VMH steroidogenic factor-1 neurons increases energy expenditure without changing food intake, resulting in a lean phenotype. Obesity is associated with many metabolic disorders, including type 2 diabetes, dyslipidemia, and cardiovascular disease. It has been shown that VMH FOXO1 signaling regulates leptin sensitivity, energy expenditure, peripheral insulin action, and glucose homeostasis, suggesting VMH as a crucial hypothalamic site of whole-body metabolism, and that transcriptional programs regulated by FOXO1 are key in regulating energy balance and glucose homeostasis. Therefore, understanding molecular mechanisms responsible for body weight regulation and glucose homeostasis in the hypothalamus will provide strategies to develop pharmacological intervention to combat metabolic disorders [105]. Further study on the role of insulin on FOXO signaling pathway in the brain will show how insulin, via the brain, regulates energy metabolism and metabolic syndrome—the killer disease that includes obesity, insulin resistance, hypertension, and dyslipidemia and affects an estimated 34% of the adult population in the United States.

In metabolism, the CNS exerts control on the endocrine pancreas and can modulate the basic feedback loop linking the concentration of the main energy-yielding substrate molecules in blood with islet beta cell functions. Thus, the elementary glucose-insulin system can be modulated under physiological conditions by both the entero-insular axis and by a brain-islet axis. Experimental data suggest intervention of the brain-islet axis under the physiological circumstances where hypothalamic factor(s) may modify the endocrine pancreatic secretion [106]. The hypothalamus of vertebrate brain is the primary output node for the limbic system. It is responsible for certain metabolic processes and other activities of the autonomic nervous system and also plays a central role in the endocrine system. Through a number of small nuclei, with a variety of functions, hypothalamus links the nervous system to the endocrine system, via the pituitary gland. In this segment, we will discuss the insulin action on the hypothalamus and also how hypothalamus controls insulin release. The central effect of insulin in brain, independent of its pleiotropic peripheral actions, has been observed in rats by administering intra-cerebroventricular agonists and antagonists of insulin signaling, in the presence of basal circulating insulin. Administration of insulin mimetic agents suppressed glucose production, but central antagonism of insulin signaling impaired the ability of circulating insulin to inhibit hepatic gluconeogenesis. This report shows the cardinal role of hypothalamus in propagating the metabolic effect of insulin and/or its signal transduction level [107].

Although glucose-stimulated insulin secretion (GSIS) is driven predominantly by direct sensing of a rise in blood glucose by pancreatic β-cells, hypothalamic glucose sensing has an important role in the integrated control of peripheral glucose homeostasis, through a novel model for the regulation of GSIS. It was shown that glucokinase (GK)-dependent glucose phosphorylation in the hypothalamus may play a facilitatory role in the regulation of the first phase of insulin secretion in response to a systemic glucose load. GK activators have been highlighted as potential therapeutic candidates for type 2 diabetes. A novel central mechanism in the control of glucose-stimulated insulin release has been suggested, which may offer a future therapeutic target for improving glycemic control in type 2 diabetes [108,109]. In the glucose-sensing brain regions, such as the hypothalamus or brainstem, both glucose-excited and glucose-inhibited neurons are present [110,111].

Available data suggest that glucose-inhibited (GI) neurons release appetite-suppressing peptides, and glucose-excited (GE) neurons, appetite-stimulating peptides [112]. Through such neuronal glucose-sensing mechanisms, the brain can constantly monitor brain glucose levels to control peripheral metabolic functions involved in energy and glucose homeostasis [113]. In a review, the recently identified roles for neuronal GK, expressed in numerous hypothalamic nuclei, in glucose homeostasis and counterregulatory responses to hypoglycemia and in regulating appetite have been detailed [114]. Allosteric activators of hepatic and pancreatic GK have been suggested as potential therapeutic candidates for type 2 diabetes mellitus (T2DM) role in diabetes therapy [109]. Based on such observations, it has been proposed that pharmacological agents targeting hypothalamic glucose-sensing pathways may also represent novel therapeutic strategies for enhancing early phase insulin secretion in T2DM. This is of significance as in T2DM, GSIS typically becomes impaired [108]. To address the mechanism of GSIS by pancreatic β-cells, two major pathways are implicated: KATP channel-dependent and KATP channel-independent pathways. In the pancreatic β-cell, prandial GSIS is biphasic and involves at least these two signaling pathways. In the KATP channel-dependent pathway, enhanced glucose metabolism increases the cellular adenosine triphosphate/adenosine diphosphate (ATP/ADP) ratio, closes KATP channels, and depolarizes the cell. Activation of voltage-dependent Ca^{+2} channels increases Ca^{+2} entry and $[Ca^{+2}]i$ and stimulates the first phase of GSIS. The KATP channel-independent pathways augment the response to increased $[Ca^{+2}]i$ and stimulates the second phase of GSIS by mechanisms that are currently unknown. However, they affect different pools of insulin-containing granules in a highly coordinated manner to restore normoglycemia [115].

Insulin plays a key role in the maintenance of nutrient homeostasis through central regulation of neuropeptides, such as neuropeptide Y (NPY) [116]. NPY, a 36-amino-acid oriexigenic, or appetite stimulant, polypeptide, like ghrelin or orexin, is widely expressed in the CNS and influences many physiological processes, including cortical excitability, stress response, food intake, circadian rhythms, and cardiovascular function. The finding that insulin, like the adipocyte-derived leptin, affects both NPY expression in the hypothalamus and food intake has led to the appreciation of the brain as an insulin target tissue, with regard to energy homeostasis in mammals [117,118]. It has been suggested that insulin may inhibit hypothalamic NPY gene expression in rats. Insulin acts in the brain to suppress feeding, whereas NPY has the opposite effect [117]. Available data suggest that fasting increases NPY biosynthesis, along an arcuate nucleus-paraventricular nucleus pathway in the hypothalamus, via a mechanism dependent on low insulin levels. As fasting lowers plasma insulin levels and increases hypothalamic synthesis of NPY, it is believed that insulin may inhibit hypothalamic NPY gene expression in rats. Understanding the role of central vs. peripherally derived NPY in whole-body energy balance could shed light on the regulation of energy metabolism and the mechanisms underlying the pathogenesis of obesity [119].

In animal models, NPY has been identified as a potential target for new drugs to treat obesity and perhaps T2DM. However, the role of human hypothalamic NPY in homeostasis and glucose regulation remains enigmatic [120]. NPY may contribute to the pathogenesis of type-1 diabetes and type-2 diabetes as a minor autoantigen [121]. The arcuate nucleus in the hypothalamus is the integrative site that receives signals from the periphery reflecting energy status and projects to other hypothalamic nuclei, such as the paraventricular nucleus [122]. NPY expressed in the arcuate nucleus is one of the most potent stimulants for food intake in the CNS such that central injection of NPY readily evokes robust feeding in satiated rats [123]. In hypothalamic organotypic cultures, it has been shown that insulin inhibits NPY gene expression in the arcuate nucleus [124]. In the hypothalamic control of food intake, insulin may regulate or potentiate this hypothalamic role. Some of the central effects of insulin converge with those of the leptin signaling machinery, resulting in the regulation of ATP-dependent potassium channels [93].

2.4   CONCLUSION

Phytoestrogens, EDCs, and insulin employ multiple peripheral and central signal transduction mechanisms to control the metabolic and reproductive functions in humans and animals. Both phytoestrogens and EDCs, the exogenous molecules, and insulin, the internally secreted endogenous hormone, are modulators of the physiological processes (Table 2.1). Plant-derived phytoestrogens can mimic, modulate, or disrupt the actions of endogenous estrogens. Many phytoestrogens and fungal estrogenic compounds may enter the food chain [125]. Due to globalization, agricultural and industrial pollution, from industrialized countries and also from developing countries like China and India, has added anthropogenic estrogenic compounds to the list of environmental estrogens. In this review, we have shed light on the central role of these substances, with an emphasis on their effects on the HPG axis. It is obvious that all three groups of substances impart a significant influence on growth and reproduction, having downstream overlapping of signal transduction mechanisms.

Available data suggest that EDCs have effects on reproduction, cancer development, metabolism, obesity, and neurovascular and cardiovascular endocrinology. Animal studies, human clinical observations, and epidemiological studies converge to implicate EDCs as a significant concern to public health. The mechanisms of EDCs action involve divergent pathways, including (but not limited to) estrogenic, antiandrogenic, thyroid, PPARγ, retinoid, and actions through other NRs; steroidogenic enzymes; neurotransmitter receptors and systems; and many other pathways [29]. Such varied mechanisms of actions have also been alluded to phytoestrogens and insulin.

The prevalence of obesity has been rapidly increasing in the United States and other countries over the past two decades. This change has involved both sexes, all age ranges, and various ethnic groups. Obesity is definitely associated with a relative increase in diabetes, cardiovascular disease, various cancers, and respiratory disorders in sleep, gallbladder disease, and osteoarthritis. Obesity is an integral component of the metabolic syndrome, which is emerging as a key constellation of risk factors for cardiovascular disease [126]. It is tempting to speculate that, in addition to reproduction, the metabolic syndrome that could be precipitated by all three modulators discussed in this review is manifested by a common central pathway, where kisspeptin pathway plays a very significant role.

The neuropeptide kisspeptin regulates reproduction by stimulating GnRH neurons via the kisspeptin receptor KISS1R. In addition to GnRH neurons, KISS1R is expressed in other brain areas and peripheral tissues, which suggests that kisspeptin has additional functions beyond reproduction. It has been reported that in mice, in addition to reproduction, kisspeptin signaling influences body weight, energy expenditure, and glucose homeostasis; therefore, alterations in kisspeptin signaling might contribute, directly or indirectly, to some facets of human obesity, diabetes, or metabolic dysfunction. This newly discovered role for the kisspeptin system extends our understanding of the relationship between reproduction and energy balance and may provide novel insight into various metabolic diseases, such as diabetes, polycystic ovary syndrome (PCOS), or obesity [127].

In peripheral tissues, kisspeptin regulates insulin release. However, neither typical protein kinase C isoforms nor p38 MAPK are involved in the potentiation of glucose-induced insulin release by kisspeptin, but intracellular signaling pathways involving phospholipase C, p42/44 MAPK, and increased intracellular calcium concentration are required for the stimulatory effects on insulin secretion. The observation that kisspeptin is also capable of stimulating insulin release *in vivo* supports the conclusion that kisspeptin is a regulator of beta cell function [128]. PCOS patients are vulnerable to develop diabetes, cardiovascular diseases, and metabolic syndrome. Insulin resistance is prevalent in women with PCOS independently of obesity and is critically involved in reproductive and metabolic complications of the syndrome. Kisspeptin, along with other proteins like leptin, RBP4, and ghrelin, has been proposed as potential new markers of IR in PCOS [129]. Lower kisspeptin serum levels in PCOS women than in controls and significantly higher kisspeptin levels in normal weight PCOS subjects than in overweight or obese women with the disease have been reported, suggesting that insulin resistance is linked to decreased kisspeptin levels [130].

Sex steroid hormones (E2) play a pivotal role in the sex-specific organization and function of the kisspeptin system. EDCs and phytoestrogens are anthropogenic or naturally occurring compounds that interact with steroid hormone signaling. Thus, these compounds have the potential to disrupt the sexually dimorphic ontogeny and function of kisspeptin signaling pathways, resulting in adverse effects on neuroendocrine physiology. Disruption of kisspeptin signaling pathways could have wide ranging effects across multiple organ systems and potentially underlies a suite of adverse human health trends including precocious female puberty, idiopathic infertility, and metabolic syndrome [131].

Recent surge in neurological disorders, obesity, T2DM, and metabolic syndrome worldwide in epidemic level could not be ascribed to nutrition or genetic predisposition alone. It is tempting to speculate that, perhaps, the anthropogenic pollutants, including phytoestrogens, and insulin are working via a hypothalamus/central pathway, involving kisspeptin and many other signal transduction systems, for the precipitation of such maladies, perhaps in synchrony with nutritional status, like excess calorie intake and sedentary lifestyle.

REFERENCES

1. Roy JR, Chakraborty S and Chakraborty TR: Estrogen-like endocrine disrupting chemicals affecting puberty in humans—A review. *Med Sci Monit* 15: RA137–145, 2009.
2. Gore AC: Neuroendocrine targets of endocrine disruptors. *Hormones* 9: 16–27, 2010.
3. Chakraborty TR, Hof PR, Ng L and Gore AC: Stereologic analysis of estrogen receptor alpha (ER alpha) expression in rat hypothalamus and its regulation by aging and estrogen. *J Comp Neurol* 466: 409–421, 2003.
4. Chakraborty TR, Ng L and Gore AC: Age-related changes in estrogen receptor beta in rat hypothalamus: A quantitative analysis. *Endocrinology* 144: 4164–4171, 2003.
5. Markou A, Duka T and Prelevic GM: Estrogens and brain function. *Hormones* 4: 9–17, 2005.
6. Bulger WH, Muccitelli RM and Kupfer D: Interactions of methoxychlor, methoxychlor base-soluble contaminant, and 2,2-bis(*p*-hydroxyphenyl)-1,1,1-trichloroethane with rat uterine estrogen receptor. *J Toxicol Environ Health* 4: 881–893, 1978.
7. Shutoh Y, Takeda M, Ohtsuka R et al: Low dose effects of dichlorodiphenyltrichloroethane (DDT) on gene transcription and DNA methylation in the hypothalamus of young male rats: Implication of hormesis-like effects. *J Toxicol Sci* 34: 469–482, 2009.
8. Huang P, Rannug A, Ahlbom E, Hakansson H and Ceccatelli S: Effect of 2,3,7,8-tetrachlorodibenzo-*p*-dioxin on the expression of cytochrome P450 1A1, the aryl hydrocarbon receptor, and the aryl hydrocarbon receptor nuclear translocator in rat brain and pituitary. *Toxicol Appl Pharmacol* 169: 159–167, 2000.
9. Hutz RJ, Carvan MJ, Baldridge MG, Conley LK and Heiden TK: Environmental toxicants and effects on female reproductive function. *Trends Reprod Biol* 2: 1–11, 2006.
10. Negri-Cesi P: Bisphenol A interaction with brain development and functions. *Dose Response* 13: 1559325815590394, 2015.
11. Krishnan AV, Stathis P, Permuth SF, Tokes L and Feldman D: Bisphenol-A: An estrogenic substance is released from polycarbonate flasks during autoclaving. *Endocrinology* 132: 2279–2286, 1993.
12. Lee YM, Seong MJ, Lee JW et al: Estrogen receptor independent neurotoxic mechanism of bisphenol A, an environmental estrogen. *J Vet Sci* 8: 27–38, 2007.
13. Salama J, Chakraborty TR, Ng L and Gore AC: Effects of polychlorinated biphenyls on estrogen receptor-beta expression in the anteroventral periventricular nucleus. *Environ Health Perspect* 111: 1278–1282, 2003.
14. Caudle WM, Richardson JR, Delea KC et al: Polychlorinated biphenyl-induced reduction of dopamine transporter expression as a precursor to Parkinson's disease-associated dopamine toxicity. *Toxicol Sci* 92: 490–499, 2006.
15. Takeuchi S, Iida M, Kobayashi S, Jin K, Matsuda T and Kojima H: Differential effects of phthalate esters on transcriptional activities via human estrogen receptors alpha and beta, and androgen receptor. *Toxicology* 210: 223–233, 2005.
16. Lim CK, Kim SK, Ko DS et al: Differential cytotoxic effects of mono-(2-ethylhexyl) phthalate on blastomere-derived embryonic stem cells and differentiating neurons. *Toxicology* 264: 145–154, 2009.
17. Wilson WW, Onyenwe W, Bradner JM, Nennig SE and Caudle WM: Developmental exposure to the organochlorine insecticide endosulfan alters expression of proteins associated with neurotransmission in the frontal cortex. *Synapse* 68: 485–497, 2014.
18. Wozniak AL, Bulayeva NN and Watson CS: Xenoestrogens at picomolar to nanomolar concentrations trigger membrane estrogen receptor-alpha-mediated Ca2+ fluxes and prolactin release in GH3/B6 pituitary tumor cells. *Environ Health Perspect* 113: 431–439, 2005.

19. Collins-Burow BM, Burow ME, Duong BN and McLachlan JA: Estrogenic and anti-estrogenic activities of flavonoid phytochemicals through estrogen receptor binding-dependent and -independent mechanisms. *Nutr Cancer* 38: 229–244, 2000.

20. File SE, Jarrett N, Fluck E, Duffy R, Casey K and Wiseman H: Eating soya improves human memory. *Psychopharmacology (Berl)* 157: 430–436, 2001.

21. Jeong EJ, Lee HK, Lee KY et al: The effects of lignan-riched extract of *Shisandra chinensis* on amyloid-beta-induced cognitive impairment and neurotoxicity in the cortex and hippocampus of mouse. *J Ethnopharmacol* 146: 347–354, 2013.

22. Kleinridders A, Ferris HA, Cai W and Kahn CR: Insulin action in brain regulates systemic metabolism and brain function. *Diabetes* 63: 2232–2243, 2014.

23. Blazquez E, Velazquez E, Hurtado-Carneiro V and Ruiz-Albusac JM: Insulin in the brain: its pathophysiological implications for States related with central insulin resistance, type 2 diabetes and Alzheimer's disease. *Front Endocrinol* 5: 161, 2014.

24. McDonnell DP and Norris JD: Connections and regulation of the human estrogen receptor. *Science* 296: 1642–1644, 2002.

25. Pettersson K and Gustafsson JA: Role of estrogen receptor beta in estrogen action. *Annu Rev Physiol* 63: 165–192, 2001.

26. Nilsson S, Makela S, Treuter E et al: Mechanisms of estrogen action. *Physiol Rev* 81: 1535–1565, 2001.

27. Katzenellenbogen BS: Estrogen receptors: Bioactivities and interactions with cell signaling pathways. *Biol Reprod* 54: 287–293, 1996.

28. Hall JM and McDonnell DP: Coregulators in nuclear estrogen receptor action: From concept to therapeutic targeting. *Mol Interv* 5: 343–357, 2005.

29. Diamanti-Kandarakis E, Bourguignon JP, Giudice LC et al: Endocrine-disrupting chemicals: An Endocrine Society scientific statement. *Endocrine Rev* 30: 293–342, 2009.

30. Toppari J, Kaleva M and Virtanen HE: Trends in the incidence of cryptorchidism and hypospadias, and methodological limitations of registry-based data. *Hum Reprod Update* 7: 282–286, 2001.

31. Parkin DM, Bray F, Ferlay J and Pisani P: Global cancer statistics, 2002. *CA Cancer J Clin* 55: 74–108, 2005.

32. Lee HR, Jeung EB, Cho MH, Kim TH, Leung PC and Choi KC: Molecular mechanism(s) of endocrine-disrupting chemicals and their potent oestrogenicity in diverse cells and tissues that express oestrogen receptors. *J Cell Mol Med* 17: 1–11, 2013.

33. Gencel VB, Benjamin MM, Bahou SN and Khalil RA: Vascular effects of phytoestrogens and alternative menopausal hormone therapy in cardiovascular disease. *Mini Rev Med Chem* 12: 149–174, 2012.

34. Shanle EK and Xu W: Endocrine disrupting chemicals targeting estrogen receptor signaling: Identification and mechanisms of action. *Chem Res Toxicol* 24: 6–19, 2011.

35. Hall JM, Couse JF and Korach KS: The multifaceted mechanisms of estradiol and estrogen receptor signaling. *J Biol Chem* 276: 36869–36872, 2001.

36. Gustafsson JA: What pharmacologists can learn from recent advances in estrogen signalling. *Trends Pharmacol Sci* 24: 479–485, 2003.

37. Swedenborg E, Ruegg J, Makela S and Pongratz I: Endocrine disruptive chemicals: Mechanisms of action and involvement in metabolic disorders. *J Mol Endocrinol* 43: 1–10, 2009.

38. Manavathi B and Kumar R: Steering estrogen signals from the plasma membrane to the nucleus: Two sides of the coin. *J Cell Physiol* 207: 594–604, 2006.

39. Harrington WR, Kim SH, Funk CC et al: Estrogen dendrimer conjugates that preferentially activate extranuclear, nongenomic versus genomic pathways of estrogen action. *Mol Endocrinol* 20: 491–502, 2006.

40. Baker ME, Medlock KL and Sheehan DM: Flavonoids inhibit estrogen binding to rat alpha-fetoprotein. *Proc Soc Exp Biol Med* 217: 317–321, 1998.
41. Boas M, Feldt-Rasmussen U, Skakkebaek NE and Main KM: Environmental chemicals and thyroid function. *Eur J Endocrinol* 154: 599–611, 2006.
42. Frye CA, Bo E, Calamandrei G et al: Endocrine disrupters: A review of some sources, effects, and mechanisms of actions on behaviour and neuroendocrine systems. *J Neuroendocrinol* 24: 144–159, 2012.
43. Patisaul HB: Endocrine disruption by dietary phyto-oestrogens: Impact on dimorphic sexual systems and behaviours. *Proc Nutr Soc* 1–15, 2016.
44. World Health Organization: *Neurological disorders: Public health challenges.* World Health Organization, Geneva, 2006.
45. Roncati L, Termopoli V and Pusiol T: Negative role of the environmental endocrine disruptors in the human neurodevelopment. *Front Neurol* 7: 143, 2016.
46. Okada M, Makino A, Nakajima M, Okuyama S, Furukawa S and Furukawa Y: Estrogen stimulates proliferation and differentiation of neural stem/progenitor cells through different signal transduction pathways. *Int J Mol Sci* 11: 4114–4123, 2010.
47. Jang YJ, Park HR, Kim TH et al: High dose bisphenol A impairs hippocampal neurogenesis in female mice across generations. *Toxicology* 296: 73–82, 2012.
48. Grandjean P and Landrigan PJ: Neurobehavioural effects of developmental toxicity. *Lancet Neurol* 13: 330–338, 2014.
49. Rossignol DA, Genuis SJ and Frye RE: Environmental toxicants and autism spectrum disorders: A systematic review. *Transl Psychiatry* 4: e360, 2014.
50. Steenland K, Hein MJ, Cassinelli RT, 2nd et al: Polychlorinated biphenyls and neurodegenerative disease mortality in an occupational cohort. *Epidemiology* 17: 8–13, 2006.
51. Gore AC: Developmental programming and endocrine disruptor effects on reproductive neuroendocrine systems. *Front Neuroendocrinol* 29: 358–374, 2008.
52. Barker DJ: The developmental origins of adult disease. *Eur J Epidemiol* 18: 733–736, 2003.
53. Chakraborty TR, Alicea E and Chakraborty S: Relationships between urinary biomarkers of phytoestrogens, phthalates, phenols, and pubertal stages in girls. *Adolesc Health Med Ther* 3: 17–26, 2012.
54. Bagchi D, Das DK, Tosaki A, Bagchi M and Kothari SC: Benefits of resveratrol in women's health. *Drugs Exp Clin Res* 27: 233–248, 2001.
55. Cheng G, Remer T, Prinz-Langenohl R, Blaszkewicz M, Degen GH and Buyken AE: Relation of isoflavones and fiber intake in childhood to the timing of puberty. *Am J Clin Nutr* 92: 556–564, 2010.
56. Patisaul HB, Todd KL, Mickens JA and Adewale HB: Impact of neonatal exposure to the ERalpha agonist PPT, bisphenol-A or phytoestrogens on hypothalamic kisspeptin fiber density in male and female rats. *Neurotoxicology* 30: 350–357, 2009.
57. Kurzer MS and Xu X: Dietary phytoestrogens. *Annu Rev Nutr* 17: 353–381, 1997.
58. Nurmi T, Mazur W, Heinonen S, Kokkonen J and Adlercreutz H: Isoflavone content of the soy based supplements. *J Pharm Biomed Anal* 28: 1–11, 2002.
59. Vergne S, Bennetau-Pelissero C, Lamothe V et al: Higher bioavailability of isoflavones after a single ingestion of a soya-based supplement than a soya-based food in young healthy males. *Br J Nutr* 99: 333–344, 2008.
60. North American Menopause Society: The role of soy isoflavones in menopausal health: Report of The North American Menopause Society/Wulf H. Utian Translational Science Symposium in Chicago, IL (October 2010). *Menopause* 18: 732–753, 2011.
61. Lee YB, Lee HJ, Won MH et al: Soy isoflavones improve spatial delayed matching-to-place performance and reduce cholinergic neuron loss in elderly male rats. *J Nutr* 134: 1827–1831, 2004.
62. Thorp AA, Sinn N, Buckley JD, Coates AM and Howe PR: Soya isoflavone supplementation enhances spatial working memory in men. *Br J Nutr* 102: 1348–1354, 2009.

63. Kuiper GG, Shughrue PJ, Merchenthaler I and Gustafsson JA: The estrogen receptor beta subtype: A novel mediator of estrogen action in neuroendocrine systems. *Front Neuroendocrinol* 19: 253–286, 1998.

64. Pfitscher A, Reiter E and Jungbauer A: Receptor binding and transactivation activities of red clover isoflavones and their metabolites. *J Steroid Biochem Mol Biol* 112: 87–94, 2008.

65. Rochester JR and Millam JR: Phytoestrogens and avian reproduction: Exploring the evolution and function of phytoestrogens and possible role of plant compounds in the breeding ecology of wild birds. *Comp Biochem Physiol A Mol Integr Physiol* 154: 279–288, 2009.

66. Borras C, Gambini J, Gomez-Cabrera MC et al: Genistein, a soy isoflavone, up-regulates expression of antioxidant genes: Involvement of estrogen receptors, ERK1/2, and NFkappaB. *FASEB J* 20: 2136–2138, 2006.

67. Borras C, Gambini J and Vina J: Mitochondrial oxidant generation is involved in determining why females live longer than males. *Front Biosci* 12: 1008–1013, 2007.

68. Raschke M, Rowland IR, Magee PJ and Pool-Zobel BL: Genistein protects prostate cells against hydrogen peroxide-induced DNA damage and induces expression of genes involved in the defence against oxidative stress. *Carcinogenesis* 27: 2322–2330, 2006.

69. Bryant M, Cassidy A, Hill C, Powell J, Talbot D and Dye L: Effect of consumption of soy isoflavones on behavioural, somatic and affective symptoms in women with premenstrual syndrome. *Br J Nutr* 93: 731–739, 2005.

70. Hooper L, Ryder JJ, Kurzer MS et al: Effects of soy protein and isoflavones on circulating hormone concentrations in pre- and post-menopausal women: A systematic review and meta-analysis. *Hum Reprod Update* 15: 423–440, 2009.

71. Basaria S, Wisniewski A, Dupree K et al: Effect of high-dose isoflavones on cognition, quality of life, androgens, and lipoprotein in post-menopausal women. *J Endocrinol Investig* 32: 150–155, 2009.

72. Kritz-Silverstein D, Von Muhlen D, Barrett-Connor E and Bressel MA: Isoflavones and cognitive function in older women: The SOy and Postmenopausal Health In Aging (SOPHIA) Study. *Menopause* 10: 196–202, 2003.

73. Huang MH, Luetters C, Buckwalter GJ et al: Dietary genistein intake and cognitive performance in a multiethnic cohort of midlife women. *Menopause* 13: 621–630, 2006.

74. Kreijkamp-Kaspers S, Kok L, Grobbee DE, de Haan EH, Aleman A and van der Schouw YT: Dietary phytoestrogen intake and cognitive function in older women. *J Gerontol A Biol Sci Med Sci* 62: 556–562, 2007.

75. White LR, Petrovitch H, Ross GW et al: Brain aging and midlife tofu consumption. *J Am Coll Nutr* 19: 242–255, 2000.

76. Borras C, Sastre J, Garcia-Sala D, Lloret A, Pallardo FV and Vina J: Mitochondria from females exhibit higher antioxidant gene expression and lower oxidative damage than males. *Free Radic Biol Med* 34: 546–552, 2003.

77. Vina J, Sastre J, Pallardo F and Borras C: Mitochondrial theory of aging: Importance to explain why females live longer than males. *Antioxid Redox Signal* 5: 549–556, 2003.

78. Pasquier J, Kamech N, Lafont AG, Vaudry H, Rousseau K and Dufour S: Molecular evolution of GPCRs: Kisspeptin/kisspeptin receptors. *J Mol Endocrinol* 52: T101–T117, 2014.

79. Richard N, Corvaisier S, Camacho E and Kottler ML: KiSS-1 and GPR54 at the pituitary level: Overview and recent insights. *Peptides* 30: 123–129, 2009.

80. Messager S, Chatzidaki EE, Ma D et al: Kisspeptin directly stimulates gonadotropin-releasing hormone release via G protein-coupled receptor 54. *Proc Natl Acad Sci U S A* 102: 1761–1766, 2005.

81. Kauffman AS, Park JH, McPhie-Lalmansingh AA et al: The kisspeptin receptor GPR54 is required for sexual differentiation of the brain and behavior. *J Neurosci* 27: 8826–8835, 2007.

82. Liu X, Lee K and Herbison AE: Kisspeptin excites gonadotropin-releasing hormone neurons through a phospholipase C/calcium-dependent pathway regulating multiple ion channels. *Endocrinology* 149: 4605–4614, 2008.

83. Xi W, Lee CK, Yeung WS et al: Effect of perinatal and postnatal bisphenol A exposure to the regulatory circuits at the hypothalamus–pituitary–gonadal axis of CD-1 mice. *Reprod Toxicol* 31: 409–417, 2011.

84. Agius L, Chowdhury MH and Alberti KG: Regulation of ketogenesis, gluconeogenesis and the mitochondrial redox state by dexamethasone in hepatocyte monolayer cultures. *Biochem J* 239: 593–601, 1986.

85. Chowdhury MH and Agius L: Epidermal growth factor counteracts the glycogenic effect of insulin in parenchymal hepatocyte cultures. *Biochem J* 247: 307–314, 1987.

86. Agius L, Chowdhury MH, Davis SN and Alberti KG: Regulation of ketogenesis, gluconeogenesis, and glycogen synthesis by insulin and proinsulin in rat hepatocyte monolayer cultures. *Diabetes* 35: 1286–1293, 1986.

87. Arias P, Rodriguez M, Szwarcfarb B, Sinay IR and Moguilevsky JA: Effect of insulin on LHRH release by perifused hypothalamic fragments. *Neuroendocrinology* 56: 415–418, 1992.

88. Miller DW, Blache D and Martin GB: The role of intracerebral insulin in the effect of nutrition on gonadotrophin secretion in mature male sheep. *J Endocrinol* 147: 321–329, 1995.

89. Gomez O, Ballester B, Romero A et al: Expression and regulation of insulin and the glucose transporter GLUT8 in the testes of diabetic rats. *Horm Metab Res* 41: 343–349, 2009.

90. Schoeller EL, Albanna G, Frolova AI and Moley KH: Insulin rescues impaired spermatogenesis via the hypothalamic-pituitary-gonadal axis in Akita diabetic mice and restores male fertility. *Diabetes* 61: 1869–1878, 2012.

91. Sliwowska JH, Fergani C, Gawalek M, Skowronska B, Fichna P and Lehman MN: Insulin: Its role in the central control of reproduction. *Physiol Behav* 133: 197–206, 2014.

92. Ferlin A, Arredi B, Zuccarello D, Garolla A, Selice R and Foresta C: Paracrine and endocrine roles of insulin-like factor 3. *J Endocrinol Invest* 29: 657–664, 2006.

93. Plum L, Schubert M and Bruning JC: The role of insulin receptor signaling in the brain. *Trends Endocrinol Metab* 16: 59–65, 2005.

94. Havrankova J, Roth J and Brownstein M: Insulin receptors are widely distributed in the central nervous system of the rat. *Nature* 272: 827–829, 1978.

95. van Houten M, Posner BI, Kopriwa BM and Brawer JR: Insulin-binding sites in the rat brain: *in vivo* localization to the circumventricular organs by quantitative radioautography. *Endocrinology* 105: 666–673, 1979.

96. Margolis RU and Altszuler N: Insulin in the cerebrospinal fluid. *Nature* 215: 1375–1376, 1967.

97. Woods SC and Porte D, Jr.: Relationship between plasma and cerebrospinal fluid insulin levels of dogs. *Am J Physiol* 233: E331–E334, 1977.

98. Diggs-Andrews KA, Zhang X, Song Z, Daphna-Iken D, Routh VH and Fisher SJ: Brain insulin action regulates hypothalamic glucose sensing and the counterregulatory response to hypoglycemia. *Diabetes* 59: 2271–2280, 2010.

99. Bruning JC, Gautam D, Burks DJ et al: Role of brain insulin receptor in control of body weight and reproduction. *Science* 289: 2122–2125, 2000.

100. Puigserver P, Rhee J, Donovan J et al: Insulin-regulated hepatic gluconeogenesis through FOXO1-PGC-1alpha interaction. *Nature* 423: 550–555, 2003.

101. Gross DN, van den Heuvel AP and Birnbaum MJ: The role of FoxO in the regulation of metabolism. *Oncogene* 27: 2320–2336, 2008.

102. Horsch D and Kahn CR: Region-specific mRNA expression of phosphatidylinositol 3-kinase regulatory isoforms in the central nervous system of C57BL/6J mice. *J Comp Neurol* 415: 105–120, 1999.

103. White MF: Insulin signaling in health and disease. *Science* 302: 1710–1711, 2003.
104. Biggs WH, 3rd, Meisenhelder J, Hunter T, Cavenee WK and Arden KC: Protein kinase B/Akt-mediated phosphorylation promotes nuclear exclusion of the winged helix transcription factor FKHR1. *Proc Natl Acad Sci U S A* 96: 7421–7426, 1999.
105. Kim KW, Donato J, Jr., Berglund ED et al: FOXO1 in the ventromedial hypothalamus regulates energy balance. *J Clin Investig* 122: 2578–2589, 2012.
106. Helman A, Marre M, Bobbioni E, Poussier P, Reach G and Assan R: The brain-islet axis: the nervous control of the endocrine pancreas. *Diabetes Metab* 8: 53–64, 1982.
107. Obici S, Zhang BB, Karkanias G and Rossetti L: Hypothalamic insulin signaling is required for inhibition of glucose production. *Nat Med* 8: 1376–1382, 2002.
108. Osundiji MA, Lam DD, Shaw J et al: Brain glucose sensors play a significant role in the regulation of pancreatic glucose-stimulated insulin secretion. *Diabetes* 61: 321–328, 2012.
109. Grimsby J, Sarabu R, Corbett WL et al: Allosteric activators of glucokinase: Potential role in diabetes therapy. *Science* 301: 370–373, 2003.
110. Kang L, Routh VH, Kuzhikandathil EV, Gaspers LD and Levin BE: Physiological and molecular characteristics of rat hypothalamic ventromedial nucleus glucosensing neurons. *Diabetes* 53: 549–559, 2004.
111. Wang R, Liu X, Hentges ST et al: The regulation of glucose-excited neurons in the hypothalamic arcuate nucleus by glucose and feeding-relevant peptides. *Diabetes* 53: 1959–1965, 2004.
112. Muroya S, Yada T, Shioda S and Takigawa M: Glucose-sensitive neurons in the rat arcuate nucleus contain neuropeptide Y. *Neurosci Lett* 264: 113–116, 1999.
113. Levin BE: Metabolic sensing neurons and the control of energy homeostasis. *Physiol Behav* 89: 486–489, 2006.
114. De Backer I, Hussain SS, Bloom SR and Gardiner JV: Insights into the role of neuronal glucokinase. *Am J Physiol Endocrinol Metab* 311: E42–E55, 2016.
115. Straub SG and Sharp GW: Glucose-stimulated signaling pathways in biphasic insulin secretion. *Diabetes Metab Res Rev* 18: 451–463, 2002.
116. Mayer CM and Belsham DD: Insulin directly regulates NPY and AgRP gene expression via the MAPK MEK/ERK signal transduction pathway in mHypoE-46 hypothalamic neurons. *Mol Cell Endocrinol* 307: 99–108, 2009.
117. Schwartz MW, Sipols AJ, Marks JL et al: Inhibition of hypothalamic neuropeptide Y gene expression by insulin. *Endocrinology* 130: 3608–3616, 1992.
118. Baskin DG, Figlewicz Lattemann D, Seeley RJ, Woods SC, Porte D, Jr. and Schwartz MW: Insulin and leptin: Dual adiposity signals to the brain for the regulation of food intake and body weight. *Brain Res* 848: 114–123, 1999.
119. Zhang W, Cline MA and Gilbert ER: Hypothalamus-adipose tissue crosstalk: Neuropeptide Y and the regulation of energy metabolism. *Nutr Metab (Lond)* 11: 27, 2014.
120. Frankish HM, Dryden S, Hopkins D, Wang Q and Williams G: Neuropeptide Y, the hypothalamus, and diabetes: Insights into the central control of metabolism. *Peptides* 16: 757–771, 1995.
121. Skarstrand H, Dahlin LB, Lernmark A and Vaziri-Sani F: Neuropeptide Y autoantibodies in patients with long-term type 1 and type 2 diabetes and neuropathy. *J Diabetes Complications* 27: 609–617, 2013.
122. Kalra SP and Kalra PS: NPY and cohorts in regulating appetite, obesity and metabolic syndrome: Beneficial effects of gene therapy. *Neuropeptides* 38: 201–211, 2004.
123. Clark JT, Kalra PS, Crowley WR and Kalra SP: Neuropeptide Y and human pancreatic polypeptide stimulate feeding behavior in rats. *Endocrinology* 115: 427–429, 1984.
124. Sato I, Arima H, Ozaki N et al: Insulin inhibits neuropeptide Y gene expression in the arcuate nucleus through GABAergic systems. *J Neurosci* 25: 8657–8664, 2005.

125. Lorand T, Vigh E and Garai J: Hormonal action of plant derived and anthropogenic non-steroidal estrogenic compounds: Phytoestrogens and xenoestrogens. *Curr Med Chem* 17: 3542–3574, 2010.
126. Rubenstein AH: Obesity: A modern epidemic. *Trans Am Clin Climatol Assoc* 116: 103–111; discussion 112–103, 2005.
127. Tolson KP, Garcia C, Yen S et al: Impaired kisspeptin signaling decreases metabolism and promotes glucose intolerance and obesity. *J Clin Investig* 124: 3075–3079, 2014.
128. Bowe JE, King AJ, Kinsey-Jones JS et al: Kisspeptin stimulation of insulin secretion: mechanisms of action in mouse islets and rats. *Diabetologia* 52: 855–862, 2009.
129. Polak K, Czyzyk A, Simoncini T and Meczekalski B: New markers of insulin resistance in polycystic ovary syndrome. *J Endocrinol Investig.* 2016. doi:10.1007/s40618-016-0523-8
130. Panidis D, Rousso D, Koliakos G et al: Plasma metastin levels are negatively correlated with insulin resistance and free androgens in women with polycystic ovary syndrome. *Fertil Steril* 85: 1778–1783, 2006.
131. Patisaul HB: Effects of environmental endocrine disruptors and phytoestrogens on the kisspeptin system. *Adv Exp Med Biol* 784: 455–479, 2013.

3 Environmental Pollution and Climate Change Impacts on Human Health with Particular Reference to Brain
A Review

Monowar Alam Khalid and
Syed Monowar Alam Shahid

CONTENTS

3.1 INTRODUCTION

The environment has a profound impact on all living beings, be it plants, animals, or humans. The physical and social environment is an important factor to shaping our lives and determining health, along with our genetic makeup and personal characteristics. A healthy environment instills confidence, safety, good health, and peace of mind, leading to a healthy growth and better contributions as individuals toward society, communities, and their ecosystems. The environment starts impacting our brain right from the beginning of the human existence, i.e., through adequate nutrition to the fetus in the womb during pregnancy, which ensures brain stimulation.

In the history of human evolution, brain enlargement was very slow during the first four million years but developed faster for the past 800,000 years (Potts, 2010), which interestingly coincided with the period of strong climate fluctuations in the world. The larger brain size enabled hominins to process and store information and plan in advance to solve abstract problems. The human brain helped hominins find new ways to cope with their surrounding environment, as evident around 280,000 years ago, which was in fact the beginning of the technological thinking of the human brain, when they made tools and spears to help in cutting and hunting, and there are evidences that prove that even grindstones were used to process plant foods (Potts, 2010). Various studies suggest that human adaptations evolved as a response to environmental changes (Potts, 2010; Potts and Sloan, 2010).

Evidences suggest that the hominid evolution occurred in East Africa, which underwent dramatic changes in its landscape during the last 10 million years from a flat homogeneous region with mountains and vegetation to numerous lake basins due to changes in local precipitation and evaporation regime. This unusual geology and climate of East Africa created periods of highly variable local climate, which could have driven hominin evolution, encephalization, and dispersal, as evident from a significant hominin speciation and brain expansion event coinciding with the occurrence of highly variable, extensive, deep water lakes (Maslin et al., 2014). Studies of Professor Mark Maslin of the University College London show that "modern humans were born out of climate change, as they had to deal with rapid switching from famine to feast, and back again, which drove the appearance of new species with bigger brains," which forced humans to disperse out of East Africa (Shultz and Maslin, 2013). Various objects found in Africa that gave a clue about the origin of human brain size and the respective tools used by humans during that period are shown in Figure 3.1.

The human brain is just about 2% of the whole body weight, but it uses 20% of the oxygen (Raichle and Gusnard, 2002). The brain is nearly 73% water and a dehydration of 2% can affect your memory and other cognitive skills (Mitchell et al., 1945; Adan, 2012). Interestingly, the brain, which weighs about three pounds, has 60% fat, making it the fattiest organ of the body, having nearly 25% of the body's cholesterol stored in it, since cholesterol is an integral part of every brain cell, without which brain cells will die (Björkhem and Meaney, 2004). The brain is the most vital organ of our body (Figure 3.2), which is the seat for consciousness, learning, memory, emotions, and detection of any changes happening around us

FIGURE 3.1   Objects found in Africa giving clues about human origin. (From Potts, R., Smithsonian Human Origins Program, 2010, http://humanorigins.si.edu/research/climate-and -human-evolution/climate-effects-human-evolution/http://humanorigins.si.edu/research/age -humans-evolutionary-perspectives-anthropocene.)

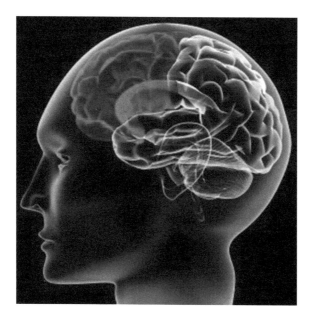

FIGURE 3.2   (See color insert.) Human brain. (From http://www.climatechanges.site/mental -health-a-factor-in-climate-change/.)

through sense organs. Our activities, beliefs, attitudes, and ideas with which we feed our brain influence our behavior through positive or negative outcome. As per World Health Organization studies, nearly 10% of the world population suffer from neuropsychiatric conditions, posttraumatic stress, obsessive compulsive disorder, panic disorder, primary insomnia, etc. Other than these, a large population of elderly people have nervous disorders like Alzheimer's disease, stroke, schizophrenia, and Parkinson's disease, which may aggravate further due to environmental pollution and climate change. More than 24 million people globally were suffering from Alzheimer's disease in 2005 (Ferri et al., 2005).

Neurodegenerative disorders, including Alzheimer's disease and Parkinson's disease, have been closely linked to disturbances in copper homeostasis in the central nervous system (CNS) and the brain (Sayre et al., 2000; Strausak et al., 2001), and Alzheimer's disease is the most common cause of dementia. Globally, more than 35 million people were suffering from dementia in 2010 (Alzheimer's Disease International, 2010). As per estimate, more than 60% of people with dementia live in least developed countries (LDCs), which is projected to increase to 71% by 2050 (Ferri et al., 2005). This increase in number is worrisome as LDCs are going to be badly impacted by environmental pollution and climate change, which will escalate already existing medical issues.

3.2 ENVIRONMENTAL POLLUTION AND ITS IMPACT ON HUMAN HEALTH

Pollution is basically an undesirable change caused by pollutants in the physical, chemical, or biological characteristics of air, water, and land, rendering it harmful on life of living beings like humans, animals, or plants. There have been environmental outbreaks and accidents happening in history that have caused neurological disorders in humans. A few prominent were when hexachlorobenzene in seed grains in South East Turkey in 1955–56 affected more than 200 people with neurological disorders and another famous neurological disorder called "Minamata" disease that occurred at Minamata, Japan, in 1956 due to methylmercury in fish, affecting more than 200 people. Lead impacted thousands of people in U.S. cities during the 1960s and 1970s, affecting them behaviorally and mentally. In 1972, more than 6500 people in Iraq were impacted with neurological disease due to methylmercury in seed grains, killing 500 of them. Carbamate pesticide in watermelons caused neurological disorder in more than a thousand reported cases in California in 1985 (Yassi and Kjellstrom, 1997). The worst industrial environmental gas tragedy happened at Bhopal, India, in December 1984, which affected 500,000 people who were exposed to methyl isocyanate (MIC) gas and other chemicals from an industrial plant, killing more than 8000 people within two weeks and the same numbers a few days later (https://en.wikipedia.org/wiki/Bhopal_disaster). The gas caused cerebral edema as found during autopsies, thus a neurological disaster. Environmental pollutants like pesticides, lead, particulate matter (PM), coal dust, bacteria, and noise have profound impact on human health. These pollutants, causing air, water, or land pollution, which are detrimental to human health are given as follows.

3.2.1 AIR POLLUTION

Air pollution is caused by a high percentage of contaminants like solid particles, liquid droplets, or gases in quantities, which is capable of causing injury to humans, animals, plant life, and even physical structures like buildings. Studies by Zanobetti et al. (2014) suggest that ambient air pollution by fine particles (PM 2.5) may be linked to the development of neurological disorders and diabetes as seen in elderly patients older than 65 years already with neurological conditions like dementia and Alzheimer's disease who were hospitalized more after a rise in fine particle pollution in their areas; however, in fewer places, hospitalization of similar patients were fewer, leaving scope for more studies to establish this association. Among the primary air pollutants that impact the health of humans and plants are the following:

Particulate matter: An increased level of suspended PM, which is fine particles in the air, causes health hazards like heart disease, reduced lung function, eye ailments, and lung cancer in humans and when settled as dust on plants causes obstruction in stomata, affecting photosynthetic functions of the leaf and thus affecting plant growth. For humans, they are detrimental as they cause wheezing, bronchitis, and asthma.

Oxides of nitrogen: The atmosphere has mainly nitrogen and nitric and nitrous oxides, formed because of bacterial decomposition and escape from automobile and industrial exhausts. These oxides of nitrogen, when combined with moisture in the atmosphere, cause smog and acidity of rain water, which adversely impacts human and plant life.

Carbon monoxide (CO): An odorless, tasteless, and invisible gas, carbon monoxide is mainly released in the atmosphere by oxidation of methane, automobile exhaust, and industries like steel and petroleum and fossil fuels' incomplete combustion. On inhalation, it decreases the oxygen-carrying capacity of the blood, causing dizziness, headache, and unconsciousness. Its prolonged exposure affects the brain and can even be fatal for humans.

Lead (Pb): Lead used as an antiknocking agent in automobiles is released in the atmosphere. It is also found in paint pigments and lead plumbing pipes, which also causes lead pollution in the environment. Once it enters the human body as air pollutant, it enters the blood and competes with calcium to reach the bone marrow, interferes in the formation of red blood corpuscles and multiplication of blood cells, leading to anemia and blood cancer. Through the blood, it affects the brain by damaging it and the kidneys by making them dysfunctional. In general, pollutants like lead are more dangerous for the brain as they damage the brain and kidney and cause trouble in learning, but others like pesticides, coal dust, and bacteria are also harmful for human health.

Ozone (O$_3$): Ozone is among the secondary pollutants found in the stratosphere as a natural protective layer of earth to save it from the harmful ultraviolet radiations, which is being depleted due to human actions. The reduced concentration of ozone allows harmful ultraviolet radiations to reach our atmosphere and thus causing skin cancer, irritation to eyes, and damage to the immune system in humans. It also adversely impacts agriculture and plants.

Toxic smog may trigger depression and damage to the nervous system along with negative impacts on other organs of the body like lungs, heart, skin, eye, etc. The brain will be impacted as toxic pollutants can cross the barrier between blood and brain, which will cause damage to vital organs (Figure 3.3). Also, metals released from fossil fuel burning will accelerate the previous process, causing depression and stroke. The nervous system may experience permanent damage due to severe exposure of carbon monoxide, which can bind with hemoglobin firmer than oxygen (Spickernell and Parmenter, 2015).

As per the Health Effects Institute, USA (Health Effects Institute Report 154, 2010), deaths in East Asia and South Asia due to leading pollutants are basically due to PM, household air pollution, and lead exposures (Figure 3.4).

There were 1.3 million deaths in East Asia and 1.1 million deaths in South Asia during 2010 due to PM alone, which was due to human activities like fossil fuel burning in vehicles and factories. Indoor or household air pollution caused 1.1

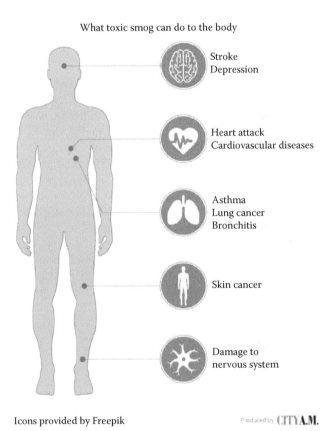

FIGURE 3.3 Impact of toxic smog on human organs. (From Spickernell, S., Parmenter, C., A toxic smog has shrouded London: Here's how it could affect our health. A City A.M. [Blog], 2015, http://www.cityam.com/212008/toxic-smog-has-shrouded-london-heres-how -it-could-affect-our-health.)

FIGURE 3.4 Death due to leading pollutants in 2010 in Asia and South Asia. (From the Health Effects Institute [HEI] Report 154, *Public Health and Air Pollution in Asia* [*PAPA*]: *Coordinated Studies of Short-term Exposure to Air Pollution and Daily Mortality in Four Cities*, HEI Public Health and Air Pollution in Asia Program, Health Effects Institute, Boston, 2010.)

million and 1.3 million deaths in East and South Asia, respectively (HEI Report 154, 2010), whereas lead pollution having fatal effects on humans and animals caused 0.2 million deaths each in East and South Asia (Figure 3.4). As per a latest report on deaths due to pollution in Asian countries (HEI Report 154, 2010), it was found that India has the highest death rate, i.e., more than 2.6 million, followed by China with 2.3 million, whereas Singapore had the lowest rate, i.e., around 3467 persons per year (Figure 3.5). The main reason was air pollution, both indoor as well as outdoor, in these already polluted cities of Asia.

3.2.2 WATER POLLUTION

Water is the most essential component on earth, without which life is not possible for any living being. The human body has about 94% of water in the fetus stage, 75% in infants, 60% in adults, and around 50% in elderly persons, and our survival is not possible if we don't take water for more than three to five days. The water in the human body is extremely important for critical body functions as it is the vital nutrient as building material for the life of every cell. It regulates the internal body temperature through sweating and transports nutrients and other essential elements to our cells. It helps in flushing out waste through urination and lubricates joints and forms saliva. Water helps absorb shocks for brain, spinal cord, and for the growing fetus (Altman, 1961).

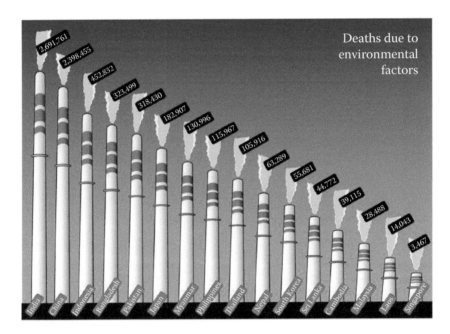

FIGURE 3.5   Deaths due to environmental factors in Asia. (From the Health Effects Institute [HEI] Report 154, *Public Health and Air Pollution in Asia [PAPA]: Coordinated Studies of Short-Term Exposure to Air Pollution and Daily Mortality in Four Cities*, HEI Public Health and Air Pollution in Asia Program, Health Effects Institute, Boston, 2010.)

Main sources of water on earth are rain water, ground water, snow, and sea water. Although 71% of our earth's surface is covered with water, only 3% of the earth's water is available to us as freshwater, among which 69% is locked in glaciers and ice caps, 30% is stored in groundwater, and a small fraction, i.e., 0.3%, is available from lakes, rivers, wetlands, and swamps all together for the use by seven billion humans and all plants and animals on this earth (Figure 3.6).

Humans are mainly responsible for pollution of water through their activities of directly discharging pollutants in the water bodies without treatment or removal of harmful compounds. Major sources of water pollution are domestic and municipal sewage, industrial waste, and radioactive and chemical wastes contaminating water bodies like rivers, lakes, ocean, aquifers, and underground water sources. It not only affects humans but equally causes damaging effects on organisms, species, both individually and at population level, or the whole communities disturbing the natural environment and balance of ecosystems. Water pollution affects human health through the food chain, as pollutants enter vegetables, fruits, grains, plants, and fodder, causing illness. Water pollution is mainly responsible for water-borne diseases like diarrhea, cholera, dysentery, typhoid fever, eye infections, malaria, schistosomiasis, sleeping sickness, and dracunchliasis. Also, vector-borne diseases due to mosquitoes and flies breeding in or near contaminated water will give rise to malaria and dengue, affecting human health.

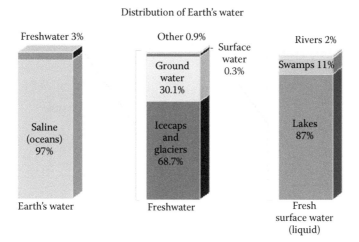

Distribution of Earth's water

FIGURE 3.6 Distribution of Earth's water. (From https://en.wikipedia.org/wiki/Water_distribution_on_Earth.)

3.2.3 LAND POLLUTION

The earth has 29% of its area under land, which is solid surfaces and exists in various forms like mountains, hills, valleys, deserts, canyons, plateaus, plains, and islands, and these geographical features influence the weather and climate. Intense anthropogenic activities have led to degradation of land and its resources directly or indirectly through destruction of the earth's surface, which reduces the productivity of land especially for agriculture, forestry, etc. The main causes of land pollution are mining activities, agriculture, landfills for wastes, deforestation and soil erosion, industrialization, construction activities, sewage treatment, and nuclear waste (Figure 3.7). Land pollution leads to contamination of soil, which renders the land unfit for cultivation, causing severe food scarcity. Chemicals and toxins are the biggest threat to soil pollution, which enter our food chain through plants and animals and affect humans, causing health problems and diseases like cancer. It affects the ecosystems by impacting local plants and animals and promoting rough and resistant exotic species. The soil with no vegetation is more prone to soil erosion, further causing erosion and loss of topsoil. The most common soil pollutants are hydrocarbons; heavy metals like cadmium, zinc, lead, chromium, copper, and mercury; pesticides; oils; tars and dioxins; etc.

3.3 HEAVY METALS AND THEIR IMPACTS ON HUMAN HEALTH

Heavy metals are naturally occurring substances present in the environment at low levels, but they become dangerous once present in large amounts. The heavy metals that are generally considered having health risks are arsenic (As), lead (Pb), cadmium (Cd), chromium (Cr) (only in the form of Cr-VI is toxic), copper (Cu), mercury (Hg), nickel (Ni), and zinc (Zn). Among these, a few are necessary for human health

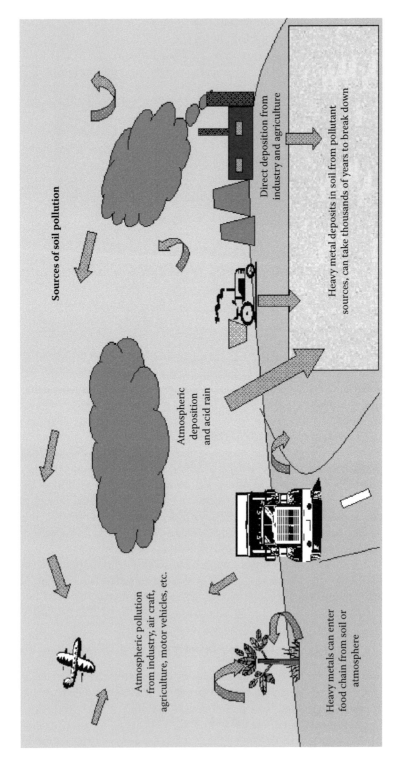

FIGURE 3.7 Sources of land pollution. (From Cooke, R., 2000, http://www.ecifm.rdg.ac.uk/landpollution.htm.)

when taken in food or as supplement at low levels. Among these, cadmium, lead, and mercury have no known biological function and are toxic to humans. Higher doses of metals like zinc, copper, iron, and cobalt, which affect humans through breathing, drinking, or eating through working or living environment, are grave risk to the health. A number of studies on metals (Martin and Griswold, 2009) have shown toxic effects of heavy metals on human health. A few of the heavy metals that impact human health especially the brain are discussed in the following.

3.3.1 ARSENIC

Arsenic is an odorless and tasteless gas, which occurs naturally in the environment and is released from volcanic activities, forest fires, rock erosion, and human activities. It is also found in paints, metals, dyes, soaps, drugs, semiconductors, etc. A large amount of arsenic can be released in the atmosphere through volcanic eruptions, lead or copper smelting, coal burning, and mining. Inorganic arsenic is carcinogenic, which can cause cancer of the lungs, liver, bladder, and skin and even at low level of exposure can cause vomiting and nausea, decrease in production of white and red blood cells, damage to blood vessels, abnormal heart rhythm, and darkening of skin (Martin and Griswold, 2009). The impact of arsenic poisoning on the human body is shown in Figure 3.8.

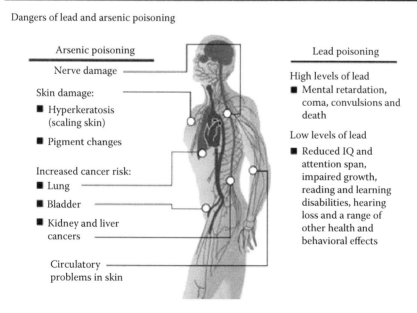

FIGURE 3.8 Arsenic poisoning and its impact on the human body. (From Aves, M., Arsenic: Physical geology. Alliances to end childhood lead poisoning and news wires, *The Denver Post*, 2015, http://skywalker.cochise.edu/wellerr/students/arsenic/project.htmAves, 2015.)

3.3.2 BARIUM

Barium is an abundant and naturally occurring metal, used extensively in industries that manufacture spark plug electrodes, vacuum tubes, oxygen removing agent, diagnostic medical equipments, etc. Barium compounds are also used in bricks, ceramics, paints, glass, rubber industries, and in manufacture of fluorescent lamps. Exposure to barium may cause difficulty in breathing, vomiting, increased or decreased blood pressure, abdominal cramps, diarrhea, and muscle weakness.

3.3.3 CADMIUM

Cadmium is an extremely toxic metal, which is found in small traces in rocks, soils, coals, and mineral fertilizers. It is widely used in batteries, plastics, pigments, and metal coatings like electroplating. Cadmium is carcinogenic for humans, and smokers are highly vulnerable due to its inhalation. If ingested at a higher level, it may cause severe damage to the lungs, and long-term low-level exposure affects the kidneys, lungs, and bones.

3.3.4 CHROMIUM

Chromium occurs in solid, liquid, and gaseous states and is found in soil, rocks, plants, and animals. Chromium is used in metal alloys in stainless steel, metal protective coatings, paints, cement, rubber, magnetic tapes, and as wood preservatives. Chromium IV compounds are known human carcinogens, whereas chromium III is an essential nutrient. If high level is inhaled through breathing, it can cause breathing problems like asthma, cough, wheezing and shortness of breath, and nose ulcers. If contacted with skin, it can cause ulcers, redness, and swelling, and long-term exposure can cause harm to the kidney, liver, and nerve tissues.

3.3.5 LEAD

Lead is present in the environment, i.e., air, soil, and water, due to human activities like fossil fuel burning, mining, and manufacturing, which are increasing its concentration in the atmosphere. Lead is used at an antiknocking agent in automobiles, as paint pigments, and in plumbing, soldering, etc., causing lead pollution in the environment, which is highly toxic. Lead affects humans by causing anemia, which affects blood and bone marrow and may lead to the dysfunction of the kidney. Lead toxicity also affects the nervous system, causing damage to the brain. Lead exposure may result in miscarriage; weakness in the fingers, wrists, or ankles; and increased blood pressure. High exposure to lead can severely damage the brain and kidneys and may cause death.

3.3.6 MERCURY

Organic and inorganic mercury compounds are formed when mercury combines with other elements. Mercury is used in thermometers, dental fillings, fluorescent bulbs, and batteries. A major chunk of mercury in the environment comes due to coal-fired power plants, whereas mercury in soil and water gets converted to methylmercury, a bio-accumulating toxin by microorganisms, which is carcinogenic to humans. The nervous system, which is highly sensitive to mercury, is negatively impacted, and its high exposure can lead to permanent damage to the brain, kidney, and fetus. Its effect on the brain leads to impairment of vision and hearing, loss of memory, and irritability and tremors (Environmental Protection Agency [EPA]). Mercury vapor may cause nausea, vomiting, diarrhea, increased heart rate/blood pressure, and eye and skin irritation.

3.3.7 SELENIUM

Selenium is widely distributed as trace mineral in most rocks and soils. In processed form, it is used mostly in industries related with electronic, glass, plastics, paints, inks, rubber, pharmaceuticals, as additive for poultry and livestock feeds, in pesticides, and antidandruff shampoos. Its radioactive form is used in diagnostic medicines. When used in traces, it is beneficial for cellular functions and is an essential trace nutrient for humans, but in large amounts, it is toxic. Its high concentration can cause vomiting, nausea, and diarrhea. Chronic oral exposure to selenium can lead to selenosis, i.e., hair loss, nail brittleness, and neurological abnormalities. High-level exposure, even for a brief period, causes respiratory tract infections, bronchitis, breathing difficulties, etc.

Humans get exposed to metal contamination through soils by use of root crops, leafy vegetables, strawberries, etc., if not washed properly. Metals may get accumulated in animals through plants and metal-laced water and get transferred to humans when eaten. In general, toxic metals like aluminum, arsenic, mercury, nickel, copper, and cadmium can interfere with glandular functions in humans and can cause neurological problems (Kaur, 2016). Among these, mercury, cadmium, and lead are hormone disruptors, causing health issues in humans (Figure 3.9).

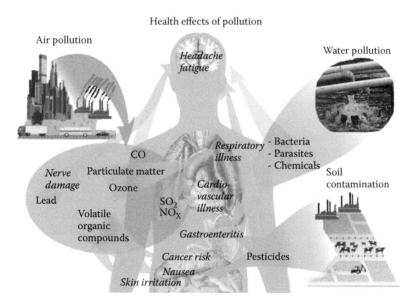

FIGURE 3.9 Effects of various pollution types on human health. (From Häggström, M., *Wikiversity J. Med.*, 1(2), 2014.)

3.4   SOUND POLLUTION

Sound pollution or noise pollution is exposure of people or animals to annoying, damaging, or stressful sounds, which can cause damage to the ears. Rapid industrialization and urbanization and use of excessive sound-making heavy machines are making disturbing sounds, which are similar to other environmental pollutants like that of air and water.

Noise directly impacts our brain as it causes irritation, which causes stress and sleeping problems, leading to anger and lack of concentration, rendering a person uninterested in his/her work. Other effects on human health include increased blood pressure or hypertension, fatigue, decreased efficiency, diverted attention, and hearing issues such as temporary or permanent deafness (Figure 3.10). In pregnant females, a sudden loud noise can cause abortion. As per a World Health Organization (2001) report, noise pollution due to unplanned urbanization may increase the prevalence of mental disorders such as depression, anxiety, chronic stress, schizophrenia, and suicide.

For humans, sounds up to 80–85 decibels (dBA) for eight hours a day are generally accepted to minimize hearing risks (Figure 3.11). Various sources of sound pollution are noise from aircrafts, road traffic, railway trains, manufacturing industries, construction and building repair activities, and noise from musical instruments. Hazards of noise pollution are sleeping disturbances, stress, increase in heart rate, hearing impairment, etc.

3.5   ELECTROMAGNETIC RADIATIONS AND ITS IMPACTS

Today, we have an almost equal number of cell phones as humans on this earth. The effects of cell phone use and impacts of its electromagnetic radiation (EMR) on humans are being studied by various researchers and agencies. Most of these studies say that EMRs are certainly harmful to humans and thus can be considered as pollutant for the neural cells, causing stress and insomnia. As per a recent report by John et al. (2012), mobile phone radiations may increase blood–brain barrier permeability by rupturing the delicate brain cell membrane (http://bebrainfit.com/cell-phone-radiation-brain/) and increases the risk of brain cancer. Studies have also proven that even a low level of radiation exposure from cell phone disrupts the production of melatonin in the body, which disturbs the sleep pattern. Melatonin, which is considered also a sleep hormone, is a potent antioxidant, and its reduced level in the human body helps cell radiations to create free radicals (unattached oxygen molecules) that cause damage to brain cells.

Over 140 proteins in the brain are negatively impacted by exposure to electromagnetic frequencies, the kind emitted by cell phones, electronics, and other electrical devices (Fragopoulou et al., 2012). Many epidemiological studies demonstrate that significant harmful biological effects occur from nonthermal radiofrequency (RF) exposure (Hill, 1965). Genetic damage, reproductive defects, cancer, neurological degeneration and nervous system dysfunction, immune system dysfunction, cognitive effects, protein and peptide damage, kidney damage, and developmental effects have all been reported in the peer-reviewed scientific literature.

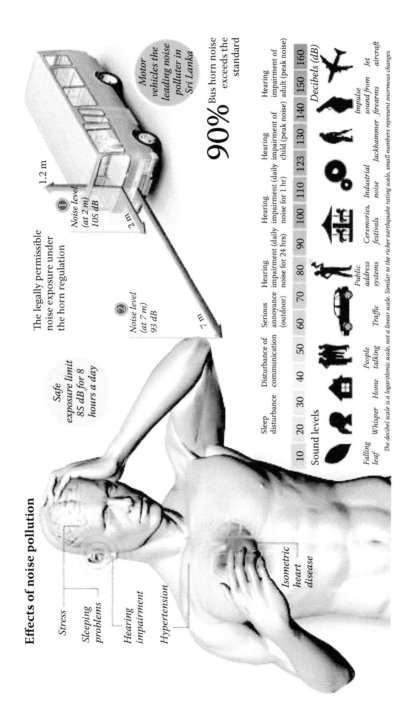

FIGURE 3.10 Effects of noise pollution and various sound levels. (Central Environmental Authority, Ear-splitting noise of the new age alarm health experts, http://www.sundaytimes.lk/151101/news/ear-splitting-noise-of-the-new-age-alarm-health-experts-169907.html.)

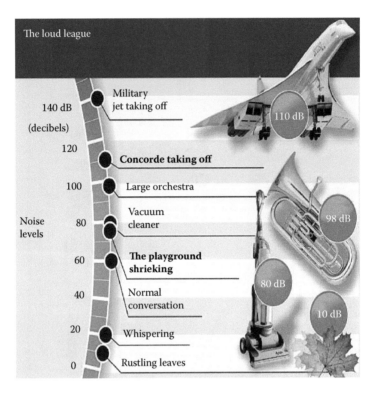

FIGURE 3.11   Noise levels as observed from different sources. (From www.dailymail.co.uk.)

3.6   CLIMATE CHANGE AND ITS IMPACT ON HUMAN HEALTH

Climate change as an environmental concern has become a global issue, which is impacting the health of ecosystems, causing ecological concerns and impacting the health of humans, animals, and plants alike. The environmental consequences of climate change, like sea level rise, flooding and drought, unpredicted and untimely precipitations, heat waves, intense hurricanes and storms, and degradation in air quality, affect directly or indirectly human health (Portier et al., 2010). The loss of life and property may cause stress, depression, and feelings of helplessness, which can influence decision making and cause neurological disorders. Extreme weather events are on the rise (Figure 3.12).

As per Intergovernmental Panel on Climate Change (IPCC) Report 2007 (Confalonieri et al., 2007), a strong consensus is emerging that human emissions of greenhouse gases are responsible for warming of our earth environment and that also is occurring at a rate greater than assessed by the IPCC (Garnaut, 2008; Spratt and Sutton, 2008). There is clear recognition of the fact that health impacts due to climate change will affect heavily on low-income or otherwise vulnerable populations (Lee, 2007; Frumkin et al., 2008). Climate change events leave a psychological impact on not only humans but also on impacted animals. These psychological events range from mild to chronic stress or other mental health disorders (Fritze et al., 2008). Weather

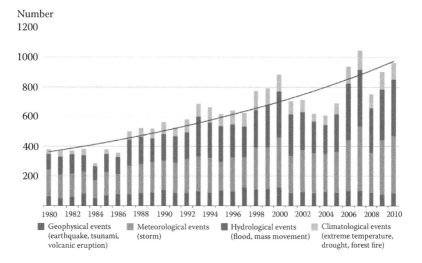

Number

FIGURE 3.12   (See color insert.) Events showing natural catastrophes worldwide, 1980–2010. (From Munich, R., *Natural Catastrophes Worldwide 1980–2011. Number of Events with Trend*, Münchener Rückversicherungs-Gesellschaft, Geo Risks Research, NatCatSERVICE, 2012.)

phenomena like El Nino are increasing events like cyclones, floods, and drought, which has started impacting children's mental health as floods ruin and destroy their schools and droughts bring misery in food availability and the extreme events cause psychological stress and trauma.

The consequences of extreme weather events may result in geographical displacement of populations and loss of life and property, which causes stress, leading to negative impacts on mental health. Other neurological disorders, due to extreme weather events, may be related to the shortage of food and water, causing food security issues, leading to malnutrition and stress (Portier et al., 2010). Also, the IPCC (2014b) 5th Assessment report clearly says with high confidence that there is risk of severe ill-health and disrupted livelihoods resulting from storm surges, sea level rise and coastal flooding, inland flooding in some urban areas, and periods of extreme heat. Also, there is risk of loss of ecosystems, biodiversity, and ecosystem goods, functions, and services, which will cause a major population to experience water scarcity and losses to life and property. The report further says that projected climate change will impact human health mainly by exacerbating health problems that already exist and climate change is expected to lead to increases in ill-health in many regions, especially in developing countries with low income. The major health impacts will be greater likelihood of injury and death due to more intense heat waves and fires; increased risks from food-borne and waterborne diseases, causing loss of work capacity and reduced labor productivity in vulnerable populations; and increased risk of undernutrition in poor regions along with risks from vector-borne diseases, which are projected to generally increase with warming, due to the extension of the infection area and season, despite reductions in some areas that become too hot for disease vectors.

Studies prove that people with mental illness will have impaired response to extreme events in case of climate change, thus rendering them unfit to take action and thus suffer more losses (USGCRP, 2016). As per the study by Sharma and Westman (1998), when body temperature rises above 40°C, it is associated with life-threatening heat stress and associated heat stroke, which may cause CNS dysfunction, causing delirium, convulsion, and coma. Nearly 50% of heat stroke victims die in a short period in spite of lowering temperature and therapeutic interventions, and those who survive are left with permanent neurological deficit, proving that heat considerably influences the structure and function of the nervous system (Sharma and Westman, 1998). The impacts of climate change on human health are shown in Figure 3.13.

In extreme events, depressed people, who are suffering a form of mental illness default in decision making, are more prone and vulnerable to losses. It is estimated that by 2050, adults with depressive disorder in the United States alone will increase by 35%, i.e., from 33.9 million to 45.8 million, and worst affected will be elderly over the age of 65 years (USGCRP, 2016). As per the World Health Organization's (2015) "The World Report on Ageing and Health," it is estimated that the world's population of elderly people over 60 years is poised to double from 12% to 22% by 2050, and among these, 80% of older people will be living in low- and middle-income countries, presenting a challenge for countries to ensure the robustness of its health and social systems. As per estimates, it is clear that the concentration of carbon dioxide is going to increase further as per current Intended Nationally Determined Contributions (INDCs) by countries and will be severely high if no action is taken (Figure 3.14).

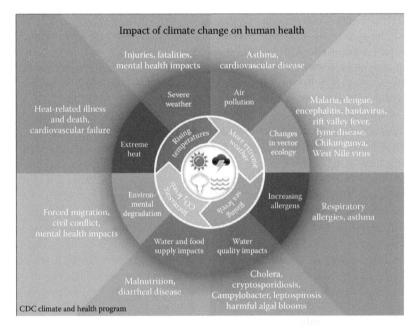

FIGURE 3.13 **(See color insert.)** Potential health effects of climate change. (From http://www.cdc.gov/climateandhealth/effects/.)

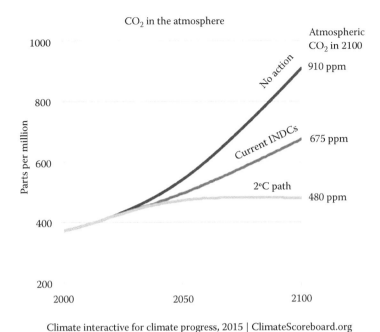

Climate interactive for climate progress, 2015 | ClimateScoreboard.org

FIGURE 3.14   Emission of CO_2 in various scenarios. (From ClimateScoreboard.org.)

Climate change projections say that many populations will age appreciably in the next 50 years, rendering the elderly more vulnerable to injury resulting from weather extremes such as heat waves, storms, and floods (Confalonieri et al., 2007), causing a heavy burden of disease and disability on the states. It is also projected that over the course of the twenty-first century, the population will grow faster in poor countries as compared to the developed world, especially in Africa, expected to rise from 26 to 60 people/km[2], whereas the density of Europe may fall from 32 to 27 people/km[2] (Cohen, 2003). Altered effects of rainfall, temperature, and weather patterns due to climate change will spread the cases of malaria from the presently 70% of total world cases occurring in Africa to newer areas altering the range and geographical distribution (World Bank et al., 2004). Another study by Hales et al. (2002) estimates that in the 2080s, 5–6 billion people would be at risk of dengue as a result of climate change and population increase, compared with 3.5 billion people if the climate remained unchanged. Heat-related morbidity and mortality are expected to increase with the evidence of the relationship between high ambient temperature and mortality, although there may be a reduction in cold-related mortality as compared to heat related as studied for the UK (Donaldson et al., 2001).

As per the IPCC Third Assessment Report (2001), climate change can directly affect human health through impacts of thermal stress, death, and injuries due to flood or storms and indirectly through changes in ranges of disease vectors like mosquitoes; water-borne pathogens; quality of air, water, and available food; and impact of socioeconomic and environmental conditions. Overall, these health consequences

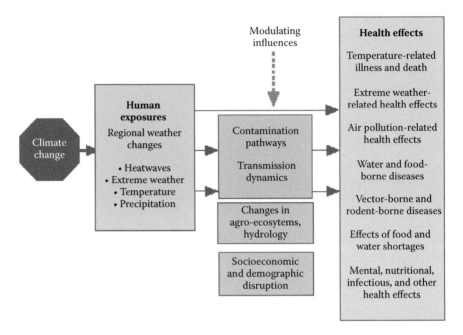

FIGURE 3.15   Pathways by which climate change affects human health. (From World Health Organization, *Climate change and human health: Risks and responses. Summary*, World Health Organization, in collaboration with UNEP and WMO, Geneva, Switzerland, 2003.)

will lead to diseases, trauma, and psychological and nutritional issues in humans, who are already depressed and demoralized due to consequences of climate change like economic loss and social displacement. The consequences of warming will encourage prolific growth of warm-loving vectors like mosquitoes and a number of insects, which will further impact more areas, affecting more humans. As per a World Health Organization (2003) study, climatic stresses like heat waves and precipitation influence contamination pathways and transmission dynamics, bringing changes in agro-ecosystems and hydrology, which finally culminates in negative health effects on humans (Figure 3.15).

As per IPCC (2014a) WGII AR5 Chapter 11, it is estimated that climate change impacts may give rise to harmful algal blooms by dinoflagellates, which may cause outbreaks of neurotoxic shellfish poisoning and cyanobacteria-producing toxins, which may cause liver, neurological, digestive, and skin diseases. Also, diatoms may produce potent neurotoxin called domoic acid, which gets bio-accumulated in shell-fish and finfish (Erdner et al., 2008). A number of studies support that increasing temperatures promote bloom formation in freshwater (Paerl et al., 2011) and marine environments (Marques et al., 2010).

3.7   ENVIRONMENTAL REFUGEES

The term "environmental refugee" was coined by El-Hinnawi in a 1985 United Nations (UN) Environmental Programme Report and is defined as "those people who have been forced to leave their traditional habitat, temporarily or permanently, because of a marked environmental disruption (natural and/or triggered by people) that jeopardized their existence and/or seriously affected the quality of their life." Among the mostly impacted environmental or climate refugees are more than 300,000 people in south and Central Somalia, who have fled into Kenya and Ethiopia in 2011 due to drought and insecurity. As per the UN's Office for the Coordination of Humanitarian Affairs, more than 10 million people (more than half of them children) in the Horn of Africa are affected due to failed rains. The story is even worse in case of a small South Pacific Island nation, Tuvalu, an LDC near New Zealand with 10,000 inhabitants affected due to climate change and sea level rise threatening their overall existence as a nation, which will become uninhabitable in the near future (Mortreux and Barnett, 2009; McAdam, 2011). As per the IPCC, factors that are most threatening in terms of creating environmental or climate refugees are sea level rise, droughts and desertification and hurricanes, torrential rains, and floods. The IPCC identified the areas most at risk as sub-Saharan Africa along with Yemen, China, Louisiana, Tuvalu, Kiribati, and Bangladesh. Currently, environmental refugees have outnumbered political refugees worldwide (UNHRC). As per an estimate by the American Association of Advancement of Science, the number of environmental refugees may reach 50 million by 2020 due to climatic factors causing agriculture disruption, deforestation, coastal flooding, shoreline erosion, industrial accidents, and pollution, whereas another estimate says that more than 22 million people are getting displaced due to climatic disasters (IDMC, 2014).

Due to natural calamities, disasters, and abrupt climatic changes, a number of people have been displaced from their original land, making them environmental refugees, who undergo a traumatic shock, feelings of insecurity, loss of jobs, and feelings of helplessness, which leave a deep impact on their minds, impacting decision making and normal action and thus making them vulnerable to depression and mental diseases. A number of authors (Trolldalen et al., 1992; Myers, 1993) agree that millions of displaced people or environmental refugees, whether because of "natural" or "man-made causes," will have a dramatic impact on host regions, threatening regional security and also overuse of the environment resource base, which may contribute to environmental collapse.

This has been shown in studies by Myers (1993), in which he shows concern about the plight of Haitian boat people, abandoning their homelands as their country has become an environmental basket case. Environmentally induced migration has already been reported across Asia, Africa, and Latin America. All this will further aggravate stress and insecurity, leading to depression and neurological issues.

Environmental pollution, once unabated, culminates in calamities like alarming respiratory problems due to air pollution, whereas polluted water causes water-borne diseases like dysentery and cholera. Noise or sound pollution causes neurological disorders, deafness, and aggravation of cardiac issues, hypertension, and high blood pressure problems. The pollution from industrial wastes has reached an enormous level, spoiling our ponds, rivers, and seas, triggering health issues, causing not only physical problems but also stress and psychological issues, which affect our brain, causing duress and insecurity and hindering decision making. Climate change is changing the environment drastically, making it nonconducive and hostile, which

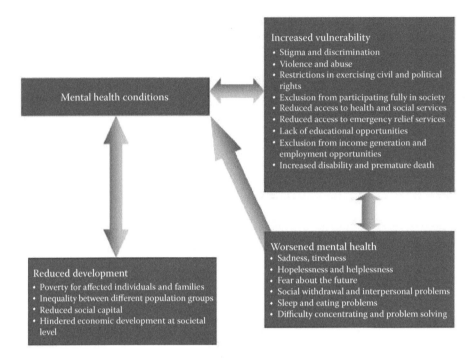

FIGURE 3.16 Mental health conditions and increased vulnerabilities. (From Funk, M., Drew, N., Freeman, M., *Mental Health and Development: Targeting People with Mental Health Conditions as a Vulnerable Group*, WHO Press, World Health Organization, Geneva, Switzerland, 2010.)

is negatively affecting the health of humans, animals, and plants and instilling in them anxiety, anger, and stress. The mental health conditions of the environmental/climatic affected people have increased vulnerability and worsened mental health effects (Figure 3.16).

3.8 WAY FORWARD

It is high time that governments all over the world rethink their strategy of materialistic development, economic boom, and carbon pathway at the cost of exploitation of natural resources, which, in fact, is threatening the existence of the whole of humanity along with the ecosystems on which our survival is so much dependent. If pollution continues to be unabated in the future and nothing is done to tackle rising green house gases, we may have more and more affected people with continued stressful conditions due to loss of property, jobs, and dear ones, developing mental health conditions. We *Homo sapiens* (Latin "Wise person") are the most intelligent of all creations, only because of the "brain," which (intelligence), if used logically, positively, and humanely to respect nature and resist greed's overexploitation and follow the sustainable development path of greener and cleaner technology, will help in reducing the impacts of pollution.

The relationship between nature and its healing impact on human mental health has been in discussion and in practice among various civilizations, cultures, and religions for hundreds of years (Bratman et al., 2012). There has been a belief that nature heals, as it provides a pollution-free and healthy environment, which de-stresses a person, making him/her calm, composed, and happy. Also, it has been observed that people living in places blessed with nature and its bounties are healthier as compared to those living in a polluted urban environment. A study reports that urban people have disadvantage in processing stress when compared to their rural counterparts (Lederbogen et al., 2011). Anthropogenic activities like land conversion, uncontrolled and unplanned urban sprawl, pollution, and, above all, greed are taking a toll on the natural environment. Writers like John Muir and the originators of the Wilderness Act in the United States have discussed nature's contributions to mental health very qualitatively (Cole and Hall, 2010), which was further supported by Marcus and Barnes (1999) in their work on the history of healing gardens in hospital settings. Studies by Cole and Hall (2010) and Hartig et al. (1991) show a decrease in self-reported stress and an increase in positive mood after prolonged experience in wilderness areas. Studies on exam-stressed students reported higher levels of positive affect and lower levels of fear after viewing slides of natural scenes than did those who viewed urban ones (Ulrich, 1979), which was also confirmed by studies of Honeyman (1992), who found the same trend when his subjects were presented with urban images containing vegetation versus urban images without vegetation. Nisbet and Zelenski (2011) demonstrated that short exposure to natural environments can increase positive mood. The study by Wells (2000) showed an increase in cognitive and functioning capacity in children who have moved to more natural surroundings versus those who had moved to more urban environment.

In the ancient Indian philosophy of Ayurveda, it is said that a healthy man is one who possesses a sound body, a sound mind, and a sound soul. It further says that man's spiritual health is dependent on his ability to live in harmony with the external universe; his mental health must depend on his ability to live in harmony with himself. Ayurveda physicians identified techniques of deep breathing and meditation for calming the mind and improving memory for practitioners of yoga. Ayurveda also uses aromatics, diet and cosmetics to calm and balance the mind (Naveen, 1993). As per Fritze et al. (2008), the Australian Psychological Society has recommended to be calm and rational to cope better with the environmental threats. Changing behaviors and optimistic about the future also help.

It is therefore suggested that we appreciate nature and its innumerable bounties in the form of natural resources bestowed on us. Indulge in nature-loving hobbies and activities and be positive toward life. Be happy and develop a lifestyle that keeps you away from stress and unnecessary exposure to pollutants. Spend time with friends and family and minimize use of cell phones and other electronic gadgets for some time, at least to relax your mind and revive concentration and peacefulness. Although income is a necessary condition for well-being, it is not a sufficient one as the World Bank's "Voices of the Poor" study says, that poor people themselves defined well-being not only in terms of income but also as "peace of mind," belonging to a community, safety, and good health among others (UNU-IHDP and UNEP, 2014).

Happiness is of great importance for the well-being of humans and also for the sustainable growth of a country like Bhutan, which assesses its people and the economy based on "Gross National Happiness Index" (GNH), a term that was coined by the fourth King of Bhutan in the 1970s, which says that "sustainable development should take a holistic approach towards notions of progress and give equal importance to non-economic aspects of wellbeing." It is based on good governance, sustainable socioeconomic development, cultural preservation, and environmental conservation, which are further classified into nine domains: psychological well-being, health, education, time use, cultural diversity and resilience, good governance, community

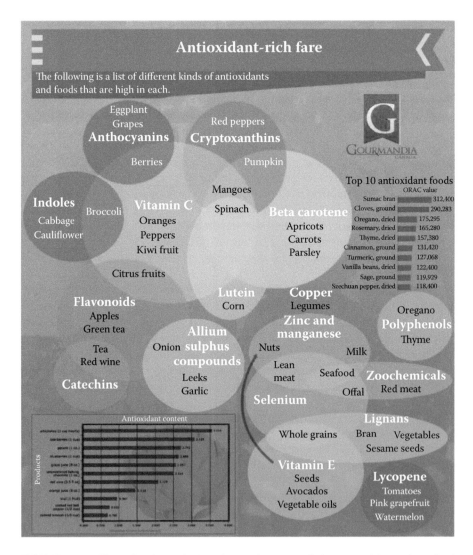

FIGURE 3.17   (See color insert.) List of some important fruits, vegetables and sea food with antioxidants. (From http://visual.ly/antioxidant-rich-fare.)

vitality, ecological diversity and resilience, and living standards. GNH is constructed through a robust multidimensional methodology known as the Alkire-Foster method (http://www.grossnationalhappiness.com/articles/).

Also, if you follow a regime of regular exercise and healthy food, you can keep yourself relaxed. As per University of Maryland Medical Center and various other studies, omega-3 fatty acid-rich diet of fishes like salmon, tuna, sardines, trout, mackerel, halibut etc.; seafood like krill, algae, and plants; and nut oils plays a crucial role in brain functions and its growth and development (http://umm.edu/health /medical/altmed/supplement/omega3-fatty-acids).

Antioxidants are present in fruits, vegetables, some seafood, and meat. It has been observed that eating such foods helps in improving cognitive skills and slows down brain aging. Few minerals and vitamins like vitamin C and zinc are antioxidants; also, the sleep hormone melatonin is a potent antioxidant. Some of the important fruits and vegetables high in antioxidants are given in Figure 3.17.

Some of the healthy foods recommended for brain health and body are those having antioxidants, which protect cells from free radical damage, slow down the cellular aging process, and help in slowing down the oxidative damage in the brain, which causes brain inflammation and contributes to anxiety and depression. Free radicals are caused by stress, lack of sleep, fried and grilled food, air pollution, and radiations from mobile phone and computers, which can be rendered harmless by binding with the antioxidants (http://bebrainfit.com). So, it is high time that good sense prevail on us, i.e., so-called civilized humans, to respect and conserve nature for the well-being of our planet earth and its living beings, as saving earth from further pollution is actually saving ourselves, the human race, from extinction faster than animals or plants!

REFERENCES

Adan, A. (2012). Cognitive performance and dehydration. *J Am Coll Nutr.* 31(2):71–78.

Alzheimer's Disease International. (2010). The global economic impact of dementia. World Alzheimer Report. Available at: http://www.alz.co.uk/research/worldreport/.

Altman, P.L. (1961). *Blood and Other Body Fluids/Analysis and Compilation.* Edited by D.S. Dittmer; prepared under the auspices of the Committee on Biological Handbooks, Washington, DC: Federation of American Societies for Experimental Biology. p. 540.

Aves, M. (2015). Arsenic: Physical geology. Alliances to end childhood lead poisoning and news wires. *The Denver Post.* Available at: http://skywalker.cochise.edu/wellerr/stu dents/arsenic/project.htm.

Björkhem, I., and Meaney, S. (2004). Brain cholesterol: Long secret life behind a barrier. *Arterioscler Thromb Vasc Biol.* 24:806–815.

Bratman, G.N., Hamilton, J.P., and Daily, G.C. (2012). The impacts of nature experience on human cognitive function and mental health. *Ann NY Acad Sci.* 1249:118–136.

Central Environmental Authority. Ear-splitting noise of the new age alarm health experts. Available at: http://www.sundaytimes.lk/151101/news/ear-splitting-noise-of-the-new-age -alarm-health-experts-169907.html.

Climatescoreboard.org. (2015). Climate scoreboard. Available at: https://www.climateinter active.org/programs/scoreboard.

Cohen, J.C. (2003). Human population: The next half century. *Science,* 302:1172–1175.

Cole, D.N., and Hall T.E. (2010). Experiencing the restorative components of wilderness environments: Does congestion interfere and does length of exposure matter? *Environ Behav.* 42: 806–823.

Confalonieri, U., Menne, B., Akhtar, R., Ebi, K.L., Hauengue, M., Kovats, R.S., Revich, B., and Woodward, A. (2007). Human health. In: *Climate Change (2007) Impacts, Adaptation and Vulnerability. Contribution of Working Group II to the Fourth Assessment Report of the Intergovernmental Panel on Climate Change.* Edited by M.L. Parry, O.F. Canziani, J.P. Palutikof, P.J. van der Linden, and C.E. Hanson. Cambridge University Press, Cambridge, UK, 391–431.

Donaldson, G.C., Kovats, R.S., Keatinge, W.R., McMichael, A. (2001). Heat-and cold-related mortality and morbidity and climate change. In: *Health Effects of Climate Change in the UK.* Edited by P. Baxtar, A. Haines, M. Hulme, R.S. Kovats, R. Maynard, D.J. Rogers, and P. Wilkinson. Department of Health, London, 70–80.

El-Hinnawi, E. (1985). Environmental Refugees United Nations Environmental Program, Nairobi.

Erdner, D.L., Dyble, J., Parsons, M.L., Stevens, R.C., Hubbard, K.A., Wrabel, M.L., Moore, S.K., Lefebvre, K.A., Anderson, D.M., Bienfang, P., Bidigare, R.R., Parker, M.S., Moeller, P., Brand, L.E., and Trainer, V.L. (2008). Centers for Oceans and Human Health: A unified approach to the challenge of harmful algal blooms. *Environ Health.* 7 Suppl 2: S2.

Ferri, C.P., Prince, M., Brayne, C., Brodaty, H., Fratiglioni, L. and Ganguli, M. (2005). Global prevalence of dementia: A Delphi consensus study. *Lancet.* 366:2112–2117.

Fragopoulou, A.F., Samara, A., Antonelou, M.H., Xanthopoulou, A., Papadopoulou, A., Vougas, K., Koutsogiannopoulou, E., Anastasiadou, E., Stravopodis, D.J., Tsangaris, G.Th., and Margaritis, L.H. (2012). Brain proteome response following whole body exposure of mice to mobile phone or wireless DECT base radiation. *Electromagn Biol Med.* 31(4): 250–274.

Fritze, J.G., Blashki, G.A., Burke, S., and Wiseman, J. (2008). Hope, despair and transformation: Climate change and the promotion of mental health and well being. *Int J Mental Health Syst.* 2:13.

Frumkin, H., Hess, J., Luber, G., Malilay, J., and McGeehin, M. (2008). Climate change: The public health response. *Am J Public Health.* 98(3):435–445.

Garnaut, R. (2008). *Garnaut Climate Change Review: Interim Report to the Commonwealth, State and Territory Governments of Australia.* Cambridge University Press, Australia.

Häggström, M. (2014). Medical gallery of Mikael Häggström 2014. *Wikiversity J Med.* 1(2): 8. doi:10.15347/wjm/2014.008. ISSN 2002-4436.

Hales, S., de Wet, N., Maindonald, J., and Woodward, A. (2002). Potential effect of population and climate changes on global distribution of dengue fever: An empirical model. *Lancet.* 360:830–834.

Hartig, T., Mang, M., Evans, G.W. (1991). Restorative effects of natural environment experiences. *Environ Behav.* 23:3–26.

Health Effects Institute (HEI) Report 154. (2010). *Public Health and Air Pollution in Asia (PAPA): Coordinated Studies of Short-term Exposure to Air Pollution and Daily Mortality in Four Cities.* HEI Public Health and Air Pollution in Asia Program, Health Effects Institute, Boston.

Hill, A.B. (1965). The environment and disease: Association or causation? *Proc R Soc Med.* 58:295–300.

Honeyman, M.K. (1992). Vegetation and stress: A comparison study of varying amounts of vegetation in countryside and urban scenes. In: *The Role of Horticulture in Human Well-Being and Social Development.* Edited by D. Relf. Timber Press, Portland, OR, 143–145.

IDMC. (2014). *Global Estimates 2014 People Displaced by Disasters.* Internal Displacement Monitoring Centre (IDMC), Norwegian Refugee Council (NRC). Edited by J. Lennard. Publication has been produced with the assistance of the European Union.

IPCC. (2001). *Synthesis Report, Third Assessment Report.* Cambridge University Press, Unied Kingdom and New York, 398p.

IPCC. (2014a). *Climate Change 2014: Synthesis Report. Contribution of Working Groups I, II and III to the Fifth Assessment Report of the Intergovernmental Panel on Climate Change.* Edited by Core Writing Team: R.K. Pachauri and L.A. Meyer. IPCC, Geneva, Switzerland.

IPCC. (2014b). Summary for policymakers In: *Climate Change 2014: Impacts, Adaptation, and Vulnerability. Part A: Global and Sectoral Aspects. Contribution of Working Group II to the Fifth Assessment Report of the Intergovernmental Panel on Climate Change.* Edited by C.B. Field, V.R. Barros, D.J. Dokken, K.J. Mach, M.D. Mastrandrea, T.E. Bilir, M. Chatterjee, K.L. Ebi, Y.O. Estrada, R.C. Genova, B. Girma, E.S. Kissel, A.N. Levy, S. MacCracken, P.R. Mastrandrea, and L.L. White. Cambridge University Press, Cambridge, United Kingdom and New York.

John, W., Hugh, S., Nancy, A., and Linda, W. (2012). *Cell Phones: Technology, Exposures, Health Effects.* Environment and Human Health Inc. North Haven, CT. www.ehhi.org.

Kaur, S.D. (2016). 5 Environmental hazards and how we can protect our health. Available at: https://www.3ho.org/3ho-lifestyle/health-and-healing/5-environmental-hazards-and -how-we-can-protect-our-health.

Lee, J. (2007). *Climate Change and Equity in Victoria Melbourne.* Friends of the Earth, Melbourne, Australia.

Lederbogen, F., Kirsch, P., Haddad, L. et al. (2011). City living and urban upbringing affect neural social stress processing in humans. *Nature* 474:498–501.

Marques, A., Nunes, M.L., Moore, S.K., and Strom, M.S. (2010). Climate change and seafood safety: Human health implications. *Food Res Int.* 43(7):1766–1779.

Marcus, C.C., and Barnes, M. (1999). *Healing Gardens: Therapeutic Benefits and Design Recommendations.* John Wiley & Sons Inc. New York.

Maslin, M.A., Brierley, C.M., Milner, A.M., Shultz, S., Trauth, M.H., and Wilson, K.E. (2014). East African climate pulses and early human evolution. *Quat Sci Rev.* 101:1–17. www.elsevier.com/locate/quascirev.

Martin, S.E., and Griswold, W. (2009). *Human Health Effects of Heavy Metals.* Center for Hazardous Substance Research (CHSR), Kansas State University, Manhattan, KS. Part of the Technical Assistance to Brownfields Communities (TAB) program.

McAdam, J. (2011). Swimming against the tide: Why a climate change displacement treaty is not the answer. *Int J Refugee Law.* 23(1):6.

Funk, M., Drew, N., and Freeman, M. (2010). *Mental Health and Development: Targeting People with Mental Health Conditions as a Vulnerable Group.* WHO Press, World Health Organization, Geneva, Switzerland.

Mitchell, H.H., Hamilton, T.S., Steggerda, F.R., and Bean, H.W. (1945). The chemical composition of the adult human body and its bearing on the biochemistry of growth. *Biol Chem.* 158:625–638.

Mortreux, C., and Barnett, J. (2009). Climate change, migration and adaptation in Funafuti, Tuvalu. *Global Environ Change.* 19:105–112.

Munich, R. (2012). *Natural Catastrophes Worldwide 1980–2011. Number of Events with Trend.* Münchener Rückversicherungs-Gesellschaft, Geo Risks Research, NatCatSERVICE, Germany.

Myers, N. (1993). *Ultimate Security: The Environmental Basis of Political Stability.* W.V. Norton, New York and London.

Naveen, P. (1993). The garden of life. An introduction to the healing plants of India. Available at: http://www.arvindguptatoys.com/arvindgupta/naveen.pdf.

Nisbet, E.K., and Zelenski, J.M. (2011). Underestimating nearby nature. *Psychol Sci.* 22:1101–1106.

Paerl, H.W., Hall, N.S., and Calandrino, E.S. (2011). Controlling harmful cyanobacterial blooms in a world experiencing anthropogenic and climatic-induced change. *Sci Total Environ.* 409(10):1739–1745.

Portier, C.J., Thigpen Tart, K., Carter, S.R., Dilworth, C.H., Grambsch, A.E., Gohlke, J., Hess, J., Howard, S.N., Luber, G., Lutz, J.T., Maslak, T., Prudent, N., Radtke, M., Rosenthal, J.P., Rowles, T., Sandifer, P.A., Scheraga, J., Schramm, P.J., Strickman, D., Trtanj, J.M., and Whung, P.-Y. (2010). *A Human Health Perspective On Climate Change: A Report Outlining the Research Needs on the Human Health Effects of Climate Change.* Environmental Health Perspectives/National Institute of Environmental Health Sciences, Research Triangle Park, NC. doi:10.1289/ehp.1002272 Available at: www.niehs .nih.gov/climatereport.

Potts, R. (2010). Smithsonian Human Origins Program. Available at: http://humanorigins .si.edu/research/climate-and-human-evolution/climate-effects-human-evolution/http:// humanorigins.si.edu/research/age-humans-evolutionary-perspectives-anthropocene.

Potts, R., and Sloan, C. (2010). What does it mean to be human? *National Geographic.* Washington, DC.

Raichle, M.E., and Gusnard, D.A. (2002). Appraising the brain's energy budget. *PNAS.* 99(16):10237–10239.

Sayre, L.M., Perry, G., Harris, P.L., Liu, Y., Schubert, K.A., and Smith, M.A. (2000). In situ oxidative catalysis by neurofibrillary tangles and senile plaques in Alzheimer's disease: A central role for bound transition metals. *J Neurochem.* 74:270–279.

Shultz, S., and Maslin, M. (2013). Early human speciation, brain expansion and dispersal influenced by African climate pulses. *PLoS ONE.* 8(10):e76750. doi:10.1371/journal.pone.0076750/ http://brainworldmagazine.com/the-impact-of-climate-change-on-our-brains/.

Spickernell, S., and Parmenter, C. (2015). A toxic smog has shrouded London: Here's how it could affect our health. A City A.M. [Blog]. Available at: http://www.cityam.com/212008 /toxic-smog-has-shrouded-london-heres-how-it-could-affect-our-health.

Sharma, H.S., and Westman, J. (1998). Brain functions in hot environment. *Prog Brain Res.* 115:1–516.

Spratt, D., and Sutton, P. (2008). *Climate Code Red: The Case for Emergency Action.* Scribe Press, Melbourne, Australia.

Strausak, D., Mercer, J.F.B., Dieter, H.H., Stremmel, W., and Multhaup, G. (2001). Copper in disorders with neurological symptoms: Alzheimer's, Menkes, and Wilson diseases. *Brain Res Bull.* 55:175–185.

Trolldalen, J.M., Birkeland, N., Borgen, J., and Scott, P.T. (1992). *Environmental Refugees: A Discussion Paper.* World Foundation for Environment and Development and Norwegian Refugee Council, Oslo, Norway.

Ulrich, R.S. (1979). Visual landscapes and psychological well being. *Landscape Res.* 4:17–23.

UNU-IHDP and UNEP. (2014). *Inclusive Wealth Report 2014. Measuring progress toward sustainability. Summary for Decision-Makers.* UNU-IHDP, Delhi, India.

USGCRP. (2016). *The Impacts of Climate Change on Human Health in the United States: A Scientific Assessment.* U.S. Global Change Research Program, Washington, DC, 307–312. Available at: http://dx.doi. org/10.7930/J02F7KCR.

Wells, N.M. (2000). At home with nature. *Environ Behav.* 32:775–795.

World Health Organization. (2001). *World Health Report 2001: Mental Health—New Understanding, New Hope.* World Health Organization, Geneva, Switzerland.

World Health Organization. (2003). Climate change and human health: Risks and responses. Summary. World Health Organization, in collaboration with UNEP and WMO, Geneva, Switzerland.

World Health Organization. (2015). The world report on ageing and health 2015. Available at: http://www.who.int/mediacentre/factsheets/fs404/en.

World Bank, African Development Bank, Asian Development Bank, DFID, Directorate-Generale for Development European Commission, Federal Ministry for Economic Cooperation and Development Germany, Ministry of Foreign Affairs Netherlands, UNDP and UNEP. (2004). *Poverty and Climate Change: Reducing the Vulnerability of the Poor through Adaptation.* World Bank, New York.

Yassi, A., and Kjellstrom, T. (1997). Linkages between environmental and occupational health, Chapter 53: Environmental Health Hazards. *Fourth Edition: International Labour Office Encyclopedia of Occupational Health and Safety.* Monograph CIS 97-2038. Available at: http://www.ilocis.org/documents/chpt53e.htm.

Zanobetti, A., Dominici, F., Wang, Y., and Schwartz, J.D. (2014). A national case-crossover analysis of the short-term effect of PM2.5 on hospitalizations and mortality in subjects with diabetes and neurological disorders. *Environ Health.* 13(1):38 DOI:10.1186/1476-069X-13-38.

WEBSITES REFERRED FOR SOURCES

http://www.climatechanges.site/mental-health-a-factor-in-climate-change/
http://humanorigins.si.edu/research/age-humans-evolutionary-perspectives-anthropocene
http://ejap.org/environmental-issues-in-asia/health-issues.html
https://en.wikipedia.org/wiki/Bhopal_disaster
https://en.wikipedia.org/wiki/Water_distribution_on_Earth
http://www.ecifm.rdg.ac.uk/landpollution.htm
http://thumbnails-visually.netdna-ssl.com/antioxidantrich-fare_5143f3e5deeab_w1500.jpg
www.dailymail.co.uk
http://bebrainfit.com/cell-phone-radiation-brain/
http://www.cdc.gov/climateandhealth/effects/
https://www.ipcc.ch/publications_and_data/ar4/wg2/en/ch8s8-4-1-2.html
http://ec.europa.eu/science-environment-policy
http://www.grossnationalhappiness.com/articles/
http://umm.edu/health/medical/altmed/supplement/omega3-fatty-acids)
http://bebrainfit.com
https://www.climateinteractive.org/programs/scoreboard

4 Metal Toxicity and Oxidative Stress in the Brain
Role of Antioxidants

Rama Nair, Manashi Bagchi, and Sreejayan Nair

CONTENTS

4.1 INTRODUCTION

Heavy metals and transition metals are ubiquitously present in the human body. Although metals are indispensable for several biological functions including gene regulation, enzyme activity, neurotransmission, and maintenance of cellular structure, excessive accumulation of metals in the central nervous system has been correlated with a wide variety of neurodegenerative diseases [1–5]. A permissive role of metal ions has been demonstrated in neurological diseases, including Alzheimer's disease [6,7], Huntington's disease [8,9], and Parkinson's disease [10,11]. Although the exact molecular mechanisms by which metals mediate neurodegeneration are yet unclear, it is well accepted that metal toxicity can lead to oxidative stress [12], mitochondrial damage [13–15], endoplasmic reticulum stress (ER stress) [16,17], autophagy [18,19], apoptosis, and epigenetic changes [20]. Despite the fact that iron, copper, manganese, and zinc are essential metals that serve a variety of biological functions, elevated levels of these metals or free (unsequestered) forms of these metals can cause neuronal damage, mainly via triggering redox reactions. Arsenic, lead, mercury, and cadmium represent the nonessential heavy metals that enter the human body as pollutants. Arsenic, lead, and mercury are the top three substances prioritized by the Agency for Toxic Substances and Disease Registry based on a combination of their frequency, toxicity, and potential for human exposure [21]. Cadmium is the seventh on the list, following vinyl chloride, polychlorinated biphenyls, and benzene [21]. This mini-review will provide a broad overview on the common biological pathways that lead to the deleterious effects of endogenous transition and heavy metals. Furthermore, in staying with the main focus of this book, natural products that can be potentially harnessed to counter these deleterious effects will also be briefly discussed.

4.2 MOLECULAR PATHWAYS MEDIATING METAL TOXICITY

4.2.1 METALS AND OXIDATIVE STRESS

Oxygen is vital for life and also leads to the production of reactive oxygen species (ROS) during basal cellular respiration, which can have detrimental effects on cellular system. Ground-state bimolecular oxygen is a "biracial" with an odd pair of electrons occupying different orbitals (Figure 4.1). Although, by the definition of "free radical," oxygen should exhibit high reactivity toward surrounding tissues— oxygen could easily extract electrons from polyunsaturated fatty acids which form tissue membranes (i.e., hypoxia, ischemia), which could lead to "autocombustion"— the reactivity of oxygen is curtailed by a spin and orbital restriction. So for oxygen to accept electrons, the electrons should occupy separate orbitals and should have opposite spin. This spin and orbital restriction can be overcome by transition metals,

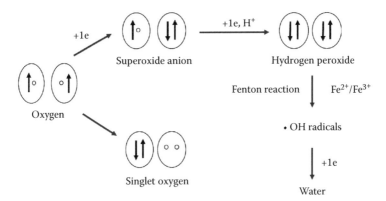

FIGURE 4.1 (See color insert.) Free radical formation from molecular oxygen. Ground-state oxygen is a biradical. It can undergo single electron reduction to form superoxide, which can be further reduced by a single electron to hydrogen peroxide. Hydrogen peroxide undergoes the classical Fenton reaction to yield hydroxyl radicals. Hydroxyl radical abstracts an electron to form water. A single electron can be promoted to the neighboring orbital to produce singlet oxygen.

which undergo single electron redox reactions. Thus, a single electro reduction of molecular oxygen leads to the formation of superoxide anion. Superoxide anion can acquire another electron forming hydrogen peroxide. Although hydrogen peroxide is not a free radical (it does not have an odd pair of electrons), it is a small molecule that can traverse cell membranes to cause alterations and damage the cellular components. More importantly, hydrogen peroxide can undergo the classical Fenton reaction in the presence of metal ions to form hydroxyl radicals, which is by far the most noxious and reactive of all free radicals. A further reduction of hydroxyl radical by acceptance of an electron forms water. Molecular oxygen can also promote one of its electrons to the nearby orbital to form singlet oxygen, another highly reactive species.

Because it is rich in polyunsaturated fatty acids and because of its high oxygen demand, the brain is extremely vulnerable to ROS. ROS can damage all types of biomolecules in the brain. For instance, it can lead to the chain reaction of lipid peroxidation, which results in the formation of neurotoxic aldehydes such as malondialdehyde and 4-hydroxynonenal [11,22]. ROS can attack proteins, resulting in the formation of protein carbonyls and protein nitration [23,24]. ROS can attack DNA by hydroxylating guanosine residues resulting in the formation of 8-hydroxy2-deoxyguanosine [25,26]. ROS can alter the expression of a variety of genes, including upregulation of cFos, cJun, and proinflammatory genes that can affect cell phenotype and function [27]. When the earth changed from the early reducing atmosphere to an oxidizing atmosphere, enzymes such as superoxide dismutase and catalases evolved to neutralize superoxide anions and hydrogen peroxide. Metal chelators such as transferrin, ferritin, ceruloplasmin, and metallothionein chelated free metal ions in the body and prevented them from being redox active. Additionally, small-molecule antioxidants such as glutathione provided a rich reducing environment in the cells. However, the presence of excessive metal ions can tip this balance toward oxidative stress, resulting in neuronal damage. Treating with antioxidants can potentially restore the balance by either scavenging the free radicals or by chelating metal ions.

4.2.2 Metals and Mitochondrial Damage

Although transition metals (copper, iron and manganese) are necessary for oxidative phosphorylation and adenosine triphosphate generation in the mitochondria, accumulation of transition metals can lead to oxidative damage and mitochondrial dysfunction. Heavy metals such as cadmium, mercury, and manganese can preferentially accumulate in the mitochondria (via calcium channels), also leading to mitochondrial dysfunction [28,29]. Brain cells, due to their increased metabolic requirement, have elevated mitochondrial content, making them highly susceptible to oxidative damage. A key pathological hallmark in Parkinson's and Huntington's disease is mitochondrial iron accumulation [30,31]. Similarly, mitochondria play a critical role in Alzheimer's disease progression via the handling of beta-amyloid protein [32]. Mitochondrial abnormalities have been cited in the hippocampus of human subjects with Alzheimer's disease [33]. Consequently, targeting excessive mitochondrial metal may represent a strategy to protect against neurodegenerative diseases.

4.2.3   Metals and Neuronal Apoptosis

Apoptosis or programmed cell death plays a significant role in cell loss during neuro-degenerative disorders [34]. Excessive accumulation of metals can trigger neuronal cell apoptosis [34,35]. Neural apoptosis has been implicated in Alzheimer's disease [36], Parkinson's disease [37], amyotrophic lateral sclerosis [38], and Huntington's disease [39]. Accumulation of iron, copper and zinc has been shown to cause apoptosis via a p53-mediated pathway [35]. Mitochondrial damage results in the release and activation of caspases, leading to apoptosis. ROS can also cause lysosomal damage, leading to the release of caustic lysosomal enzymes, resulting in cell death. Based on this role of metal ions in neurodegenerative disease, metal chelators are now being designed to protect against neuronal apoptosis [7,40].

4.2.4   Metals and ER Stress

The ER is the site for protein folding, handling, posttranslational modification, and delivery of proteins to their final destination [41]. When the ER is overwhelmed by either excessive synthesis of proteins or by slow breakdown, it accumulates unfolded proteins, leading to ER stress. ER stress triggers an unfolded protein response via upregulation of glucose-regulated protein (GRP) and heat shock proteins, which counteract ER stress by decreasing protein synthesis and upregulating both the protein folding and the degradation pathways, leading to cellular recovery [42]. Heavy metals such as lead, cadmium, and mercury have been shown to modulate ER stress by inhibiting protein folding, reducing GRP78 and disrupting calcium homeostasis [43–45]. The aforementioned studies suggest that ER is a target for metal toxicity and ER-chaperones such as tauroursodeoxycholic acid [46] may be harnessed to treat neurodegenerative disease caused by excessive metal accumulation.

4.2.5   Metals and Autophagy

Autophagy is a cellular breakdown pathway whereby cytoplasmic components are delivered to lysosomes for degradation [47]. Under normal conditions, autophagy plays a significant role in maintaining cellular homeostasis by providing nutrients to the cells in the time of need and aiding neuronal survival. However, like oxidative stress and ER stress, unregulated autophagy can function as a double-edged sword and prove to be detrimental. Accumulating evidence suggests that autophagy mediates neurotoxicity induced by exposure to essential metals, such as manganese, copper, and iron, and other heavy metals, such as cadmium, lead, and mercury [18].

4.2.6   Metals and Epigenetic Alterations

Epigenetics regulates the dynamic nature of chromatin structure and can lead to heritable changes in gene expression without alterations in the DNA sequence [48]. Epigenetic changes that are widely studied include DNA methylation and histone-posttranslational modifications such as histone acetylation. Heavy metals such as nickel, chromium, lead, cadmium, and arsenic have been shown to cause DNA hypermethylation [49–51]. Metal-induced alterations in methylation can lead to a cascade of events including aberrant gene expression, which can result in altered cellular phenotype [52]. Thus, agents that reduce DNA hypermethylation or alter histone deacetylases may have potential therapeutic value in treating neuronal metal toxicity.

4.3 NATURAL PRODUCTS TO COUNTER METAL-INDUCED NEURONAL DAMAGE

Naturally occurring antioxidants have received growing attention as neuroprotective agents to prevent oxidative damage. As it is beyond the scope of this article to discuss all the natural products that have been studied for their neuroprotective properties against metal damage, this section is by no means comprehensive and will focus only on those molecules/natural products that have been extensively studied.

4.3.1 RESVERATROL

Resveratrol is a polyphenolic compound with strong antioxidant properties that is found abundantly in grapes, berries, nuts, and red wine [53]. Several studies describe the salutary role of resveratrol in neurodegenerative disease [54,55]. Resveratrol exhibits strong antioxidant, antiinflammatory, and metal chelation properties as shown by in vitro and in vivo studies [56,57]. Resveratrol has been shown to inhibit iron-induced mitochondrial dysfunction by inhibiting glycogen synthase kinase-3 beta activity [58] and reduces lipid peroxidation by scavenging free radicals [59]. Accumulation of aluminum in the brain has been reported to be linked to neuropathological conditions [60]. Aluminum has been shown to reduce aluminum-induced ROS production and the activation of neuroinflammatory response in vivo [61]. Elevated levels of labile copper in the brain has been observed in Alzheimer's disease [62]. By virtue of its ability to chelate copper [63], resveratrol, in theory, may also benefit Alzheimer's disease [64]. However, resveratrol has been shown to reduce cupric to cuprous iron, triggering a pro-oxidant status rather than an antioxidant status, which may prove to be detrimental rather than protective [54]. Nubling and coworkers reported that the interaction between phosphorylated tau and metal ions can lead to the formation of potentially toxic oligomer species on tau oligomer formation and its coaggregation with α-synuclein at the level of individual oligomers [65]. Interestingly, resveratrol-mediated activation of sirtuin-1 (SIRT1) can lead to direct deacetylation of acetylated tau, thereby promoting its proteasomal degradation [66], which may represent yet another mechanism by which resveratrol may attenuate neuronal metal toxicity.

4.3.2 FLAVONOIDS

Flavonoids are natural antioxidant substances present in several plant-derived products including green tea, raspberries, blueberries, and red wine. Green tea is rich in the flavonol catechins, the predominant one being (-)-epigallocatechin-3-gallate (EGCG) [67]. EGCG was most potent among different polyphenols, including

(-)-epicatechin, vanillic acid, gallic acid, quercetin, and myricetin, in protecting hydroxyl-radical induced DNA damage [68]. The neuroprotective effects of EGCG have been documented in various models of neurodegenerative diseases [69–71]. Green tea catechins improve neuronal cell survival and mitochondrial function via their antioxidant/metal chelation activities and by modulating genes involved in the cell survival/death pathway [69,72]. A recent study revealed that green tea treatment reduced the accumulation of lead in the brain in mice subjected to lead treatment, suggesting that green tea can compete with intestinal absorption of heavy metals and/or cause the efflux of heavy metals from the brain [73]. EGCG has been shown to regulate amyloid precursor protein through the iron responsive elements and reduce the toxic levels of amyloid beta peptide in Parkinson's and Alzheimer's disease models [74]. EGCG inhibits the nuclear factor kappa B, the oxidative stress-induced gene, and suppresses cytokine expression by chelating iron [75]. In addition, EGCG inhibits iron-dependent lipid peroxidation and increases glutathione peroxidase activity and reduced glutathione concentrations [76,77], indicating that EGCG is a potent metal chelating antioxidant.

4.3.3 CURCUMIN

Curcumin (diferuloyl methane) is the yellow coloring pigment of the root turmeric. Curcumin is a potent scavenger of superoxide anion [78], nitric oxide [79], and singlet oxygen [80] and an inhibitor of lipid peroxidation [81]. By virtue of its diketone moiety (which can exist as a keto-enol tautomer), curcumin also functions as a potent metal chelator [81]. Several recent studies suggest that curcumin exhibits neuroprotective properties [82]. Inhibition of metal-ion induced amyloid-beta aggregation has been suggested as a potential mechanism involved in the neuroprotection by curcumin [83–86]. Additionally, curcumin has been shown to protect nigral dopaminergic neurons by iron chelation in a rat model of Parkinson's disease [87]. Curcumin and its natural derivatives have also been shown to alleviate lead-induced memory dysfunction [88], methyl-mercury-induced cognitive and motor dysfunction [89], and memory deficits associated with cadmium treatment [90].

4.3.4 OTHER NATURAL PRODUCTS

Several other natural products have been shown to exhibit neuroprotection either by reducing metal-induced oxidative stress in the brain or by functioning as a chain-breaking antioxidant. Ferulic acid (4-hydroxy-3-methoxycinnamic acid), which is widely distributed in the plant kingdom, exhibits neuroprotective effect via its antioxidant properties [91]. Recent studies have demonstrated the protective effects of *Bacopa monnieri* extract against aluminum, lead, and mercury-induced neurotoxicity [92–95]. Naringenin, a flavonoid abundant in the peels of citrus fruit, has been recently shown to protect against iron-overload-induced cerebral cortex neurotoxicity [96]. Rutin (quercetin-3-O-rutinoside) is another natural flavonoid glycoside that has been studied for its neuroprotective properties based on its antioxidant effect [97].

4.4 CONCLUSION

The apparent link between metal toxicity and neurodegenerative diseases has given rise to the notion of metal chelation and/or antioxidant therapy as a therapeutic option. Several phyto-products and herbs have demonstrated neuroprotective efficacy in preclinical studies. A major challenge of using phytochemicals, mainly the ones with phenolic groups, is its poor bioavailability and rapid metabolism. Additionally, an antioxidant can function as a pro-oxidant after donating an electron to neutralize free radicals and/or if it has a coordination site following complex formation with metal ions. As free radicals are highly reactive for an antioxidant to be viable therapeutically, it should reach the target site at high concentration and should have a greater rate constant for reaction with the free radicals than the endogenous ligands. Novel targeting mechanism such as the use of nanoparticles may to some extent circumvent these challenges. Also, structure activity relationship studies may help unveil the part of the molecule responsible for activity, and such knowledge will be useful in optimizing the molecule for maximum efficacy.

REFERENCES

1. P. Chen, M.R. Miah, M. Aschner, Metals and neurodegeneration, *F1000Res*, 5 (2016).
2. E. Aizenman, P.G. Mastroberardino, Metals and neurodegeneration, *Neurobiol Dis*, 81 (2015) 1–3.
3. K.A. Jellinger, The relevance of metals in the pathophysiology of neurodegeneration, pathological considerations, *Int Rev Neurobiol*, 110 (2013) 1–47.
4. A.R. White, K.M. Kanninen, P.J. Crouch, Editorial: Metals and neurodegeneration: restoring the balance, *Front Aging Neurosci*, 7 (2015) 127.
5. K.J. Barnham, A.I. Bush, Biological metals and metal-targeting compounds in major neurodegenerative diseases, *Chem Soc Rev*, 43 (2014) 6727–6749.
6. E.J. McAllum, D.I. Finkelstein, Metals in Alzheimer's and Parkinson's disease: Relevance to dementia with Lewy bodies, *J Mol Neurosci*, 60 (2016) 279–298.
7. J.S. Cristovao, R. Santos, C.M. Gomes, Metals and neuronal metal binding proteins implicated in Alzheimer's disease, *Oxid Med Cell Longev*, 2016 (2016) 9812178.
8. M. Muller, B.R. Leavitt, Iron dysregulation in Huntington's disease, *J Neurochem*, 130 (2014) 328–350.
9. J.F. Dominguez, A.C. Ng, G. Poudel, J.C. Stout, A. Churchyard, P. Chua, G.F. Egan, N. Georgiou-Karistianis, Iron accumulation in the basal ganglia in Huntington's disease: cross-sectional data from the IMAGE-HD study, *J Neurol Neurosurg Psychiatry*, 87 (2016) 545–549.
10. A.R. White, K.M. Kanninen, P.J. Crouch, Editorial: Metals and neurodegeneration: restoring the balance. *Front Aging Neurosci*, 7 (2015) 127.
11. P.A. Adlard, R.S. Chung, Editorial: The molecular pathology of cognitive decline: Focus on metals, *Front Aging Neurosci*, 7 (2015) 116.
12. S. Caito, M. Aschner, Neurotoxicity of metals, *Handb Clin Neurol*, 131 (2015) 169–189.
13. S.A. Schneider, Neurodegeneration with brain iron accumulation, *Curr Neurol Neurosci Rep*, 16 (2016) 9.
14. J.R. Liddell, Targeting mitochondrial metal dyshomeostasis for the treatment of neurodegeneration, *Neurodegener Dis Manag*, 5 (2015) 345–364.
15. M. Aoun, V. Tiranti, Mitochondria: A crossroads for lipid metabolism defect in neurodegeneration with brain iron accumulation diseases, *Int J Biochem Cell Biol*, 63 (2015) 25–31.

16. Y. Liu, J.R. Connor, Iron and ER stress in neurodegenerative disease, *Biometals*, 25 (2012) 837–845.
17. Y. Qian, E. Tiffany-Castiglioni, Lead-induced endoplasmic reticulum (ER) stress responses in the nervous system, *Neurochem Res*, 28 (2003) 153–162.
18. Z. Zhang, M. Miah, M. Culbreth, M. Aschner, Autophagy in neurodegenerative diseases and metal neurotoxicity, *Neurochem Res*, 41 (2016) 409–422.
19. S. Krishan, P.J. Jansson, E. Gutierrez, D.J. Lane, D. Richardson, S. Sahni, Iron metabolism and autophagy: A poorly explored relationship that has important consequences for health and disease, *Nagoya J Med Sci*, 77 (2015) 1–6.
20. M. Caffo, G. Caruso, G.L. Fata, V. Barresi, M. Visalli, M. Venza, I. Venza, Heavy metals and epigenetic alterations in brain tumors, *Curr Genom*, 15 (2014) 457–463.
21. Agency for Toxic Substances and Disease Registry. Priority list of hazardous substances. http://www.atsdr.cdc.gov/SPL/index.html, accessed on September 16, 2016.
22. J. Long, C. Liu, L. Sun, H. Gao, J. Liu, Neuronal mitochondrial toxicity of malondialdehyde: Inhibitory effects on respiratory function and enzyme activities in rat brain mitochondria, *Neurochem Res*, 34 (2009) 786–794.
23. M.A. Korolainen, T. Pirttila, Cerebrospinal fluid, serum and plasma protein oxidation in Alzheimer's disease, *Acta Neurol Scand*, 119 (2009) 32–38.
24. T.T. Reed, W.M. Pierce, Jr., D.M. Turner, W.R. Markesbery, D.A. Butterfield, Proteomic identification of nitrated brain proteins in early Alzheimer's disease inferior parietal lobule, *J Cell Mol Med*, 13 (2009) 2019–2029.
25. M. Bogdanov, W.R. Matson, L. Wang, T. Matson, R. Saunders-Pullman, S.S. Bressman, M. Flint Beal, Metabolomic profiling to develop blood biomarkers for Parkinson's disease, *Brain*, 131 (2008) 389–396.
26. N. Aguirre, M.F. Beal, W.R. Matson, M.B. Bogdanov, Increased oxidative damage to DNA in an animal model of amyotrophic lateral sclerosis, *Free Radic Res*, 39 (2005) 383–388.
27. X. Wang, A. Zaidi, R. Pal, A.S. Garrett, R. Braceras, X.W. Chen, M.L. Michaelis, E.K. Michaelis, Genomic and biochemical approaches in the discovery of mechanisms for selective neuronal vulnerability to oxidative stress, *BMC Neurosci*, 10 (2009) 12.
28. A. Grubman, A.R. White, J.R. Liddell, Mitochondrial metals as a potential therapeutic target in neurodegeneration, *Br J Pharmacol*, 171 (2014) 2159–2173.
29. J.N. Meyer, M.C. Leung, J.P. Rooney, A. Sendoel, M.O. Hengartner, G.E. Kisby, A.S. Bess, Mitochondria as a target of environmental toxicants, *Toxicol Sci*, 134 (2013) 1–17.
30. P.G. Mastroberardino, E.K. Hoffman, M.P. Horowitz, R. Betarbet, G. Taylor, D. Cheng, H.M. Na, C.A. Gutekunst, M. Gearing, J.Q. Trojanowski, M. Anderson, C.T. Chu, J. Peng, J.T. Greenamyre, A novel transferrin/TfR2-mediated mitochondrial iron transport system is disrupted in Parkinson's disease, *Neurobiol Dis*, 34 (2009) 417–431.
31. H.D. Rosas, Y.I. Chen, G. Doros, D.H. Salat, N.K. Chen, K.K. Kwong, A. Bush, J. Fox, S.M. Hersch, Alterations in brain transition metals in Huntington disease: An evolving and intricate story, *Arch Neurol*, 69 (2012) 887–893.
32. C.M. Pinho, P.F. Teixeira, E. Glaser, Mitochondrial import and degradation of amyloid-beta peptide, *Biochim Biophys Acta*, 1837 (2014) 1069–1074.
33. P.I. Moreira, C. Carvalho, X. Zhu, M.A. Smith, G. Perry, Mitochondrial dysfunction is a trigger of Alzheimer's disease pathophysiology, *Biochim Biophys Acta*, 1802 (2010) 2–10.
34. K.P. Loh, S.H. Huang, R. De Silva, B.K. Tan, Y.Z. Zhu, Oxidative stress: Apoptosis in neuronal injury, *Curr Alzheimer Res*, 3 (2006) 327–337.
35. C.W. Levenson, Trace metal regulation of neuronal apoptosis: From genes to behavior, *Physiol Behav*, 86 (2005) 399–406.

36. B.O. Popescu, M. Ankarcrona, Mechanisms of cell death in Alzheimer's disease: Role of presenilins, *J Alzheimers Dis*, 6 (2004) 123–128.
37. W.G. Tatton, R. Chalmers-Redman, D. Brown, N. Tatton, Apoptosis in Parkinson's disease: Signals for neuronal degradation, *Ann Neurol*, 53 Suppl 3 (2003) S61–S70; discussion S70–S62.
38. S. Przedborski, Programmed cell death in amyotrophic lateral sclerosis: A mechanism of pathogenic and therapeutic importance, *Neurologist*, 10 (2004) 1–7.
39. M.A. Hickey, M.F. Chesselet, Apoptosis in Huntington's disease, *Prog Neuropsychopharmacol Biol Psychiatry*, 27 (2003) 255–265.
40. R.B. Mounsey, P. Teismann, Chelators in the treatment of iron accumulation in Parkinson's disease, *Int J Cell Biol*, 2012 (2012) 983245.
41. M.J. Berridge, The endoplasmic reticulum: A multifunctional signaling organelle, *Cell Calcium*, 32 (2002) 235–249.
42. M. Ni, A.S. Lee, ER chaperones in mammalian development and human diseases, *FEBS Lett*, 581 (2007) 3641–3651.
43. Y. Qian, Y. Zheng, K.S. Ramos, E. Tiffany-Castiglioni, GRP78 compartmentalized redistribution in Pb-treated glia: Role of GRP78 in lead-induced oxidative stress, *Neurotoxicology*, 26 (2005) 267–275.
44. Y. Shinkai, C. Yamamoto, T. Kaji, Lead induces the expression of endoplasmic reticulum chaperones GRP78 and GRP94 in vascular endothelial cells via the JNK-AP-1 pathway, *Toxicol Sci*, 114 (2010) 378–386.
45. S.K. Sharma, P. Goloubinoff, P. Christen, Heavy metal ions are potent inhibitors of protein folding, *Biochem Biophys Res Commun*, 372 (2008) 341–345.
46. R.J. Viana, C.J. Steer, C.M. Rodrigues, Amyloid-beta peptide-induced secretion of endoplasmic reticulum chaperone glycoprotein GRP94, *J Alzheimers Dis*, 27 (2011) 61–73.
47. S. Nair, J. Ren, Autophagy and cardiovascular aging: Lesson learned from rapamycin, *Cell Cycle*, 11 (2012) 2092–2099.
48. D.C. Dolinoy, J.R. Weidman, R.L. Jirtle, Epigenetic gene regulation: Linking early developmental environment to adult disease, *Reprod Toxicol*, 23 (2007) 297–307.
49. J.F. Reichard, M. Schnekenburger, A. Puga, Long term low-dose arsenic exposure induces loss of DNA methylation, *Biochem Biophys Res Commun*, 352 (2007) 188–192.
50. M. Takiguchi, W.E. Achanzar, W. Qu, G. Li, M.P. Waalkes, Effects of cadmium on DNA-(cytosine-5) methyltransferase activity and DNA methylation status during cadmium-induced cellular transformation, *Exp Cell Res*, 286 (2003) 355–365.
51. R. Martinez-Zamudio, H.C. Ha, Environmental epigenetics in metal exposure, *Epigenetics*, 6 (2011) 820–827.
52. R.O. Wright, A. Baccarelli, Metals and neurotoxicology, *J Nutr*, 137 (2007) 2809–2813.
53. E. Wenzel, V. Somoza, Metabolism and bioavailability of trans-resveratrol, *Mol Nutr Food Res*, 49 (2005) 472–481.
54. I. Muqbil, F.W. Beck, B. Bao, F.H. Sarkar, R.M. Mohammad, S.M. Hadi, A.S. Azmi, Old wine in a new bottle: The Warburg effect and anticancer mechanisms of resveratrol, *Curr Pharm Des*, 18 (2012) 1645–1654.
55. A. Granzotto, P. Zatta, Resveratrol and Alzheimer's disease: Message in a bottle on red wine and cognition, *Front Aging Neurosci*, 6 (2014) 95.
56. C.D. Venturini, S. Merlo, A.A. Souto, C. Fernandes Mda, R. Gomez, C.R. Rhoden, Resveratrol and red wine function as antioxidants in the nervous system without cellular proliferative effects during experimental diabetes, *Oxid Med Cell Longev*, 3 (2010) 434–441.
57. A.Y. Sun, Q. Wang, A. Simonyi, G.Y. Sun, Resveratrol as a therapeutic agent for neurodegenerative diseases, *Mol Neurobiol*, 41 (2010) 375–383.

58. S.M. Shin, I.J. Cho, S.G. Kim, Resveratrol protects mitochondria against oxidative stress through AMP-activated protein kinase-mediated glycogen synthase kinase-3beta inhibition downstream of poly(ADP-ribose)polymerase-LKB1 pathway, *Mol Pharmacol*, 76 (2009) 884–895.

59. B. Tadolini, C. Juliano, L. Piu, F. Franconi, L. Cabrini, Resveratrol inhibition of lipid peroxidation, *Free Radic Res*, 33 (2000) 105–114.

60. J.R. Walton, Aluminum involvement in the progression of Alzheimer's disease, *J Alzheimers Dis*, 35 (2013) 7–43.

61. A. Zaky, B. Mohammad, M. Moftah, K.M. Kandeel, A.R. Bassiouny, Apurinic/apyrimidinic endonuclease 1 is a key modulator of aluminum-induced neuroinflammation, *BMC Neurosci*, 14 (2013) 26.

62. S.A. James, I. Volitakis, P.A. Adlard, J.A. Duce, C.L. Masters, R.A. Cherny, A.I. Bush, Elevated labile Cu is associated with oxidative pathology in Alzheimer disease, *Free Radic Biol Med*, 52 (2012) 298–302.

63. V. Tamboli, A. Defant, I. Mancini, P. Tosi, A study of resveratrol-copper complexes by electrospray ionization mass spectrometry and density functional theory calculations, *Rapid Commun Mass Spectrom*, 25 (2011) 526–532.

64. N.G. Faux, C.W. Ritchie, A. Gunn, A. Rembach, A. Tsatsanis, J. Bedo, J. Harrison, L. Lannfelt, K. Blennow, H. Zetterberg, M. Ingelsson, C.L. Masters, R.E. Tanzi, J.L. Cummings, C.M. Herd, A.I. Bush, PBT2 rapidly improves cognition in Alzheimer's Disease: Additional phase II analyses, *J Alzheimers Dis*, 20 (2010) 509–516.

65. G. Nubling, B. Bader, J. Levin, J. Hildebrandt, H. Kretzschmar, A. Giese, Synergistic influence of phosphorylation and metal ions on tau oligomer formation and coaggregation with alpha-synuclein at the single molecule level, *Mol Neurodegener*, 7 (2012) 35.

66. S.W. Min, S.H. Cho, Y. Zhou, S. Schroeder, V. Haroutunian, W.W. Seeley, E.J. Huang, Y. Shen, E. Masliah, C. Mukherjee, D. Meyers, P.A. Cole, M. Ott, L. Gan, Acetylation of tau inhibits its degradation and contributes to tauopathy, *Neuron*, 67 (2010) 953–966.

67. N.T. Zaveri, Green tea and its polyphenolic catechins: Medicinal uses in cancer and noncancer applications, *Life Sci*, 78 (2006) 2073–2080.

68. N. Salah, N.J. Miller, G. Paganga, L. Tijburg, G.P. Bolwell, C. Rice-Evans, Polyphenolic flavanols as scavengers of aqueous phase radicals and as chain-breaking antioxidants, *Arch Biochem Biophys*, 322 (1995) 339–346.

69. S.A. Mandel, T. Amit, O. Weinreb, L. Reznichenko, M.B. Youdim, Simultaneous manipulation of multiple brain targets by green tea catechins: A potential neuroprotective strategy for Alzheimer and Parkinson diseases, *CNS Neurosci Ther*, 14 (2008) 352–365.

70. S.A. Mandel, O. Weinreb, T. Amit, M.B. Youdim, Molecular mechanisms of the neuroprotective/neurorescue action of multi-target green tea polyphenols, *Front Biosci (Schol Ed)*, 4 (2012) 581–598.

71. N.A. Singh, A.K. Mandal, Z.A. Khan, Potential neuroprotective properties of epigallocatechin-3-gallate (EGCG), *Nutr J*, 15 (2016) 60.

72. T. Amit, Y. Avramovich-Tirosh, M.B. Youdim, S. Mandel, Targeting multiple Alzheimer's disease etiologies with multimodal neuroprotective and neurorestorative iron chelators, *FASEB J*, 22 (2008) 1296–1305.

73. A. Winiarska-Mieczan, The potential protective effect of green, black, red and white tea infusions against adverse effect of cadmium and lead during chronic exposure—A rat model study, *Regul Toxicol Pharmacol*, 73 (2015) 521–529.

74. Y. Avramovich-Tirosh, L. Reznichenko, T. Mit, H. Zheng, M. Fridkin, O. Weinreb, S. Mandel, M.B. Youdim, Neurorescue activity, APP regulation and amyloid-beta peptide reduction by novel multi-functional brain permeable iron-chelating-antioxidants, M-30 and green tea polyphenol, EGCG, *Curr Alzheimer Res*, 4 (2007) 403–411.

75. J. Li, L. Ye, X. Wang, J. Liu, Y. Wang, Y. Zhou, W. Ho, (-)-Epigallocatechin gallate inhibits endotoxin-induced expression of inflammatory cytokines in human cerebral microvascular endothelial cells, *J Neuroinflammation*, 9 (2012) 161.
76. I. Dobrzynska, A. Sniecinska, E. Skrzydlewska, Z. Figaszewski, Green tea modulation of the biochemical and electric properties of rat liver cells that were affected by ethanol and aging, *Cell Mol Biol Lett*, 9 (2004) 709–721.
77. J. Ostrowska, W. Luczaj, I. Kasacka, A. Rozanski, E. Skrzydlewska, Green tea protects against ethanol-induced lipid peroxidation in rat organs, *Alcohol*, 32 (2004) 25–32.
78. N. Sreejayan, M.N. Rao, Free radical scavenging activity of curcuminoids, *Arzneimittelforschung*, 46 (1996) 169–171.
79. Sreejayan, M.N. Rao, Nitric oxide scavenging by curcuminoids, *J Pharm Pharmacol*, 49 (1997) 105–107.
80. M. Subramanian, Sreejayan, M.N. Rao, T.P. Devasagayam, B.B. Singh, Diminution of singlet oxygen-induced DNA damage by curcumin and related antioxidants, *Mutat Res*, 311 (1994) 249–255.
81. Sreejayan, M.N. Rao, Curcuminoids as potent inhibitors of lipid peroxidation, *J Pharm Pharmacol*, 46 (1994) 1013–1016.
82. D. Chin, P. Huebbe, K. Pallauf, G. Rimbach, Neuroprotective properties of curcumin in Alzheimer's disease—Merits and limitations, *Curr Med Chem*, 20 (2013) 3955–3985.
83. A. Kochi, H.J. Lee, S.M. Vithanarachchi, V. Padmini, M.J. Allen, M.H. Lim, Inhibitory activity of curcumin derivatives towards metal-free and metal-induced amyloid-beta aggregation, *Curr Alzheimer Res*, 12 (2015) 415–423.
84. S. Chan, S. Kantham, V.M. Rao, M.K. Palanivelu, H.L. Pham, P.N. Shaw, R.P. McGeary, B.P. Ross, Metal chelation, radical scavenging and inhibition of Abeta(4)(2) fibrillation by food constituents in relation to Alzheimer's disease, *Food Chem*, 199 (2016) 185–194.
85. V.S. Mithu, B. Sarkar, D. Bhowmik, A.K. Das, M. Chandrakesan, S. Maiti, P.K. Madhu, Curcumin alters the salt bridge-containing turn region in amyloid beta(1–42) aggregates, *J Biol Chem*, 289 (2014) 11122–11131.
86. T. Jiang, X.L. Zhi, Y.H. Zhang, L.F. Pan, P. Zhou, Inhibitory effect of curcumin on the Al(III)-induced Abeta(4)(2) aggregation and neurotoxicity in vitro, *Biochim Biophys Acta*, 1822 (2012) 1207–1215.
87. X.X. Du, H.M. Xu, H. Jiang, N. Song, J. Wang, J.X. Xie, Curcumin protects nigral dopaminergic neurons by iron-chelation in the 6-hydroxydopamine rat model of Parkinson's disease, *Neurosci Bull*, 28 (2012) 253–258.
88. A. Dairam, J.L. Limson, G.M. Watkins, E. Antunes, S. Daya, Curcuminoids, curcumin, and demethoxycurcumin reduce lead-induced memory deficits in male Wistar rats, *J Agric Food Chem*, 55 (2007) 1039–1044.
89. F. Zahir, S.J. Rizvi, S.K. Haq, R.H. Khan, Effect of methyl mercury induced free radical stress on nucleic acids and protein: Implications on cognitive and motor functions, *Indian J Clin Biochem*, 21 (2006) 149–152.
90. P. da Costa, J.F. Goncalves, J. Baldissarelli, T.R. Mann, F.H. Abdalla, A.M. Fiorenza, M.M. da Rosa, F.B. Carvalho, J.M. Gutierres, C.M. de Andrade, M.A. Rubin, M.R. Schetinger, V.M. Morsch, Curcumin attenuates memory deficits and the impairment of cholinergic and purinergic signaling in rats chronically exposed to cadmium, *Environ Toxicol* (2015).
91. A. Sgarbossa, D. Giacomazza, M. di Carlo, Ferulic acid: A hope for Alzheimer's disease therapy from plants, *Nutrients*, 7 (2015) 5764–5782.
92. J.S. Nannepaga, M. Korivi, M. Tirumanyam, M. Bommavaram, C.H. Kuo, Neuroprotective effects of *Bacopa monniera* whole-plant extract against aluminum-induced hippocampus damage in rats: Evidence from electron microscopic images, *Chin J Physiol*, 57 (2014) 279–285.

93. M.K. Velaga, C.K. Basuri, K.S. Robinson Taylor, P.R. Yallapragada, S. Rajanna, B. Rajanna, Ameliorative effects of *Bacopa monniera* on lead-induced oxidative stress in different regions of rat brain, *Drug Chem Toxicol*, 37 (2014) 357–364.

94. T. Sumathi, C. Shobana, J. Christinal, C. Anusha, Protective effect of *Bacopa monniera* on methyl mercury-induced oxidative stress in cerebellum of rats, *Cell Mol Neurobiol*, 32 (2012) 979–987.

95. B. Mahitha, B. Deva Prasad Raju, K. Mallikarjuna, N. Durga Mahalakshmi Ch, N.J. Sushmal, *Bacopa monniera* stabilized silver nanoparticles attenuates oxidative stress induced by aluminum in Albino mice, *J Nanosci Nanotechnol*, 15 (2015) 1101–1109.

96. Y. Chtourou, H. Fetoui, R. Gdoura, Protective effects of naringenin on iron-overload-induced cerebral cortex neurotoxicity correlated with oxidative stress, *Biol Trace Elem Res*, 158 (2014) 376–383.

97. S. Habtemariam, Rutin as a natural therapy for Alzheimer's disease: Insights into its mechanisms of action, *Curr Med Chem*, 23 (2016) 860–873.

Section II

Phytopharmaceuticals

5 Effects of Dietary Supplements and Nutraceuticals of Plant Origin on Central and Peripheral Nervous System Pathology

Odete Mendes

CONTENTS

Published data support the potential beneficial effects of dietary supplements, nutraceuticals, and phytopharmaceuticals on central and peripheral nervous system pathology. Data on the neuroprotective effects of such compounds have been reported for multiple disease conditions including neurodegenerative and cognition disorders, Alzheimer's disease (AD), ischemic injury associated with stroke and trauma, neurotoxicity, and primary or metastatic cancer.

AD is one of the most common neurodegenerative diseases. It manifests itself in the form of dementia. Dietary regimens associated with the use of dietary supplementation may be a relevant tool in moderating the risk of AD in aged populations. AD pathology is characterized by amyloid plaque deposits in the cerebral cortex and the hippocampus. The key pathological features of AD include the presence of intracellular neurofibrillary tangles of hyperphosphorilated tau protein, the presence of extracellular amyloid deposits observable in plaques, and the presence of dystrophic or degenerating neurites (Grossi et al., 2013). Observable behavioral dysfunctions of AD are believed to be related to the presence of these plaques. Oxidative stress is one of the main factors involved in the pathogenesis of this disease. The generation of reactive oxygen species (ROS) can induce both functional and structural defects in cell membranes due to lipid peroxidation and carbonyl modification of proteins, which can contribute to the lesions observed with AD.

Plants and vegetables with high levels of vitamins, such as vitamin C, as part of the diet, in the form of dietary supplements or phytopharmaceuticals may be able to delay the onset or development of AD by acting as antioxidative agents. Another important factor in AD pathogenesis is the integrity of the blood–brain barrier (BBB). Compounds with strong antioxidative properties may also play an important role in improving BBB integrity and thus impact AD pathology. Polyphenols may act as ROS scavengers, impacting oxidative stress balance and the development of disease. Diets rich in polyphenols such as extra virgin olive oil and red wine may contribute to learning improvements and may have behavioral benefits associated with the onset and development of AD. Another main component of the pathogenicity of neurodegenerative diseases is inflammation associated with both astrocyte and microglial activation. Inflammatory mediators such as interleukin 6 (IL6) and tumor necrosis factor-alpha (TNFα) have been associated with the presence of neuritic plaques. IL1β and IL6 have also been related to AD impairments (Essa et al., 2015). Plant extracts that have high levels of vitamins or are rich in flavonoids, such as *Ginkgo biloba*, may have both antioxidant and anti-inflammatory properties that may be protective to AD development (Joseph et al., 2003). Also, dietary supplements rich in pomegranate, fig, and dates may impact inflammation favorably by modulating inflammatory mediators such as cytokines (IL1β, IL2–6 and IL9, TNFα) and may be of neuroprotective interest. Increments of acetylcholinesterase (AChE) activity have also been observed around amyloid plaques (Subash et al., 2014), and these increases have been associated with dementia. Diets rich in figs and fig leaf extracts potentially have an impact in AChE activity and may have a meaningful impact in disease manifestation.

Ischemic injury to the brain and spinal cord occurs when there is impairment of nervous system vasculature. Any event that markedly reduces the blood flow to the brain causes a decrease in oxygen delivery to brain tissues and may result in ischemia. Common causes of brain or spinal cord ischemia include traumatic events and vascular occlusion that lead to stroke. Ischemia and oxygen deprivation lead to lipid peroxidation and formation of ROS that cause tissue damage. Plant-derived supplements or phytopharmaceuticals that have high levels of polyphenols or other constituents with free radical scavenging properties and antioxidative properties may impact the development of neural tissue damage associated with reperfusion that occurs during ischemic injury. Cardiovascular disease, associated with vascular impairment, is one of the leading causes of stroke. Pathological lesions of partial or complete luminal occlusion of brain vasculature result in low oxygen tension or severe hypoxia, respectively, and may lead to severe brain injury or death. The pathogenesis of stroke is intimately involved with reperfusion events that occur after vascular occlusion and subsequent hypoxia. During brain reperfusion, oxidative stress is the main pathogenic pathway associated with tissue damage. Stroke outcomes may therefore also be impacted by diets rich in antioxidants, such as vitamin E (VE) and vitamin C, carotenoids, and polyphenolic phytochemicals that are widely present in fruits, vegetables, herbal teas, and wine (Sweeney et al., 2002). Polyphenols are products of plant metabolism that are comprised of molecules defined by the presence of an aromatic ring that has one or more hydroxyl substitutes. They are abundant in multiple fruits, including pomegranate. Pomegranate juice has protective effects in oxidative stress and may also help cardiovascular disease by having antiatherosclerotic and anti-atherogenic properties (Loren et al., 2005).

Dietary supplementation also has a role in increasing brain neuronal plasticity that may impact the development of slow-onset pathologies and impact inflammatory events associated with the establishment of post ischemic or partial occlusion etiologies. Brain plasticity is associated with areas of stem cell activity located in the subventricular zone (SVZ) and the subgranular layer (SGZ) of the dentate gyrus of the hippocampus. Neural stem cells located in these areas may migrate and differentiate into interneurons, granular cells, and astrocytes. During stroke, brain injury may trigger limited or transient neurogenesis; however, strategies that may be impacted by phytopharmaceuticals and dietary supplementation could potentially enhance neurogenesis and neuronal plasticity and thus may have beneficial outcomes in the treatment of these types of brain pathologies (Kaneko et al., 2012).

Another area of potential impact of phytopharmaceuticals and dietary supplementation in the development of brain pathology associated with stroke or trauma is the observed dysfunction of brain neurotransmitter systems. A transient period of increased cholinergic activity follows acute brain injury and associated excitotoxicity via nicotinic and muscarinic receptors. In chronic injury, a decline in cholinergic functions is observed. Activation of postsynaptic glutamate receptors, leading to glutamate excitotoxicity linked to lethal influx of calcium via cell membrane N-methyl-D-aspartate (NMDA) receptors, also has a relevant effect of posthypoxic injury. Supplementation with choline or polyunsaturated fatty acids (PUFAs) may have a positive impact in improving neurotransmitter balance and thus support brain health and promote neuroprotective events (Guseva et al., 2008). By affecting microgial and astrogial activity, plant-derived supplements may also have a role in establishing of both acute and chronic inflammatory processes associated with traumatic and ischemic brain pathologies. Strong anti-inflammatory properties have been ascribed to ginseng. In addition to repressing microglial activation, it also impacts postischemic inflammation resolution by decreasing proinflammatory cytokines such as IL6 and suppression of phosphorylation of nuclear factor-kappa B (NF-kB) and the cyclooxygenase 2 pathway (Rastogi et al., 2015).

Prevention of primary and secondary neuronal damage associated with toxic insult and related cognitive impairment has been ascribed to alpha-linoleic acid (ALA), an essential omega 3 PUFA found in green leaves, flaxseed oil, pumpkin seeds, beans, and walnuts. ALA has an important role in brain function and protection and has both anti-inflammatory and neuroplastic properties. Alpha-linolenic acid (LIN), a PUFA found in vegetable products, has an impact on activated NF-kB levels in the hippocampus that may have neuroprotective effects against neuronal cell death. Vitamin supplements such as VE have also a neuroprotective effect in nervous tissue toxic injury associated with ROS formation. A main pathway of toxic insult is the formation of ROS and reactive nitrogen species (RNS) such as super peroxide and nitric oxide. The protective effects of VE are thought to be due to its capacity to scavenge these toxic agents, thus protecting the neuronal cells from oxidative stress. These effects are associated with the activation or downregulation with many injury pathways that involve blc2 and Ras–mitogen-activated protein kinase (MAPK). Protective effects mediated via antioxidative properties are directly linked to increased phospholipid membrane stability (Galal et al., 2014).

Breast cancer brain metastases are associated with very poor prognosis and represent a challenging therapeutic target. Polyphenolic flavonoids have demonstrated potential efficacy in impacting cancer cell migration and invasive phenotype. Malignant brain tumors, such as gliomas, are also associated with a low median survival time (Vidak et al., 2015). One of the most important risk factors associated with brain cancer progression is oxidative stress. Flavonoids mediate reduction in oxidative stress by direct scavenging of ROS and RNS, prevention of calcium influx, increased levels of endogenous ROS/RNS scavengers such as glutamate, alteration of mitochondrial function, and decrease in enzymes associated with the generation of ROS/RNS. Flavonoids may also mitigate tumor progression via anti-inflammatory properties such as suppression of COX 2 expression, cancer apoptotic pathways, IL secretion, and direct and proliferative effects. Herbal supplements containing flavonoids can exert neuroprotective actions by modulation of intracellular mechanisms. They interact with multiple signaling pathway cascades including PI3 kinase, tyrosine kinase, protein kinase C, and MAPK. Inhibitory actions of kinase cascades may be beneficial in the treatment of cancer and proliferative diseases. Flavonoids have the potential to bind to the ATP-binding sites of a large number of proteins including mitochondrial ATPase and other ATPases that may modulate the mitochondrial transition pore and impact the apoptotic process. They may also impact gene expression regulation by exerting pharmacological activity in transcription factor regulation (Spencer, 2007). Flavonoids such as epigallocatechin 3-gallate may increase the chemosensitivity of tumor cells to a variety of chemotherapeutic agents and improve chemosensitizations across the BBB as well as having anti-inflammatory, antimutagenic, antiproliferative, and anticarcinogenic effects by impacting brain tumor invasiveness properties and cancer cell apoptosis and proliferative characteristics.

The aim of this chapter is to present a brief review of data that may support the neuroprotective effect of phytopharmaceuticals in different types of neuropathology.

5.1   NEURODEGENERATIVE AND COGNITION PATHOLOGY

Neurodegenerative diseases are a broad spectrum of diseases characterized by functional or sensory impairment of the central and/or peripheral nervous system. Numerous fruits have been reported to have potential impact on brain degenerative diseases (Keservani et al., 2016). These diseases are characterized by loss of integrity of the neuron cell body or the axon. The manifestation of each disease depends on the type of neuron that is affected and the specific neuronal lesion that occurs. The causes of these diseases may also vary among the different conditions.

Oxidative stress is one of the most common pathogenic events associated with neuronal degeneration. Numerous phytopharmaceuticals and nutraceuticals that have antioxidant properties may function as neuroprotective agents. One such phytopharmaceutical is *Curculigo orchioides*. It is a plant whose biologically active ingredient is curculigoside (CUR). CUR has been found to have neuroprotective properties (Tian et al., 2012). It is also thought to improve learning and memory due to a decrease in cerebral AChE activity and inhibition of the expression of beta-site APP cleaving enzyme 1 in the hippocampus. In animal behavior models, CUR improves latency and decreases the number of errors in behavioral studies with aged rats. This is associated with a decrease in activity of AChE and down regulation of β-site APP (amyloid-β precursor protein) cleaving enzyme 1 in the hippocampus. Overall, decreases in these molecules and activities resulted in improved cognitive function in aged rats (Wu et al., 2012). CUR has also been associated with the prevention of spatial memory deficit in APP/PS1 mutated transgenic mice and has a positive impact in Aβ deposition in AD. These effects have been attributed to CUR antioxidative properties (Zhao et al., 2015).

In addition to impacting pathogenic processes associated with oxidative pathways, phytopharmaceuticals may also impact levels of neurotransmitters such as dopamine and other signaling molecules related to brain function. Parkinson's disease (PD) occurs when there is degeneration of the dopaminergic neurons of the substantia nigra pars compacta. Spirulina (*Spirulina fusiform*) is a blue-green algae that contains high levels of antioxidant and anti-inflammatory molecules such as c-phycocyanin and phenolic compounds. Administration of spirulina to rats decreased PD endpoints such as the number of body rotations, improved body movements and locomotor activity, improved the balance of antioxidant brain levels, and significantly increased the levels of dopamine (Chattopadhyaya et al., 2015).

VE dietary supplementation was able to correct metabolic abnormalities associated with neuronal ceroid lipofuscinosis in the mnd (motor neuron disease) transgenic mouse, contributing to a positive impact in the concentration of phenylalanine in extracts of cerebral tissue, an increase in glutamate and *N*-acetyl-ʟ-aspartate, and decreased creatinine and glutamine in extracts of cortex tissue (Griffin et al., 2002). ALA is an essential omega-3 polyunsaturated fatty acid found in plants and markedly decreases neuronal cell death on hippocampal CA1 and CA3 subfields in a model of epilepsy induced by kianic acid. The neuroprotective effects of the polyunsaturated fatty acids may be linked, at least in part, to potassium channel opening (Lauritzen et al., 2000). ALA also increases levels of brain-derived neurotrophic factor (BDNF). Given that BDNF also has an antidepressant activity, administration of ALA is also associated with reduced measures of depressive-like behavior (Blondeau et al., 2009).

5.2 ALZHEIMER'S DISEASE

There are numerous reports that dietary supplements of plant and fruit origin, such as figs, may impact the development of AD. AD is a progressive neurodegenerative disorder that is characterized by the presence of amyloid (Aβ) plaques and neurofibrillary tangles in the brain cortex and limbic regions associated with chronic inflammation and neuronal damage. Aβ plaques activate inflammatory cells such as microglial cells and astrocytes and increase levels of certain cytokines and acute phase proteins.

Polyphenols present in figs, dates, and pomegranates are thought to have levels of antioxidants that may positively impact the development of AD pathology. Supplementation given to transgenic mice that are models of AD, APPsw/Tg2576, caused a significant reduction in Aβ peptide levels, the main component of Aβ plaques in both the hippocampus and cortex. Levels of ATP were also increased with dietary supplementation, suggesting an improvement of cerebral energy production in this model of AD (Essa et al., 2015).

In chronic AD, the deposition of senile Aβ plaques and neurofibrillary tangles leads to hyperphosphorylation of tau protein and progressive loss of neurons and neuronal processes, leading to cognitive impairment. One of the proposed mechanisms of pathogenicity of AD is the increase in ROS and free radicals associated with the Aβ peptide. These may cause structural damage to the cell membranes, leading to lipid peroxidation. Fig-rich diets, with their high concentrations of polyphenols, are a good source of antioxidants that act as scavengers. Administration of a 4% fig rich diet in APPsw/Tg2576 mice caused attenuation of the lipid peroxidation associated with AD development, observed as an increase in antioxidant enzyme activity in both cortex and hippocampus (Subash et al., 2014).

Resveratrol (3,5,4′-trihydroxy-trans-stilbene) is a polyphenol produced naturally by several plants such as grapes, peanuts, blueberries, raspberries, and mulberries. It is believed to have a potentially beneficial impact in neurodegenerative disease. Tg19959 mice develop an AD phenotype with formation of amyloid plaques. Administration of a diet with 0.2% resveratrol reduced plaque pathology with a reduction of plaque counts and plaque burden in the medial cortex, striatum and hypothalamus (Karuppagounder et al., 2009). Red wine also has relevant content in polyphenols including phenolic acids, flavonols, proanthocyanidins, and anthocyanins. Cabernet sauvignon (generated from *Vitis vinifera*)-derived anthocyanins reduce the generation of Aβ peptides in primary cortico-hippocampal neurons from Tg2576 mice. Quercetin-3-O-glucoronide and malvidin-3-O-glucoronide are cabernet sauvignon polyphenol metabolites that attenuated the generation of Aβ peptides in primary cortico-hippocampal neurons from Tg2576 mice treated with Quercetin. They have an effect on the amyloid peptide assembly and may improve AD-type deficits in hippocampal basal synaptic transmission via the MAPK signaling pathway (Ho et al., 2013). Blueberry-containing diets also show promise in the therapy of AD disease (Joseph et al., 2003). Extra-virgin olive oil also contains a number of polyphenols and secoridoids that include oleuropein aglycone (OLE). In TgCRND8 mice, a model for Aβ deposition administration of OLE improved memory deficits associated with AD and changes the number, size, and shape of cortex and hippocampus Aβ plaques (Figure 5.1) in both early and late stages of the disease associated with a reduction in migration of microglia to the plaque area, a reduction in astrocyte reactivity, and an increase in autophagic marker expression and lysosomal activity (Grossi et al., 2013).

Vitamin C has been shown to prevent or ameliorate AD cognitive-enhancing behavior. Administration of vitamin C to APP/PSEN1 mice that are a model for AD showed an improvement in behavior by Y-maze spontaneous alternation and swim accuracy in the water maze for very old mice; however, these were not accompanied by significant changes in Aβ protein deposition (Harrison et al., 2009). In order to

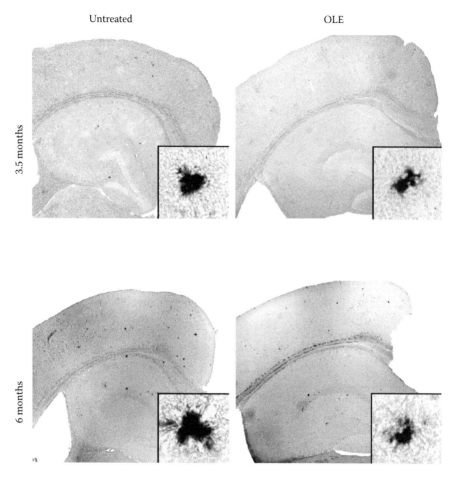

FIGURE 5.1   Treatment with OLE has a protective effect in TgCRND8 mice that are models of AD by decreasing the number, size, and shape of Aβ plaques in the cortex and hippocampus in both the early and late stages of the disease. Immunopositive deposits of AB protein can be observed in the untreated animals. However, animals treated with OLE show a significant reduction of these immunopositive deposits. (Reproduced from Grossi, C., Rigacci, S., Ambrosini, S., Ed Dami, T., Luccarini, I., Traini, C. et al., *PLoS ONE*, 8, e71702, 2013.)

evaluate the impact of vitamin C in the pathogenesis of AD, KO-Transgenic mice were generated by the crossing of 5 familial AD mutation (5XFAD) and mice lacking ʟ-gulono-gamma-lactone oxidase, the enzyme that synthesizes ascorbic acid. Vitamin C supplementation significantly reduced amyloid deposition in the cortex and the hippocampus, prevented disruption of the BBB by improving tight junction structure (TJ), and prevented mitochondrial alterations. Gliosis was also decreased in brains of mice supplemented with high doses of vitamin C. Ultrastructural evaluation showed that high supplementation with vitamin C (Figure 5.2) resulted in longer TJs and therefore improved BBB integrity. Abnormal mitochondrial morphology

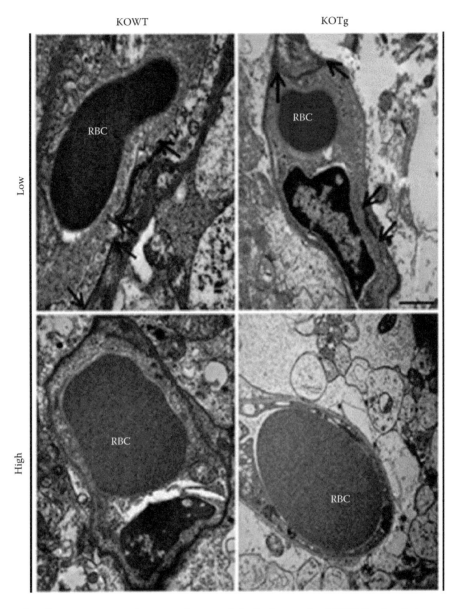

FIGURE 5.2 Electron microscopy images of the brain vessels of 5XFAD mice that are a model of AD. TJs were altered in these mice. Supplementation with high doses of vitamin C increased the length of these TJs (1225.6 nm versus 624.2 nm in standard vitamin C-treated-KO-Tg mice), demonstrating an improvement of the integrity of the BBB. Arrow, TJs; RBC, red blood cells. Scale bar = 2 μm. (Reproduced from Kook, S.Y., Lee, K.M., Kim, Y., Cha, M.Y., Kang, S., Baik, S.H., Lee, H., Park, R., Mook-Jung, I., *Cell Death Dis.*, 5, e1083, 2014.)

was also prevented by high supplementation and KO-JT mice showed no alterations of mitochondrial structure (Kook et al., 2014).

5.3 STROKE AND TRAUMA

There are continuous research efforts to develop dietary supplementation that may positively impact vascular injury associated with ischemic damage and subsequent neurodegeneration and nervous system pathology.

Mechanisms of neuroprotection include antioxidative effects observable with the administration of polyphenols. Diets supplemented with blueberry, spinach, and spirulina had an impact on both behavioral and histopathological endpoints in mice where stroke was induced by surgical ligation of the middle cerebral artery (Wang et al., 2005), with an observable increase in locomotor activity and reduction of the infraction volume (Figure 5.3). Polyphenols provided to pregnant mice as dietary supplements have been shown to protect against neonatal neurodegeneration when mice were subjected to hypoxic-ischemic brain injury. Rapid development of both necrotic and delayed apoptotic injury occurs after neonatal ischemic injury. Dietary supplementation to the mother with foods that are rich in polyphenols present in pomegranate juice, blueberries, green tea, and apple juice have shown protective in infantile rats with hydrocephalus and white matter degeneration (Etus et al., 2003) by reducing ischemic brain damage (Sweeney et al., 2002; Dajas et al., 2003).

Pomegranate juice administered to dams decreased brain tissue loss in pups that had ligation of the carotid for 45 minutes on postnatal day 7. This event was associated with a decrease in the activation of apoptosis-mediator caspase 3 (Loren et al., 2005). NT-20 is a formulated mixture that contains polyphenols from blueberry extracts, catechins from tea, amino acids such as carnosine, and vitamin D3. One hour after permanent middle cerebral artery occlusion (MCAO), NT-20-treated animals showed fewer abnormalities in motor and neurologic tests. Administration of NT-20 14 days after stroke significantly reduced the postischemic glial scar evaluated by means of glial fibrillary acidic protein (GFAP) staining. In addition, an increase in cell proliferation indicated by 5′-Bromo-2-deoxyuridine (BrdU) labeling was observed at the neurogenic (SVZ) and nonneurogenic striatum, potentially impacting proper clustering of newly formed cells. By double labeling with BrdU and GFAP, there was evidence that the induced neuronal differentiation had an increased tendency towards neuronal versus glial lineage (Kaneko et al., 2012).

Compounds that show effects of mitigation of apoptotic and excitatory pathways, implement neuro stem cell activity, or exert anti-inflammatory activity may also act as neuroprotectants in the development of ischemic brain injury. In the transient period after traumatic brain injury (TBI), an excess of cholinergic activity may contribute to excitotoxicity via effects mediated by nicotinic and/or muscarinic receptor subtypes. Conversely, there is a decrease in brain cholinergic function in the chronic phase of TBI that may be associated with ischemic brain injury. Also associated with ischemic events, microglial activation may have an important role in the development of brain pathology associated with TBI, specifically by impacting ROS formation, cytokine production, and proteinase or complement activity. Dietary supplementation with

Control   Blueberry   Spinach   Spirulina

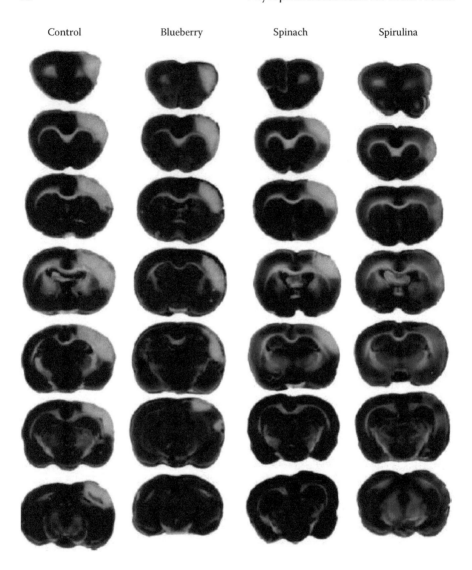

FIGURE 5.3 **(See color insert.)** Diets supplemented with blueberry, spinach, and spirulina had an impact in both behavioral and histopathological endpoints in mice where stroke was induced by surgical ligation of the middle cerebral artery, with an observable reduction of the infarction volume. Infarction areas can be seen in the coronal brain sections as light/white colored areas in the outer aspect of the brain sections. Brains from animals treated with spirulina have fewer and smaller infarction areas when compared to control animals. (Reproduced from Wang, Y., Chang, C.F., Chou, J., Chen, H.L., Deng, X., Harvey, B.K., Cadet, J.L., Bickford, P.C., *Exp. Neurol.*, 193, 75–84, 2005.)

choline was associated with cortical tissue preservation, consistent with a neuroprotective action (Guseva et al., 2008). In induced ischemic injury, microglial activation is observed after 12 days post injury in the hippocampus and the ventral posterior medial thalamus. Supplementation with choline decreases microglial activation in the cerebral cortex, hippocampus dentate gyrus, and thalamus.

ALA, a PUFA, exhibits both brain anti-inflammatory and neuroplastic properties. PUFAs found in vegetable oils may have neuroprotectant effects if administered 30 minutes prior to brain ischemic injury, with an observable decrease in neuronal cell death (Lauritzen et al., 2000). ALA also increases levels of vesicle-associated membrane protein 2, synaptosomal-associated protein 25, BDNF, and V-glutathione 1 and 2. These may impact glutamate transmission, affecting the brain's response to excitotoxicity (Blondeau et al., 2009), reduce the post ischemic infarct volume after 24 hours of occlusion of the middle cerebral artery, and increase survival rates 10 days after the ischemic insult. Ginseng is a term that encompasses 11 species of plants of the genus Panax. It is known to have extensive neuroprotective effects (Rastogi et al., 2015). Ginsenoside Rb1 represses microgial activation and decreases cytokine activity, such as IL6, in models of transient middle artery occlusion and models of ischemic injury (Park et al., 2004; Zhu et al., 2012). Gensenosides also have antiapoptotic effects in ischemic injury associated with regulation of apoptotic and antiapoptotic effectors such as an increase in bcl 2 and a decrease in BAX, respectively (Yuan et al., 2007). These effectors impact the release of cytochrome c and apoptosis inducing factor from the mitochondria and therefore mediate apoptotic events (Liang et al., 2013). Ginseng saponins have been shown to induce proliferation of neural stem cells and their differentiation into functional neurons, causing an increase in neurological scores in rats (Zheng et al., 2011) in permanent MCAO. Also, *G. biloba* may positively impact heme-oxigenase 1 during oxidative stress associated with MCAO, affecting excitotoxicity and thus conferring neuroprotection (Saleem et al., 2008). Tea catechins may impact the extent of damage after ischemic injury associated with stroke (Arab and Liebeskind, 2010). Spirulina, a blue-green algae, may impact hippocampal dentate gyrus response to lipopolysaccharide (LPS)-induced damage by decreasing stem cell/progenitor proliferation in the subgranular zone cells and decreasing astrocyte response to LPS, thus protecting neural stem cells from insult (Bachstetter et al., 2010).

5.4 TOXICITY PROTECTION

Another area in which phytopharmaceuticals may demonstrate neuroprotective effects is pathology associated with neurotoxicants. Examples of this type of protection have been observed with pesticide, such as organophosphates (OPs) and deltametrin (DM), neurotoxicity.

OPs are toxicants to the peripheral and central nervous systems. They act by inhibiting AChE, thus leading to the accumulation of acetylcholine, causing cholinergic respiratory and cardiovascular impairment; overactivation of muscarinic receptors that may cause seizures and secondary increases in glutamate and γ-aminobutyricacid. Glutamate receptors, such as the NMDA receptor, are involved in the maintenance of nerve potential and may be associated with agent-induced seizures and the excitotoxic-mediated brain damage related to OP nervous system toxicity. Brain regions affected by OP toxicity include the amygdala, hippocampus, and the piriform cortex. Lesions associated with these events consist of necrosis and/or apoptosis as well as neuronal, axonal, or dendrite degeneration. ALA may prevent neuronal damage associated with OP brain toxicity by being protective against OP induce neuropathology

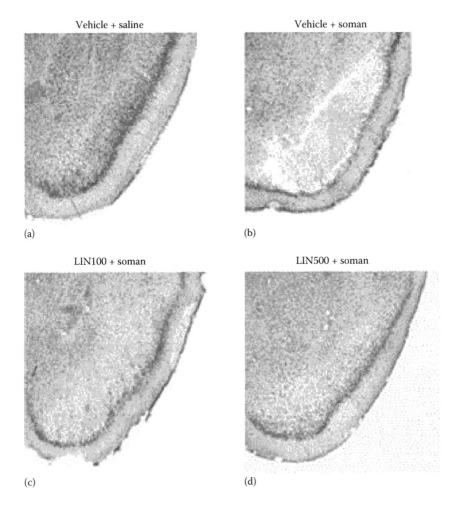

Vehicle + saline   Vehicle + soman

(a)   (b)

LIN100 + soman   LIN500 + soman

(c)   (d)

FIGURE 5.4 (See color insert.) Single intravascular treatment with LIN at 500 nmol/kg three days prior to soman injection has a marked effect in ameliorating the tissue friability, edema, and extensive cell death and disruption in the pyriform cortex. Animals were perfused after 24 hours after soman exposure and the brains were processed and stained with cresyl violet. A section of the piriform cortex is presented in the images. A reduction of neurons compared to control (a) was observed with soman treatment (b). Treatment with LIN at 500 nmol/kg (d) showed a marked reduction in neuronal cell loss when compared with the non-LIN treated soman exposed animals (b) and treatment with LIN at 100 nmol/kg (c). (Reproduced from Pan, H., Hu, X.Z., Jacobowitz, D.M., Chen, C., McDonough, J., Van Shura, K., Lyman, M., Marini, A.M., *Neurotoxicology*, 33, 1219–1229, 2012.)

(Piermartiri et al., 2015). Intravenous administration of ALA caused an increase in BDNF in the cortex and hippocampus (Blondeau et al., 2009). ALA also has effects in protecting against memory deficits observable with OP toxicity and reduction in neuronal degeneration (Pan et al., 2015). Additionally, ALA may also have a role in the neurogenesis via an mTORC1 (mammalian target of rapamycin kinase)-mediated mechanism, which may impact the Akt pathway and BDNF secretion, potentially impacting the brain plasticity response after toxic brain injury by serving as a blue-print for restoring brain function. LIN, the only omega-3 fatty acid found in vegetable products, may almost completely abolish neuronal cell death in CA1 and CA3 sub-fields in the hippocampus induced by kainic acid and OP nerve agent soman. A single intravascular treatment with LIN at 500 nmol/kg three days prior to soman injection has a marked effect in ameliorating the tissue friability, edema, and extensive cell death and disruption observed in the piriform cortex (Pan et al., 2012). Treatment with LIN at 500 nmol/kg showed a marked reduction in neuronal cell loss when compared with the non-LIN-treated soman exposed animals (Figure 5.4). LIN also increases neuronal survival, enhances cyto-architectural integrity, and reduces brain friability in soman-induced neuropathology. Administration of LIN over a seven-day period after exposure to soman reduced neuronal cell death in the cingulate cortex, the amygdala, the hippocampus, and piriform cortex up to 21 days (Pan et al., 2015).

DM accumulates in fatty tissues such as the brain, where it may increase ROS formation, leading to oxidative stress and apoptotic cell death, and is believed to be associated with neurodegenerative diseases. VE is thought to have a neuroprotec-tive effect by neutralizing compounds that negatively affect cell membranes, caus-ing lipid peroxidation, and leading to a decrease in neuronal cell death. Neuronal cell damage may also be decreased by VE-associated increased expression of anti-apoptotic mediator bcl 2 (Numakawa et al., 2006). Administration of VE caused a significant decrease in DM-induced oxidative stress and a marked decrease in apop-tosis indicators such as DNA fragmentation (Galal et al., 2014).

5.5 CANCER

Dietary supplements may decrease the development of primary brain cancer such as glioblastoma and secondary carcinoma metastasis.

Plant extracts may directly impact cancer cell regulatory pathways associated with both migration and invasiveness and may have an indirect impact via anti-oxidant and anticarcinogenic activities. Moreover, they have shown to have positive impacts when administered as coadjuvant therapies. Sibilinin is a natural polypheno-lic flavonoid isolated from seed extracts of the herb milk thistle (*Silybium mariatum*). It has been suggested that it may be a natural inhibitor of signal transducers and acti-vator of transduction 3. Investigational studies have demonstrated that sibilinin has efficacious anticancer properties associated with cancer cell migration and invasive-ness. Patients with non-small cell lung cancer have a very poor prognosis once brain metastases are established. Administration of sibilinin has shown in clinical reports to be associated with marked reduction of peritumoral brain edema and activity against progressive brain metastasis. When used in a joint therapy, sibilinin may be associated with the decrease in tumor volume (Bosch-Barrera et al., 2016).

Flavonoids are polyphenolic compounds of plant origin that may cross the BBB, where they may be active in direct scavenging of ROS and RNS, and impact Ca2+ influx and the levels of endogenous scavengers such as glutathione. Reduction of oxidative stress may be both neuroprotective and anticarcinogenic. Quercetin and its derivatives are the most abundant flavonoids in the Western diet as they are present in apples, lemons, lettuce, and numerous other fruits and vegetables (Vidak et al., 2015). Quercetin coadministration with temozilode may impact apoptosis in human glioblastoma multiforme T98G cells via activation of the mitochondrial death pathway in association with the inhibition of heat shock proteins HSP27 and HSP72, activation of caspases 3 and 9, cytochrome c release from the mitochondria, and decrease in mitochondrial membrane potential (Jakubowicz-Gil et al., 2013). Quercetin decreases U138MG glioma cell line proliferation and viability induced by cell death and arrested cell cycle. It also may impact the mitotic index (Braganhol et al., 2006).

Derivatives of flavonoid catechins may appear in food in the form of glicosydes. They are very potent antioxidants and scavengers of ROS and RNS. Catechin derivatives may impact brain inflammation by modulating the MAPK pathway and its downstream effectors such as inducible nitric oxide synthase (iNOS) and TNFα (Spencer, 2007). They may also impact excitotoxicity by increasing cell glutamate uptake, and impact S100A9, a calcium binding protein that may have neutrophic effects, in both neuronal and glial cells (Abib et al., 2008). In glioblastoma multiformis, catechins may impact the sensitivity of cancer cells to therapeutic agents (Chen et al., 2011) and increase cancer cell apoptosis, altered expression of matrix metalloproteinases, and cancer cell proliferations rate (Li et al., 2014). Proanthocyanins, an example of which is grape seed extract (GSE), are also antioxidants and impact lipid peroxidation. Additionally, GSE improves rat glial viability after exposure to oxidative stress and increased iNOS expression and subsequent low-level nitric oxide production (Fujishita et al., 2009).

5.6   CONCLUSION

There are numerous data reports that demonstrate that phytopharmaceuticals and dietary supplements of plant origin may impact multiple pathologic pathways that affect the development of nervous tissue lesions. The effects of these types of compounds are mediated by their actions on multiple pathways relevant to the establishment of these pathological lesions.

The antioxidative properties of this type of compounds are paramount in the progression of brain toxic and ischemic injury. Also important are anti-inflammatory and neuroplastic characteristics that are associated with neuroprotection and progression of neurodegenerative diseases. Additional modes of action for phytopharmaceuticals include antiapoptotic and antiproliferative properties that may affect the development of primary or metastatic cancer.

These mechanisms of action make these molecules relevant in multiple areas of central and peripheral nervous system disease that include, but are not limited to, neurodegeneration, neoplasia, and ischemic or toxic injury.

LIST OF ABBREVIATIONS

AChE	Acetylcholinesterase
AD	Alzheimer's disease
AIF	Apoptosis inducing factor
ALA	Alpha-linoleic acid
Aβ	Beta-amyloid plaques
BACE 1	β-Site APP (amyloid-β precursor protein) cleaving enzyme 1
BBB	Blood–brain barrier
BDNF	Brain derived neurotrophic factor
BrdU	5′-Bromo-2-deoxyuridine
COX2	Cyclooxygenase 2
CUR	Curculigoside
DM	Deltametrin
GABA	Glutamate and γ-aminobutyricacid
GFAP	Glial fibrillary acidic protein
GSE	Grape seed extract
HSP	Heat shock proteins
IL	Interleukin
iNOS	Inducible nitric oxide synthase
LIN	Alpha-linolenic acid
LPS	Lipopolysaccharide
MAPK	Ras–mitogen-activated protein kinase
MCAO	Middle cerebral artery occlusion
mnd	Motor neuron disease
mTORC	Mammalian target of rapamycin kinase
NCL	Neuronal ceroid lipofuscinosis
NF-kB	Nuclear Factor-kappa B
NMDA	N-methyl-D-aspartate
NO	Nitric oxide
OLE	Oleuropein aglycone
OP	Organophosphates
PD	Parkinson's disease
PUFAs	Polyunsaturated fatty acids
RNS	Reactive nitrogen species
ROS	Reactive oxygen species
SGZ	Subgranular layer
SNAP-25	Synaptosomal-associated protein 25
STAT 3	Signal transducers and activator of transduction 3
SVZ	Subventricular zone
TBI	Traumatic brain injury
TJ	Tight junction
TNF	Tumor necrosis factor
VAMP-2	Vesicle-associated membrane protein 2
VE	Vitamin E
V-GLUT	V-Glutathione 1 and 2

REFERENCES

Abib, R.T., Quincozes-Santos, A., Nardin, P., Wofchuk, S.T., Perry, M.L., Gonçalves, C.A., and C. Gottfried. 2008. *Epicatechin gallate* increases glutamate uptake and S100B secretion in C6 cell lineage. *Mol Cell Biochem.* 310(1–2):153–8.

Arab, L. and Liebeskind, D.S. 2010. Tea flavonoids and stroke in man and mouse. *Arch Biochem Biophys.* 501(1):31–6.

Bachstetter, A.D., Jernberg, J., Schlunk, A., Vila, J.L., Hudson, C., Cole, M.J., Shytle, R.D., Tan, J., Sanberg, P.R., Sanberg, C.D., Borlongan, C., Kaneko, Y., Tajiri, N., Gemma, C., and Bickford, P.C. 2010. Spirulina promotes stem cell genesis and protects against LPS induced declines in neural stem cell proliferation. *PLoS One.* 5(5):e10496. doi: 10.1371 /journal.pone.0010496.

Blondeau, N., Nguemeni, C., Debruyne, D.N., Piens, M., Wu, X., Pan, H., Hu, X., Gandin, C., Lipsky, R.H., Plumier, J.C., Marini, A.M, and Heurteaux, C. 2009. Subchronic alpha-linolenic acid treatment enhances brain plasticity and exerts an antidepressant effect: A versatile potential therapy for stroke. *Neuropsychopharmacology.* 34(12):2548–59.

Bosch-Barrera, J., Sais, E., Cañete, N., Marruecos, J., Cuyàs, E., Izquierdo, A., Porta, R., Haro, M., Brunet, J., Pedraza, S., and Menendez, J.A. 2016. Response of brain metastasis from lung cancer patients to an oral nutraceutical product containing silibinin. *Oncotarget.* doi: 10.18632/oncotarget.7900. Epub ahead of print.

Braganhol, E., Zamin, L.L., Canedo, A.D., Horn, F., Tamajusuku, A.S., Wink, M.R., Salbego, C., and Battastini, A.M. 2006. Antiproliferative effect of quercetin in the human U138MG glioma cell line. *Anticancer Drugs.* 17(6):663–71.

Chattopadhyaya, I., Gupta, S., Mohammed, A., Mushtaq, N., Chauhan, S., and Ghosh, S. 2015. Neuroprotective effect of *Spirulina fusiform* and amantadine in the 6-OHDA induced Parkinsonism in rats. *BMC Complement Altern Med.* 15(296):1–11.

Chen, T.C., Wang, W., Golden, E.B., Thomas, S., Sivakumar, W., Hofman, F.M., Louie, S.G., and Schönthal, A.H. 2011. Green tea epigallocatechin gallate enhances therapeutic efficacy of temozolomide in orthotopic mouse glioblastoma models. *Cancer Lett.* 302(2):100–8. doi: 10.1016/j.canlet.2010.11.008. Epub January 22, 2011.

Dajas, F., Rivera, F., Blasina, F., Arredondo, F., Echeverry, C., Lafon, L., Morquio, A., and Heinzen, H. 2003. Cell culture protection and in vivo neuroprotective capacity of flavonoids. *Neurotox Res.* 5(6):425–32.

Essa, M.M., Subash, S., Akbar, M., Al-Adawi, S., and Guillemin, G.J. 2015. Long-term dietary supplementation of pomegranates, figs and dates alleviate neuroinflammation in a transgenic mouse model of Alzheimer's disease. *PLoS One.* 10(3):e0120964. doi: 10.1371/journal.pone.0120964. eCollection.

Etus, V., Altug, T., Belce, A., and Ceylan, S. 2003. Green tea polyphenol-epigallocatechin gallate prevents oxidative damage on periventricular white matter of infantile rats with hydrocephalus. *Tohoku J Exp Med.* 200(4):203–9.

Fujishita, K., Ozawa, T., Shibata, K., Tanabe, S., Sato, Y., Hisamoto, M., Okuda, T., and Koizumi, S. 2009. Grape seed extract acting on astrocytes reveals neuronal protection against oxidative stress via interleukin-6-mediated mechanisms. *Cell Mol Neurobiol.* 29(8):1121–9.

Galal, M.K., Khalaf, A.A., Ogaly, H.A., and Ibrahim, M.A. 2014. Vitamin E attenuates neurotoxicity induced by deltamethrin in rats. *BMC Complement Altern Med.* 14(458):1–7.

Griffin, J.L., Muller, D., Woograsingh, R., Jowatt, V., Hindmarsh, A., Nicholson, J.K., and Martin, J.E. 2002. Vitamin E deficiency and metabolic deficits in neuronal ceroid lipofuscinosis described by bioinformatics. *Physiol Genom.* 3;11(3):195–203.

Grossi, C., Rigacci, S., Ambrosini, S., Ed Dami, T., Luccarini, I., Traini, C., Failli, P., Berti, A., Casamenti, F., and Stefani, M. 2013. The polyphenol oleuropein aglycone protects TgCRND8 mice against Aß plaque pathology. *PLoS One.* 8(8):e71702. doi: 10.1371 /journal.pone.0071702. eCollection.

Guseva, M.V., Hopkins, D.M., Scheff, S.W., and Pauly, J.R. 2008. Dietary choline supplementation improves behavioral, histological, and neurochemical outcomes in a rat model of traumatic brain injury. *J Neurotrauma*. 25(8):975–83.

Harrison, F.E., Hosseini, A.H., McDonald, M.P., May, J.M. Kitt-Hale, B., and Morgan, D. 2009. Vitamin C reduces spatial learning deficits in middle-aged and very old APP/PSEN1 transgenic and wild-type mice. *Pharmacol Biochem Behav*. 93(4):443–50.

Ho, L., Ferruzzi, M.G., Janle, E.M., Wang, J., Gong, B., Chen, T.Y., Lobo, J., Cooper, B., Wu, Q.L., Talcott, S.T., Percival, S.S., Simon, J.E., and Pasinetti, G.M. 2013. Identification of brain-targeted bioactive dietaryquercetin-3-*O*-glucuronide as a novel intervention for Alzheimer's disease. *FASEB J*. 27(2):769–81.

Jakubowicz-Gil, J., Langner, E., Bądziul, D., Wertel, I., and Rzeski, W. 2013. Apoptosis induction in human glioblastoma multiforme T98G cells upon temozolomide and quercetin treatment. *Tumour Biol*. 34(4):2367–78.

Joseph, J.A., Denisova, N.A., Arendash, G., Gordon, M., Diamond, D., Shukitt-Hale, B. and Morgan, D. 2003. Blueberry supplementation enhances signaling and prevents behavioral deficits in an Alzheimer disease model. *Nutr Neurosci*. 6(3):153–62.

Kaneko, Y., Cortes, L., Sanberg, C., Acosta, S., Bickford, P.C., and Borlongan, C.V. 2012. Dietary supplementations as neuroprotective therapies: Focus on NT-020 diet benefits in a rat model of stroke. *Int J Mol Sci*. 13(6):7424–44.

Karuppagounder, S.S., Pinto, J.T., Xu, H., Chen, H.L., Beal, M.F., and Gibson, G.E. 2009. Dietary supplementation with resveratrol reduces plaque pathology in a transgenic model of Alzheimer's disease. *Neurochem Int*. 54(2):111–8.

Keservani, R.K., Sharma, A.K., and Kesharwani, R.K. 2016. Medicinal effect of nutraceutical fruits for the cognition and brain health. *Scientifica (Cairo)*. 2016:3109254. doi: 10.1155/2016/3109254. Epub February 4, 2016.

Kook, S.Y., Lee, K.M., Kim, Y., Cha, M.Y., Kang, S., Baik, S.H., Lee, H., Park, R., and Mook-Jung, I. 2014. High-dose of vitamin C supplementation reduces amyloid plaque burden and ameliorates pathological changes in the brain of 5XFAD mice. *Cell Death Dis*. 5:e1083. doi: 10.1038/cddis.2014.26.

Lauritzen, I., Blondeau, N., Heurteaux, C., Widmann, C., Romey, G., and Lazdunski, M. 2000. Polyunsaturated fatty acids are potent neuroprotectors. *EMBO J*. 19(8):1784–93.

Li, H., Li, Z., Xu, Y.M., Wu, Y., Yu, K.K., Zhang, C., Ji, Y.H., Ding, G., and Chen, F.X. 2014. Epigallocatechin-3-gallate induces apoptosis, inhibits proliferation and decreases invasion of glioma cell. *Neurosci Bull*. 30(1):67–73.

Liang, J., Yu, Y., Wang, B., Lu, B., Zhang, J., Zhang, H., and Ge, P. 2013. Ginsenoside Rb1 attenuates oxygen-glucose deprivation-induced apoptosis in SH-SY5Y cells via protection of mitochondria and inhibition of AIF and cytochrome c release. *Molecules*. 18(10):12777–92.

Loren, D.J., Seeram, N.P., Schulman, R.N., and Holtzman, D.M. 2005. Maternal dietary supplementation with pomegranate juice is neuroprotective in an animal model of neonatal hypoxic-ischemic brain injury. *Pediatr Res*. 57(6):858–64.

Numakawa, Y., Numakawa, T., Matsumoto, T., Yagasaki, Y., Kumamaru, E., Kunugi, H., Taguchi, T., and Niki, E. 2006. Vitamin E protected cultured cortical neurons from oxidative stress-induced cell death through the activation of mitogen-activated protein kinase and phosphatidylinositol 3-kinase. *J Neurochem*. 97(4):1191–202.

Pan, H., Hu, X.Z., Jacobowitz, D.M., Chen, C., McDonough, J., Van Shura, K., Lyman, M., and Marini, A.M. 2012. Alpha-linolenic acid is a potent neuroprotective agent against soman-induced neuropathology. *Neurotoxicology*. 33(5):1219–29.

Pan, H., Piermartiri, T.C., Chen, J., McDonough, J., Oppel, C., Driwech, W., Winter, K., McFarland, E., Black, K., Figueiredo, T., Grunberg, N., and Marini, A.M. 2015. Repeated systemic administration of the nutraceutical alpha-linolenic acid exerts neuroprotective efficacy, an antidepressant effect and improves cognitive performance when given after soman exposure. *Neurotoxicology*. 51:38–50.

Park, E.K., Choo, M.K., Oh, J.K., Ryu, J.H., and Kim, D.H. 2004. Ginsenoside Rh2 reduces ischemic brain injury in rats. *Biol Pharm Bull.* 27(3):433–6.

Piermartiri, T., Pan, H., Figueiredo, T.H. and Marini, A.M. 2015. α-Linolenic acid, A nutraceutical with pleiotropic properties that targets endogenous neuroprotective pathways to protect against organophosphate nerve agent-induced neuropathology. *Molecules.* 20(11):20355–80.

Rastogi, V., Santiago-Moreno, J., and Doré S. 2015. Ginseng: A promising neuroprotective strategy in stroke. *Front Cell Neurosci.* 2015 Jan 20;8:457. eCollection 2014.

Saleem, S., Zhuang, H., Biswal, S., Christen, Y., and Doré, S. 2008. *Ginkgo biloba* extract neuroprotective action is dependent on heme oxygenase 1 in ischemic reperfusion brain injury. *Stroke.* 39(12):3389–96.

Spencer, J.P. 2007. The interactions of flavonoids within neuronal signaling pathways. *Genes Nutr.* 2(3):257–73.

Subash, S., Essa, M.M., Al-Asmi, A., Al-Adawi, S., and Vaishnav, R. 2014. Chronic dietary supplementation of 4% figs on the modification of oxidative stress in Alzheimer's disease transgenic mouse model. *Biomed Res Int.* 2014:546357. doi: 10.1155/2014/546357. Epub June 19, 2014.

Sweeney, M.I., Kalt, W., MacKinnon, S.L., Ashby, J., and Gottschall-Pass, K.T. 2002. Feeding rats diets enriched in lowbush blueberries for six weeks decreases ischemia-induced brain damage. *Nutr Neurosci.* 5(6):427–31.

Tian, Z., Yu, W., Liu, H.B., Zhang, N., Li, X.B., Zhao, M.G., and Liu, S.B. 2012. Neuroprotective effects of curculigoside against NMDA-induced neuronal excitoxicity in vitro. *Food Chem Toxicol.* 50(11):4010–5.

Vidak, M., Rozman, D., and Komel, R. 2015. Effects of flavonoids from food and dietary supplements on glial and glioblastoma multiforme cells. *Molecules.* 20:19406–32.

Wang, Y., Chang, C.F., Chou, J., Chen, H.L., Deng, X., Harvey, B.K., Cadet, J.L., and Bickford, P.C. 2005. Dietary supplementation with blueberries, spinach, or spirulina reduces ischemic brain damage. *Exp Neurol.* 193(1):75–84.

Wu, X.Y., Li, J.Z., Guo, J.Z., and Hou, B.Y. 2012. Ameliorative effects of curculigoside from *Curculigo orchioides* Gaertn on learning and memory in aged rats. *Molecules.* 17(9):10108–18.

Yuan, Q.L., Yang, C.X, Xu, P., Gao, X.Q., Deng, L., Chen, P., Sun, Z.L., and Chen, Q.Y. 2007. Neuroprotective effects of ginsenoside Rb1 on transient cerebral ischemia in rats. *Brain Res.* 1167:1–12.

Zhao, L., Liu, S., Wang, Y., Zhang, Q., Zhao, W., Wang, Z., and Yin, M. 2015. Effects of curculigoside on memory impairment and bone loss via anti-oxidative character in APP/PS1 mutated transgenic mice. *PLoS One.* 10(7):1–13.

Zheng, G.Q., Cheng, W., Wang, Y., Wang, X.M., Zhao, S.Z., Zhou, Y., Liu, S.J., and Wang, X.T. 2011. Ginseng total saponins enhance neurogenesis after focal cerebral ischemia. *J Ethnopharmacol.* 133(2):724–8.

Zhu, J., Jiang, Y., Wu, L., Lu, T., Xu, G., and Liu, X. 2012. Suppression of local inflammation contributes to the neuroprotective effect of ginsenoside Rb1 in rats with cerebral ischemia. *Neuroscience.* 202:342–51.

6 Huperzine A and Shankhapushpi in Brain Health

*Debasis Bagchi, Manashi Bagchi,
Anand Swaroop, and Hiroyoshi Moriyama*

CONTENTS

6.1 INTRODUCTION

Globally, there is a need to explore the ability of natural products to boost mental health, focus and concentration, and cognitive function. Studies have demonstrated that proper nutrition in conjunction with physical, mental, and social activities may have a greater benefit in maintaining or improving brain health and function (Hugel, 2015; Kim et al., 2010). Balanced diet, appropriate nutrition, cognitive activity, social engagement, and regular physical exercise can significantly attenuate brain health and function with advancing age and potentially reduce the risk of cognitive decline (Kelsey et al., 2010; Shetty and Bates, 2015). Advancing age can exhibit health-related challenges that take a toll emotionally, financially, and physically. Moreover, regular stress is a challenging problem (Kelsey et al., 2010; Qian and Ke, 2014). Novel nutraceuticals exhibit significant protection against neurodegenerative diseases associated with exacerbated oxidative stress in the central nervous system (CNS) (Kelsey et al., 2010; Kim et al., 2010; Shetty and Bates, 2015; Subedee et al., 2015).

6.2 BENEFITS OF NUTRACEUTICALS AND FUNCTIONAL FOODS FOR AGING BRAIN

Research studies on a number of phytopharmaceuticals and medicinal plants have demonstrated the efficacy of huperzine A (HupA), berry anthocyanins, *trans*-resveratrol, *Ginkgo biloba*, *Bacopa monniera*, *Centella asiatica*, ginseng, vitamin B12, alpha lipoic acid, vinpocetine, tocotrienols and palm oil, selenium, black pepper, acetylcholine, and gamma aminobutyric acid (GABA) in boosting brain function and physical well-being (Bobade et al., 2015; Kelsey et al., 2010; Polotow et al., 2015; Rajan et al., 2015; Shetty and Bates, 2015; Shin et al., 2010; Subedee et al., 2015; Zhang and Yang, 2014). Turmeric and curcumin have a long history as a healing herb for cognitive decline (Lee et al., 2013). Research studies have shown that turmeric and curcumin have the amazing ability to effectively cross the blood–brain barrier and work as an antioxidant to scavenge noxious free radicals that can damage healthy cells as well as lipids, proteins, and DNA in aging brain (Hugel, 2015; Kim et al., 2010). These novel nutraceuticals, especially resveratrol, curcumin, and green tea catechins, have the potential to prevent Alzheimer's disease (AD) because of their antiamyloidogenic, antioxidative, and anti-inflammatory properties (Hugel 2015). It is important to mention that medicinal herbs and structurally diverse botanicals have been widely used in Asia for more than 2000 years. Botanical extracts with antiamyloidogenic activity, including green tea catechins, turmeric, *Salvia miltiorrhiza*, berry anthocyanins, and *Panax ginseng*, have demonstrated significant efficacy in AD (Hugel 2015; Kim et al., 2010; Shin et al., 2010; Subedee et al., 2015). Furthermore, Indian ginger, rosemary, sage, salvia herbs, black pepper, as well as Chinese celery, singly and in combination, exhibited highly promising neuroprotective efficacy against AD (Hugel, 2015; Kelsey et al., 2010; Qian and Ke, 2014; Subedee et al., 2015; Zhang, 2012; Zhang and Tang, 2006; Zhu and Tang, 1988).

Consumption of marine fishes and general seafood has been recommended for long-term nutritional intervention to preserve mental health, hinder neurodegenerative processes, and sustain cognitive capacities in humans. Omega-3 and omega-6 polyunsaturated fatty acids, n-3/n-6 PUFAs, phosphatidyl serine, and marine antioxidants prevent the initiation and progression of many neurological disorders (Polotow et al., 2015). Marine antioxidant carotenoid astaxanthin have shown promising results against free radical-promoted neurodegenerative processes and cognition loss (Kelsey et al., 2010).

Studies have demonstrated that adults at risk of cognitive impairment demonstrated that a combination of physical activity, proper nutrition, cognitive training, social activities, and management of heart health risk factors slowed cognitive decline (Kim et al., 2010) and activate adaptive cellular stress responses, called "neurohormesis," and suppress disease processes (Hugel, 2015). Furthermore, amyloid-β-induced pathogenesis of AD can also be partially protected or retarded by these novel dietary phytochemicals (Subedee et al., 2015; Zhang, 1986; Zhang and Yang, 2014).

6.3 ALZHEIMER'S DISEASE

AD is an age-related neurodegenerative disease affecting the elderly and there is hardly any effective therapy (Hugel, 2015; Kim et al., 2010). AD is recognized as one of the most complicated neurodegenerative diseases and is a major social problem (Venkatesan et al., 2015). It is estimated that 26.6 million patients with AD are reported worldwide. Furthermore, this number is estimated to increase to 102.6 million by 2050 (Venkatesan et al., 2015). It is important to mention that patients with diabetes mellitus have an approximate double risk for the generation of dementia with advancing age (Kodl and Seaquist, 2008).

6.4 HUPERZINE A

6.4.1 PHYSICOCHEMICAL CHARACTERISTICS

HupA [IUPAC name: (1R,9S,13E)-1-amino-13-ethylidene-11-methyl-6-azatricyclo-[7.3.1.02,7]-trideca-2(7),3,10-trien-5-one; commercially known as CogniUp] is a naturally occurring firmoss, a sesquiterpene alkaloid, an acetylcholinesterase (AChE) inhibitor, and *N*-methyl-D-aspartate receptor (also known as glutamate receptor) antagonist (Table 6.1; Bagchi and Barilla, 1998; Wang et al., 2006, 2008). It is found in an extract from the Chinese club moss botanically known as *Huperzia serrata* (also known as *Lycopodium serratum*). HupA grows at high alleviations and in cold climates. It has been used for centuries in Chinese Folk Medicine (known as *Qian Ceng Ta*). HupA is also available in *Huperzia elmeri*, *Huperzia carinat*, and *Huperzia aqualupian*, and its highest content is found in *Huperzia pinifolia* (Bagchi and Barilla, 1998; Ishiuchi et al., 2013; Wang et al., 2006, 2008, 2011). The chemical stability of HupA is very good, and it is resistant to structural changes in acidic and alkaline solution, which indicates that HupA has a longer shelf life (Figure 6.1).

TABLE 6.1
Physicochemical Characteristics of HupA

Physicochemical Characteristics	Data
Molecular formula	$C_{15}H_{18}N_2O$
Molar mass	242.32 g/mol
CAS ID	102518-79-6
Melting point	422.6°F (217°C)
Solubility	DMSO, ethanol

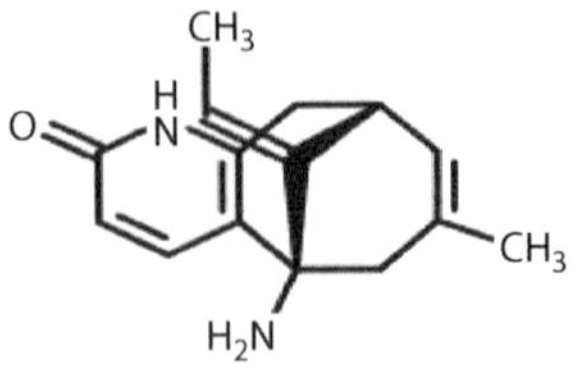

FIGURE 6.1   Structure of HupA.

6.4.2 SAFETY

The acute oral toxicity (LD_{50}) of HupA (CogniUp) was found to be 4.6 mg/kg body weight in mice. Considering a therapeutic oral dose of 0.2 mg HupA/kg body weight, it demonstrates a wide margin of safety (Bagchi and Barilla, 1998). Extensive laboratory testing of HupA has been done, demonstrating its non-mutagenic property as demonstrated by Ames' bacterial reverse mutation assay. Teratology studies in mice and rabbits or off-springs of animals exhibited no external or internal organ or skeleton deformity or toxicity (Bagchi and Barilla, 1998). HupA didn't cause hepatotoxicity in dogs and rabbits or any other side effects such as nausea, vomiting, gastrointestinal upset, depression, and other commonly observed adverse effects. In the clinical studies conducted so far, no significant adverse events were observed (Bai et al., 2000, 2013; Yue et al., 2012). HupA also displayed good pharmacokinetics with a rapid absorption and a wide distribution in the body at a low to moderate rate of elimination (Ha et al., 2011).

6.4.3 HUPA AND THERAPEUTIC USE

In earlier days, it was used to treat fever and inflammation, and for several decades, it was used for the treatment of dementia especially for AD, diabetes-induced neuronal impairment, chronic inflammation, and cognitive decline in elderly individuals (Bai et al., 2000; Damar et al., 2016; Venkatesan et al., 2015; Yue et al., 2012). Other benefits of HupA are shown in Table 6.2.

6.4.4 LEARNING ABILITY, MEMORY ENHANCEMENT, AND COGNITIVE PERFORMANCE

HupA (CogniUp) was found to improve cognitive performance in a broad range of animal models involving mice, rats, and monkeys with induced amnesia. HupA also markedly improved the retention of a learned task when tested 24 hours later in aged mice. Enhancement of learning and memory performance, increased retention, and faster retrieval processes were observed. The loss of cholinergic neurons in the brain has been demonstrated to be a part of the aging process itself. This loss is considered important in the process of memory impairment, including dementia. HupA improves cholinergic function by inhibiting acetylcholine degradation in the brain (Sun et al., 1999; Tang et al., 1994; Vincent et al., 1987; Zhu and Tang, 1988).

TABLE 6.2
Diverse Health Benefits of HupA

1. Superior safety and efficacy profile as compared to other cholinesterase inhibitors
2. Learning and memory retention
3. Improve focus and concentration
4. Treatment of cognition and memory impairment
5. Improve nerve transmission to muscles
6. Attenuate muscle fatigue
7. Powerful and reversible long-term inhibitor of AChE activity in the brain
8. Dementia resulting from strokes, and senile or presenile dementia
9. Improved clinical picture for patients with myasthenia gravis
10. Improved short-term and long-term memory in patients with cerebral arteriosclerosis (hardening of arteries in the brain)
11. Alleviation of symptoms related to glaucoma
12. Prevention against organophosphate pesticide toxicity
13. Prevents nerve gas toxicity
14. A novel psychotherapeutic agent for improving cognitive function in Alzheimer's patients
15. Osteoarthritis and joint pain
16. Boosting of immune function

HupA (CogniUp) has demonstrated to improve performance in mice and rats in running through mazes and protected young and aged animals against sodium nitrite, cycloheximide, carbon dioxide-treated, and electroconvulsive shock-induced passive response. HupA improved the memory in squirrel monkeys in another set of experiments. It is important to note that the duration of improved effects on learning and memory retention process with oral HupA was longer as compared to other existing AChE inhibitors (Tang et al., 1986; Vincent et al., 1987). HupA targets different sites of AChE, and its ability to inhibit AChE is eightfold and twofold more effective than donepezil and rivastigmine, respectively (Jia et al., 2013). It also prevents the selective degeneration of acetylcholine-producing neurons in the brain and enhances the availability of acetylcholine in the brain of patients suffering from neurological disorders, muscle contraction, and dementia between nerves and muscle to function better. The improvement in neuromuscular cholinergic transmission leads to an improvement of patient memory. It has been demonstrated repeatedly that memory loss or impairments and cognitive dysfunctions are accompanied by a dramatic reduction in acetylcholine synthesis and/or release in the nerve cells. HupA works by a unique mechanism that has been reported (Lin et al., 1996). Acetylcholine is the neurotransmitter in the brain that is responsible for carrying electrical impulses from one nerve to another. Acetylcholine at the end section of nerve fibers converts into small vesicles, where it is stored until released. Once acetylcholine has been secreted by the nerve ending, it persists for a few seconds. In a normal brain, the enzyme AChE serves a housekeeping function by breaking down the acetylcholine. Acetylcholine breaks down into an acetate moiety and choline. The choline is then transmitted back into the nerve ending to be used again to make acetylcholine. People with AD exhibit a deficiency of acetylcholine because of the damaged brain cells. HupA stops AChE from breaking down acetylcholine and prevents deficiency, thus improving mental function (Ha et al., 2011; Lunardi et al., 2013; Malkova et al., 2011).

The finding of a severely damaged and underactive cholinergic system in the brains of patients with memory impairments and AD has led to clinical trials of new cholinomimetics, including AChE inhibitors (Malkova et al., 2011; Wang et al., 2008, 2011). Results demonstrated that memory loss or impairments and cognitive dysfunctions are accompanied by dramatic reduction in acetylcholine synthesis or release in the nerve cells. Investigating acetylcholine release is one way of testing the function of cholinergic synapses (the transmission of acetylcholine across the gap to the next nerve cell). This can be accomplished with different drugs or nutraceuticals in animal or human autopsy tissue slices using techniques that measure extracellular acetylcholine (Yue et al., 2012). Acetylcholine release is governed by complex factors such as membrane integrity and cholinergic receptor biochemistry. Receptors for acetylcholine usually function to different degrees depending on factors such as heredity (as for example, fewer insulin receptors increase the probability of getting diabetes) and adverse conditions during early childhood when the brain is still developing (some types of schizophrenia) (Jia et al., 2013; Sun et al., 1999; Xing et al., 2014; Zhang, 2012).

6.4.5 Molecular Mechanisms of Neuroprotection

The loss of cholinergic neurons in the brain has been demonstrated to be a part of the aging process itself (Zhang, 2012). This loss is considered an important element in the process of memory loss or memory impairment including dementia. HupA improved cholinergic function by inhibiting acetylcholine degradation in the brain (Wang et al., 2008).

HupA has demonstrated to exert its effects via α7nAChRs and α4β2nAChRs, thereby inducing a potent anti-inflammatory response by reducing interleukin (IL)-1β and tumor necrosis factor (TNF)-α expression and suppressing transcriptional activation of nuclear factor-kappa B (NF-κB) (Ishiuchi et al., 2013). Thus, it provides protection from excitotoxicity and neuronal death as well as increase in GABAergic transmission associated with anticonvulsant activity (Bai et al., 2013; Damar et al., 2016; Lunardi et al., 2013; Yu et al., 2013; Zhang and Tang, 2006).

HupA exerts a neuroprotective effect against AD by inhibiting AChE, altering Aβ peptide processing, reducing oxidative stress, and promoting the expression of antiapoptotic protein and nerve growth factor (NGF; Zhang and Tang, 2006). The memory deficits in transient cerebral ischemia and reperfusion mouse models were reversed by increasing the expression levels of NGF, brain-derived neurotrophic factor (BDNF), and transforming growth factor-beta (TGF-β) through mitogen-activated protein kinase (MAPK)/extracellular signal-regulated kinase (ERK)-mediated neuroprotection (Jia et al., 2013; Malkova et al., 2011; Xing et al., 2014). In SHSY5Y neuroblastoma cells, HupA treatment reversed the reduction in NGF level that was caused by H_2O_2-induced oxidative stress; this effect was due to the activation of p75NTR and TrkA receptors and the upstream MAP/ERK signaling pathway (Tang et al., 2005a). Furthermore, HupA promotes neurite outgrowth in rat PC12 cells and in rat cortical astrocyte cells by inhibiting AChE and upregulating the expression levels of NGF and p75NTR (Tang et al., 2005b). HupA attenuates cognitive defects in streptozotocine-induced diabetic rats by increasing the levels on the activity of acetyltransferase (ChAT), BDNF, superoxide dismutase (SOD), glutathione peroxidase, and catalase (CAT), while simultaneously inhibiting AChE, malondialdehyde (MDA), CAT, NF-κB, TNF-α, IL-1β, IL-6, and caspase-3 (Venkatesan et al., 2015).

6.4.6 HUMAN STUDY

Studies demonstrated the improvement in muscle function using HupA (CogniUp) and favorable effects in the treatment of age-related memory impairment (Zhang, 1986). In another comparative study, hydergine (a vasodilator, 600 μg) and HupA (30 μg, intramuscularly) appeared to improve memory for one to four hours in 100 elderly individuals (age: 46–82 years; 54 males; 46 females) suffering from memory impairment. Among these 100 subjects, 83 had no serious brain disease but were suffering from age-related amnesia or memory dysfunction, and the remaining 17 subjects had probable AD (Bagchi and Barilla, 1998). This study was very encouraging. Minimal or no significant adverse effects were observed.

Another study was conducted by Zhang et al. (1991). The therapeutic effect of HupA was investigated in a randomized, double-blind trial in 56 subjects (age: 64 ± 7 years; 52 males; 4 females) suffering from multi-infarct dementia (resulting from repeated small strokes). Also studied were patients with senile dementia and 104 patients of senile and presenile simple memory disorders (age: 63 ± 7 years; 58 males; 46 females). Each group was divided into two smaller groups, the placebo and the HupA-treatment group. The control group was treated with saline (intramuscularly). HupA dose for the multi-infarct dementia was 50 μg (intramuscularly) twice a day over a period of four weeks, while for the senile and presenile memory disorders, it was 30 μg (intramuscularly) twice a day over a period of two weeks. The Weschler memory scale (WMS) was employed to determine the improvement in memory function. HupA treatment significantly improved the memory of the HupA treatment groups with minimal observed side effects.

Another randomized, clinical study was conducted in the treatment of 202 patients (15 centers) with mild to moderate AD to evaluate the efficacy and safety of HupA over a period of 12 weeks (Zhang et al., 2002). A total of 100 subjects were given HupA (400 μg/day), while 100 subjects received the placebo. Different scales were used to evaluate the cognitive function, activity of daily life (ADL), noncognitive disorders, and overall clinical efficacy. Safety evaluation was conducted every six weeks. Significant improvement was observed in the HupA group as compared to the placebo group. Mild and transient adverse events (edema of bilateral ankles and insomnia) were observed in 3% of HupA-treated patients (Zhang et al., 2002).

A multicenter, randomized, double-blind, parallel, placebo-controlled study was conducted in 103 patients with AD. HupA was administered orally 0.2 mg (four tablets) in 50 patients, while 53 patients receive placebo over a period of eight weeks (Xu et al., 1995). All patients were evaluated with WMS, Hasegawa dementia scale, Mini-Mental State Examination (MMSE) Scale, activity of daily living scale, treatment emergency symptom scale, and total blood and urine chemistry. About 58% (29/50) of patients treated with HupA showed improvements in their memory ($P < 0.01$), cognitive ($P < 0.01$), and behavioral ($P < 0.01$) functions. The efficacy of HupA was better than that of placebo (36%, 19/53) ($P < 0.05$). No severe side effect was observed (Xu et al., 1995).

Another randomized, double-blind, placebo-controlled study was conducted with 78 patients with mild to moderate vascular dementia (Xu et al., 2012). The placebo group (n = 39) received vitamin C (100 mg bid), while the treatment group (n = 39) received HupA (0.1 mg bid) for 12 consecutive weeks. The MMSE, clinical dementia rating (CDR), and ADL scores were used for the assessment of cognition. The assessments were made prior to treatment and 4, 8, and 12 weeks of the treatment. After 12 weeks of treatment, the MMSE, CDR, and ADL scores significantly improved in the HupA-treated group ($P < 0.01$), whereas the placebo group did not show any such improvement. No serious adverse events were recorded during the treatment (Xu et al., 2012).

The efficacy of HupA was assessed on memory and learning performance of adolescent students using double-blind and matched pair method. Thirty-four pairs of junior middle school students complaining of memory inadequacy were divided into two groups by normal psychological health inventory, similar memory quotient (MQ), same sex, and class. The treatment group orally received two capsules (each containing 50 μg) bid, while the placebo group received two capsules of placebo (starch and lactose) bid over a period of four weeks. At the end of trial, the HupA group's MQ (115 ± 6) was more than that of the placebo group (104 ± 9; $P < 0.01$), and the scores of Chinese language lesson in the HupA group were elevated significantly. The researchers concluded that HupA enhances memory and learning performance of adolescent students.

6.4.7 GLAUCOMA

Glaucoma is a disease in which there is an increase in pressure within the eye, which ultimately causes damage to the optic nerve and irreversible blindness can occur. AChE inhibitors are generally used to alleviate this condition. Since HupA can inhibit the transport of choline in the brain, this allows more to be available for the synthesis of acetylcholine, which is important in maintaining normal eye function. HupA has been suggested as a better therapeutic agent than other AChE inhibitors including physostigmine, neostigmine, or tacrine for the treatment of glaucoma (Ashani et al., 1994; Xing et al., 2014).

6.4.8 MYASTHENIA GRAVIS: A NEUROMUSCULAR DISORDER

This is a neuromuscular disease characterized by weakness and fatigue of the skeletal muscle. This involves nerve transmission at the neuromuscular junction. Generally, when a nerve activates a muscle at the neuromuscular junction, the muscle should contract or move. Patients with myasthenia gravis can initially move the muscle, repeated activation causes a diminished response, and muscle movement can't be maintained. AChE inhibitors, by maintaining the acetylcholine levels high, can increase the muscle response. Cheng et al. (1986) assessed the comparative clinical efficacy of HupA with prostigmine, another AChE inhibitor, in 128 patients suffering from myasthenia gravis. Approximately 99% demonstrated controlled or improved clinical symptoms in patients treated with HupA. The duration of action of HupA was 7 ± 6 hours. Nausea and other side effects were significantly less when compared with those induced by prostigmine (Cheng et al., 1986).

6.4.9 NERVE AGENT AND PESTICIDE TOXICITY

Organophosphates (OPS), widely used as pesticides and nerve agents during wartime, are well known for entering the nervous system, reacting with cholinesterase, irreversibly inhibiting it, and inducing potential brain injury leading to coma and death. In the eyes, the pupil will markedly constrict with pain and spasm of muscles; effects on the lungs include "tightness" in the chest and wheezing due to constriction of bronchial tubes, as well as increased mucus. In the gastrointestinal tract, nausea, vomiting, cramps, diarrhea, and extreme salivation are produced. Death usually occurs by respiratory failure, often accompanied by a heart attack (Grunwald et al., 1994; Jia et al., 2013; Wang et al., 2008; Yu et al., 2013).

Research demonstrated that HupA has a remarkable selectivity for AChE and has superior blood–brain barrier penetration ability (Lin et al., 1996). HupA is chemically stable and safe and exhibits long-lasting prophylactic treatment against nerve agent toxicity in humans. A comparative study against OPS toxicity using the nerve agent soman was conducted in mice, which were pretreated with HupA or physostigmine. HupA dramatically prevented irreversible phosphorylation of AChE by soman. The effect of HupA against soman toxicity lasted a long time. It is important to mention that HupA levels persist in the brain for a longer period than physostigmine does. The researchers demonstrated that HupA might provide significantly longer therapeutic activity than other AChE inhibitors used to manage diseases where there is a deficiency in cholinergic neurons (Grunwald et al., 1994).

Overall, HupA demonstrated the ability to reduce dementia and neurodegeneration effectively as well as noxious OPS-induced neurotoxicity in numerous species of animals and humans. There are enough evidences that HupA boosts learning ability, memory enhancement, and cognitive performance. Moreover, it has tremendous potential as a therapeutic intervention in patients suffering from myasthenia gravis and AD with minimal side effects. More ongoing research studies are in progress around the world to further establish its eminent neuroprotective ability in animals and humans.

6.5 SHANKHAPUSHPI

Shankhapushpi (SP, botanically known as *Convolvulus pluricaulis*, also known as morning glory) is a perennial herb whose branches are spread on the ground and can be more than 30 cm long (Bhowmik et al., 2012; Sethiya et al., 2009; Shiksharthi et al., 2011). The flowers are deep blue or purple in color, about 5 mm in size, and shaped like a "*Sankha*" (means conch, which is blown to make sound), while the leaves are elliptic in shape. It widely grows in India in sandy and stony areas and prefers dry climate (Amin et al., 2014; Amin and Sharma, 2015; Mudgal, 1975). It is important to indicate that SP has been used in India effectively over the centuries due to its unique chemical constituents, which helps the brain to calm down, relieve tension, and attenuate focus. According to Ayurveda, it is considered as a remarkable *"medhya drug"* (one that improves mental abilities, works on rejuvenating nervous function, and acts as brain stimulator, memory enhancer and ability to recall) (Amin et al., 2014; Beg et al., 2011). The herb produces its action by modulation of neuro-chemistry of the brain and keeps brain cells active and healthy. It is beneficial in mental weakness, forgetfulness, memory loss, low retention power, anti-ageing, and diverse central nervous system (CNS) disorders, including insanity, epilepsy, and nervous debility (Amin et al., 2014; Amin and Sharma, 2015).

It has been reported that the whole plant including the flower is used in Ayurvedic medical treatment for memory loss, long-term memory enhancement, insomnia, use for mental stimulation, and rejuvenation therapy (Kumar, 2006; Sethiya et al., 2009). It also calms the mind and reduces anxiety, neurosis, mental stress, work-related stress, and depression (Sharma et al., 2009). SP has been reported to regulate the production of the body's stress hormones like adrenaline and cortisol and thus helps in reducing stress and anxiety (Nahata et al., 2009; Patel et al., 2012; Pawar et al., 2001a, 2001b; Sharma et al., 2009; Singh and Mehta, 1977). SP also works as a tranquilizer and psychostimulant (Gupta et al., 1981; Singh and Mehta, 1977). Daily consumption of SP extract prevents memory loss and also helps to decrease the cholesterol level in the blood including triglycerides and phospholipids (Sethiya et al., 2009).

Furthermore, its consumption retards memory loss and has been successful in improving memory in neurodegerative diseases such as Parkinson's syndrome and AD (Sethiya et al., 2009). It slows the progression of memory loss in these diseases. It is also very popular to treat insomnia effectively. Research has demonstrated that SP is effective in counteracting the stress and memory retention of diabetic patients (Patel et al., 2012).

6.5.1 Safety and Storage

The herb (SP) is nontoxic and doesn't produce any adverse effects. The SP extract has been reported to be stable and light brown in color and the oral LD_{50} is 1250 mg/kg body weight in mice (Pawar et al., 2001a, 2001b; Sethiya et al., 2009). Treated mice showed a sedative effect at doses greater than 200 mg/kg body weight and reflected a moderate to marked decrease in locomotor activity which lasted for nearly 12 hours (Pawar et al., 2001a, 2001b; Sethiya et al., 2009).

It is recommended to store SP in a cool dry place away from direct sunlight.

6.5.2 Chemical Constituents

Chemical investigations of SP demonstrate the presence of glycosides, coumarins, flavonoids, and alkaloids. The active chemical constituents and phytonutrients are β-sitosterol, ceryl alcohol, hydroxycinnamic acid, octacosanol, octacosanoltetraco-sane, tetracosane, scopoline, scopoletin, convolvidine, subhirsine, convolvine, and phyllabine, along with glucose and sucrose, which have been identified as part of glycosides, as well as 20-oxodotriacontanol, tetratriacontanoic acid, 29-oxodotriac-ontanol (Deshpande and Srivastava, 1969; Srivastava and Deshpande, 1975; Sethiya et al., 2009) (Figure 6.2).

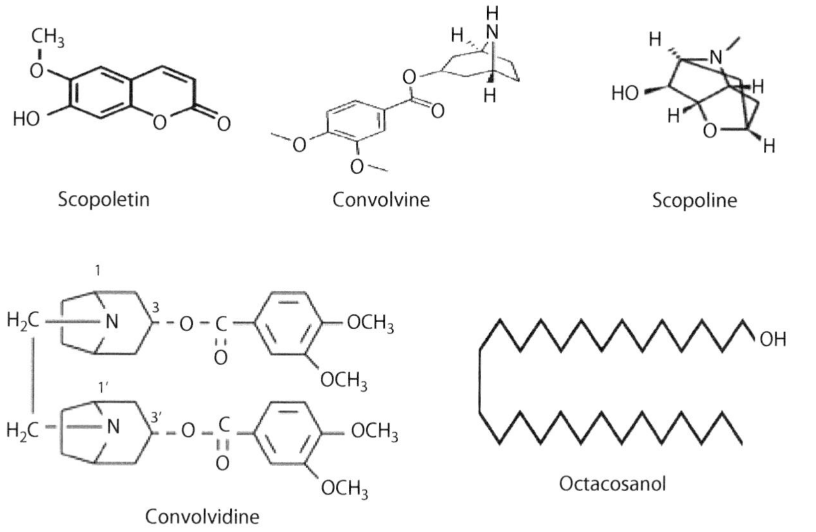

FIGURE 6.2 Chemical structures of scopoletin, convolvine, convolvidine, and scopoline.

6.5.3 RESEARCH AND CLINICAL STUDIES

The detailed research and phytochemical evaluation of SP by high performance liquid chromatography has been reported (Sethiya et al., 2010). It is worthwhile to mention that in this study, all the parameters measured followed the guidelines of World Health Organization and Pharmacopeial guidelines.

Preclinical studies on various extract of SP reported significant improvement on learning behavior and memory enhancement, and hence, it is used as a brain tonic to promote intellect and memory and to alleviate nervous disorder and hypertension (Bagchi and Barilla, 1998).

Clinical studies have demonstrated beneficial effects in patients with anxiety, neurosis and nervousness. It induces a feeling of calm and peace; good sleep; and relief from anxiety, panic attack, stresses, and mental fatigue; and reduces the level of anxiety significantly. In the future, randomized double-blind clinical studies are required in patients suffering from diverse neurological disorders, anxiety, and stress (Amin et al., 2014; Amin and Sharma, 2015; Sethiya et al., 2009).

In a human clinical study, 30 subjects (male: 12; female: 18; age 16–25 years) who were suffering from lack of concentration (100%), disturbed sleep (80.6%), inability to relax (92.7%), depression (55.6%), fear (47.4%), and anxiety (78.4%) were administered four tablets of 500 mg of SP tablet each/day two hours after food with milk over a period of 60 consecutive days (Amin et al., 2014). All these volunteers had educational and professional stress. Paired t test was used for analyzing the data. This study obtained institutional review board approval and was registered at CTRI (CTRI/2013/10/004100). All volunteers signed the consent forms. WMS was adapted to collect data pre and post treatment. Total effect of therapy was assessed as follows:

Complete improvement: 100% improvement (score between 32 and 40)
Marked improvement: Score between 24 and 31
Moderate improvement: Score between 16 and 23
Mild improvement: Score between 8 and 15
Unchanged: Score between 0 and 8

It is important to mention that patients with any type of known psychiatric disorders or any systemic diseases were excluded from this study.

Overall, results have shown significant increase in long-term memory retention, recollection of memory, and control over senses and anxiety. The study provides a safer and affordable therapeutic option in augmenting memory. No adverse events were reported.

Nurturing this type of study on all age groups will help bring solution in public health initiative to boost health care delivery system in the areas of mental health care with minimal or no side effects (Amin et al., 2014).

In another clinical study, the efficacy of SP was compared with Yogic procedure according to Ayurveda (Amin et al., 2015). In this study, 102 physically healthy volunteers (age: 16–25 years, without any pathological conditions) were randomly divided into two groups (Amin et al., 2015). Group A received Yogic procedure (*asana*, *pranayama*, and *chanting* were followed with counseling) and were given placebo tablets, and Group B was given SP tablets, made with whole part of SP plant. Written consent was obtained from all subjects. This study also obtained institutional review board approval and was registered at CTRI (CTRI/2013/10/004100). WMS was used to compare data before and after intervention over a period of two months. Paired and unpaired t tests were used for analysis of data using Sigmastat software. Group B showed significant improvement as compared to the Yoga placebo group. SP dramatically improved the memory process, improved mental alertness and long-term memory retention power, and increased self-esteem level, thereby enhancing their academic and professional performances. Improvement in short- and long-term memory was observed in the SP group. Visual recognition was also significantly improved in the SP group (Amin et al., 2015).

In another clinical study, a total of 160 male and female patients with thyrotoxicosis, a disease involving various psychological and emotional factors, which are considered as an important factor for the development of the disease, were chosen (Gupta et al., 1981). Since the emotional factor is considered as a causative factor, SP was used as a brain tonic and a tranquilizer; SP was used in this study. The subjects were divided into three groups—Group 1: neomercazole + diazepam; Group 2: SP syrup alone; and Group 3: SP syrup + neomercazole. SP treatment was found to be more effective than the tranquilizer. In addition, no side effects were noted in the SP group (Gupta et al., 1981).

6.5.4 Mechanisms of Action

It is important to mention that SP is nootropic, which is referred to as a brain tonic as well as memory and cognitive enhancer, which increases concentration and focus.

SP works as nootropic by altering the availability of brain supply of neurochemicals (neutrotransmitters, enzymes, and hormones) by improving the brain's oxygen supply and optimizing blood flow to the brain, thereby increasing nutrient supply and stimulating nerve growth, which improves brain function and memory (Amin et al., 2014; Beg et al., 2011; Sethiya et al., 2009).

Several mechanisms have been postulated for the nootropic activity of SP as follows:

1. Increases brain protein content (Dandiya, 1990)
2. Enhances neuropeptide synthesis (Dandiya, 1990)
3. Scavenges noxious free radicals and ameliorates oxidative stress (Nahata et al., 2009)
4. Antistress activity due to reduction of behavioral pattern and suppression of aggressive behavior and alteration and improvement in sleep pattern (Dhingra and Valecha, 2007)
5. Antidepressant activity due to its action with the adrenergic, dopaminergic, and serotonergic systems (Dhingra and Valecha, 2007; Nahata et al., 2009).
6. Anti-AChE and neuroprotective activity (Amin et al., 2015).
7. Nootropic activity of flavonoids (Shiksharthi et al., 2011).
8. Potent anxiolytic, anti-AChE, and neuroprotective activities (Barar and Sharma, 1965; Mudgal, 1975; Pawar et al., 2001a, 2001b; Sharma et al., 2009).

Overall, preclinical and clinical studies of SP demonstrated remarkable short-term memory loss and long-term memory enhancement, antistress effects, improvement in sleep pattern, antidepressant activity, decrease in anxiety and anti-AChE activity, and dramatic neuroprotective ability (Barar and Sharma, 1965; Pawar et al., 2001a, 2001b; Sethiya et al., 2009).

6.6   CONCLUSION

Although the use of HupA and SP has shown promising results in patients with memory dysfunctions, data supporting their use are limited by weak study design. Systematic randomized, placebo-controlled clinical studies are warranted in larger study population.

REFERENCES

Amin H, Sharma R, Vyas H, Vyas M, Prajapati PK, Dwivedi R. 2014. Nootropic (medhya) effect of Bhavita Sankhapuspi tablets: A clinical appraisal. *Anc Sci Life* 34: 109–12.

Amin H, Sharma R. 2015. Nootropic efficacy of Satvavaajaya Chikitsa and Ayurvedic drug therapy: A comparative clinical exposition. *Int J Yoga* 8: 109–16.

Ashani Y, Grunwald J, Kronman C, Velan B, Shafferman A. 1994. Role of tyrosine 337 in the binding of huperzine A to the active site of human acetylcholinesterase. *Mol Pharmacol* 45: 555–60.

Bagchi D, Barilla J. 1998. *Huperzine A: Boost your brain power.* Keats Publishing Inc., New Canaan, CT, pp 1–43.

Bai DL, Tang XC, He XC. 2000. Huperzine A, a potential therapeutic agent for treatment of Alzheimer's disease. *Curr Med Chem* 7: 355–74.

Bai F, Xu Y, Chen J, Liu Q, Gu J, Wang X, Ma J, Li H, Onuchic JN, Jiang H. 2013. Free energy landscape for the binding process of huperzine A to acetylcholinesterase. *PNAS USA* 110: 4273–8.

Barar FS, Sharma VN. 1965. Preliminary pharmacological studies on *Convolvulus pluricaulis* chois—An Indian indigenous herb. *Indian J Physiol Pharmacol* 9: 99–102.

Beg S, Swain S, Hasan H, Barkat MA, Hussain MS. 2011. Systematic review of herbals as potential anti-inflammatory agents: Recent advances, current clinical status and future perspectives. *Pharmacogn Rev* 5: 120–37.

Bhowmik D, Kumar KPS, Paswan S, Srivastava S, Yadav AP, Dutta A. 2012. Traditional Indian herbs *Convolvulus pluricaulis* and its medicinal importance. *J Pharmacogn Phytochem* 1: 44–51.

Bobade V, Bodhankar SL, Aswar U, Vishwaraman M, Thakurdesai P. 2015. Prophylactic effects of asiaticoside-based standardized extract of *Centella asiatica* (L.) urban leaves on experimental migraine: Involvement of 5HT1A/1B receptors. *Chin J Nat Med* 13(4): 274–82. doi: 10.1016/S1875-5364(15)30014-5.

Cheng YS, Lu CZ, Ying ZL, Ni WY, Zhang CJ, Sang GW. 1986. 128 cases of myasthenia gravis treated with huperzine A. *New Drugs Clin Remed* 5: 197–9.

Damar U, Gersner R, Johntone J, Schacter S, Rotenberg A. 2016. Huperzine A as a neuroprotective and antiepileptic drug: A review of preclinical research. *Expert Rev Neurother.* [Epub ahead of print].

Dandiya PC. 1990. The pharmacological basis of herbal drugs acting on CNS. *East Pharm* 33: 39–47.

Deshpande SM, Srivastava DN. 1969. Chemical composition of the fatty acids of *Convolvulus pluricaulis. Indian Oil Soap J* 34: 217–8.

Grunwald J, Raveh L, Doctor BP, Ashani Y. 1994. Huperzine A as a pretreatment candidate drug agent against nerve agent toxicity. *Life Sci* 54: 991–7.

Gupta RC, Singh PM, Prasad GC, Udupa KN. 1981. Probable mode of action of Shankhapuspi in the management of thyrotoxicosis. *Anc Sci Life* 1: 49–57.

Ha GT, Wong RK, Zhang Y. 2011. Huperzine A as potential treatment of Alzheimer's disease: An assessment of chemistry, pharmacology, and clinical studies. *Chem Biodivers* 8: 1189–204.

Hugel HM. 2015. Brain food for Alzheimer-free ageing: Focus on herbal medicines. *Adv Exp Med Biol* 863: 95–116.

Ishiuchi K, Park JJ, Long RM, Gang DR. 2013. Production of huperzine A and other Lycopodium alkaloids in Huperzia species grown under controlled conditions and in vitro. *Phytochemistry* 91: 208–19.

Jia JY, Zhao QH, Liu Y, Gui YZ, Liu GY, Zhu DY, Yu C, Hong Z. 2013. Phase I study on the pharmacokinetics and tolerance of ZT-1, a prodrug of huperzine A, for the treatment of Alzheimer's disease. *Acta Pharmacol Sin* 34: 976–82.

Kelsey NA, Wilkins HM, Linseman DA. 2010. Nutraceutical antioxidants as novel neuroprotective agents. *Molecules* 15: 7792–814.

Kim J, Lee HJ, Lee KW. 2010. Naturally occurring phytochemicals for the prevention of Alzheimer's disease. *J Neurochemistry* 112: 1415–30.

Kodl CT, Seaquist ER. 2008. Cognitive dysfunction and diabetes mellitus. *Endocrine Rev* 29: 494–511.

Kumar V. 2006. Potential medicinal plants for CNS disorders: An overview. *Phytother Res* 20: 1023–35.

Lee WH, Loo CY, Bebawy M, Luk F, Mason RS, Rohanizadeh R. 2013. Curcumin and its derivatives: Their application in neuropharmacology and neuroscience in the 21st century. *Curr Neuropharmacol* 11: 338–78.

Lin JH, Hu GY, Tang XC. 1996. Facilitatory effect of huperzine-A on mouse neuromuscular transmission in vitro. *Acta Pharmacol Sin* 17: 299–301.

Lunardi P, Nardin P, Guerra MC, Abib R, Leite MC, Goncalves CA. 2013. Huperzine A, but not tacrine, stimulates S100B secretion in astrocyte cultures. *Life Sci* 92: 701–7.

Malkova L, Kozikowski AP, Gale K. 2011. The effects of huperzine A and IDRA 21 on visual recognition memory in young macaques. *Neuropharmacology* 60: 1262–8.

Mudgal V. 1975. Studies on medicinal properties of *Convolvulus pluricaulis* and *Boerhaavia diffusa*. *Planta Med* 28:62–8.

Nahata A, Patil UK, Dixit VK. 2009. Anxiolytic activity of *Evolvuvus alsinoides* and *Convulvulus pluricaulis* in rodents. *Pharmaceutical Biol* 47: 444–51.

Patel DV, Chandola H, Baghel MS, Joshi JR. 2012. Clinical efficacy of shankhapushpi and a herbo-mineral compound in type II diabetes. *Ayu* 33: 230–7.

Pawar SA, Dhuley JN, Naik SR. 2001a. Neuropharmacology of an extract derived from *Convolvulus pluricaulis*. *Pharm Biol* 39: 4.

Pawar SA, Dhuley JN, Naik SR. 2001b. Neuropharmacology of an extract derived from *Convolvulus microphyllus*. *Pharmaceutical Biol* 39; 253–8.

Polotow TG, Poppe SC, Vardaris CV, Ganini D, Guariroba M, Mattei R, Hatanaka E, Martins MF, Bondan EF, Barros MP. 2015. Redox status and neuro inflammation indexes in cerebellum and motor cortex of Wistar rats supplemented with natural sources of omega-3 fatty acids and astaxanthin: Fish oil, krill oil, and algal biomass. *Mar Drugs* 13: 6117–37.

Qian ZM, Ke Y. 2014. Huperzine A: Is it an effective disease-modifying drug for Alzheimer's disease? *Front Aging Neurosci* 6: 216. doi: 10.3389/fnagi.2014.00216. eCollection 2014.

Rajan KE, Preethi J, Singh HK. 2015. Molecular and functional characterization of *Bacopa monniera*: A retrospective review. *Evid Based Complement Alternat Med* 2015: 945217. doi: 10.1155/2015/945217. Epub August 27, 2015.

Sethiya NK, Nahata A, Mishra SH, Dixit VK. 2009. An update on shankhapushpi, a cognition-boosting Ayurvedic medicine. *J Chin Integr Med* 7: 1001–22.

Sharma K, Arora V, Rana AC, Bhatnagar M. 2009. Anxiolytic effect of *Convolvulus pluricaulis* petals on elevated plus maize model of anxiety in mice. *J Herb Med Toxicol* 3: 41–6.

Shetty AK, Bates A. 2015. Potential of GABA-ergic cell therapy for schizophrenia, neuropathic pain, and Alzheimer's and Parkinson's diseases. *Brain Res.* pii: S0006-8993(15)00717-9. doi: 10.1016/j.brainres.2015.09.019. Epub September 28, 2015.

Shiksharthi AR, Mittal S, Ramana J. 2011. Systematic review of herbals as potential memory enhancers. *Int J Res Pharm Biomed Sci* 3: 918–25.

Shin PH, Chan YC, Liao JW, Wang MF, Yen GC. 2010. Antioxidant and cognitive promotion effects of anthocyanin-rich mulberry (*Morus atropurpurea* L.) on senescence-accelerated mice and prevention of Alzheimer's disease. *J Nutr Biochem* 21: 598–605.

Singh RH, Mehta AK. 1977. Studies on the anti-anxiety effect of the *Medhya Rasayana* drug "Shakhapushpi" (*C. pluricaulis* Chois). *J Res Ind Med* 13, 1–7.

Srivastava DN, Deshpande SM. 1975. Gas chromatographic identification of fatty acids, fatty alcohols, and hydrocarbons of *Convolvulus pluricaulis* (Choisy). *J Am Oil Chem Soc* 52: 318–9.

Subedee L, Suresh RN, Mk J, Hi K, Am S, Vh P. 2015. Preventive role of Indian black pepper in animal models of Alzheimer's diseases. *J Clin Diagn Res* 9(4): FF01–4. doi: 10.7860/JCDR/2015/8953.5767. Epub April 1, 2015.

Sun QQ, Xu SS, Pan JL, Guo HM, Cao WQ. 1999. Huperzine A capsules enhance memory and learning performance in 34 pairs of matched adolescent students. *Zhongguo Yao Li Xue Bao* 20: 601–3.

Tang XC, Han YF, Chen XP, Zhu XD. 1986. Effects of huperzine A on learning and retrieval process of discrimination performance in rats. *Acta Pharmacol Sin* 7: 507–11.

Tang LL, Wang R, Tang XC. 2005a. Huperzine A protects SHSY5Y neuroblastoma cells against oxidative stress damage via nerve growth factor production. *Eur J Pharmacol* 519: 9–15.

Tang LL, Wang R, Tang XC. 2005b. Effects of huperzine A on secretion of nerve growth factor in cultured rat cortical astrocytes and neurite outgrowth in rat PC12 cells. *Acta Pharmacol Sin* 26: 673–8.

Tang XC, Ziong ZQ, Qian BC, Zhou ZF, Zhang CL. 1994. Cognition improvement by oral huperzine A: A novel acetylcholinesterase inhibitor. *Alzheimer's therapy: Therapeutic strategies.* Editors: Giacobini E, Becker R, Birkhauser Publications, Boston, MA, pp. 113–9.

Vincent GP, Rumennik L, Cumin R, Martin J, Spinwall J. 1987. The effects of huperzine A, an acetyl cholinesterase inhibitor, on the enhancement of memory in mice, rats and monkeys. *Neuroscience Abs* 13: 844.

Venkatesan R, Ji E, Kim SY. 2015. Phytochemicals that regulate neurodegenerative disease by targeting neurotrophins: A comprehensive review. *BioMed ResInt* 2015: Article ID 814068 http://dx.doi.org/10.1155/2015/814068.

Wang CY, Zheng W, Wang T, Xie JW, Wang SL, Zhao BL, Teng WP, Wang ZY. 2011. Huperzine A activates Wnt/β-catenin signaling and enhances the nonamyloidogenic pathway in an Alzheimer transgenic mouse model. *Neuropsychopharmacology* 36: 1073–89.

Wang R, Yan H, Tang X-C. 2006. Progress in studies of huperzine A, a natural cholinesterase inhibitor from Chinese herbal medicine. *Acta Pharmacol Sin* 27: 1–26.

Wang ZF, Wang J, Zhang HY, Tang XC. 2008. Huperzine A exhibits anti-inflammatory and neuroprotective effects in a rat model of transient focal cerebral ischemia. *Journal of Neurochemistry* 106: 1594–603.

Xing SH, Zhu CX, Zhang R, An L. 2014. Huperzine A in the treatment of Alzheimer's disease and vascular dementia: A meta-analysis. *Evidence-Based Complementary and Alternative Medicine* Volume 2014, Article ID 363985, 10 pages http://dx.doi.org/10.1155/2014/363985.

Xu SS, Gao ZX, Weng Z, Du ZM, Xu WA, Yang JS, Zhang ML, Tong ZH, Fang YS, Chai XS. 1995. Efficacy of tablet huperzine-A on memory, cognition, and behavior in Alzheimer's disease. *Zhongguo Yao Li Xue Bao* 16: 391–5.

Xu ZQ, Liang XM, Juan W, Zhang YF, Zhu CX, Jiang XJ. 2012. Treatment with Huperzine A improves cognition in vascular dementia patients. *Cell Biochem Biophys* 62: 55–8.

Yu D, Thakor DK, Han I, Ropper AE, Haragopal H, Sidman RL, Zafonte R, Schachter SC, Teng YD. 2013. Alleviation of chronic pain following rat spinal cord compression injury with multimodal actions of huperzine A. *PNAS USA* 110: E746–55.

Yue J, Dong BR, Lin X, Yang M, Wu HM, Wu T. 2012. Huperzine A for mild cognitive impairment. *Cochrane Database Syst Rev* 12: CD008827. doi: 10.1002/14651858 .CD008827.pub2.

Zhang HY. 2012. New insights into huperzine A for the treatment of Alzheimer's disease. *Acta Pharmacol Sin* 33: 1170–5.

Zhang HY, Tang XC. 2006. Neuroprotective effects of huperzine A: New therapeutic targets for neurodegenerative disease. *Trends Pharmacol Sci* 27: 619–26.

Zhang RW, Tang XC, Han YY, Sang GW, Zhang YD, Ma YX, Zhang CI, Yang RM. 1991. Drug evaluation of huperzine A in the treatment of senile memory disorders. *Acta Pharmacol Sin* 12: 250–2.

Zhang L, Yang L. 2014. Anti-inflammatory effects of vinpocetine in atherosclerosis and ischemic stroke: A review of the literature. *Molecules* 20: 335–47.

Zhang SL. 1986. Therapeutic effects of huperzine A on the aged humans with memory impairment. *New Drugs Clin Remed* 5: 260–2.

Zhang Z, Wang X, Chen Q, Shu L, Wang J, Shan G. 2002. Clinical efficacy and safety of huperzine alpha in treatment of mild to moderate Alzheimer disease, a placebo-controlled, double-blind, randomized trial. *Zhonghua Yi Xue Za Zhi* 82: 941–4.

Zhu XD, Tang XC. 1988. Improvement of impaired memory in mice by huperzine A and huperzine B. *Acta Pharmacol Sin* 9: 492–7.

7 Resveratrol and Depression

Jui-Hu Shih, Chien-Fu Fred Chen, and I-Hsun Li

CONTENTS

7.1 INTRODUCTION

Depression is a common mental disorder and a major cause of morbidity and mortality in children and adolescents, with an estimated 350 million people affected globally. Depression is projected to become the second leading cause of disability worldwide by 2030 (Mathers et al. 2008). The major hypothesis of depression was proposed that the main symptoms of depression are due to a functional disorder of the brain monoamine neurotransmitters (particularly adrenaline [NA] and 5-hydroxytryptamine [5-HT]), their presynaptic transporters, postsynaptic receptors, and monoamine oxidase (MAO) for neurotransmitter metabolism (Brigitta 2002, Blier and de Montigny 1994). Recent studies also suggested that major depression is associated with hyperactivity of the hypothalamic–pituitary–adrenal (HPA) axis, resulting in hypersecretion of cortisol (corticosterone in rodents), adrenocorticotropin, and corticotropin-releasing hormone (CRH) (Dinan 1994, Juruena et al. 2004). Besides, previous studies have pointed to a positive correlation between interleukin

119

(IL)-1β, IL-6, and tumor necrosis factor alpha (TNF-α) levels and depressive symptoms (Hannestad et al. 2011). Traditionally, the first-line therapeutic antidepressants are tricyclic antidepressants, selective serotonin reuptake inhibitors (SSRIs), serotonin, and noradrenaline reuptake inhibitors and MAO inhibitors. Clinically, these drugs have many limitations, including insufficiency of efficacy, undesirable adverse side effects, and a long latency to onset (Blier and El Mansari 2013, Rush et al. 2003, 2006). Thus the development of safe and effective pharmacotherapeutics remains necessary. Recently, the use of herbal medicines (e.g., St. John's wort and resveratrol) has also provided a useful approach in the management of depression (Sarris et al. 2012).

Resveratrol (3,5,4′-trihydroxy-*trans*-stilbene; Figure 7.1) is a natural polyphenolic phytoalexin produced naturally by several plants when they were exposed to injury or pathogen infection, such as bacteria or fungi (Langcake and Pryce 1977). Sources of resveratrol include the skin of grapes, blueberries, raspberries, red wine, peanuts, and medicinal herbs, such as *Polygonum cuspidatum* (Zhang et al. 2013). Interest in resveratrol began with the "French Paradox" describing the lower incidence of cardiovascular events by consuming red wine, which contains high concentrations of nonflavonoid resveratrol, even in some of French population with a high-fat intake (Lippi et al. 2010). Since then, more studies have demonstrated that resveratrol exerts a variety of biological effects, including antioxidant, antiapoptotic, anti-inflammatory, antiaging, antidepressant, and antiangiogenic properties (Kasiotis et al. 2013, Li et al. 2014, Lin et al. 2014, Liu et al. 2016, Zhang et al. 2010). Furthermore, there is also evidence suggesting that resveratrol can function as a neuroprotective agent to ameliorate symptoms associated with neuronal damage caused by cerebral ischemia (Lopez et al. 2015, Wan et al. 2016) and a host of neurodegenerative diseases such as Alzheimer's disease (Feng et al. 2009, Gertz et al. 2012), Parkinson's disease (Blanchet et al. 2008, Ferretta et al. 2014), and Huntington's disease (Kumar et al. 2007).

However, the obvious and important limitation of resveratrol for therapeutic application in brain illness is its low bioavailability in brain due to its short biological half-life, rapid metabolism, and elimination (Baur and Sinclair 2006). Previous pharmacokinetic studies indicate that the oral bioavailability of resveratrol is almost zero, which casts doubt on the physiological relevance of the high concentrations typically used for *in vitro* experiments (Walle et al. 2004, Wenzel and Somoza 2005). Despite these disadvantages, resveratrol has still attracted great interest in the research community, with 7984 publications referenced on the U.S. National

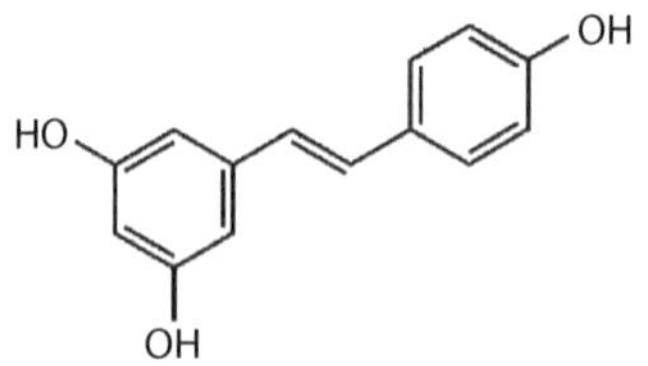

FIGURE 7.1 Chemical structure of resveratrol (3,5,4′-trihydroxy-*trans*-stilbene).

Library of Medicine's PubMed database between 1981 and 2015, of which 7.6% studied brain-related illness. It is implying that trace amounts of resveratrol in the brain are enough to regulate neurophysiological actions and treat brain disease.

This review provides a brief overview of current scientific literature on resveratrol, its plausible mechanisms of action, and its potential use as a therapeutic intervention in depressive-like disorders. A literature search was performed using PubMed for articles published up to 2016. Specific keywords, such as *resveratrol, depression, depressive-like disorders, serotonin, monoamine oxidase, serotonin transporter,* and *anti-depressant,* were used in combinations to filter the searches. Relevant information pertaining to the focus of this review was included.

7.2 ABSORPTION, DISTRIBUTION, METABOLISM, AND ELIMINATION OF RESVERATROL

Pharmacokinetic studies in humans found that resveratrol appears to be well absorbed (~75%) primarily through transepithelial diffusion after oral administration of the drug (Walle et al. 2004). When resveratrol was absorbed into the blood circulation, more than 90% of free resveratrol would be bound to human plasma and 50% of its plasma metabolites were bound to the proteins (Burkon and Somoza 2008). The apparent volume of distribution after intravenous administration was about 1.8 L/kg, indicating a fair amount of extravascular distribution (Walle et al. 2004). As shown in Figure 7.2, free resveratrol usually was metabolized rapidly to the 3-O-sulfated, 3-O-glucuronidated, and 4'-O-glucuronidated conjugates in the intestine and liver (Burkon and Somoza 2008, Goldberg et al. 2003, Walle 2011). After metabolism by the glucuronidation pathway, 87% of resveratrol conjugates with better polarity and solubility were rapidly eliminated via the kidney (Marier et al. 2002). Renal

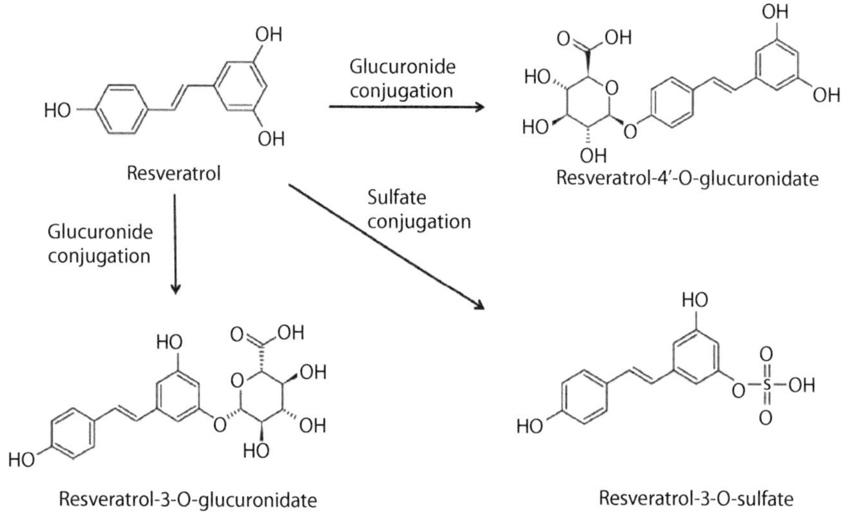

FIGURE 7.2 Predominant metabolites of postulated metabolic pathway of resveratrol.

excretion was the major route of elimination in animals and humans (Goldberg et al. 2003, Meng et al. 2004, Soleas et al. 2001, Yu et al. 2002), and fecal excretion was responsible for elimination of the rest of metabolites (Walle et al. 2004). For both total unchanged resveratrol and its metabolites, the first peak concentrations in serum occurred 30–60 minutes after oral administration, at about 1.8–2 µmoles/L, but the serum level of free unmetabolized resveratrol was a small fraction of the total resveratrol concentration, no more than 1.7% to 1.9%, indicating very low oral bioavailability for resveratrol (Goldberg et al. 2003, Soleas et al. 2001, Walle et al. 2004). A second peak at about 1.3 µmoles/L was observed at six hours, suggesting enteric recirculation of conjugated metabolites by reabsorption following intestinal hydrolysis, and the plasma half-life was about 9.2 hours (Walle et al. 2004).

Even though accumulating studies have pointed low bioavailability of resveratrol, a study by Juan and colleagues (2010) demonstrated that there were still detectable levels of resveratrol and its metabolites in various tissues by high-performance liquid chromatography method. At 90 minutes after intravenous administration of 15 mg/kg *trans*-resveratrol in rats, the higher amounts of *trans*-resveratrol and its metabolites were detected in the kidney (*trans*-resveratrol: 1.45 nmol/g; glucuronidated conjugate: 2.91 nmol/g; sulfated conjugate: not detected) and lung (*trans*-resveratrol: 1.13 nmol/g; glucuronidated conjugate: 0.28 nmol/g; sulfated conjugate: 0.42 nmol/g), whereas the tissue extracts from liver (*trans*-resveratrol: 0.35 nmol/g; glucuronidated conjugate: 0.58 nmol/g; sulfated conjugate: 0.11 nmol/g), brain (*trans*-resveratrol: 0.17 nmol/g; glucuronidated conjugate: not detected; sulfated conjugate: 0.04 nmol/g), and testes (*trans*-resveratrol: 0.05 nmol/g; glucuronidated conjugate: 0.70 nmol/g; sulfated conjugate: 0.23 nmol/g) contained only moderate amounts of *trans*-resveratrol (Juan et al. 2010). Lou and colleagues (2014) also investigated levels of resveratrol and its metabolites in various tissues by stable isotope-dilution ultra-high performance liquid chromatography-tandem mass analysis and found that resveratrol and its metabolites were detectable in brain with pmol/g ranges (resveratrol: 15.55 pmol/g; 3-O-glucuronidated conjugate: 34.59 pmol/g; 4′-O-glucuronidated conjugate: 4.40 pmol/g; sulfated conjugates: 171.32 pmol/g) at 30 minutes after intravenous administration of 20 mg/ kg resveratrol in rats (Lou et al. 2014). These results showed that resveratrol and its metabolites can pass through the blood–brain barrier and have chances to regulate the microenvironment of neurons in the central nervous system.

7.3 MODE OF ACTION IN DEPRESSION

In the 1990s, resveratrol was only reported for its cardioprotective property (Bertelli et al. 1995). Since Jang and colleagues demonstrated chemopreventive activity of resveratrol (Jang et al. 1997), however, many studies have been published with varying degrees of evidence that resveratrol was linked to many health benefits in a wide range of diseases, including cancer, central nervous system injury, diabetes, cardiovascular, and neurodegenerative diseases through different mechanisms of action which are briefly summarized in Table 7.1. In recent years, increased attention has been given to the antidepressive effect of resveratrol. Therefore, in the following sections, we will mainly review researches about mechanisms of antidepressant action of resveratrol.

TABLE 7.1
Biological Activity, Effects, and Plausible Mechanisms of Action of Resveratrol

Biological Activity	Effect	Plausible Mechanism(s)	Reference
Antioxidative	Inhibition of reactive oxygen species (ROS)	Induction of SOD, CAT, and GSH peroxidase-1	Xia et al. (2010)
	Reduction of free radicals	Activation of PPARγ coactivator 1α	Lagouge et al. (2006)
Anti-inflammatory	Attenuated DNA damage and upregulation of IL-6 and TNF-a	Via sirtuin-1 activation	Csiszar et al. (2008)
	Reduced proinflammatory arachadonic acid metabolism	Inhibition of cyclooxygenase-1 and -2	Mohamed et al. (2014), Simao et al. (2012)
Antiproliferative	Induction of apoptosis	Downregulation of the PI3K/Akt/mTOR pathway	Roy et al. (2009)
Antiangiogenesis	Inhibition of tumor growth, cell migration, invasion, and metastasis	Downregulation of MMP-2 and MMP-9 through the p38 kinase and PI3K pathways	Gweon and Kim (2013, 2014)
Antiplatelet	Inhibition of platelet activation and aggregation	Inhibition of protein kinase C activation and protein tyrosine phosphorylation	Wu et al. (2007)
Antidiabetic	Improvement of glucose homeostasis, decreased insulin resistance, protection of pancreatic β-cells, increased insulin secretion	Increased expression/ activity of AMPK and SIRT1	Szkudelski and Szkudelska (2015)

7.3.1 INHIBITION OF SEROTONIN TRANSPORTERS AND MAO-A

The serotonin transporter, known as the sodium-dependent serotonin transporter, is a type of monoamine transporter protein that terminates the action of serotonin and transports serotonin from the synaptic cleft to the presynaptic neuron in a sodium-dependent manner (Owens and Nemeroff 1994). Alternatively, serotonin is broken down by the enzyme MAO into the inactive metabolite 5-hydroxyindole acetaldehyde. There are two subtypes of MAO. MAO-A has a high affinity for serotonin and MAO-B has a low affinity for serotonin. MAO-B is present inside serotonergic neurons but degrades serotonin only at high concentrations (Alves et al. 2007, Nagatsu 2004). Both serotonin transporters and MAO play important roles in mood regulation.

The first *ex vivo* study for screening the potential antidepressive effects of resveratrol on the uptake of [³H]noradrenaline ([³H]NA) and [³H]5-hydroxytryptamine ([³H]5-HT) by synaptosomes from rat brain was conducted by Yanez et al. (2006). This study demonstrated that both *cis*-resveratrol and *trans*-resveratrol (5–200 μM) inhibited the uptake of [³H]NA and [³H]5-HT in a concentration-dependent manner. *Trans*-resveratrol was more slightly sufficient than *cis*-resveratrol but less selective than fluoxetine, an SSRI (Yanez et al. 2006). These *ex vivo* effects of resveratrol on the uptake of [³H]5-HT implies that serotonin transporters may be a potential biological target of resveratrol. Therefore, our groups conducted an *in vivo* study of serotonin transporter occupancy of resveratrol using small-animal positron emission tomography with the serotonin transporter radioligand, *N,N*-dimethyl-2-(2-amino-4-[18F] fluorophenylthio) benzylamine (Shih et al. 2016). We found that resveratrol exhibits binding potentials to serotonin transporter *in vivo* in a dose-dependent manner with variation among brain regions (Shih et al. 2016). In addition, both *cis*-resveratrol and *trans*-resveratrol (5–200 μM) concentration dependently inhibited the enzymatic activity of human recombinant MAO, especially in MAO-A isoform (Yanez et al. 2006). An *in vivo* study by Xu and colleagues also demonstrated that *trans*-resveratrol (10–80 mg/kg) dose dependently inhibited MAO-A activity in mouse brain (Xu et al. 2010). Xu et al. (2010) further showed that *trans*-resveratrol (40 or 80 mg/kg) increased the 5-HT and NA levels in the frontal cortex, hippocampus, and hypothalamus. In accordance with these findings, Huang and colleagues (2013) also found that low doses of *trans*-resveratrol (10 and 20 mg/kg) combined with piperine (2.5 mg/kg) produced significant synergistic effects to increase monoamines and inhibit MAO-A activity (Huang et al. 2013). In summary, resveratrol exerts inhibitory effects on serotonin transporters and MAO activity to increase the levels of 5-HT in the synaptic cleft.

7.3.2 REGULATION OF cAMP RESPONSE ELEMENT-BINDING PROTEIN/BRAIN-DERIVED NEUROTROPHIC FACTOR/ EXTRACELLULAR SIGNAL-REGULATED KINASE SIGNALING

Recently, numerous studies have pointed that neurotropic factors might be associated with pathophysiology of mental disorders (Castren et al. 2007). Growing evidence also indicated that brain-derived neurotrophic factor (BDNF), regulated by upstream transcription factors (e.g., cAMP response element-binding protein [CREB]), plays essential roles in the differentiation of new neurons and synaptogenesis (Huang and Reichardt 2001). Additionally, BDNF expression or polymorphism of this protein is involved with depression as shown in both patients and in the animal model of depression (Duman and Monteggia 2006, Karege et al. 2005). Usually, chronic administration of antidepressant drugs upregulates the expression of BDNF (Calabrese et al. 2007). Inversely, acute treatment of antidepressant drugs will enhance the function of serotonin transporters via the activation of tropomyosin receptor kinase B, which leads to a decrease in extracellular levels of 5-HT and is responsible for the long latency to onset (Benmansour et al. 2008). In addition, extracellular signal-regulated kinase (ERK) is a downstream signal transduction protein activated by BDNF, which is related to cell proliferation and neuroprotection (Mebratu and Tesfaigzi 2009).

It has been demonstrated that the CREB/BDNF/ERK signaling pathway can be upregulated by resveratrol in the different depressive-like animal models. Wang and colleagues (2013) treated normal mice intraperitoneally with 20–80 mg/kg resveratrol for 21 days and found that resveratrol increased BDNF mRNA levels, BDNF protein, and ERK phosphorylation (pERK) levels in the prefrontal cortex and hippocampus. This effect was similar to that seen with the antidepressant drug fluoxetine (10 mg/kg for 21 days), an SSRI, in this study (Wang et al. 2013). The same research team further used the five-week chronic unpredictable mild stress (CUMS) model to explore the effects of chronic administration of resveratrol (20–80 mg/kg for five consecutive weeks) on the CREB/BDNF/ERK signaling pathway in rats (Liu et al. 2014). The results shown by Liu et al. (2014) also revealed that five weeks of CUMS exposure significantly decreased BDNF, pERK, and pCREB levels in the hippocampus and amygdala, while the resveratrol treatment normalized these levels. In a study of lipopolysaccharide (LPS)-induced depressive-like behavior, Ge and colleagues (2015) reported that pretreatment with resveratrol (80 mg/kg, i.p.) for seven consecutive days ameliorated the expression levels of proinflammatory cytokines induced by a single dose of LPS (0.83 mg/kg) and upregulated pCREB/BDNF expression in prefrontal cortex and hippocampus in mice (Ge et al. 2015). In addition, whether in the repeated corticosterone-induced depression model or the spontaneous depressive-like strain (Wistar-Kyoto [WKY] rats), it is proven again that chronic administration of resveratrol exerts antidepressant effects, which is associated with an increased BDNF level in hippocampus (Ali et al. 2015, Hurley et al. 2014).

7.3.3 MODULATION OF HPA AXIS

HPA activity is governed by the secretion of CRH from the hypothalamus. CRH activates the secretion of adrenocorticotropic hormone (ACTH) from the pituitary, which in turn stimulates the secretion of the glucocorticoids (cortisol in human and corticosterone in rodents) from the adrenal cortex. The circulating glucocorticoids exert the effect of negative feedback through binding mineralocorticoid, and glucocorticoid receptors to maintain the internal homeostasis of HPA hormones. The activated HPA axis regulates peripheral physiologic functions such as gluconeogenesis and the regulation of the immune system. However, the overactivation of HPA axis may also lead to deleterious effects on the brain. Recently, numerous studies have indicated that the long-term increased activity of HPA axis is associated with psychiatric disorders, particularly in major depression (Juruena 2014, Pariante and Lightman 2008, Vreeburg et al. 2009). As the hippocampus is a brain region with a high concentration of glucocorticoid receptors, several studies have reported that a significant reduction in hippocampal volume in patients with major depressive disorders contributes to cognitive impairment and memory performance (Rubinow et al. 1984, Wolkowitz et al. 1990). Therefore, in the search for antidepressant drugs with enhanced efficacy, targeting the HPA axis may be a valid strategy.

In recent years, accumulating evidence suggested that resveratrol exerts the inhibitory effect on the activity of HPA axis in the different animal models of stress. Wang and colleagues (2013) showed that the chronic administration of 80 mg/kg resveratrol for 21 consecutive days significantly reduced the serum corticosterone levels compared with the control group in the mice at 30 minutes after a forced swim test (FST). It is also proven that rats treated only with the low dose of resveratrol (15 mg/kg) in the three-week CUMS model have lower serum corticosterone concentrations (Ge et al. 2013). However, a higher dose of resveratrol (80 mg/kg) is needed in the five-week CUMS model to produce a similar inhibitory effect on the HPA axis (Liu et al. 2014). Another study showed that the low dose of resveratrol (20 mg/kg) could not ameliorate the increased levels of serum corticosterone in rats exposed to four-week CUMS (Sakr et al. 2015). These results implied that the dose of resveratrol for inhibiting the activity of HPA axis should be increased proportionally with the extent of stress. In addition, Ali and colleagues (2015) orally treated mice with 80 mg/kg resveratrol prior to the subcutaneous administration of 40 mg/kg corticosterone for 21 consecutive days. It is interesting that resveratrol also can reduce exogenous corticosterone of mice in the model of repeated corticosterone-induced depression (Ali et al. 2015), while the mechanism is unclear. To date, however, only one study has pointed that resveratrol suppresses steroidogenesis of rat adrenocortical cells by inhibiting cytochrome P450 c21-hydroxylase (Supornsilchai et al. 2005). Further, in another study, resveratrol successfully ameliorated middle cerebral artery occlusion-induced series abnormalities related to depressive-like behaviors, including an increased expression of the CRH and the differential expression of glucocorticoid receptor in the frontal cortex, hippocampus, and hypothalamus in a dose-dependent manner (Pang et al. 2015). Taken together, these data suggest that resveratrol not only inhibits steroidogenesis but also regulates the expression of CRH and glucocorticoid receptor in brain to improve the neuroendocrine system disorder induced by stress or inflammation.

7.3.4 MODULATION OF HYPOTHALAMIC–PITUITARY–THYROID AXIS

The association between thyroid function and psychiatric disorders particularly mood disorders has long been recognized. Disturbances in thyroid hormone homeostasis may impair both function and structure of the brain, resulting in neurobehavioral alterations in emotion and cognition (He et al. 2011, Zhu et al. 2006). Both excess and insufficient thyroid hormones can cause depression, generally reversible with adequate treatment (Hage and Azar 2012). Clinical studies have revealed that the treatment of levothyroxine improves mood and normalizes the elevated relative cerebral glucose metabolism in several brain regions of the depression patients (Bauer et al. 2005, Uhl et al. 2014). However, the correct dosage of levothyroxine remains elusive. Thus, there is still an urgent clinical need for development of new drugs to alleviate depression in patients with thyroid disorders.

With respect to the thyroid, few data are available on the effects of resveratrol on thyroid function in depression patients or the animal models. Recently, only some studies suggest that resveratrol can act as a thyroid regulator. Resveratrol has been reported to exert the apoptotic effect to arrest the cell growth of human papillary and follicular thyroid cancer via the mitogen-activated protein kinase and p53 signal transduction pathway (Shih et al. 2002). In an experimental study designed to assess the effects over three months of resveratrol on the hypothalamic–pituitary–thyroid (HPT) axis in ovariectomized rats, it was demonstrated that serum levels of 1.0 and

8.1 μM resveratrol lead to a significant increase in total serum triiodothyronine levels, while no thyroid morphological changes were observed (Bottner et al. 2006). In addition, in the rat thyroid cell line RFTL-5, resveratrol could increase iodide trapping and the expression of the natrium/iodide symporter after 6 to 12 hours of treatment. But this increase was transient, as the increase was no longer detectable after 24 hours (Sebai et al. 2010). However, Giuliani and colleagues (2014) proved that resveratrol decreased the iodide uptake and the sodium/iodide symporter protein expression in FRTL-5 cell line after 48 hours of the treatment, and the *in vivo* inhibitory effect of resveratrol on iodide uptake was also further confirmed in the rats (Giuliani et al. 2014). These results suggest that resveratrol can act as a dual-directional regulator on iodide uptake of thyroid cells at different time points. Moreover, Sarkar and colleagues demonstrated that resveratrol can exert the protective effects against fluoride-induced thyroidal dysfunction including reduction of nucleic acids, thyroid hormones, and metabolic enzyme activities like Na^+-K^+-ATPase, thyroidal peroxidase, and 5,5'-deiodinase (Sarkar and Pal 2014). In spite of the multiple beneficial effects of resveratrol in thyroid gland, there has been relatively little research on the effects of resveratrol on the HPT axis of depression patients or rodents.

Until now, only one study has been conducted by Ge and colleagues (2016) to assess the potential antidepressant-like effect of resveratrol in subclinical hypothyroidism (SCH) rats, which exhibit depressive-like behaviors, induced by hemithyroid electrocauterization. This study showed that resveratrol administration (15 mg/kg/day for 16 days) decreased the elevated plasma thyroid-stimulating hormone and the mRNA expression of hypothalamic thyrotropin-releasing hormone and simultaneously regulated both the HPA axis and the Wnt/β-catenin pathway to manifest the antidepressant effect in SCH rats (Ge et al. 2016). Therefore, research in other depression models is obviously required and will help to differentiate effects of resveratrol on various potential causes of depression.

7.3.5 REGULATION OF OXIDATIVE STRESS

The etiology of depression remains uncertain and complicated. Within the last decade, a growing number of literature in both humans and preclinical animal models proposed the oxidative stress hypothesis of depressive disorder (Michel et al. 2007, 2010). Several environmental and genetic factors such as psychosocial stress, BDNF, and superoxide dismutase (SOD) lead to increased induction of oxidative-stress-related membrane-lipid peroxidation and DNA damage. Enhanced oxidative stress was also observed in the chronic-stress-induced animal depression model. These animals expressed increased lipid peroxidation and decreased endogenous antioxidant defense such as SOD, glutathione (GSH) peroxidase, and catalase (CAT) in the cerebral cortex, striatum, and hippocampus (Che et al. 2015). These changes are responsible for structural and functional alterations of neurons in specific brain regions, especially in the prefrontal cortex and the hippocampus, known to play a role in the development of depression (Michel et al. 2012).

In order to simulate the oxidative-stress-induced cell injury induced by HPA activation, Ge and colleagues (2013) stimulated human neuroblastoma SK-N-BE(2)C cells with 10 μmoles/L corticosterone and treated with resveratrol (10^{-5}–10^{-10} mole/L) two hours later *in vitro*. They also injected 15 mg/kg resveratrol into the rats exposed to three-week CUMS to explore the effect of resveratrol on malonaldehyde (MDA) levels, an indicator for lipid peroxidation. The results of this study showed that resveratrol could remarkably reduce the elevation of MDA level in the supernatant of the corticosterone-stimulated cells. The chronic administration of resveratrol also significantly decreased the serum MDA level in CUMS rats, indicating the antioxidant effect of resveratrol (Ge et al. 2013). Liu and colleagues also demonstrated the fact that resveratrol treatment (80 mg/kg i.p. for four consecutive weeks) significantly restored the CUMS-induced lipid peroxidation and SOD level in the prefrontal cortex and the hippocampus of the rat brain (Liu et al. 2016). Furthermore, Sakr and colleagues examined the effects of resveratrol (20 mg/kg for 28 days, through gavage) and fluoxetine (10 mg/kg, through gavage) on the four-week CUMS-induced oxidative stress in the rat testes, responsible for steroidogenesis. Interestingly, these results showed that resveratrol significantly decreased MDA level and increased antioxidants (GSH, SOD, and CAT) in the testicular tissue of the rats exposed to CUMS, while fluoxetine worsened testicular oxidative stress and functions (Sakr et al. 2015). In summary, resveratrol could exert antidepressant properties by modulating antioxidant enzymes and antioxidant responsive elements of the brain and had fewer adverse effects on testicular function than fluoxetine did.

7.3.6 Upregulation of Mammalian Target of Rapamycin Pathway

In the last few years, a special attention has been given to the role of mammalian target of rapamycin (mTOR) signaling in major depression. Several studies have reported decreased brain mTOR activation in humans diagnosed with major depression disorder and in animal models of depression. A postmortem study showed robust deficits in mTOR signaling in the prefrontal cortex of patients who suffered from major depression (Jernigan et al. 2011). Rodents exposed to CUMS also exhibit depressive-like behaviors associated with a reduction in phosphorylation levels of mTOR and its downstream signaling components in the prefrontal cortex, hippocampus and amygdala (Chandran et al. 2013, Zhong et al. 2014, Zhu et al. 2013). Furthermore, some drugs such as ketamine, glutamatergic agents, muscarinic receptor antagonists, and 5-HT$_{2C}$ receptor antagonists were reported to exert antidepressant effects via the activation of mTOR (Ignacio et al. 2015). Therefore, the upregulation of mTOR pathway may be an effective strategy for management of major depression.

While it has been reported that resveratrol exerts antiproliferative effects and induces apoptosis in cancer cells via inhibiting mTOR pathway (Roy et al. 2009), to date, there is only one study aiming to explore the role of resveratrol in modulation of mTOR pathway in animal models of depression (Liu et al. 2016). Liu and colleagues found that four-week CUMS was able to decrease phosphorylation of mTOR and its upstream serine/threonine kinase Akt in hippocampus and prefrontal cortex, and these biochemical abnormalities were ameliorated by chronic resveratrol treatment (80 mg/kg/i.p. four weeks). These data suggested that resveratrol plays dual roles in modulation of mTOR pathway in the different disease models (activation in depression but inhibition in cancer).

7.4 EFFECTS OF RESVERATROL ON DEPRESSION-LIKE BEHAVIORAL PHENOTYPES

7.4.1 FST/TAIL SUSPENSION TEST

The FST, a despair-based test, was developed in 1977 by Posolt and colleagues to measure depressant-like behavior in rats and mice (Porsolt et al. 1977, 1978). The FST is one of the most commonly used behavioral paradigms for assessing anti-depressant-like activity of drugs. The test is based on the observation that animals develop a passive immobile behavior in an inescapable transparent cylinder filled with water. After antidepressant administration, the rodents will actively perform escape-directed behaviors, such as swimming and climbing, with longer duration than animals with vehicle treatment. In the FST, subacute antidepressant (including SSRIs) administration decreases immobility, with a corresponding increase in climbing or swimming behavior (Lucki 1997). It is well accepted that increased time of swimming behavior in rats relates to an elevation of brain 5-HT levels, whereas an increase in climbing behavior indicates an upregulated release of norepinephrine or dopamine (Bogdanova et al. 2013, Javelot et al. 2014, Vieira et al. 2008).

The tail suspension test (TST), which was first introduced in 1985 to measure the potential effectiveness of antidepressants (Steru et al. 1985), shares a common theoretical basis and behavioral measure with the FST. In this procedure, tails of rodents are suspended using adhesive tape to a horizontal bar for six minutes, and the time of immobility is recorded. Typically, the suspended rodents are immediately engaged in several agitation or escape-like behaviors, followed temporally by developing an immobile posture (Cryan et al. 2005). If antidepressants at doses are given prior to the test, the subjects will be actively engaged in escape-directed behaviors for longer periods of time than after vehicle treatment, exhibiting a decrease in duration of immobility.

Xu and colleagues (2010) used the FST and TST as screening platforms for assessment of the antidepressive effect of resveratrol in mice and found that a single dose of resveratrol (20–80 mg/kg) significantly decreased immobility time in both despair tests in a dose-dependent manner. Wang and colleagues (2013) further assessed the effect of the chronic administration of resveratrol (20–80 mg/kg for 21 consecutive days) on the changes in depression-related behavior of mice in the FST and TST and pointed that the chronic administration of resveratrol induced a significant decrease in immobility time, a complementary increase in swimming time, but no difference in climbing time of the FST (Wang et al. 2013). The chronic administration of resveratrol was also shown to decrease the immobility time in the TST (Wang et al. 2013). Moreover, resveratrol has robustly been demonstrated to ameliorate effectively the depressive-like behaviors induced by different models such as CUMS (Ge et al. 2013, Liu et al. 2014, 2016), post-stroke (Pang et al. 2015), repeated injection of corticosterone (Ali et al. 2015), chronic constriction injury of sciatic nerves (Zhao et al. 2014), and noninduced (WKY rats) (Hurley et al. 2014) and LPS-induced depression model (Ge et al. 2015). Interestingly, these results showed consistently that resveratrol increased significantly the swimming time but did not influence the climbing time, implying that catecholaminergic neurons may not be regulated by resveratrol.

This phenomenon could be explained by the fact that resveratrol inhibited the uptake of [^{3}H]5-HT more potently than [^{3}H]NA by synaptosomes from rat brain (IC$_{50}$ for inhibiting [^{3}H]5-HT uptake: 32.5 μM; IC$_{50}$ for inhibiting [^{3}H]5-NA uptake: 69.7 μM) (Yanez et al. 2006). In summary, resveratrol exerts antidepressant effects by mainly regulating serotonergic neurons to improve depressive-like behaviors.

7.4.2 SUCROSE PREFERENCE TEST

The sucrose preference test (SPT) is a reward-based test used as an indicator for anhedonia, a lack of interest in rewarding stimuli present in some forms of affective disorder including depression. In the SPT, rodents usually favor seeking out a sweet rewarding drink over plain drinking water. A bias toward the sweetened drink is typical; failure to do so is indicative of anhedonia or depression (Brenes Saenz et al. 2006). Resveratrol has also been reported to reverse successfully the reduction of sucrose preference or intake induced by various depression models such as CUMS, post-stroke, repeated injection of corticosterone, noninduced (WKY rats), and LPS-induced models (Ali et al. 2015, Ge et al. 2013, 2015, Hurley et al. 2014, Liu et al. 2014, 2016, Pang et al. 2015, Sakr et al. 2015, Yu et al. 2013). These findings, which are consistent with the results of despair-based FST or TST, suggest that resveratrol could exert antidepressant-like effects in the reward-based test.

7.4.3 OPEN FIELD TEST

The open field test is an anxiety-based test that provides simultaneous measures of locomotion, exploration, and anxiety of the animals. The behaviors scored included line crossing, center square entries/durations, rearing, grooming, and so on. The more frequent line crossing and rearing events usually indicate a better locomotor activity and a lower anxiety level of the animal. The higher number of central entries and the longer duration of time spent in central area represent higher exploratory behavior and lower anxiety levels (Walsh and Cummins 1976). While chronic mild stress has been shown to reduce rodents' grooming in the open field test, the observation of grooming may serve as a useful measure of stress and anxiety in laboratory animals (Gould 2009). Recently, it has been reported that resveratrol could ameliorate the reduction of crossing number and grooming number induced by CUMS (Liu et al. 2014, 2016). However, some studies showed that both acute and chronic administration of resveratrol had no effect on the locomotor activity in the normal mice, noninduced (WKY rats) or LPS-induced model of depression (Ge et al. 2015, Hurley et al. 2014, Wang et al. 2013). Therefore, further research is needed to confirm the antianxiety activity of resveratrol.

7.5 CONCLUSIONS

Depression is not a homogeneous disorder, but a complex phenomenon, which has many subtypes and probably more than one etiology. Thus, evaluating the safety and efficacy of drugs that exhibit remarkable multipotent ability to regulate serotonin transporters, MAO-A, CREB/BDNF/ERK signaling pathway, HPA axis, HPT axis, oxidative stress, and mTOR pathway may be important elements in the development of new therapeutic strategies to treat major depressive disorder. Despite the fact that many beneficial actions of resveratrol have been demonstrated to improve the depression-like behaviors of rodents (Figure 7.3), poor solubility and brain bioavailability of resveratrol are the major obstacles in translating preclinical findings into meaningful clinical trials. However, clinical trials are underway to assess the efficacy and safety of resveratrol in brain diseases such as dementia and Alzheimer's disease, and it has been demonstrated that resveratrol (0.5–2 g orally for 52 weeks) could penetrate the blood–brain barrier at nM levels to have central effects on modulating the levels of amyloid beta and be reported to have mild gastrointestinal side

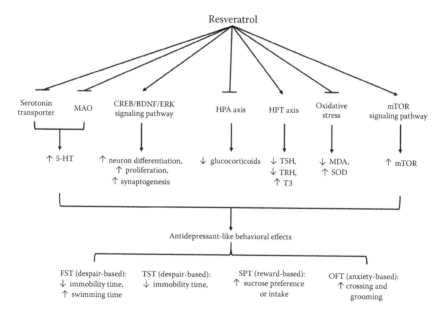

FIGURE 7.3 Various important actions of resveratrol that form the basis of its benefit in depression. Resveratrol inhibits serotonin transporters and MAO activity to increase the level of serotonin in synaptic cleft. Resveratrol activates both CREB/BDNF/ERK and mTOR signaling pathways to enhance the differentiation and proliferation of neurons. Resveratrol modulates the HPA and HPT axis of the neuroendocrine system responsible for regulating hormones and using hormone-based signals to change the body's responses to stress. Resveratrol ameliorates oxidative stress by decreasing the MDA level of lipid peroxidation and increasing antioxidant defense SOD. On the basis of these multiple beneficial actions, resveratrol exerts the antidepressant-like behavioral effects in despair-, reward- and anxiety-based tests. Abbreviations: MAO, monoamine oxidase; CREB, cAMP response element-binding protein; BDNF, brain-derived neurotrophic factor; ERK, extracellular signal-regulated kinase; HPA axis, hypothalamic–pituitary–adrenal axis; HPT axis, hypothalamic–pituitary–thyroid axis; mTOR, mammalian target of rapamycin; 5-HT, 5-hydroxytryptamine; TSH, thyroid-stimulating hormone; TRH, thyrotropin-releasing hormone; T3, triiodothyronine; MDA, malonaldehyde; SOD, superoxide dismutase; FST, forced swim test; TST, tail suspension test; SPT, sucrose preference test; OFT, open field test.

effects, including nausea and diarrhea (Turner et al. 2015). In addition, some short-term studies have also reported no apparent adverse effects but only mild to moderate gastrointestinal side effects at concentrations of about 2.5 to 5 g doses (Brown et al. 2010, la Porte et al. 2010). These results of clinical studies suggest the potential of resveratrol as promising bioactive agents with only minor adverse effects to improve brain health. To date, clinical studies of resveratrol in depression have yet to be conducted to prove its effectiveness of anti-depression in real life. Therefore, there is a need for future clinical studies to determine whether resveratrol may be beneficial in depression.

ACKNOWLEDGMENTS

This work was supported by Tri-Service General Hospital, Taipei, Taiwan (TSGH-C102-130, TSGH-C104-138, and TSGH-C105-137) and Ministry of Science and Technology, Taiwan (MOST 105-2314-B-016-027-MY2).

REFERENCES

Ali, S. H., R. M. Madhana, V. A. K, E. R. Kasala, L. N. Bodduluru, S. Pitta, J. R. Mahareddy, and M. Lahkar. 2015. "Resveratrol ameliorates depressive-like behavior in repeated corticosterone-induced depression in mice." *Steroids* 101:37–42. doi: 10.1016/j.steroids.2015.05.010.

Alves, E., T. Summavielle, C. J. Alves, J. Gomes-da-Silva, J. C. Barata, E. Fernandes, L. Bastos Mde, M. A. Tavares, and F. Carvalho. 2007. "Monoamine oxidase-B mediates ecstasy-induced neurotoxic effects to adolescent rat brain mitochondria." *J Neurosci* 27 (38):10203–10. doi: 10.1523/JNEUROSCI.2645-07.2007.

Bauer, M., E. D. London, N. Rasgon, S. M. Berman, M. A. Frye, L. L. Altshuler, M. A. Mandelkern, J. Bramen, B. Voytek, R. Woods, J. C. Mazziotta, and P. C. Whybrow. 2005. "Supraphysiological doses of levothyroxine alter regional cerebral metabolism and improve mood in bipolar depression." *Mol Psychiatry* 10 (5):456–69. doi: 10.1038/sj.mp.4001647.

Baur, J. A., and D. A. Sinclair. 2006. "Therapeutic potential of resveratrol: The in vivo evidence." *Nat Rev Drug Discov* 5 (6):493–506. doi: 10.1038/nrd2060.

Benmansour, S., T. Deltheil, J. Piotrowski, L. Nicolas, C. Reperant, A. M. Gardier, A. Frazer, and D. J. David. 2008. "Influence of brain-derived neurotrophic factor (BDNF) on serotonin neurotransmission in the hippocampus of adult rodents." *Eur J Pharmacol* 587 (1–3):90–8. doi: 10.1016/j.ejphar.2008.03.048.

Bertelli, A. A., L. Giovannini, D. Giannessi, M. Migliori, W. Bernini, M. Fregoni, and A. Bertelli. 1995. "Antiplatelet activity of synthetic and natural resveratrol in red wine." *Int J Tissue React* 17 (1):1–3.

Blanchet, J., F. Longpre, G. Bureau, M. Morissette, T. DiPaolo, G. Bronchti, and M. G. Martinoli. 2008. "Resveratrol, a red wine polyphenol, protects dopaminergic neurons in MPTP-treated mice." *Prog Neuropsychopharmacol Biol Psychiatry* 32 (5):1243–50. doi: 10.1016/j.pnpbp.2008.03.024.

Blier, P., and C. de Montigny. 1994. "Current advances and trends in the treatment of depression." *Trends Pharmacol Sci* 15 (7):220–6.

Blier, P., and M. El Mansari. 2013. "Serotonin and beyond: Therapeutics for major depression." *Philos Trans R Soc Lond B Biol Sci* 368 (1615):20120536. doi: 10.1098/rstb.2012.0536.

Bogdanova, O. V., S. Kanekar, K. E. D'Anci, and P. F. Renshaw. 2013. "Factors influencing behavior in the forced swim test." *Physiol Behav* 118:227–39. doi: 10.1016/j.physbeh.2013.05.012.

Bottner, M., J. Christoffel, G. Rimoldi, and W. Wuttke. 2006. "Effects of long-term treatment with resveratrol and subcutaneous and oral estradiol administration on the pituitary-thyroid-axis." *Exp Clin Endocrinol Diabetes* 114 (2):82–90. doi: 10.1055/s-2006-923888.

Brenes Saenz, J. C., O. R. Villagra, and J. Fornaguera Trias. 2006. "Factor analysis of forced swimming test, sucrose preference test and open field test on enriched, social and isolated reared rats." *Behav Brain Res* 169 (1):57–65. doi: 10.1016/j.bbr.2005.12.001.

Brigitta, B. 2002. "Pathophysiology of depression and mechanisms of treatment." *Dialogues Clin Neurosci* 4 (1):7–20.

Brown, V. A., K. R. Patel, M. Viskaduraki, J. A. Crowell, M. Perloff, T. D. Booth, G. Vasilinin, A. Sen, A. M. Schinas, G. Piccirilli, K. Brown, W. P. Steward, A. J. Gescher, and D. E. Brenner. 2010. "Repeat dose study of the cancer chemopreventive agent resveratrol in healthy volunteers: Safety, pharmacokinetics, and effect on the insulin-like growth factor axis." *Cancer Res* 70 (22):9003–11. doi: 10.1158/0008-5472.CAN-10-2364.

Burkon, A., and V. Somoza. 2008. "Quantification of free and protein-bound trans-resveratrol metabolites and identification of trans-resveratrol-C/O-conjugated diglucuronides—Two novel resveratrol metabolites in human plasma." *Mol Nutr Food Res* 52 (5):549–57. doi: 10.1002/mnfr.200700290.

Calabrese, F., R. Molteni, P. F. Maj, A. Cattaneo, M. Gennarelli, G. Racagni, and M. A. Riva. 2007. "Chronic duloxetine treatment induces specific changes in the expression of BDNF transcripts and in the subcellular localization of the neurotrophin protein." *Neuropsychopharmacology* 32 (11):2351–9. doi: 10.1038/sj.npp.1301360.

Castren, E., V. Voikar, and T. Rantamaki. 2007. "Role of neurotrophic factors in depression." *Curr Opin Pharmacol* 7 (1):18-21. doi: 10.1016/j.coph.2006.08.009.

Chandran, A., A. H. Iyo, C. S. Jernigan, B. Legutko, M. C. Austin, and B. Karolewicz. 2013. "Reduced phosphorylation of the mTOR signaling pathway components in the amygdala of rats exposed to chronic stress." *Prog Neuropsychopharmacol Biol Psychiatry* 40:240–5. doi: 10.1016/j.pnpbp.2012.08.001.

Che, Y., Z. Zhou, Y. Shu, C. Zhai, Y. Zhu, S. Gong, Y. Cui, and J. F. Wang. 2015. "Chronic unpredictable stress impairs endogenous antioxidant defense in rat brain." *Neurosci Lett* 584:208–13. doi: 10.1016/j.neulet.2014.10.031.

Cryan, J. F., C. Mombereau, and A. Vassout. 2005. "The tail suspension test as a model for assessing antidepressant activity: Review of pharmacological and genetic studies in mice." *Neurosci Biobehav Rev* 29 (4–5):571–625. doi: 10.1016/j.neubiorev.2005.03.009.

Csiszar, A., N. Labinskyy, A. Podlutsky, P. M. Kaminski, M. S. Wolin, C. Zhang, P. Mukhopadhyay, P. Pacher, F. Hu, R. de Cabo, P. Ballabh, and Z. Ungvari. 2008. "Vasoprotective effects of resveratrol and SIRT1: Attenuation of cigarette smoke-induced oxidative stress and proinflammatory phenotypic alterations." *Am J Physiol Heart Circ Physiol* 294 (6):H2721–35. doi: 10.1152/ajpheart.00235.2008.

Dinan, T. G. 1994. "Glucocorticoids and the genesis of depressive illness. A psychobiological model." *Br J Psychiatry* 164 (3):365–71.

Duman, R. S., and L. M. Monteggia. 2006. "A neurotrophic model for stress-related mood disorders." *Biol Psychiatry* 59 (12):1116–27. doi: 10.1016/j.biopsych.2006.02.013.

Feng, Y., X. P. Wang, S. G. Yang, Y. J. Wang, X. Zhang, X. T. Du, X. X. Sun, M. Zhao, L. Huang, and R. T. Liu. 2009. "Resveratrol inhibits beta-amyloid oligomeric cytotoxicity but does not prevent oligomer formation." *Neurotoxicology* 30 (6):986–95. doi: 10.1016/j.neuro.2009.08.013.

Ferretta, A., A. Gaballo, P. Tanzarella, C. Piccoli, N. Capitanio, B. Nico, T. Annese, M. Di Paola, C. Dell'aquila, M. De Mari, E. Ferranini, V. Bonifati, C. Pacelli, and T. Cocco. 2014. "Effect of resveratrol on mitochondrial function: Implications in parkin-associated familiar Parkinson's disease." *Biochim Biophys Acta* 1842 (7):902–15. doi: 10.1016/j.bbadis.2014.02.010.

Ge, J. F., L. Peng, J. Q. Cheng, C. X. Pan, J. Tang, F. H. Chen, and J. Li. 2013. "Antidepressant-like effect of resveratrol: Involvement of antioxidant effect and peripheral regulation on HPA axis." *Pharmacol Biochem Behav* 114–115:64–9. doi: 10.1016/j.pbb.2013.10.028.

Ge, J. F., Y. Y. Xu, G. Qin, J. Q. Cheng, and F. H. Chen. 2016. "Resveratrol ameliorates the anxiety- and depression-like behavior of subclinical hypothyroidism rat: Possible involvement of the HPT axis, HPA axis, and Wnt/beta-catenin pathway." *Front Endocrinol (Lausanne)* 7:44. doi: 10.3389/fendo.2016.00044.

Ge, L., L. Liu, H. Liu, S. Liu, H. Xue, X. Wang, L. Yuan, Z. Wang, and D. Liu. 2015. "Resveratrol abrogates lipopolysaccharide-induced depressive-like behavior, neuroinflammatory response, and CREB/BDNF signaling in mice." *Eur J Pharmacol* 768:49–57. doi: 10.1016/j.ejphar.2015.10.026.

Gertz, M., G. T. Nguyen, F. Fischer, B. Suenkel, C. Schlicker, B. Franzel, J. Tomaschewski, F. Aladini, C. Becker, D. Wolters, and C. Steegborn. 2012. "A molecular mechanism for direct sirtuin activation by resveratrol." *PLoS One* 7 (11):e49761. doi: 10.1371/journal.pone.0049761.

Giuliani, C., I. Bucci, S. Di Santo, C. Rossi, A. Grassadonia, M. Mariotti, M. Piantelli, F. Monaco, and G. Napolitano. 2014. "Resveratrol inhibits sodium/iodide symporter gene expression and function in rat thyroid cells." *PLoS One* 9 (9):e107936. doi: 10.1371/journal.pone.0107936.

Goldberg, D. M., J. Yan, and G. J. Soleas. 2003. "Absorption of three wine-related polyphenols in three different matrices by healthy subjects." *Clin Biochem* 36 (1):79–87.

Gould, T. D. 2009. *Mood and anxiety related phenotypes in mice: characterization using behavioral tests.* 2 vols, Neuromethods. New York: Humana Press.

Gweon, E. J., and S. J. Kim. 2013. "Resveratrol induces MMP-9 and cell migration via the p38 kinase and PI-3K pathways in HT1080 human fibrosarcoma cells." *Oncol Rep* 29 (2):826–34. doi: 10.3892/or.2012.2151.

Gweon, E. J., and S. J. Kim. 2014. "Resveratrol attenuates matrix metalloproteinase-9 and -2-regulated differentiation of HTB94 chondrosarcoma cells through the p38 kinase and JNK pathways." *Oncol Rep* 32 (1):71–8. doi: 10.3892/or.2014.3192.

Hage, M. P., and S. T. Azar. 2012. "The link between thyroid function and depression." *J Thyroid Res* 2012:590648. doi: 10.1155/2012/590648.

Hannestad, J., N. DellaGioia, and M. Bloch. 2011. "The effect of antidepressant medication treatment on serum levels of inflammatory cytokines: A meta-analysis." *Neuropsychopharmacology* 36 (12):2452–9. doi: 10.1038/npp.2011.132.

He, X. S., N. Ma, Z. L. Pan, Z. X. Wang, N. Li, X. C. Zhang, J. N. Zhou, D. F. Zhu, and D. R. Zhang. 2011. "Functional magnetic resource imaging assessment of altered brain function in hypothyroidism during working memory processing." *Eur J Endocrinol* 164 (6):951–9. doi: 10.1530/EJE-11-0046.

Huang, E. J., and L. F. Reichardt. 2001. "Neurotrophins: Roles in neuronal development and function." *Annu Rev Neurosci* 24:677–736. doi: 10.1146/annurev.neuro.24.1.677.

Huang, W., Z. Chen, Q. Wang, M. Lin, S. Wu, Q. Yan, F. Wu, X. Yu, X. Xie, G. Li, Y. Xu, and J. Pan. 2013. "Piperine potentiates the antidepressant-like effect of trans-resveratrol: Involvement of monoaminergic system." *Metab Brain Dis* 28 (4):585–95. doi: 10.1007/s11011-013-9426-y.

Hurley, L. L., L. Akinfiresoye, O. Kalejaiye, and Y. Tizabi. 2014. "Antidepressant effects of resveratrol in an animal model of depression." *Behav Brain Res* 268:1–7. doi: 10.1016/j.bbr.2014.03.052.

Ignacio, Z. M., G. Z. Reus, C. O. Arent, H. M. Abelaira, M. R. Pitcher, and J. Quevedo. 2015. "New perspectives on the involvement of mTOR in depression as well as in the action of antidepressant drugs." *Br J Clin Pharmacol.* doi: 10.1111/bcp.12845.

Jang, M., L. Cai, G. O. Udeani, K. V. Slowing, C. F. Thomas, C. W. Beecher, H. H. Fong, N. R. Farnsworth, A. D. Kinghorn, R. G. Mehta, R. C. Moon, and J. M. Pezzuto. 1997. "Cancer chemopreventive activity of resveratrol, a natural product derived from grapes." *Science* 275 (5297):218–20.

Javelot, H., M. Messaoudi, C. Jacquelin, J. F. Bisson, P. Rozan, A. Nejdi, C. Lazarus, J. C. Cassel, C. Strazielle, and R. Lalonde. 2014. "Behavioral and neurochemical effects of dietary methyl donor deficiency combined with unpredictable chronic mild stress in rats." *Behav Brain Res* 261:8–16. doi: 10.1016/j.bbr.2013.11.047.

Jernigan, C. S., D. B. Goswami, M. C. Austin, A. H. Iyo, A. Chandran, C. A. Stockmeier, and B. Karolewicz. 2011. "The mTOR signaling pathway in the prefrontal cortex is compromised in major depressive disorder." *Prog Neuropsychopharmacol Biol Psychiatry* 35 (7):1774–9. doi: 10.1016/j.pnpbp.2011.05.010.

Juan, M. E., M. Maijo, and J. M. Planas. 2010. "Quantification of trans-resveratrol and its metabolites in rat plasma and tissues by HPLC." *J Pharm Biomed Anal* 51 (2):391–8. doi: 10.1016/j.jpba.2009.03.026.

Juruena, M. F. 2014. "Early-life stress and HPA axis trigger recurrent adulthood depression." *Epilepsy Behav* 38:148–59. doi: 10.1016/j.yebeh.2013.10.020.

Juruena, M. F., A. J. Cleare, and C. M. Pariante. 2004. "[The hypothalamic pituitary adrenal axis, glucocorticoid receptor function and relevance to depression]." *Rev Bras Psiquiatr* 26 (3):189–201. doi:/S1516-44462004000300009.

Karege, F., G. Bondolfi, N. Gervasoni, M. Schwald, J. M. Aubry, and G. Bertschy. 2005. "Low brain-derived neurotrophic factor (BDNF) levels in serum of depressed patients probably results from lowered platelet BDNF release unrelated to platelet reactivity." *Biol Psychiatry* 57 (9):1068–72. doi: 10.1016/j.biopsych.2005.01.008.

Kasiotis, K. M., H. Pratsinis, D. Kletsas, and S. A. Haroutounian. 2013. "Resveratrol and related stilbenes: Their anti-aging and anti-angiogenic properties." *Food Chem Toxicol* 61:112–20. doi: 10.1016/j.fct.2013.03.038.

Kumar, P., S. S. Padi, P. S. Naidu, and A. Kumar. 2007. "Cyclooxygenase inhibition attenuates 3-nitropropionic acid-induced neurotoxicity in rats: Possible antioxidant mechanisms." *Fundam Clin Pharmacol* 21 (3):297–306. doi: 10.1111/j.1472-8206.2007.00485.x.

la Porte, C., N. Voduc, G. Zhang, I. Seguin, D. Tardiff, N. Singhal, and D. W. Cameron. 2010. "Steady-state pharmacokinetics and tolerability of trans-resveratrol 2000 mg twice daily with food, quercetin and alcohol (ethanol) in healthy human subjects." *Clin Pharmacokinet* 49 (7):449–54. doi: 10.2165/11531820-000000000-00000.

Lagouge, M., C. Argmann, Z. Gerhart-Hines, H. Meziane, C. Lerin, F. Daussin, N. Messadeq, J. Milne, P. Lambert, P. Elliott, B. Geny, M. Laakso, P. Puigserver, and J. Auwerx. 2006. "Resveratrol improves mitochondrial function and protects against metabolic disease by activating SIRT1 and PGC-1alpha." *Cell* 127 (6):1109–22. doi: 10.1016/j.cell.2006.11.013.

Langcake, P., and R. J. Pryce. 1977. "A new class of phytoalexins from grapevines." *Experientia* 33 (2):151–2.

Li, J., L. Feng, Y. Xing, Y. Wang, L. Du, C. Xu, J. Cao, Q. Wang, S. Fan, Q. Liu, and F. Fan. 2014. "Radioprotective and antioxidant effect of resveratrol in hippocampus by activating Sirt1." *Int J Mol Sci* 15 (4):5928–39. doi: 10.3390/ijms15045928.

Lin, C. J., T. H. Chen, L. Y. Yang, and C. M. Shih. 2014. "Resveratrol protects astrocytes against traumatic brain injury through inhibiting apoptotic and autophagic cell death." *Cell Death Dis* 5:e1147. doi: 10.1038/cddis.2014.123.

Lippi, G., M. Franchini, E. J. Favaloro, and G. Targher. 2010. "Moderate red wine consumption and cardiovascular disease risk: Beyond the 'French paradox'." *Semin Thromb Hemost* 36 (1):59–70. doi: 10.1055/s-0030-1248725.

Liu, D., K. Xie, X. Yang, J. Gu, L. Ge, X. Wang, and Z. Wang. 2014. "Resveratrol reverses the effects of chronic unpredictable mild stress on behavior, serum corticosterone levels and BDNF expression in rats." *Behav Brain Res* 264:9–16. doi: 10.1016/j.bbr.2014.01.039.

Liu, S., T. Li, H. Liu, X. Wang, S. Bo, Y. Xie, X. Bai, L. Wu, Z. Wang, and D. Liu. 2016. "Resveratrol exerts antidepressant properties in the chronic unpredictable mild stress model through the regulation of oxidative stress and mTOR pathway in the rat hippocampus and prefrontal cortex." *Behav Brain Res* 302:191–9. doi: 10.1016/j.bbr.2016.01.037.

Lopez, M. S., R. J. Dempsey, and R. Vemuganti. 2015. "Resveratrol neuroprotection in stroke and traumatic CNS injury." *Neurochem Int* 89:75–82. doi: 10.1016/j.neuint.2015.08.009.

Lou, B. S., P. S. Wu, C. W. Hou, F. Y. Cheng, and J. K. Chen. 2014. "Simultaneous quantification of trans-resveratrol and its sulfate and glucuronide metabolites in rat tissues by stable isotope-dilution UPLC-MS/MS analysis." *J Pharm Biomed Anal* 94:99–105. doi: 10.1016/j.jpba.2014.01.039.

Lucki, I. 1997. "The forced swimming test as a model for core and component behavioral effects of antidepressant drugs." *Behav Pharmacol* 8 (6–7):523–32.

Marier, J. F., P. Vachon, A. Gritsas, J. Zhang, J. P. Moreau, and M. P. Ducharme. 2002. "Metabolism and disposition of resveratrol in rats: Extent of absorption, glucuronidation, and enterohepatic recirculation evidenced by a linked-rat model." *J Pharmacol Exp Ther* 302 (1):369–73.

Mathers, C., D. M. Fat, J. T. Boerma, and World Health Organization. 2008. *The global burden of disease: 2004 update*. Geneva, Switzerland: World Health Organization.

Mebratu, Y., and Y. Tesfaigzi. 2009. "How ERK1/2 activation controls cell proliferation and cell death: Is subcellular localization the answer?" *Cell Cycle* 8 (8):1168–75. doi: 10.4161/cc.8.8.8147.

Meng, X., P. Maliakal, H. Lu, M. J. Lee, and C. S. Yang. 2004. "Urinary and plasma levels of resveratrol and quercetin in humans, mice, and rats after ingestion of pure compounds and grape juice." *J Agric Food Chem* 52 (4):935–42. doi: 10.1021/jf030582e.

Michel, T. M., S. Camara, T. Tatschner, S. Frangou, A. J. Sheldrick, P. Riederer, and E. Grunblatt. 2010. "Increased xanthine oxidase in the thalamus and putamen in depression." *World J Biol Psychiatry* 11 (2 Pt 2):314–20. doi: 10.3109/15622970802123695.

Michel, T. M., S. Frangou, D. Thiemeyer, S. Camara, J. Jecel, K. Nara, A. Brunklaus, R. Zoechling, and P. Riederer. 2007. "Evidence for oxidative stress in the frontal cortex in patients with recurrent depressive disorder—A postmortem study." *Psychiatry Res* 151 (1–2):145–50. doi: 10.1016/j.psychres.2006.04.013.

Michel, T. M., D. Pulschen, and J. Thome. 2012. "The role of oxidative stress in depressive disorders." *Curr Pharm Des* 18 (36):5890–9.

Mohamed, H. E., S. E. El-Swefy, R. A. Hasan, and A. A. Hasan. 2014. "Neuroprotective effect of resveratrol in diabetic cerebral ischemic-reperfused rats through regulation of inflammatory and apoptotic events." *Diabetol Metab Syndr* 6 (1):88. doi: 10.1186/1758-5996-6-88.

Nagatsu, T. 2004. "Progress in monoamine oxidase (MAO) research in relation to genetic engineering." *Neurotoxicology* 25 (1–2):11–20. doi: 10.1016/S0161-813X(03)00085-8.

Owens, M. J., and C. B. Nemeroff. 1994. "Role of serotonin in the pathophysiology of depression: Focus on the serotonin transporter." *Clin Chem* 40 (2):288–95.

Pang, C., L. Cao, F. Wu, L. Wang, G. Wang, Y. Yu, M. Zhang, L. Chen, W. Wang, W. Lv, L. Chen, J. Zhu, J. Pan, H. Zhang, Y. Xu, and L. Ding. 2015. "The effect of trans-resveratrol on post-stroke depression via regulation of hypothalamus–pituitary–adrenal axis." *Neuropharmacology* 97:447–56. doi: 10.1016/j.neuropharm.2015.04.017.

Pariante, C. M., and S. L. Lightman. 2008. "The HPA axis in major depression: Classical theories and new developments." *Trends Neurosci* 31 (9):464–8. doi: 10.1016/j.tins.2008.06.006.

Porsolt, R. D., G. Anton, N. Blavet, and M. Jalfre. 1978. "Behavioural despair in rats: A new model sensitive to antidepressant treatments." *Eur J Pharmacol* 47 (4):379–91.

Porsolt, R. D., A. Bertin, and M. Jalfre. 1977. "Behavioral despair in mice: A primary screening test for antidepressants." *Arch Int Pharmacodyn Ther* 229 (2):327–36.

Roy, P., N. Kalra, S. Prasad, J. George, and Y. Shukla. 2009. "Chemopreventive potential of resveratrol in mouse skin tumors through regulation of mitochondrial and PI3K/AKT signaling pathways." *Pharm Res* 26 (1):211–7. doi: 10.1007/s11095-008-9723-z.

Rubinow, D. R., R. M. Post, R. Savard, and P. W. Gold. 1984. "Cortisol hypersecretion and cognitive impairment in depression." *Arch Gen Psychiatry* 41 (3):279–83.

Rush, A. J., M. E. Thase, and S. Dube. 2003. "Research issues in the study of difficult-to-treat depression." *Biol Psychiatry* 53 (8):743–53.

Rush, A. J., M. H. Trivedi, S. R. Wisniewski, A. A. Nierenberg, J. W. Stewart, D. Warden, G. Niederehe, M. E. Thase, P. W. Lavori, B. D. Lebowitz, P. J. McGrath, J. F. Rosenbaum, H. A. Sackeim, D. J. Kupfer, J. Luther, and M. Fava. 2006. "Acute and longer-term outcomes in depressed outpatients requiring one or several treatment steps: A STAR*D report." *Am J Psychiatry* 163 (11):1905–17. doi: 10.1176/ajp.2006.163.11.1905.

Sakr, H. F., A. M. Abbas, A. Z. Elsamanoudy, and F. M. Ghoneim. 2015. "Effect of fluoxetine and resveratrol on testicular functions and oxidative stress in a rat model of chronic mild stress-induced depression." *J Physiol Pharmacol* 66 (4):515–27.

Sarkar, C., and S. Pal. 2014. "Ameliorative effect of resveratrol against fluoride-induced alteration of thyroid function in male Wistar rats." *Biol Trace Elem Res* 162 (1–3):278–87. doi: 10.1007/s12011-014-0108-3.

Sarris, J., M. Fava, I. Schweitzer, and D. Mischoulon. 2012. "St John's wort (*Hypericum perforatum*) versus sertraline and placebo in major depressive disorder: Continuation data from a 26-week RCT." *Pharmacopsychiatry* 45 (7):275–8. doi: 10.1055/s-0032-1306348.

Sebai, H., S. Hovsepian, E. Ristorcelli, E. Aouani, D. Lombardo, and G. Fayet. 2010. "Resveratrol increases iodide trapping in the rat thyroid cell line FRTL-5." *Thyroid* 20 (2):195–203. doi: 10.1089/thy.2009.0171.

Shih, A., F. B. Davis, H. Y. Lin, and P. J. Davis. 2002. "Resveratrol induces apoptosis in thyroid cancer cell lines via a MAPK- and p53-dependent mechanism." *J Clin Endocrinol Metab* 87 (3):1223–32. doi: 10.1210/jcem.87.3.8345.

Shih, J. H., K. H. Ma, C. F. Chen, C. Y. Cheng, L. H. Pao, S. J. Weng, Y. S. Huang, C. Y. Shiue, M. K. Yeh, and I. H. Li. 2016. "Evaluation of brain SERT occupancy by resveratrol against MDMA-induced neurobiological and behavioral changes in rats: A 4-[(18) F]-ADAM/small-animal PET study." *Eur Neuropsychopharmacol* 26 (1):92–104. doi: 10.1016/j.euroneuro.2015.11.001.

Simao, F., A. Matte, A. S. Pagnussat, C. A. Netto, and C. G. Salbego. 2012. "Resveratrol preconditioning modulates inflammatory response in the rat hippocampus following global cerebral ischemia." *Neurochem Int* 61 (5):659–65. doi: 10.1016/j.neuint.2012.06.009.

Soleas, G. J., J. Yan, and D. M. Goldberg. 2001. "Measurement of trans-resveratrol, (+)-catechin, and quercetin in rat and human blood and urine by gas chromatography with mass selective detection." *Methods Enzymol* 335:130–45.

Steru, L., R. Chermat, B. Thierry, and P. Simon. 1985. "The tail suspension test: A new method for screening antidepressants in mice." *Psychopharmacology (Berl)* 85 (3):367–70.

Supornsilchai, V., K. Svechnikov, D. Seidlova-Wuttke, W. Wuttke, and O. Soder. 2005. "Phytoestrogen resveratrol suppresses steroidogenesis by rat adrenocortical cells by inhibiting cytochrome P450 c21-hydroxylase." *Horm Res* 64 (6):280–6. doi: 10.1159/000089487.

Szkudelski, T., and K. Szkudelska. 2015. "Resveratrol and diabetes: From animal to human studies." *Biochim Biophys Acta* 1852 (6):1145–54. doi: 10.1016/j.bbadis.2014.10.013.

Turner, R. S., R. G. Thomas, S. Craft, C. H. van Dyck, J. Mintzer, B. A. Reynolds, J. B. Brewer, R. A. Rissman, R. Raman, P. S. Aisen, and Study Alzheimer's Disease Cooperative. 2015. "A randomized, double-blind, placebo-controlled trial of resveratrol for Alzheimer disease." *Neurology* 85 (16):1383–91. doi: 10.1212/WNL.0000000000002035.

Uhl, I., J. A. Bez, T. Stamm, M. Pilhatsch, H. J. Assion, C. Norra, U. Lewitzka, F. Schlagenhauf, M. Bauer, and G. Juckel. 2014. "Influence of levothyroxine in augmentation therapy for bipolar depression on central serotonergic function." *Pharmacopsychiatry* 47 (4–5):180–3. doi: 10.1055/s-0034-1383654.

Vieira, C., T. C. De Lima, P. Carobrez Ade, and C. Lino-de-Oliveira. 2008. "Frequency of climbing behavior as a predictor of altered motor activity in rat forced swimming test." *Neurosci Lett* 445 (2):170–3. doi: 10.1016/j.neulet.2008.09.001.

Vreeburg, S. A., W. J. Hoogendijk, J. van Pelt, R. H. Derijk, J. C. Verhagen, R. van Dyck, J. H. Smit, F. G. Zitman, and B. W. Penninx. 2009. "Major depressive disorder and hypothalamic–pituitary–adrenal axis activity: Results from a large cohort study." *Arch Gen Psychiatry* 66 (6):617–26. doi: 10.1001/archgenpsychiatry.2009.50.

Walle, T. 2011. "Bioavailability of resveratrol." *Ann N Y Acad Sci* 1215:9–15. doi: 10.1111/j.1749-6632.2010.05842.x.

Walle, T., F. Hsieh, M. H. DeLegge, J. E. Oatis, Jr., and U. K. Walle. 2004. "High absorption but very low bioavailability of oral resveratrol in humans." *Drug Metab Dispos* 32 (12):1377–82. doi: 10.1124/dmd.104.000885.

Walsh, R. N., and R. A. Cummins. 1976. "The Open-Field Test: A critical review." *Psychol Bull* 83 (3):482–504.

Wan, D., Y. Zhou, K. Wang, Y. Hou, R. Hou, and X. Ye. 2016. "Resveratrol provides neuroprotection by inhibiting phosphodiesterases and regulating the cAMP/AMPK/SIRT1 pathway after stroke in rats." *Brain Res Bull* 121:255–62. doi: 10.1016/j.brainresbull.2016.02.011.

Wang, Z., J. Gu, X. Wang, K. Xie, Q. Luan, N. Wan, Q. Zhang, H. Jiang, and D. Liu. 2013. "Antidepressant-like activity of resveratrol treatment in the forced swim test and tail suspension test in mice: The HPA axis, BDNF expression and phosphorylation of ERK." *Pharmacol Biochem Behav* 112:104–10. doi: 10.1016/j.pbb.2013.10.007.

Wenzel, E., and V. Somoza. 2005. "Metabolism and bioavailability of trans-resveratrol." *Mol Nutr Food Res* 49 (5):472–81. doi: 10.1002/mnfr.200500010.

Wolkowitz, O. M., V. I. Reus, H. Weingartner, K. Thompson, A. Breier, A. Doran, D. Rubinow, and D. Pickar. 1990. "Cognitive effects of corticosteroids." *Am J Psychiatry* 147 (10):1297–303. doi: 10.1176/ajp.147.10.1297.

Wu, C. C., C. I. Wu, W. Y. Wang, and Y. C. Wu. 2007. "Low concentrations of resveratrol potentiate the antiplatelet effect of prostaglandins." *Planta Med* 73 (5):439–43. doi: 10.1055/s-2007-967173.

Xia, N., A. Daiber, A. Habermeier, E. I. Closs, T. Thum, G. Spanier, Q. Lu, M. Oelze, M. Torzewski, K. J. Lackner, T. Munzel, U. Forstermann, and H. Li. 2010. "Resveratrol reverses endothelial nitric-oxide synthase uncoupling in apolipoprotein E knockout mice." *J Pharmacol Exp Ther* 335 (1):149–54. doi: 10.1124/jpet.110.168724.

Xu, Y., Z. Wang, W. You, X. Zhang, S. Li, P. A. Barish, M. M. Vernon, X. Du, G. Li, J. Pan, and W. O. Ogle. 2010. "Antidepressant-like effect of trans-resveratrol: Involvement of serotonin and noradrenaline system." *Eur Neuropsychopharmacol* 20 (6):405–13. doi: 10.1016/j.euroneuro.2010.02.013.

Yanez, M., N. Fraiz, E. Cano, and F. Orallo. 2006. "Inhibitory effects of cis- and trans-resveratrol on noradrenaline and 5-hydroxytryptamine uptake and on monoamine oxidase activity." *Biochem Biophys Res Commun* 344 (2):688–95. doi: 10.1016/j.bbrc.2006.03.190.

Yu, C., Y. G. Shin, A. Chow, Y. Li, J. W. Kosmeder, Y. S. Lee, W. H. Hirschelman, J. M. Pezzuto, R. G. Mehta, and R. B. van Breemen. 2002. "Human, rat, and mouse metabolism of resveratrol." *Pharm Res* 19 (12):1907–14.

Yu, Y., R. Wang, C. Chen, X. Du, L. Ruan, J. Sun, J. Li, L. Zhang, J. M. O'Donnell, J. Pan, and Y. Xu. 2013. "Antidepressant-like effect of trans-resveratrol in chronic stress model: Behavioral and neurochemical evidences." *J Psychiatr Res* 47 (3):315–22. doi: 10.1016/j.jpsychires.2012.10.018.

Zhang, F., J. Liu, and J. S. Shi. 2010. "Anti-inflammatory activities of resveratrol in the brain: Role of resveratrol in microglial activation." *Eur J Pharmacol* 636 (1–3):1–7. doi: 10.1016/j.ejphar.2010.03.043.

Zhang, H., C. Li, S. T. Kwok, Q. W. Zhang, and S. W. Chan. 2013. "A review of the pharmacological effects of the dried root of *Polygonum cuspidatum* (Hu Zhang) and its constituents." *Evid Based Complement Alternat Med* 2013:208349. doi: 10.1155/2013/208349.

Zhao, X., C. Yu, C. Wang, J. F. Zhang, W. H. Zhou, W. G. Cui, F. Ye, and Y. Xu. 2014. "Chronic resveratrol treatment exerts antihyperalgesic effect and corrects co-morbid depressive like behaviors in mice with mononeuropathy: Involvement of serotonergic system." *Neuropharmacology* 85:131–41. doi: 10.1016/j.neuropharm.2014.04.021.

Zhong, P., W. Wang, B. Pan, X. Liu, Z. Zhang, J. Z. Long, H. T. Zhang, B. F. Cravatt, and Q. S. Liu. 2014. "Monoacylglycerol lipase inhibition blocks chronic stress-induced depressive-like behaviors via activation of mTOR signaling." *Neuropsychopharmacology* 39 (7):1763–76. doi: 10.1038/npp.2014.24.

Zhu, D. F., Z. X. Wang, D. R. Zhang, Z. L. Pan, S. He, X. P. Hu, X. C. Chen, and J. N. Zhou. 2006. "fMRI revealed neural substrate for reversible working memory dysfunction in subclinical hypothyroidism." *Brain* 129 (Pt 11):2923–30. doi: 10.1093/brain/awl215.

Zhu, W. L., S. J. Wang, M. M. Liu, H. S. Shi, R. X. Zhang, J. F. Liu, Z. B. Ding, and L. Lu. 2013. "Glycine site N-methyl-D-aspartate receptor antagonist 7-CTKA produces rapid antidepressant-like effects in male rats." *J Psychiatry Neurosci* 38 (5):306–16. doi: 10.1503/jpn.120228.

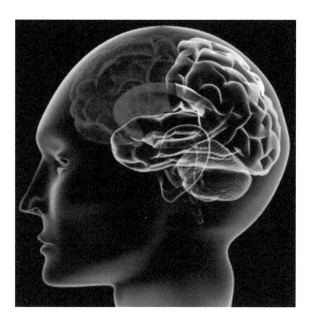

FIGURE 3.2 Human brain. (From http://www.climatechanges.site/mental-health-a-factor -in-climate-change/.)

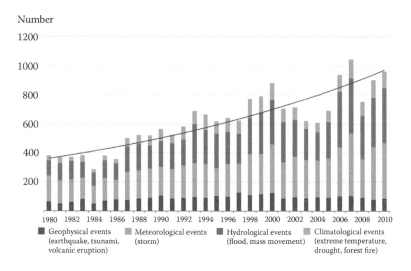

FIGURE 3.12 Events showing natural catastrophes worldwide, 1980–2010. (From Munich, R., *Natural Catastrophes Worldwide 1980–2011. Number of Events with Trend*, Münchener Rückversicherungs-Gesellschaft, Geo Risks Research, NatCatSERVICE, 2012.)

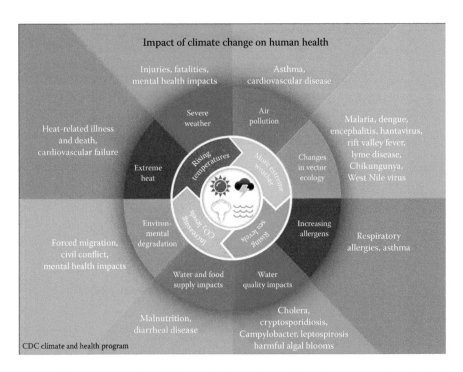

FIGURE 3.13 Potential health effects of climate change. (From http://www.cdc.gov/climate andhealth/effects/.)

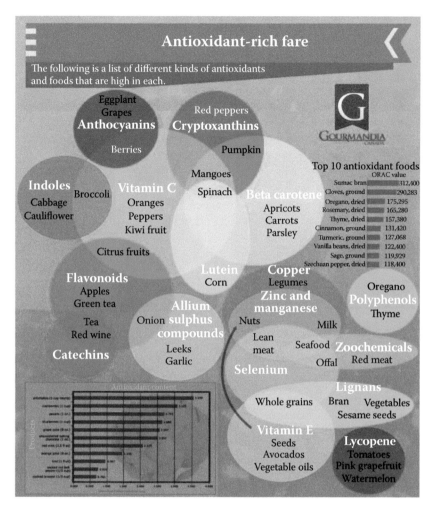

FIGURE 3.17 List of some important fruits, vegetables and sea food with antioxidants. (From http://visual.ly/antioxidant-rich-fare.)

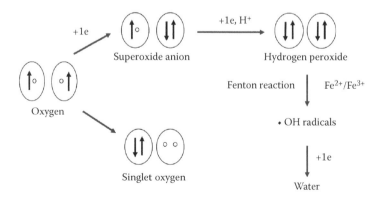

FIGURE 4.1 Free radical formation from molecular oxygen. Ground-state oxygen is a biradical. It can undergo single electron reduction to form superoxide, which can be further reduced by a single electron to hydrogen peroxide. Hydrogen peroxide undergoes the classical Fenton reaction to yield hydroxyl radicals. Hydroxyl ratdical abstracts an electron to form water. A single electron can be promoted to the neighboring orbital to produce singlet oxygen.

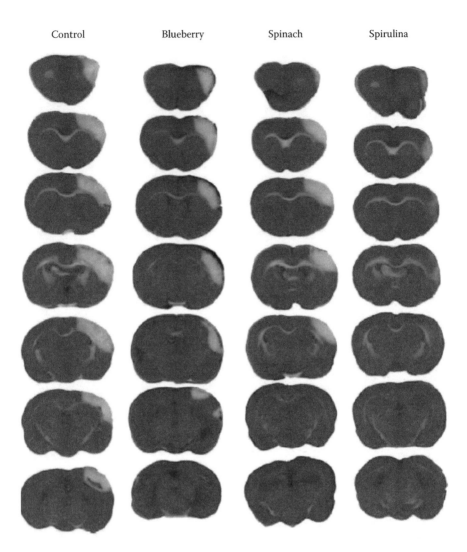

FIGURE 5.3 Diets supplemented with blueberry, spinach, and spirulina had an impact in both behavioral and histopathological endpoints in mice where stroke was induced by surgical ligation of the middle cerebral artery, with an observable reduction of the infarction volume. Infarction areas can be seen in the coronal brain sections as light/white colored areas in the outer aspect of the brain sections. Brains from animals treated with spirulina have fewer and smaller infarction areas when compared to control animals. (Reproduced from Wang, Y., Chang, C.F., Chou, J., Chen, H.L., Deng, X., Harvey, B.K., Cadet, J.L., Bickford, P.C., *Exp. Neurol.*, 193, 75–84, 2005.)

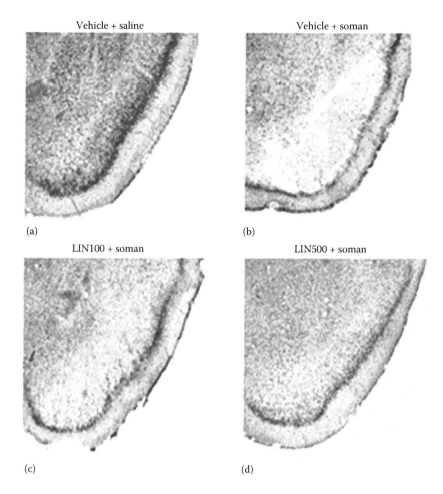

FIGURE 5.4 Single intravascular treatment with LIN at 500 nmol/kg three days prior to soman injection has a marked effect in ameliorating the tissue friability, edema, and extensive cell death and disruption in the pyriform cortex. Animals were perfused after 24 hours after soman exposure and the brains were processed and stained with cresyl violet. A section of the pyriform cortex is presented in the images. A reduction of neurons compared to control (a) was observed with soman treatment (b). Treatment with LIN at 500 nmol/kg (d) showed a marked reduction in neuronal cell loss when compared with the non-LIN treated soman exposed animals (b) and treatment with LIN at 100 nmol/kg (c). (Reproduced from Pan, H., Hu, X.Z., Jacobowitz, D.M., Chen, C., McDonough, J., Van Shura, K., Lyman, M., Marini, A.M., *Neurotoxicology*, 33, 1219–1229, 2012.)

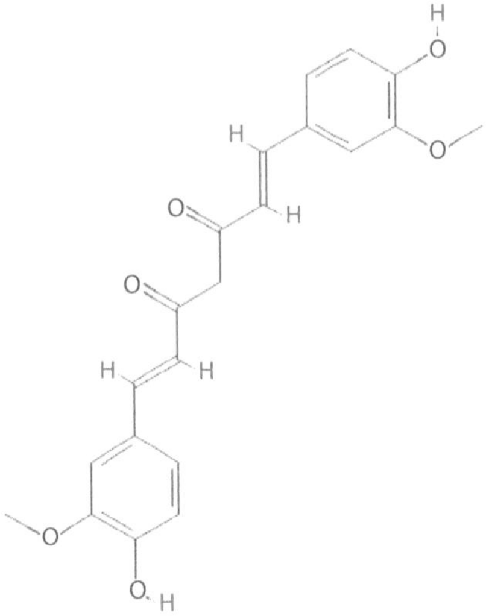

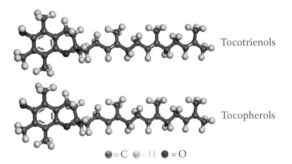

FIGURE 8.1 Curcumin. Chemical names: curcumin; diferuloylmethane; natural yellow 3; 458-37-7; turmeric yellow; turmeric. Molecular formula: C21H20O6. Molecular weight: 368.3799 g/mol.

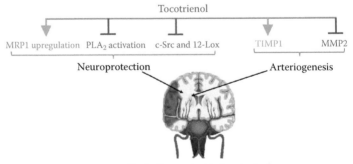

FIGURE 9.1 Chemical structure of TCT and TCP. TCTs and TCPs consist of a chromanol head attached to a 16-carbon side chain. TCPs have a saturated side chain and TCTs have an unsaturated side chain containing three *trans* double bonds.

Tocotrienol

| MRP1 upregulation | PLA$_2$ activation | c-Src and 12-Lox | TIMP1 | MMP2 |

Neuroprotection · Arteriogenesis

Protection against ischemic stroke

FIGURE 9.3 Specific molecular targets of TCT for protection against stroke. TCT confers protection against stroke through two main pathways: neuroprotection and arteriogenesis. Molecular targets of TCT for neuroprotection include upregulating MRP1 synthesis and inhibiting PLA$_2$ activation, c-Src, and 12-LOX. Arteriogenesis targets are upregulating TIMP1 expression and inhibiting MMP2 activity.

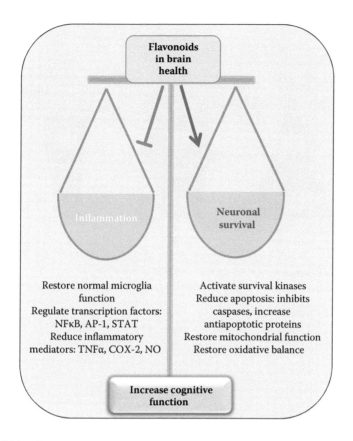

FIGURE 10.2 Schematic model of the role of flavonoids in brain function.

FIGURE 12.1 Aerial parts of *Bacopa monnieri*.

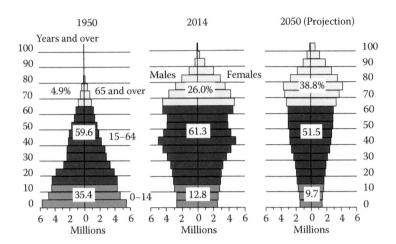

FIGURE 15.1 Changes in the population pyramid in Japan. (From Statistics Bureau, Japan. *Statistical Handbook of Japan 2016*. http://www.stat.go.jp/english/data/handbook/index.htm.)

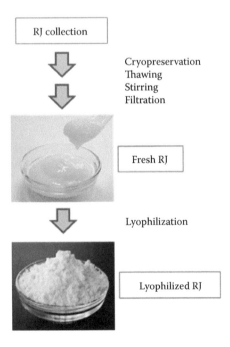

FIGURE 15.2 Processing of RJ for commercial applications.

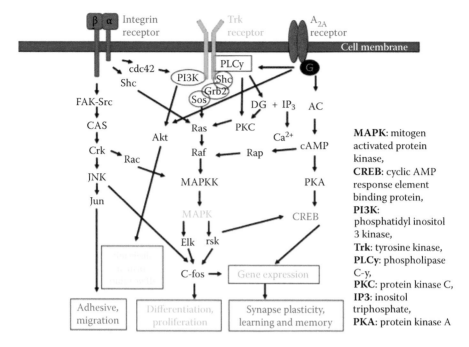

FIGURE 15.6 Intracellular signal transduction pathway.

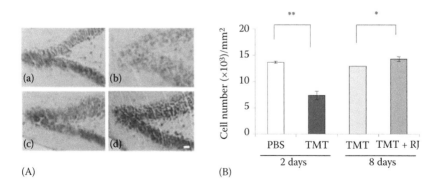

FIGURE 15.8 The ameliorative effect of RJ on the TMT-induced neurodegeneration in the hippocampal dentate gyrus of mice. (A) The Nissl-stained hippocampal dentate gyrus at two days after treatment with PBS (a) or TMT (b) and at eight days after treatment with TMT (c) or TMT + six-day feeding of 5% RJ diet (d). Scale bar: 25 μm. (B) The number of granule cells in the hippocampal dentate gyrus at two days after treatment with PBS or TMT and at eight days after treatment with TMT or TMT + six-day feeding of 5% RJ diet. The values represent mean ± SE (n = 4). A statistical difference between the two values was determined by Student's t-test to be significant at $**p < .001$ and $*p < .05$.

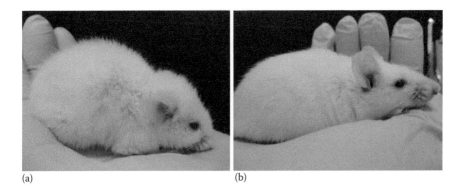

FIGURE 18.1 (a) SAMP8 mouse, 42-week-old male. (b) SAMR1 mouse, 42-week-old male.

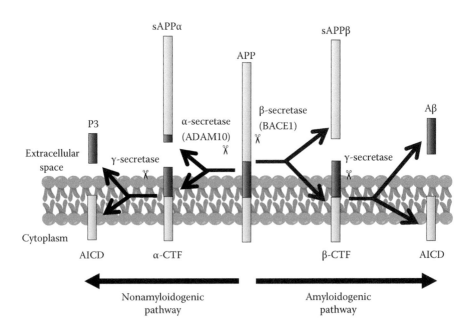

FIGURE 18.2 The cleavage of amyloid beta-protein precursor (APP) by the secretase enzymes. Aβ, beta-amyloid; AICD, APP intracellular domain; ADAM10, a disintegrin and metalloprotease domain; BACE1, Beta site APP cleaving enxyme I; CTF, C-terminal fragment; sAPP, soluble APP.

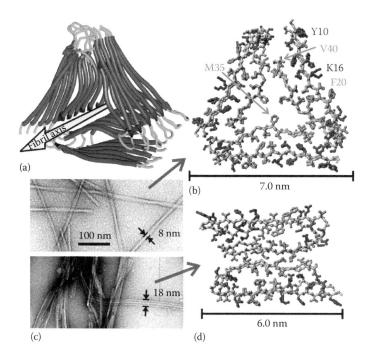

FIGURE 19.1 Amyloid fibrils formed by amyloid β peptide.[18,19] (a) Ribbon representation of fibrils with twisted morphology. (b) Atomic representation of fibrils with twisted morphology viewed down the fibril axis. Hydrophobic, polar negatively charged, and positively charged amino acids are green, magenta, red, and blue, respectively. Unstructured residues 1–8 omitted. (c) Comparison of twisted (upper) and striated ribbon (lower) fibril morphologies by TEM. (d) Atomic representation of fibrils with striated ribbon morphology viewed down the fibril axis.

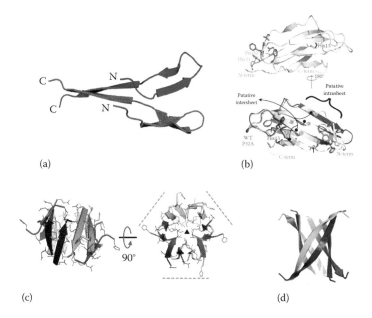

FIGURE 19.2 Structures of amyloid oligomers.[20] (a) Globulomer, $A\beta_{42}$[34]; (b) hexamer, β2-microglobulin[35]; (c) hexamer, PrP fragment[36]; (d) hexamer, αB-crystallin fragment.[24]

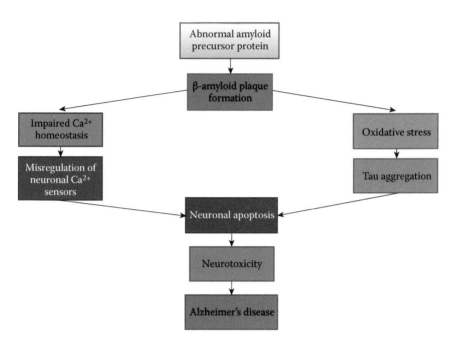

FIGURE 21.1 Pathology of Alzheimer's disease.

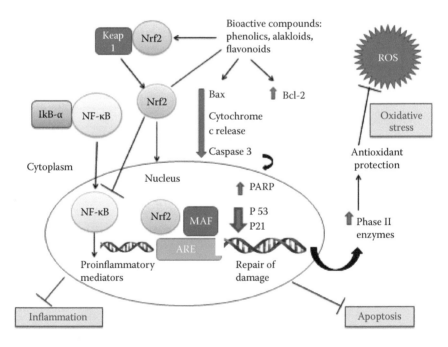

FIGURE 21.3 Summary of the mechanisms underlying neuroprotective effects of bioactive compounds. (Adapted from Giacoppo, S., Galuppo, M., Montaut, S., Iori, R., Rollin, P., Bramanti, P., Mazzon, E., *Fitoterapia*, 106, 12–21, 2015.)

8 Current Perspectives on the Beneficial Role of *Curcuma longa* and *Centella asiatica* Linn. in Neurological Disorders

Shahnaz Subhan, Manashi Bagchi, and Abdul Ilah

CONTENTS

8.1 INTRODUCTION

Neurodegenerative disorders represent clusters of serious diseases that results in progressive deterioration of the normal structure and physiology of the central nervous system. The pathophysiology of Alzheimer's, Parkinson's, or other neurodegenerative disorders involves multifaceted permutation of genetic and environmental factors. Combinations of lifestyle modification linked with environmental factors, jointly or alone, represent the largest share of cases of these disorders [1]. Neurodegenerative diseases are more common and have a disproportionate impact on countries with longer life expectancies and represent the fourth highest source of overall disease burden in high-income countries. Oxidative stress can lead to neuronal death via several mechanisms such as mitochondrial dysfunction, DNA damage, membrane permeability loss, protein aggregation, and apoptosis [2]. Keeping in view the complex etiology and pathophysiology of these neurodegenerative disorders, along with a miniscule of available treatment options associated with them, the role of natural agents and herbal extracts as therapeutic alternatives, alone or in combination with synthetic drugs, could not be ruled out [3]. In the same context, the present chapter has been aimed to investigate the role of selected natural plants like *Curcuma longa* and *Centella asiatica* Linn. in various neurodegenerative disorders and to provide an updated overview on the mechanism of action in various pathophysiologies and recent progress in clinical biology. The rationale for the selection of these plants was based on their strong anti-inflammatory and anti-oxidant potential and the large body of evidence that suggests their efficacy in preclinical as well as in clinical studies.

8.2 CURCUMA LONGA

The concept of using phytochemicals has ushered in a new revolution in pharmaceuticals. Naturally occurring polyphenols (like curcumin, morin, resveratrol, etc.) have gained importance because of their minimal side effects, low cost and abundance. Curcumin (diferuloylmethane) is a component of turmeric isolated from the rhizome of *Curcuma longa*, with a variety of pharmacologic properties. Research for more than two decades has revealed the pleiotropic nature of the biological effects of this molecule [4,5]. More than 7000 published articles have shed light on the various aspects of curcumin including its antioxidant, hypoglycemic, anti-inflammatory and anticancer activities. Apart from these well-known activities, this natural polyphenolic compound also exerts its beneficial effects by modulating different signaling molecules, including transcription factors, chemokines, cytokines, tumor suppressor genes, adhesion molecules, microRNAs, etc. Oxidative stress and inflammation play a pivotal role in various diseases like diabetes, cancer, arthritis, Alzheimer's disease (AD), and cardiovascular diseases [4–8].

8.2.1 MECHANISM OF ACTION IN THE BRAIN

Curcumin, a polyphenolic natural product, exhibits therapeutic activity against a number of diseases, attributed mainly to its chemical structure (Figure 8.1) and unique physical, chemical, and biological properties. It is a diferuloyl methane molecule [1,7-bis (4-hydroxy-3-methoxyphenyl)-1,6-heptadiene-3,5-dione)] containing two ferulic acid residues joined by a methylene bridge. It has three important functionalities: an aromatic *o*-methoxy phenolic group, alpha, beta-unsaturated beta-diketo moiety, and a seven-carbon linker. Extensive research in the last two decades has provided evidence for the role of these different functional groups in its crucial biological activities. A few highlights of chemical structural features associated with the biological activity of curcumin are the *o*-methoxyphenol group and methylenic hydrogen, which are responsible for the antioxidant activity of curcumin, and curcumin donates an electron/hydrogen atom to reactive oxygen species. Curcumin interacts with a number of biomolecules through noncovalent and covalent binding. The hydrogen bonding and hydrophobicity of curcumin, arising from the aromatic and tautomeric structures along with the flexibility of the linker group, are responsible for the noncovalent interactions. The alpha, beta-unsaturated beta-diketone moiety covalently interacts with protein thiols, through Michael reaction. The beta-diketo group forms chelate with transition metals, thereby reducing the metal induced toxicity [9–11].

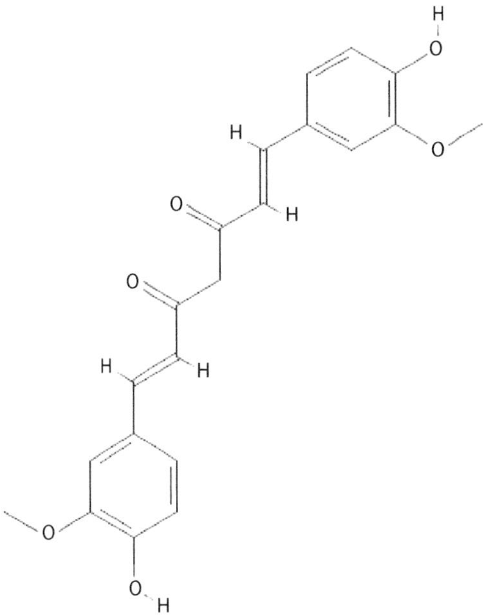

FIGURE 8.1   **(See color insert.)** Curcumin. Chemical names: curcumin; diferuloylmethane; natural yellow 3; 458-37-7; turmeric yellow; turmeric. Molecular formula: C21H20O6. Molecular weight: 368.3799 g/mol.

Curcumin mediates its neuroprotective effects not only in neurotraumatic disorders (stroke, spinal cord injury, traumatic brain injury, and epilepsy) but also in Parkinson disease, Huntington disease, and prion diseases. In addition, curcumin also promotes its beneficial effects in neuropsychological disorders (depression, bipolar disorders, and tardive dyskinesia). The mechanism associated with the neuroprotective action of curcumin is not fully understood. However, it is becoming increasingly evident that anti-inflammatory and antioxidant properties of curcumin may be responsible for neuroprotective effects [1,6]. At the molecular level, the neuroprotective effects of curcumin are accompanied by downregulating activities of phospholipases, lipooxygenase, and cyclooxygenase (COX)-2, which lead to low levels of leukotrienes, thromboxanes, and prostaglandins. In addition, curcumin also inhibits the expression of tumor necrosis factor-alpha (TNF-α), interleukin (IL)-12, monocyte chemoattractant protein-1 (MCP-1)-1, and interferon-inducible protein. Curcumin also modulates various neurotransmitter levels in the brain [12]. Curcumin is an excellent antioxidant anti-inflammatory agent. It can cross the blood–brain barrier (BBB). Curcumin mediates its beneficial effect not only by modulating APP processing and downregulating gene expression of proteins associated by apoptosis and neuroinflammation but also by increasing Aβ uptake by macrophages. Curcumin also induces heat shock proteins and reduces protein misfolding and aggregation [1].

Curcumin blocks the formation of reactive-oxygen species, possesses anti-inflammatory properties as a result of inhibition of COXs and other enzymes involved in inflammation, and disrupts cell signal transduction by various mechanisms including inhibition of protein kinase C. These effects may play a role in the agent's observed antineoplastic properties, which include inhibition of tumor cell proliferation and suppression of chemically induced carcinogenesis and tumor growth in animal models of cancer [9,10].

8.2.2 CLINICAL STUDIES AND RESEARCH

C. longa or turmeric has been commonly used in Ayurvedic medicine since at least 3000 BC. The first study for its use in curing human disease was observed in 1937 [13]. *C. longa*, or turmeric, possessed several pharmaceutically important chemical agents, such as terpenoids [14], volatile oils, sugar, proteins, and curcuminoids [15]. Curcumin or diferuloyl methane (which gives the characteristic bright yellow color to turmeric), demethoxycurcumin, bisdemethoxycurcumin, and cyclocurcumin are the four most important curcuminoid compounds and are found to play a significant role in curing several human as well as animal diseases. Among all these four curcuminoids, curcumin is well studied so far and has a wide spectrum of pharmacological actions on diverse human diseases [16]. Several beneficial effects of curcumin for the nervous system (at least 10 known neuroprotective actions) have been reported [6]. To date, approximately 3000 studies have been accomplished on curcumin (diferuloyl methane). Some reports suggested that curcumin has several positive effects on many functions of the body, such as cardiovascular, endocrine, musculoskeletal, respiratory, gastrointestinal, and most importantly neurological systems by acting on around 100 genetic pathways [17]. Curcumin is also well known for its anti-inflammatory, anti-hyperlipidemic, antihypertensive, antitumor, anticancer, antiphlogistic, antidiabetic, antipsoriasis, antithrombotic, antihepatotoxic, and neuroprotective properties [6,18]. Curcumin is a light molecular weight polar compound and has the potential to cross BBB efficiently. Thus, curcumin could enhance mature hippocampus neurogenesis activities by generating new cells in the region of the hippocampus [19]. And at the same time, curcumin is found to inhibit reactive astrocyte production and to protect hippocampal cells from its death by kainic acids [20]. Beta amyloid plaques are one of the main factors for AD. Recently, it was observed that use of curcumin has effectively disaggregated beta amyloid plaques and helped to ameliorate the recovery pathway [7,18]. Curcumin also inhibits immunostimulatory function of dendritic cells: MAPKs and translocation of NF-B as potential targets [21].

Depression is a serious neurological disorder. Irritable mood, loss of interest and concentration, feeling of extreme guilt, significant body weight instability, hypersomnia or insomnia, and suicidal tendencies are the characteristic features of depression and are the result of neurological dysfunction. Approximately 20% of people all over the world have been suffering from these neurological problems. Despite several antidepressant drugs being available for these problems, most drugs are associated with severe side effects. Curcumin has been found to be the sole alternative drug of the disease without side effects. Curcumin found to have potential antidepressant activity, and it was experimentally proved in an animal model by using forced swim test and chronic unpredictable stress [22,23]. A scientific report suggested that curcumin is the effective natural medicine to treat neurological disorders like tardive dyskinesia, a kind of neurological disorder and characterized by uncontrolled movement of the jaw and face [24]. Further study has shown that curcumin increased the antimobility effects of two monoamine oxidase inhibitors, selegiline (5 mg/kg, intraperitoneal route [i.p.]) and tranylcypromine (5 mg/kg, i.p.). These experiments showed the role of monoamine oxidase enzyme in the antidepressant property of curcumin.

Indeed, accumulating cell culture and animal model data show that dietary curcumin is a strong candidate for use in the prevention or treatment of major disabling age-related neurodegenerative diseases like AD, Parkinson's disease, and stroke. In an animal model of stroke, curcumin treatment protected neurons against ischemic cell death and ameliorated behavioral deficits. Moreover, curcumin has been shown to reverse chronic stress-induced impairment of hippocampal neurogenesis and increase the expression of brain-derived neurotrophic factor in an animal model of depression [25]. Adriana et al. reviewed the evidence from several animal models that curcumin improves health span by preventing or delaying the onset of various neurodegenerative diseases [26].

Recent studies have indicated that curcumin can target newly identified signaling pathways, including those associated with microRNA, cancer stem cells, and autophagy. Extensive research from preclinical and clinical studies has delineated the molecular basis for the pharmaceutical uses of this polyphenol against cancer, pulmonary diseases, neurological diseases, liver diseases, metabolic diseases, autoimmune diseases, cardiovascular diseases, and numerous other chronic diseases. Multiple studies have indicated the safety and efficacy of curcumin in numerous animals including rodents, monkeys, horses, rabbits, and cats and have provided a solid basis for evaluating its safety and efficacy in humans. To date, more than 65 human clinical trials of curcumin, which included more than 1000 patients, have been completed, and as many as 35 clinical trials are underway. Curcumin is now used as a supplement in several countries including the United States, India, Japan, Korea, Thailand, China, Turkey, South Africa, Nepal, and Pakistan [5].

8.2.3 Metabolism

The bioavailability of any chemical agents in the human body usually depends on absorptions, strong intrinsic activity, and low rate of metabolism and reduced elimination from the body. It is reported that curcumin possesses strong intrinsic activity and, thus, is a strong therapeutic agent for several diseases. Three decades of epidemiological experimental output on curcumin suggest that curcumin has poor absorption and rapid metabolism rate that extremely reduced its bioavailability in the blood serum and in tissues [27]. Wahlstrom and Blennow (1978) reported and recorded an insignificant amount of curcumin in blood plasma by feeding 1 g/kg of curcumin to Sprague-Dawley rats [28]. Similarly, Ravindranath and Chandrasekhara showed that after orally feeding of 400 mg of curcumin to rats, no curcumin was traced in heart blood serum and very little amount was found in portal blood within 15 to 24 hours [29]. Later, it was revealed that tritium-labeled curcumin was better absorbed and detected in rat blood [30]. Interestingly, when curcumin was administered to humans at a dose of 2 g/kg, the availability of curcumin was extremely less in amount in blood serum [31].

Pan et al. examined the pharmacokinetic actions of curcumin by administrating either orally or through intraperitoneal route in mice. It was observed that after oral administration of 1.0 g/kg of curcumin, little plasma curcumin levels of 0.13 µg/mL appeared in blood plasma after 15 minutes, whereas a highest plasma curcumin level of 0.22 µg/mL was obtained after one hour interval and plasma curcumin concentrations declined below the detection limit after six hours. On the other hand, different plasma curcumin levels were observed after i.p. administration of 0.1 g/kg curcumin. Plasma curcumin levels were found to be high (2.25 µg/mL) after 15 minutes of administration and declined speedily by one hour [32]. In another report (Yang et al.), curcumin was administered at the rate 10 mg/kg via i.p. route into rats and a maximum serum curcumin level (0.36 µg/mL) was observed while on the other side when a 50-fold higher curcumin dose was administered orally that showed only 0.06 µg/mL highest serum level in rat [33]. These observations clearly indicated the role of route of administration on serum levels concentration of curcumin and further showed that the serum levels of curcumin in rats and in humans were not directly comparable.

The poor absorptions and poor bioavailability of curcumin in the blood plasma level has been overcome by special scientific techniques such as use of adjuvant that can block metabolic cascade of curcumin. This is one of the most important means by which the bioavailabilities of curcumin are improved. There are many other means that have been adapted to enhance the bioavailability of curcumin, such as (1) use of nanoparticles, (2) liposomes, (3) phospholipids, and (4) micelle formulations. These formulations have supported for better permeability, longer circulation, and resistance to metabolic processes. Adjuvants like piperine, a known inhibitor of hepatic and intestinal glucuronidation, were combined with curcumin and administered in rats and healthy human volunteers by Shoba et al. It is noticed that in rats, 2 g/kg of curcumin alone raised the maximum serum curcumin level up to 1.35 µg/mL at 0.83-hour interval, while combined administration of piperine and curcumin (20 mg/kg) increased the serum concentration of curcumin within a very short period of time and elimination and clearance of curcumin were significantly decreased, consequently increasing curcumin bioavailability by 154% in the blood serum. On the other hand, in humans, when 2 g curcumin alone was administered, serum levels were found to be very low. However, combined administration of piperine enhanced by 2000% the bioavailability of curcumin serum level in human blood [31].

Nanoparticles targeted and triggered efficient drug delivery systems, and along with nanoparticle technology, the possibility of bioavailability of therapeutic agents has widened more. It has been reported that polymer-based nanoparticle of curcumin, namely, "nanocurcumin," with less than 100 nm size, was effectively used for curcumin delivery, and it was found to have similar in vitro activity as that of free curcumin in pancreatic cell lines [34].

8.3 *CENTELLA ASIATICA* LINN.

Centella asiatica (CA) L. Urban (syn. *Hydrocotyle asiatica* L.), locally well known as Gotu Kola belonging to family Apiaceae (Umbelliferae), is a psychoactive traditional medicinal plant which has been used for centuries in Ayurvedic system and traditional Chinese medicine as a medhya rasayana [35,36]. Although consumption

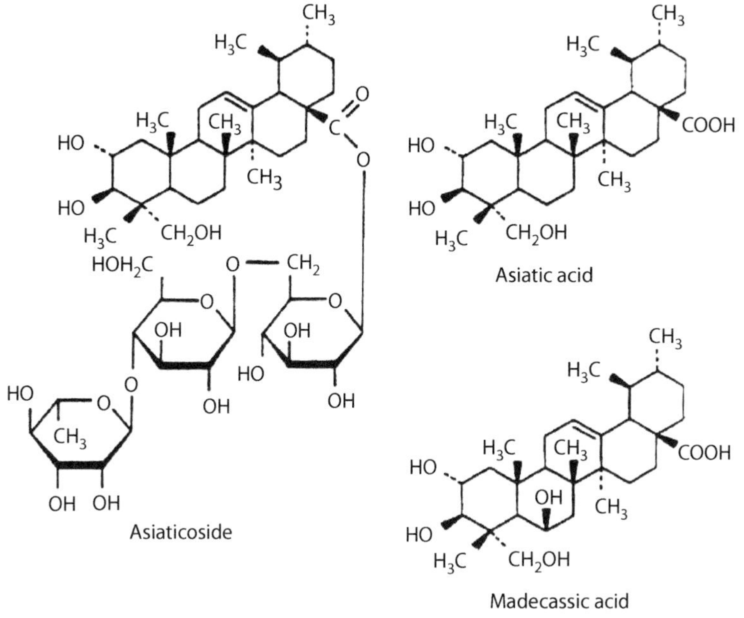

FIGURE 8.2   Bioactive compounds of *Centella asiatica*. (From Bonte, F., Dumas, M., Chaudagne, C., Meybeck, A., *Planta Med.*, 60, 33–135, 1994 [51].)

of the plant is indicated for various illnesses, its potential neuroprotective properties have been well studied and documented. CA acts as an antioxidant, reducing the effect of oxidative stress in vitro and in vivo [37,38]. At the in vitro level, CA promotes dendrite arborisation and elongation and also protects the neurons from apoptosis [37,39]. In vivo studies have shown that the whole extract and also individual compounds of CA have a protective effect against various neurological diseases [37,40]. Most of the in vivo studies on neuroprotective effects have focused on AD [41], Parkinson's disease [42], learning and memory enhancement [43], neurotoxicity [44,45], and other mental illnesses such as depression and anxiety and epilepsy [46–48].

Major bioactive compounds of this plant contain highly variable triterpenoid saponins, including asiaticoside, madecassoside [35,49,50] (Figure 8.2), oxyasiaticoside, centelloside, brahmoside, brahminoside, thankunoside, isothankunoside, and related sapogenins. It also contains triterpenoid acids viz. asiatic acid (AA), madecassic acid, brahmic acid, isobrahmic acid, betulic acid, etc. However, its exact mechanism of action in the treatment and management of neurodisorders has not been fully understood [35]. The market value of AA is US$31,455 per gram, of asiaticoside is US$15,020 per gram, and of madecassoside is US$44,367 per gram (Extrasynthase, Genax Cedex, France).

8.3.1   Molecular Mechanism and Clinical Studies of Neuroprotection

Cognitive dysfunction is a major health problem in the twenty-first century, and many neuropsychiatric disorders and neurodegenerative disorders, such as schizophrenia, depression, AD dementia, cerebrovascular impairment, seizure disorders, head injury, and Parkinsonism, can be severely functionally debilitating in nature. In course of time, a number of neurotransmitters and signaling molecules have been identified that have been considered as therapeutic targets. Conventional as well as newer molecules have been tried against these targets. Phytochemicals from medicinal plants play a vital role in maintaining the brain's chemical balance by influencing the function of receptors for the major inhibitory neurotransmitters [52]. Active constituents of these herbals might play an important role in preserving the integrity of various neurotransmitters and their receptor in the brain, influencing its functions at the molecular level [35]. Recent studies have also embarked on finding the molecular mechanism of neuroprotection by CA by decreasing the oxidative stress parameters [37,53].

Water extract of CA (CAW) improved performance in the Morris water maze (MWM) test in aged animals and had a modest effect on the performance of young animals. CAW also increased the expression of mitochondrial and antioxidant response genes in the brain and liver of both young and old animals. The expression of synaptic markers was also increased in the hippocampus and frontal cortex of CAW-treated animals [54]. In another study, the data indicated that CA extract can impact the amyloid cascade altering amyloid β pathology in the brains of PSAPP mice and modulating components of the oxidative stress response that has been implicated in the neurodegenerative changes that occur with AD [55]. The effects of a water extract of CA (GKW) in the Tg2576 mouse, a murine model of AD with high β-amyloid burden showed that CA offers a unique therapeutic mechanism and novel active compounds of potential relevance to the treatment of AD [56]. The effect of chloroform:methanolic (80:20) extract of CA (CA; 100 and 200 mg/kg) was evaluated on the course of free radical generation and excitotoxicity in monosodiumglutamate (MSG)-treated female Sprague-Dawley rats. It can be concluded that CA protected MSG-induced neurodegeneration attributed to its antioxidant and behavioral properties. This activity of CA can be explored in epilepsy, stroke, and other degenerative conditions in which the role of glutamate is known to play vital role in the pathogenesis [46].

An investigation of the differential effects of ursane triterpenoids from CA, and their semisynthetic analogues, on GABAA receptors suggests that AA 1 may be a lead compound for the enhancement of cognition and memory [43]. The effect of AA on the treatment of spinal cord injury in rats attenuated the levels of lipid peroxidation products (MDA) and proinflammatory cytokines (TNF-α, IL1β). It also increased the Tarlov functional recovery scores of the rats [57]. AA is a triterpene extracted from CA and has been reported as an antioxidant and anti-inflammatory agent that offers neuroprotection against glutamate toxicity. This study investigated the effect of AA in a rotenone (an inhibitor of mitochondrial complex I) induced in an in vitro model of Parkinson's disease. Following the exposure of SH-SY5Y cells to rotenone, there was a marked overproduction of reactive oxygen species (ROS), mitochondrial dysfunction (as indexed by the decrease in mitochondrial membrane potential [MMP]), and apoptosis (Hoechst and dual staining, comet assay; expressions of proapoptotic and antiapoptotic indices). Pretreatment with AA reversed these changes, which might be due to its antioxidant, mitoprotective, and antiapoptotic properties [44].

AA attenuates glutamate-induced cognitive deficits of mice and protects human neuroblastoma SH-SY5Y cells against glutamate-induced apoptosis in vitro. Pretreatment of SH-SY5Y cells with AA (0.1–100 nmol/L) attenuated toxicity induced by 10 mmol/L glutamate in a concentration-dependent manner. AA 10 nmol/L significantly decreased apoptotic cell death and reduced ROS, stabilized the MMP, and promoted the expression of PGC-1α and Sirt1. In mice models, oral administration of AA (100 mg/kg) significantly attenuated cognitive deficits in the MWM test and restored lipid peroxidation and glutathione and the activity of SOD in the hippocampus and cortex to the control levels. AA (50 and 100 mg/kg) also attenuated neuronal damage of the pyramidal layer in the CA1 and CA3 regions [39]. Axonal regeneration is important for functional recovery following nerve damage. CA ethanolic extract (100 µg mL^{-1}) elicits a marked increase in neurite outgrowth in human SH-SY5Y cells in the presence of nerve growth factor. AA showed marked activity at 1 µm (0.5 µg mL^{-1}). Hence, Centella ethanolic extract may be useful for accelerating repair of damaged neurons [58].

Sleep deprivation (SD) is an experience of inadequate or poor quality of sleep that may produce significant alterations in multiple neural systems. CA is a psychoactive medicinal herb with immense therapeutic potential. The present study suggests that the possible nitric oxide modulatory mechanism could be involved in the neuroprotective effect of CA against SD-induced anxiety-like behavior, oxidative damage, and neuroinflammation [59].

8.4   CONCLUSIONS

Exploring alternative sources for neurological degenerative therapy has led researchers to set eyes on herbal medicine since most herbal compounds have antioxidant and anti-inflammatory properties. Phytodrugs, mainly from *C. longa* and *C. asiatica*, can prevent this neuronal damage and, therefore, cellular death. However, information is still missing on relevant aspects such as metabolism, pharmacokinetics, and bioavailability in the brain as well as any changes that they may have in the central nervous system. Nevertheless, these natural compounds can be used in the treatment of neurodegenerative diseases and also could serve as models for developing new specific drugs against these pathologies.

REFERENCES

1. Farooqui AA. 2016. Effects of curcumin on neuroinflammation in animal models and in patients with Alzheimer disease. *Therapeutic Potentials of Curcumin for Alzheimer Disease*. pp 259–296. Springer International Publishing, Switzerland.
2. Pérez-Hernández J, Zaldívar-Machorro VJ, Villanueva-Porras D, Vega-Ávila E, Chavarría A. 2016. *A Potential Alternative against Neurodegenerative Diseases: Phytodrugs*. Hindawi Publishing Corporation. Oxidative Medicine and Cellular Longevity. Volume 2016, Article ID 8378613. http://dx.doi.org/10.1155/2016/8378613.

3. Srivastava P, Singh Yadav R. 2016. Efficacy of natural compounds in neurodegenerative disorders. In Schousboe A (ed). *Advances in Neurobiology. Volume 12. The Benefits of Natural Products for Neurodegenerative Diseases.* Springer International Publishing, Switzerland, pp 107–123.

4. Ghosh S, Banerjee S, and Sil PC. 2015. The beneficial role of curcumin on inflammation, diabetes and neurodegenerative disease: A recent update. *Food Chem Toxicol.* 83:111–124.

5. Gupta SC, Kismali G, Aggarwal BB. 2013. Curcumin, a component of turmeric: From farm to pharmacy. *Biofactors.* 39(1):2–13. doi: 10.1002/biof.1079.

6. Aggarwal, BB, Harikumar, KB. 2009. Potential therapeutic effects of curcumin, the anti-inflammatory agent, against neurodegenerative, cardiovascular, pulmonary, metabolic, autoimmune and neoplastic diseases. *Int J Biochem Cell Biol.* 41:40–59.

7. Mishra S, Palanivelu K. 2008. The effect of curcumin (turmeric) on Alzheimer's disease: An overview. *Ann Indian Acad Neurol.* 11(1):13–19. doi: 10.4103/0972-2327.40220.

8. Witkin JM, Li X. 2013. Curcumin, an active constituent of the ancient medicinal herb *Curcuma longa* L.: Some uses and the establishment and biological basis of medical efficacy. *CNS Neurol Disord Drug Targets.* 12(4):487–497.

9. Priyadarsini KI. 2013. Chemical and structural features influencing the biological activity of curcumin. *Curr Pharm Des.* 19(11):2093–2100.

10. PubChem. https://pubchem.ncbi.nlm.nih.gov

11. NIST Standard Reference Database 1A v14. http://www.nist.gov/srd/nist1a.cfm

12. Farooqui AA. 2016. Therapeutic importance of curcumin in neurological disorders other than Alzheimer disease. *Therapeutic Potentials of Curcumin for Alzheimer Disease.* pp 297–334. Springer International Publishing, Switzerland.

13. Albert O. 1937. Turmeric (curcumin) in biliary diseases. *Lancet.* 229:619–621.

14. Afzal A, Oriqat G, Akram Khan M, Jose J, Afzal M. 2013. Chemistry and biochemistry of terpenoids from curcuma and related species. *J Biol Active Prod Nature.* 3(1). http://dx.doi.org/10.1080/22311866.2013.782757

15. Jurenka JS. 2009. Anti-inflammatory properties of curcumin, a major constituent of *Curcuma longa*: A review of preclinical and clinical research. *Altern Med Rev.* 14(3):277.

16. Ahsan H, Parveen N, Khan NU et al. 1999. Pro-oxidant, anti-oxidant and cleavage activities on DNA of curcumin and its derivatives demethoxycurcumin and bisdemethoxycurcumin. *Chem Biol Interact.* 121:161–175.

17. Zhou H, Beevers C, Huang S. 2011. Targets of curcumin. *Curr Drug Targets.* 12(3):332–347.

18. Kulkami S, Dhir A, Akula KK. 2009. Potentials of curcumin as an antidepressant. *Sci World J.* 9:1233–1241.

19. Kim SJ, Son TG, Park HR et al. 2008. Curcumin stimulates proliferation of embryonic neural progenitor cells and neurogenesis in the adult hippocampus. *J Biol Chem.* 283:14497–14505.

20. Shin HI, Lee JY, Son E et al. 2007. Curcumin attenuates the kainic acid-induced hippocampal cell death in the mice. *Neurosci Lett.* 416:49–54.

21. Kim GY, Kim KH, Lee SH, Yoon MS, Lee HJ, Moon DO. 2005. Curcumin inhibits immunostimulatory function of dendritic cells: MAPKs and translocation of NF-B as potential targets. *J Immunol.* 174:8116–8124.

22. Bhutani MK, Bishnoi M, Kulkarni SK. 2009. Anti-depressant like effect of curcumin and its combination with piperine in unpredictable chronic stress-induced behavioral, biochemical and neurochemical changes. *Pharmacol Biochem Behav.* 92:39–43.

23. Kulkarni SK, Bhutani MK, Bishnoi M. 2008. Antidepressant activity of curcumin: Involvement of serotonin and dopamine system. *Psychopharmacology (Berl).* 201:435–442.

24. Kulkarni SK, Dhir A. 2010. An overview of curcumin in neurological disorders. *Indian J Pharm Sci.* 72(2):149–154. doi: 10.4103/0250-474X.65012.

25. Xu Y, Ku B, Cui L et al. 2007. Curcumin reverses impaired hippocampal neurogenesis and increases serotonin receptor 1A mRNA and brain-derived neurotrophic factor expression in chronically stressed rats. *Brain Res*. 1162:9–18.

26. Monroy A, Lithgow GJ, Alavez S. 2013. Curcumin and neurodegenerative diseases. *Biofactors*. 39(1):122–132. doi: 10.1002/biof.1063.

27. Anand P, Kunnumakkara AB, Newman RA, Aggarwal BB. 2007. Bioavailability of curcumin: Problems and promises. *Mol Pharmaceutics*. 4(6):807–818. doi: 10.1021/mp700113r.

28. Wahlstrom B, Blennow G. 1978. A study on the fate of curcumin in the rat. *Acta Pharmacol Toxicol (Copenhagen)*. 43(2):86–92.

29. Ravindranath V, Chandrasekhara N. 1980. Absorption and tissue distribution of curcumin in rats. *Toxicology*. 16(3):259–265.

30. Ravindranath V, Chandrasekhara N. 1981. Metabolism of curcumin—Studies with [3H]curcumin. *Toxicology*. 22(4):337–344.

31. Shoba G, Joy D, Joseph T, Majeed M, Rajendran R, Srinivas PS. 1998. Influence of piperine on the pharmacokinetics of curcumin in animals and human volunteers. *Planta Med*. 64(4):353–356.

32. Pan MH, Huang TM, Lin JK. 1999. Biotransformation of curcumin through reduction and glucuronidation in mice. *Drug Metab Dispos*. 27(4):486–494.

33. Yang KY, Lin LC, Tseng TY, Wang SC, Tsai TH. 2007. Oral bioavailability of curcumin in rat and the herbal analysis from *Curcuma longa* by LC-MS/MS. *J Chromatogr B Anal Technol Biomed Life Sci*. 853(1–2):183–189.

34. Karikar C, Maitra A, Bisht S, Feldmann G, Soni S, Ravi R. 2007. Polymeric nanoparticle-encapsulated curcumin ("nanocurcumin"): A novel strategy for human cancer therapy. *J Nanobiotechnol*. 5:3.

35. Nalini K, Aroor A, Karanth K, Rao A. 1992. *Centella asiatica* fresh leaf aqueous extract on learning and memory and biogenic amine turnover in albino rats. *Fitoterapia*. 63:232–237.

36. Anand T, Naika M, Kumar PG, Khanum F. 2011. Antioxidant and DNA damage preventive properties of *Centella asiatica* (L). Urb., *Pharmacognosy Mag*. 2:53–58.

37. Lokanathan Y, Omar N, Ahmad Puzi NN, Saim A, Hj Idrus R. 2016. Recent updates in neuroprotective and neuroregenerative potential of *Centella asiatica*. *Malays J Med Sci*. 23(1):4–14.

38. Herbs 2000.com. 2010. Source or traditional and nutritional health care. http://www.herbs2000.com

39. Xu M-F, Xiong Y-Y, Liu J-K, Qian J-J, Zhu L, Gao J. 2012. Asiatic acid, a pentacyclic triterpene in *Centella asiatica*, attenuates glutamate-induced cognitive deficits in mice and apoptosis in SH-SY5Y cells. *Acta Pharmacol Sin*. 33:578–587.

40. Kumar MHV, Gupta YK. 2002. Effect of different extracts of *Centella asiatica* on cognition and markers of oxidative stress in rats. *J Ethnopharmacol* 79:253–260.

41. Veerendra Kumar MH, Gupta YK. 2003. Effect of *Centella asiatica* on cognition and oxidative stress in an intracerebroventricular streptozotocin model of Alzheimer's disease in rats. *Clin Exp Pharmacol Physiol*. 30(5–6):336–342.

42. Haleagrahara N, Ponnusamy K. 2010. Neuroprotective effect of *Centella asiatica* extract (CAE) on experimentally induced parkinsonism in aged Sprague-Dawley rats. *J Toxicol Sci*. 35(1):41–47.

43. Hamid K, Ng I, Tallapragada VJ, Váradi L, Hibbs DE, Hanrahan J, Groundwater PW. 2016. An investigation of the differential effects of ursane triterpenoids from *Centella asiatica*, and their semisynthetic analogues, on GABAA receptors. *Chem Biol Drug Des*. 88(3):386–397.

44. Nataraj J, Manivasagam T, Thenmozhi AJ, Essa MM. 2016. Neuroprotective effect of asiatic acid on rotenone-induced mitochondrial dysfunction and oxidative stress-mediated apoptosis in differentiated SH-SYS5Y cells. *Nutr Neurosci*. 1–9.

45. Jew SS, Yoo CH, Lim DY, Kim H, Jung MI, Jung MW, Choi H, Jung YH, Kim H, Park HG. 2000. Structure–activity relationship study of asiatic acid derivatives against beta amyloid (A beta)-induced neurotoxicity. *Bioorg Med Chem Lett.* 10(2):119–121.

46. Ramanathan M, Sivakumar S, Anandvijayakumar PR, Saravanababu C, Rathinavel PP. 2007. Neuroprotective evaluation of standardized extract of *Centella asiatica* in monosodium glutamate treated rats. *Indian J Exp Biol.* 45(5):425–431.

47. Roa KG, Rao SM, Rao SG. 2005. *Centella asiatica* (Linn.) induced behaviour changes during growth spurt period in neonatal rats. *Nueroanatomy.* 4:18–23.

48. Wijeweera P, Arnason JT, Koszycki D, Merali Z. 2006. Evaluation of anxiolytic properties of gotu kola—(*Centella asiatica*) extracts and asiaticoside in rat behavioural models. *Phytomedicine.* 13:668–676.

49. Harborne JB, Baxter H, Moss GP. 1999. *Phytochemical Dictionary: A Handbook of Bioactive Compounds from Plants.* Taylor & Francis Ltd., London, pp 41–89.

50. Mangas S, Moyano E, Hernández-Vázquez L, Bonfill M. 2009. *Centella asiatica* (L) Unban: An updated approach. In: Palazón JB, Cusidó RM (eds). *Plant Secondary Terpenoids.* Research Signpost, Trivandrum, India, pp 55–74.

51. Bonte F, Dumas M, Chaudagne C, Meybeck A. 1994. Influence of asiatic acid, madecassic acid and asiaticoside on human collagen I synthesis. *Planta Med.* 60:33–135.

52. Kumar GP, Khanum F. 2012. Neuroprotective potential of phytochemicals. *Pharmacogn Rev.* 6(12):81–90. doi: 10.4103/0973-7847.99898.

53. Agrawal V, Subhan S. 2003. *In vitro* plant regeneration and protein profile analysis in *Centella asiatica* (Linn.) Urban: A medicinal plant. Pl. *Cell Biotech Mol Biol.* 4:83–90.

54. Gray NE, Harris CJ, Quinn JF, Soumyanath A. 2016. *Centella asiatica* modulates antioxidant and mitochondrial pathways and improves cognitive function in mice. *J Ethnopharmacol.* 180:78–86.

55. Dhanasekaran M, Holcomb LA, Hitt AR, Tharakan B, Porter JW, Young KA, Manyam BV. 2009. *Centella asiatica* extract selectively decreases amyloid β levels in hippocampus of Alzheimer's disease animal model. *Phytother Res.* 23:14–19. doi: 10.1002/ptr.2405.

56. Soumyanath A, Zhong YP, Henson E et al. 2012. *Centella asiatica* extract improves behavioral deficits in a mouse model of Alzheimer's disease: Investigation of a possible mechanism of action. *Int J Alzheimer Dis.* 2012:Article ID 381974. doi: 10.1155/2012/381974.

57. Gurcan O, Gurcay AG, Kazanci A, Senturk S, Bodur E, Karaca EU, Turkoglu OF, Bavbek M. 2017. Effect of asiatic acid on the treatment of spinal cord injury: An experimental study in rats. *Turk Neurosurg.* 27(2):259–264. doi: 10.5137/1019-5149.JTN.15747-15.2.

58. Soumyanath A, Zhong Y-P, Yu X, Bourdette D, Koop DR, Gold SA, Gold BG. 2005. *Centella asiatica* accelerates nerve regeneration upon oral administration and contains multiple active fractions increasing neurite elongation in-vitro. *J Pharm Pharmacol.* 57(9):1221–1229.

59. Chanana P, Kumar A. 2016. Possible involvement of nitric oxide modulatory mechanisms in the neuroprotective effect of *Centella asiatica* against sleep deprivation induced anxiety like behaviour, oxidative damage and neuroinflammation. *Phytother Res.* 30:671–680. doi: 10.1002/ptr.5582.

9 Tocotrienol Vitamin E in Brain Health

Richard Stewart and Savita Khanna

CONTENTS

9.1 INTRODUCTION TO NATURAL VITAMIN E

By examining the history of vitamin E, one can fully appreciate the advances made in vitamin E research. Vitamin E was discovered in 1922 by Herbert McLean Evans and Katharine Scott Bishop at the University of California, Berkeley.[1] In the experiments that led to the discovery of vitamin E, rats were fed a special semipurified diet, which led to a very interesting observation—that all rats (except pregnant females) would grow very well. In the case of pregnant females, the pups would die in the womb while maintained on this diet, a phenomenon that appeared to be rescued when lettuce or wheat germ was added to the diet. Dr. Evans and Dr. Bishop initially decided to call the missing factor in the diet "Factor X." This Factor X was in the lipid extract of lettuce, so they recognized that the mysterious factor must be fat soluble.[1] In 1924, Dr. Bennett Sure independently identified a missing factor in the diet that was making rats sterile and proposed the name vitamin E to the missing factor.[2]

The natural vitamin E family is made up of eight fat-soluble molecules: α-, β-, γ-, and δ-tocopherol (TCP) and α-, β-, γ-, and δ-tocotrienol (TCT).[3] Structurally, each member of the vitamin E family consists of a chromanol (a chromanol ring with an alcohol hydroxyl group) with an attached 16-carbon phytyl-like side chain. The side chain of the TCP analogue is saturated and the side chain of the TCT is unsaturated and possesses an isoprenoid side chain[4] (Figure 9.1). The various TCP and TCT analogues differ in the methylation status of their chromanol ring. It is believed that the difference in methylation status of the chromanol ring explains the preferential uptake of some vitamin E analogues over others.[5] As a result of their structure, all members of the vitamin E family are lipid soluble and amphipathic, which allows them to integrate with membrane lipids.[6,7] It is within the cell membrane that vitamin E performs its well-known role as an antioxidant,[6] but it also contributes to regulating gene expression and signal transduction.[7,8] While the structures of the vitamin E family molecules are similar, they do not have redundant biological activities.[6,8]

Vitamin E is a highly potent and widely accepted lipophilic, phenolic antioxidant consumed in our diet.[6,9] The strong antioxidant capability of all TCPs and TCTs stems from the chromanol ring, which can readily donate hydrogen to neutralize free radicals.[4] TCTs have been suggested to be more efficient scavengers of free radicals than TCPs in the lipid membrane due to a more even distribution when compared to TCPs.[4] Vitamin E is a chain-breaking antioxidant that prevents the propagation of free radical damage in biological membranes by scavenging a potent peroxyl radical.[10–13] Being a chain-breaking antioxidant, vitamin E preserves the activity of several signal transduction enzymes in the cell membranes.[6] Since excessive free radical damage has been implicated in numerous neurological, cardiovascular, cancer, pulmonary diseases, and rheumatoid arthritis, vitamin E has been shown to have beneficial effects through its antioxidative properties.[6]

Antioxidant-independent mechanisms of vitamin E action have also been well studied in a variety of models. Vitamin E molecules can interact with various

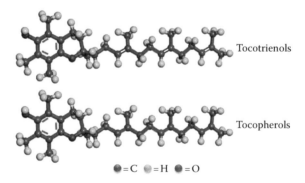

FIGURE 9.1   **(See color insert.)** Chemical structure of TCT and TCP. TCTs and TCPs consist of a chromanol head attached to a 16-carbon side chain. TCPs have a saturated side chain and TCTs have an unsaturated side chain containing three *trans* double bonds.

signaling molecules, such as protein kinases, lipid kinases, and specific enzymes to modulate signal transduction and gene expression.[7]

9.2 BIOSYNTHESIS OF TCPs AND TCTs

The biosynthesis of TCTs and TCPs is initiated by prenylation of homogentisate (HGA). For TCT synthesis, geranylgeranyl diphosphate is the prenyl donor, and for TCP, phytyl diphosphate (PDP) is the prenyl donor. TCPs consist of a chromanol ring attached to a 16-carbon side chain. The chromanol ring is derived from HGA and the carbon side chain is derived from PDP. The enzyme HGA phytyltransferase catalyzes the condensation reaction between HGA and the 16-carbon side chain to form TCP.[14] TCP is abundant in photosynthetic tissue and helps to maintain optimal photosynthesis rate during periods of high-light stress.[15]

Structurally, the difference between TCPs and TCTs is in their carbon side chain. The unsaturated side chain with three trans double bonds in TCTs appears to be much more flexible, which can increase curvature stress on phospholipid membranes,[16] and this has been confirmed by Atkinson using scanning calorimetry.[17] The TCT biosynthesis is carried out by the enzyme homogentisic acid transferase (HGGT). Transgenic expression of barley HGGT in *Arabidopsis thaliana* leaves, typically devoid of any TCTs, resulted in detectable TCT content and a 10- to 15-fold increase of total vitamin E content. Transgenic expression of barley HGGT was also tested in corn seeds, which led to detectable TCT content and an approximate sixfold increase in total vitamin E content.[18] Genetically engineering metabolic pathways has also been shown to be an effective platform to increase TCT biosynthesis.[19] Upregulating the biosynthesis of *p*-hydroxyphenylpyruvate and HGA, the aromatic precursors of vitamin E, was achieved by the expression of yeast prephenate dehydrogenase in tobacco plants that overexpress *A. thaliana p*-hydroxyphenylpyruvate dioxygenase. This resulted in high concentrations of TCT in tobacco leaves that are absent in wild-type leaves. These results demonstrated that the transgenic expression of HGGT alone was sufficient to induce TCT biosynthesis in plant organs and sesame cells.[19]

9.3 NATURAL SOURCES OF TCTs

TCTs are synthesized by plants in both inedible and edible forms. Rubber latex is an example of an inedible plant product that is rich in TCTs.[20] Palm oil has been discovered to be a major source of dietary TCTs, with a total vitamin E content of 70% TCTs and 30% TCPs. Oil is extracted from the mesocarp of the palm fruit and the endosperm or kernel. Palm oil has long been a component of the West African diet, where the palm (*Elaeis guineensis*) is native and inhabitants of West Africa use the oil from the palm fruit for culinary purposes.[21] Large-scale palm farming operations in Asia, Africa, and Latin America primarily focus on palm oil production, making it readily available for dietary purposes.[21]

Other major dietary sources of TCTs include rice bran oil, which is rich in γ-TCT but contains only trace amounts of α-TCT. Newly discovered forms of TCT, desmethyl TCTs, were discovered in stabilized and heated rice bran. Rice bran oil is a major by-product of the rice milling industry, also making it readily available for dietary purposes.[22] Although there is not a significant amount of scientific evidence supporting the nutritional benefits of rice bran oil, it is believed to be a healthy and safe oil for culinary purposes.[23]

Other common dietary grains, such as oats, barley, and rye, contain small amounts of TCTs. In oat and barley, α-TCT is the major form. In wheat, β-TCT is the major form of TCT. Processing of these grains for consumption affects the levels of TCTs and TCPs present in these grains. Autoclaving grains increases the levels of TCTs and TCPs, with the exception of β-TCT, which remained unaffected. Drum drying of steamed, rolled oats causes a significant loss in both TCTs and TCPs. Steaming and flaking dehulled oat groats decreases the levels of TCTs, but not TCPs.[24] It is important to note that the amount of palm oil and wheat germ that one would have to consume in order to match published literature claiming a biological effect is approximately 200 g of palm or rice bran oil and 1.5–4 kg of cereals.[8] Other plant products, such as sunflower seeds, walnuts, and olive oil, contain exclusively TCP, as they are products of dicot plants.[25]

9.4   BIOAVAILABILITY OF ORAL TCTs

Research efforts to understand the transport of dietary vitamin E in the past two decades have focused mainly on α-TCP[26–29] (Figure 9.2). These efforts toward understanding the transport of vitamin E led to the discovery of α-TCP transfer protein (TTP). TTP has been shown to be involved in the secretion of α-TCP into blood plasma.[27] TTP readily binds α-TCP, hence its name, and it also has been shown to bind α-TCT but at an order of magnitude lower than α-TCP.[30] Studies to determine the extent to which α-TCT is delivered to tissue have employed the use of a TTP knockout mouse model. Female TTP knockout mice are infertile due to a deficiency

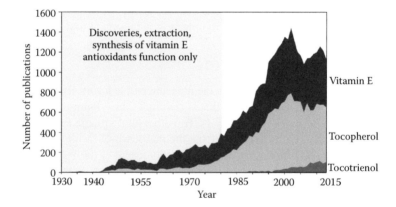

FIGURE 9.2   Vitamin E research publications. Graph showing the number of vitamin E publications per year. TCT research is a small minority of the vitamin E research conducted. TCPs are by far the most studied group, with a significant proportion of vitamin E research not directly specifying the form of vitamin E that is being studied. Frequently, "vitamin E" is a catch-all term for a mixture of various isoforms of TCPs and TCTs.

in α-TCP. Placenta in TTP knockout mice exhibited a decrease in labyrinthine trophoblasts and death of embryos during mid-gestation phase.[31] In a study by Khanna *et al.* in 2005, oral supplementation of α-TCT rescued infertility in female TTP knockout mice, while oral supplementation of α-TCP did not.[32] This study clearly demonstrated that in addition to restoring fertility to TTP knockout mice orally supplemented with TCT, there appeared to be effective delivery of TCTs to vital organs, including the liver, brain, heart, skin, lungs, adipose tissue, and whole blood. This observation suggested that TCTs could be delivered to tissue via a TTP-independent mechanism.

Studies have shown that orally supplemented TCTs not only reach the brain but also accumulate in concentrations that are high enough to protect against brain injury such as stroke.[33,34] This could be achieved by long-term supplementation of TCT. These observations indicates that there could be competition between TCPs and TCTs for specific transport mechanisms, as TCT transport is optimal when dietary TCP uptake is minimized.[30] In a 2005 study by Khanna *et al.,* five generations of rats were studied for 60 weeks.[32] TCTs were found to be delivered to all the vital organs studied, and in some cases, the levels of TCTs were higher than the levels of TCPs. In skin samples collected from rats that were cosupplemented with TCP and TCT, the levels of TCT in the skin were lower compared to the group that was supplemented with TCT only. This study suggested that there is an efficient transport mechanism for TCT and that TCT can be retained in skin tissue over time.[32]

After vitamin E is consumed, it enters blood circulation via intestinal chylomicrons. Chylomicrons are converted to remnant particles, which allows the distribution of vitamin E to circulating lipoproteins, which then facilitates vitamin E distribution to tissues.[29] In the liver, newly absorbed lipids are incorporated into new very low-density lipoproteins (LDL), which places the liver in charge of blood plasma α-TCP levels. In TTP knockout mice, the liver does not release vitamin E into the blood plasma, but rather the liver metabolizes and releases vitamin E in bile.[29] In TTP knockout mice supplemented with α-TCT, TCT levels in the liver were lower than those in peripheral tissue.[32] This suggests that the liver is not in control of delivering TCT to peripheral tissue.

Human clinical trials investigating oral supplementation of TCTs have been conducted to determine bioavailability and pharmacokinetics. A study conducted by Yap *et al.* in 2001 sought to determine if there was a difference in TCT uptake in fasted and fed subjects. Volunteers in the study were assigned to two groups, fed or fasted, and given a single dose of 300 mg of mixed TCTs. TCT concentrations in blood plasma peaked at over 1 μM when volunteers were in the fed state, which was significantly higher than the volunteers in the fasted state. The fed state increased the onset of TCT absorption; however, the elimination half-life for TCTs was still significantly shorter than for TCPs.[35] Sex-based differences in the transport of vitamins are known to exist,[36] and in a rodent model of TCT supplementation, it was reported that female rodents exhibited more efficient tissue delivery of TCT.[32] Based on the information resulting from this rodent study, the outcomes of TCT supplementation in women were investigated by Khosla *et al.,* in which healthy female volunteers were given a single dose of 400 mg of mixed TCTs. Blood plasma, LDL, triglyceride-rich lipoprotein (TGRL), and high-density lipoprotein (HDL) were analyzed for TCT content. TCT content in blood plasma concentrations peaked at 3 μM, while concentrations of 1.7 μM were measured in LDL, 0.9 μM in TGRL, and 0.5 μM in HDL.[37] In the rodent model, it was reported that nM α-TCT concentration in brain tissue was sufficient to protect against stroke.[33] The human trial showing μM range of TCT after supplementation in plasma is encouraging and suggests a beneficial effect in patients with stroke.

9.5   SAFETY OF TCTs

The U.S. Food and Drug Administration (FDA) classifies α-TCT as Generally Recognized as Safe (GRN307) and recognizes that TCT is not a drug with adverse side effects.[38] The palm fruit of *E. guineensis* is the richest natural source of TCTs and has been a part of the human diet for thousands of years.[39] The earliest estimation for the domestication of palm is believed to have occurred in West Africa approximately 5000 years ago.[21] It is therefore not surprising that clinical studies investigating the effects of supplementing patients with TCTs have reported an absence of adverse outcomes due to the supplementation.[40–43] In a clinical trial conducted by Patel et al., 16 healthy subjects received 400 mg of mixed (α-, γ-, and δ-) TCTs daily for 12 weeks with no reported adverse side effects.[44] In a separate clinical trial conducted by Gopalan et al., 121 volunteers aged 35 years or older were randomized to receive placebo or 200 mg of TCTs twice daily for a period of two years. This study also reported no adverse side effects that required a patient to be excluded from the study.[41]

Supplemental antioxidants, particularly vitamin E, has become popular as a method to fight LDL oxidation and curb atherogenesis.[45] Later meta-analyses of these supplemental studies have surprisingly revealed that vitamin E increased all-cause mortality[45] and had a nonsignificant effect in reducing the risk of cardiovascular diseases and cancer.[46] A meta-analysis of α-TCP supplementation clinical studies conducted by Gee *et al.* in 2011 revealed that the supplementation of naturally sourced α-TCP significantly increased all-cause mortality. It was suggested that this higher incidence of all-cause mortality is due to the suppression of the bioavailability of other forms of vitamin E.[47] This study also showed the absence of statistical differences in clinical trials when synthetically sourced α-TCP was used as the vitamin E source.[47] Continued clinical trials investigating the supplementation of the other forms of vitamin E, with increased emphasis on TCTs, still need to be performed.

9.6   FUNCTIONAL DISTINCTIVENESS
OF VITAMIN E FAMILY MEMBERS

All eight members of vitamin E share close structural similarities and common biological importance as lipophilic antioxidants at higher concentrations; however, each member of the vitamin E family also exhibits unique biological functions independent of their role as antioxidant molecules[48]. The new biological functions of vitamin E have been shown to affect cell signal transduction and gene expression, both *in vitro* and *in vivo*.

One of the first characterized antioxidant-independent functions of vitamin E involved the role of α-TCP in inhibiting platelet adhesion. It was discovered that α-TCP specifically inhibited platelet adhesion by decreasing the extent of pseudopodia protrusions in activated platelets.[49,50] Further research also identified biochemical pathways that were shown to be modulated by α-TCP. Protein kinase C activity is specifically inhibited by α-TCP.[51]

α-TCP is effective in posttranscriptional inhibition of phospholipase A2 (PLA$_2$) and 5-lipoxygenase but activates diacylglycerol kinase and phosphatase 2A. α-TCP also transcriptionally regulates the genes that code for α-TTP,[52] matrix metalloprotease, and α-tropomycin.[53] As a result of the biochemical pathways that α-TCP affects, certain cell behaviors such as cell proliferation and monocyte adhesion are directly modulated by α-TCP.[51] The different forms of TCP also possess unique biological activities. γ-TCP metabolites, but not α-TCP metabolites, exhibit increased natriuretic activity.[54]

TCTs differ from TCPs by the degree of unsaturation of the hydrocarbon tail. TCPs have a saturated tail, whereas TCTs have an unsaturated tail consisting of three trans double bonds. It has also been reported that unsaturated hydrocarbon tail of TCTs allows efficient penetration into tissue that has saturated fatty acid layers, such as the brain.[55]

Specifically, α-TCT has been shown to protect against inducible neurodegeneration by regulating modulators of cell death such as 12-lipoxygenase in nanomolar concentrations.[56–58] Oral supplementation of TCTs in humans has been shown to increase blood plasma TCT concentrations to levels that are significantly greater than required for neuroprotection.[43,57] α-TCT concentrations in the micromolar range have been shown to regulate 3-hydroxy-3-methylglutaryl coenzyme A (HMG-CoA) reductase, an enzyme that regulates cholesterol synthesis.[59] TCTs also increased the mean lifespan of *Caenorhabditis elegans* by decreasing oxidative protein damage.[59]

9.7 NEUROPROTECTIVE FUNCTIONS OF TCT

Vitamin E is essential for normal neurological structure and function,[60,61] and its critical significance in neurological health and diseases was identified several decades ago (Figure 9.2). Recent studies on the neuroprotective properties of TCTs have recognized the significance of α-TCT as the most potent form of vitamin E. It has been reported that α-TCT, but not α-TCP, protects brain against stroke.[33,38,62,63]

9.7.1 STROKE

The World Health Organization reported in 2015 that stroke, which is also known as brain attack, is the second leading cause of death in the age group 60 years and above and a fifth leading cause in people aged 15 to 59 years old around the world. There are three major types of stroke: ischemic (the most common, accounting for about 87% of all strokes), hemorrhagic, and transient ischemic attack (TIA). Ischemic strokes occur when a clot blocks a blood vessel in the brain and prevents the delivery of blood containing glucose and oxygen, which are essential for brain tissue. Hemorrhagic stroke occurs when a blood vessel within the brain ruptures, leaking blood onto surrounding tissue. TIA, or mini strokes, occurs when a vessel blockage occurs for a short period of time, typically less than five minutes. TIA usually precedes a major stroke. More than 33% of people who experience a TIA event will experience a major ischemic stroke within one year.[64]

Current therapeutics for stroke patients are limited. The most widely used treatment is intravenous thrombolysis. However, this treatment is not ideal and must be administered within a specific time frame after stroke onset, which limits its use to approximately 10% of ischemic stroke cases.[65,66] If we explore the stroke-related research, most of the research has focused on the identification of neuroprotective agents, which typically antagonize a molecular pathway or cellular event that contributes to neuronal cell death. Numerous targets have been studied over the years, such as ion channel antagonists, calcium blockers, and antioxidants, which are often met with promising preclinical outcomes followed by failure in a clinical setting.[65] This failure of clinical trials in stroke research suggests that there is an imminent need to change the current strategy when looking for therapeutic options. Based on the complex pathophysiological cascade associated with stroke, targeting a single mechanism alone is likely to have a limited effect on stroke outcome. Therefore, multimodal mechanisms (i.e., neuro and vascular) may be more effective to attenuate stroke.[66] The identification of therapies that combine neuroprotective and vascular protective properties may be critical in the fight against stroke. TCT is readily available in natural food sources and has been shown to have both neuroprotective and vascular protective properties and therefore is a viable option as potent stroke therapeutics.

TCT confers neuroprotection through both its antioxidant properties and modulation of molecular pathways, as described in the following.

9.7.1.1 Antioxidative Neuroprotection

The brain tissue contains a high concentration of polyunsaturated fatty acids that are sensitive to oxidative stress.[67] Approximately 20% of all brain polyunsaturated fatty acids are arachidonic acid (AA) and docosahexaenoic acid (DHA), and both are important components of phospholipid membranes. Neither of these molecules are synthesized *de novo* in the brain. They are either obtained directly from diet or can be synthesized in the liver from the precursor molecules linoleic and α-linoleic acid.[68] Both AA and DHA play a major role in membrane fluidity, gene transcription, and signal transduction.[69] Free radical insult resulting from brain injury, i.e., stroke, can lead to an increase in lipid peroxidation products, which are known to be neuromodulators of the cell death cascade.[70] Free radicals react with the double bonds of polyunsaturated fatty acids to produce alkyl radicals, which react with molecular oxygen to produce a peroxyl radical. These peroxyl radicals can then abstract hydrogen from adjacent polyunsaturated fatty acids, leading to the production of another alkyl radical and creating a chain reaction of lipid oxidation.[71] The degradation of lipid peroxides leads to increased concentrations of α,β-unsaturated aldehydes that can bind to proteins and alter their function. Vitamin E is known to be the chain breaking antioxidant that protects lipid membranes from peroxidation.[72]

TCTs are thought to be more effective antioxidants than TCPs. This is due to the unsaturated isoprenoid tail component of TCTs that allows it to assume unique three-dimensional structures and greater penetration into tissue with a saturated fatty layer.[73] α-TCT exhibits significantly higher peroxyl radical scavenging activity and greater protection against reactive oxygen species (ROS) compared to α-TCPs in liposomes and liver microsomes. TCTs were also more effective at combating cytochrome P-450-induced oxidation.[74] It has been demonstrated that the TCT-rich fraction of palm oil was more effective than α-TCP alone in preventing lipid oxidative damage to rat brain mitochondria.[75]

9.7.1.2 Regulation of pp60c-Src Tyrosine Kinase

α-TCT has shown to be neuroprotective against glutamate-induced neurotoxicity at nanomolar concentrations,[58,76] and this nanomolar concentration of TCT is far below the required concentration for effective antioxidant activity, which suggests neuroprotection independent of antioxidant properties. Nanomolar amounts of α-TCT blocked glutamate-induced cell death by suppressing glutamate-induced early activation of c-Src kinase.[58] Activation of pp60c-Src (c-Src) tyrosine kinase as a result of pathological glutamate release contributes to neurodegeneration. Inhibition or deficiency of c-Src in mice has been shown to have a beneficial effect in stroke.[77] In an *in vitro* model of glutamate-induced cell death employing HT4 neuronal cells, α-TCT exhibited dose-dependent protection against viability loss from the range of 50 to 250 nM. It was found that α-TCT strongly inhibited the induction of c-Src tyrosine kinase activity, resulting in neuroprotection.[58] The involvement of c-Src tyrosine kinase activity in glutamate-induced cell death was determined using experiments in which cells overexpressed either catalytically active or inactive c-Src. α-TCT was able to prevent glutamate-induced death in cells that were overexpressing active c-Src, suggesting that α-TCT either inhibited an event that leads to c-Src activation or c-Src kinase activity.[58] Further experiments have shown that α-TCT inhibits c-Src activation by preventing the necessary reorganization of the SH domains but does not affect c-Src kinase activity. Glutamate-induced c-Src activation represents a major check-point in oxytosis (oxidative stress-induced programmed cell death). α-TCT-mediated inhibition of c-Src activation prevents oxytosis[78] and causes neuroprotection[58] (Figure 9.3).

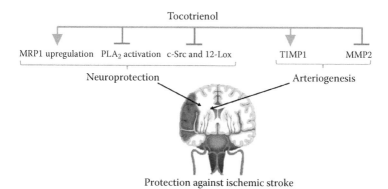

Tocotrienol

MRP1 upregulation   PLA$_2$ activation   c-Src and 12-Lox   TIMP1   MMP2

Neuroprotection   Arteriogenesis

Protection against ischemic stroke

FIGURE 9.3 **(See color insert.)** Specific molecular targets of TCT for protection against stroke. TCT confers protection against stroke through two main pathways: neuroprotection and arteriogenesis. Molecular targets of TCT for neuroprotection include upregulating MRP1 synthesis and inhibiting PLA$_2$ activation, c-Src, and 12-LOX. Arteriogenesis targets are upregulating TIMP1 expression and inhibiting MMP2 activity.

9.7.1.3   Inhibition of 12-LOX

In response to the depletion of glutathione, neural cells activate 12-lipoxygenase, which upon activation orchestrates a production of peroxides and an accumulation of intracellular Ca^{2+}, which eventually leads to cell death.[78,79] There are three main lipoxygenases expressed in the central nervous system: 5-, 12-, and 15-LOX. Each LOX inserts molecular oxygen into the 5-, 12-, or 15-carbon atoms of AA, hence the name of each protein. Studies have shown that 12-LOX plays a central role in executing glutamate-induced neurodegeneration.[57,78,79] 12-LOX inhibitor, BL15, inhibited glutamate-induced neurotoxicity in HT4 and immature primary cortical neurons, and interestingly, neurons isolated from 12-LOX-deficient mice were observed to be resistant to glutamate-induced death. These observations confirmed the role of 12-LOX in neurodegeneration.[57] Studies done in a cell free system with α-TCT and pure 12-LOX indicated that α-TCT was directly involved in suppressing 12-LOX-mediated AA metabolism.[57] These findings suggest that α-TCT, present in physiologically relevant concentrations, is a potent inhibitor of 12-LOX-mediated AA metabolism and neurodegeneration (Figure 9.3).

9.7.1.4   Phospholipase A2

PLA_2 isoenzymes are a large group of enzymes that are responsible for the hydrolysis of the sn-2 ester bond of glycerophospholipids. This hydrolysis produces a free fatty acid, such as AA, and a lysophospholipid. Two categories of PLA_2 enzymes, $sPLA_2$ (secreted small molecular weight) and $cPLA_2$ (cytosolic calcium dependent), have defined roles in pathologies that involve AA metabolism.[81] Following release from the cell membrane lipid bilayer by PLA_2, the free AA has three possible outcomes: (i) reincorporation into the membrane phospholipids, (ii) diffuse out of the cell, or (iii) be metabolized.

In pathological conditions, free AA undergoes oxidative metabolism via nonenzymatic and enzymatic pathways. The uncontrolled oxidative metabolism of AA is known as "AA cascade." In pathological conditions such as stroke, significant increases in PLA_2 and AA levels have been shown in stroke affected brain.[63,81] α-TCT at nanomolar concentrations has been shown to inhibit glutamate-induced $cPLA_2$ activity[82] in neuronal cells. TCT has also been reported to disrupt AA cascade on the basis of both antioxidant-dependent and independent mechanism[33,57,63] (Figure 9.3).

9.7.1.5   Multidrug Resistance Protein 1

Multidrug Resistance Protein 1 (MRP1) is a transport protein that is involved in the export of glutathione disulfide (GSSG) from cells during oxidative stress.[83] Increased oxidative stress resulting from brain injury, such as ischemic stroke, causes the rapid oxidation of GSH to GSSG and also impairs GSSG reductase activity, resulting in an increase in the GSSG/GSH ratio within the cell.[81,84] GSSG has been identified as a potent inducer of GSH-deficient neural cell death via the 12-LOX pathway.[84] Strategies directed at improving cellular GSSG clearance may be effective in minimizing oxidative stress-related tissue injury such as stroke. Orally supplemented α-TCT was found to enhance MRP1 expression in the context of stroke-induced brain injury and protects the brain against stroke in mice[38] (Figure 9.3).

9.7.1.6 microRNA Targets of TCT

MicroRNAs (miRNAs) are ~21–23-nucleotide single-stranded RNAs, involved in posttranscriptional gene silencing that alter protein expression.[85–87] A growing body of publications in miRNA research confirm its critical role in almost every aspect of biology, including development as well as pathological conditions. These findings suggest that modulation of miRNA expression may serve as a novel therapeutic tool.[87] Major changes in the miRNA transcriptomes have also been observed in various pathological conditions related to the brain.[88]

The miR-29 family has four closely related miR-29 precursors: hsa-miR-29a, hsa-miR-29b-1, hsa-miR-29b-2, and hsa-miR-29c. miR-29-b-1 and miR-29b-2 share identical mature sequences and are collectively referred to as miR-29b. The mature forms of miR-29 family members are highly conserved in humans, mouse, and rats. miR-29b has recently been recognized as a neuronal cell survival factor by silencing proapoptotic BH3-only family genes.[89] In another study, functional significance of miR-29b in stroke was explored. Loss of miR-29b was reported in stroke-affected mouse brain. Such loss contributed by the activity of the 12-LOX pathway linked AA metabolism to miR-dependent mechanisms in stroke.[63] To investigate the specific significance of miR-29b loss in mouse brain, rescue experiments were conducted where miR-29b mimic was delivered to the area at risk for stroke. Delivery of miR-29b mimic to the brain improves post-stroke sensorimotor function and attenuates stroke-induced neurodegeneration. Oral supplementation of TCT, a 12-LOX inhibitor, rescued stroke-induced loss of miR-29b and minimized lesion size.[63]

miR-199a-5p has also been recognized as neuronal cell survival factor and TCT sensitive.[38] miR-199a-5p is a hypoxia-sensitive miRNA and its role in regulating hypoxia-inducible factor 1α has been established.[90] The hypoxia aspect of ischemic stroke may be responsible for the downregulation of miR199a-5p post-stroke. Park *et al.* showed involvement of miR-199a-5p in regulating MRP1. In this study, α-TCT supplementation decreased the miR-199a-5p levels and thereby increased the expression of MRP1, allowing enhanced neuroprotection.[38]

9.7.2 CEREBROVASCULAR PROTECTION

It has been argued that neuroprotection alone is not an effective strategy to mitigate damage from ischemic stroke.[66] Within the brain, there exists a network of small blood vessels that enables collateral circulation that can stabilize cerebral blood flow when the major vessels are obstructed.[91] These collateral vessels occur through leptomeningeal arterioles that overlap and provide secondary routes for blood flow from the middle cerebral artery to the anterior and posterior cerebral arteries.[92] Within 24 hours of blood vessel occlusion, arteriogenic collateral growth occurs.[93] Arteriogenesis in response to a pathology is marked by early expression of tissue inhibitor of metalloproteinase 1 (TIMP1) in growing collateral vessels.[94] TIMP1 is a glycoprotein responsible for the inhibition of matrix metalloproteinase 2 (MMP2). TIMP1 binds to MMP2 in a 1:1 stoichiometric ratio to disable it and provide control of MMP2 activity. TIMP1 upregulation correlates with increased angiogenic activity by locally inhibiting MMP2.[66] TCT supplementation in canines was shown to improve blood flow in collateral vessels and improve post-stroke arteriogenesis[34] by upregulating TIMP1 expression[34] and inhibiting MMP2 activity.[34]

9.7.3 COGNITIVE DEFICITS

A progressive decline in cognitive function is one of the many side effects of aging and other pathologies. As the brain ages, it undergoes many changes, such as loss of volume, weight, neuronal cell death, and neurochemical alterations.[95–97] Sustained, elevated levels of ROS within the brain are also thought to contribute to neurodegeneration through lipid peroxidation, protein oxidation, and DNA damage.[98] Indeed, increased levels of lipid peroxidation products, for instance malondialdehyde (MDA), have been correlated with memory deficits.[99] The primary risk factor for Alzheimer's disease (AD) is age.[100] Excessive alcohol consumption may also contribute to cognitive decline through increased ROS production.[101] Nutritional antioxidants, TCT, are unique molecules to combat cognitive decline as they function not only as potent antioxidants but also as biochemically active molecules that inhibit lipid peroxidation and cell death pathways.

9.7.3.1 Aging

In an aged brain, there are increases in oxidative stress and accumulation of amyloid beta peptides, as well as decreases in proteasomal activity and mitochondrial function. Aging also brings about a loss of efficiency for endogenous antioxidant systems, which contributes to increased oxidative stress.[102] These factors all contribute to age-related decline in cognitive function.

In an *in vivo* rat model, TCT was able to reduce the cognitive deficit of aged rats.[103] Taridi *et al.* specifically investigated the effect of TCT-rich fraction (TRF) on reference memory and working memory in aged rats. Their studies have shown that aged rats that are supplemented with TRF fared better in place navigation tests, determined by a Morris water maze, suggesting increased reference memory capability. Rats supplemented with TRF also performed better in a working memory task that directly measured the path that the rats took, after being trained, to find a hidden platform.[103] These behavior tests provide evidence of rescue of cognitive impairment in TRF supplemented animals.

DNA damage and lipid peroxidation in the brains of aged rodents are increased when compared to younger rodents. Supplementation with TRF was able to decrease the amount of DNA damage in older rodents as well as decrease lipid peroxidation. Levels of lipid peroxidation were measured by determining the concentration of MDA in blood plasma. The activity of antioxidant enzymes such as superoxide dismutase, catalase, and glutathione were all increased in the TRF supplemented mice when compared to age-matched controls.[103] Taken together, the decrease in DNA damage and lipid peroxidation, the increase in antioxidant enzyme activity, and improvement in behavioral function provide promising evidence for the ability of TCT to attenuate age-related cognitive deficits, warranting further clinical research.

9.7.3.2 Alzheimer's Disease

AD has two major components in its pathology: neuritic plaques (extracellular deposits of amyloid beta in the brain gray matter) and neurofibrillary tangles (aggregates of hyperphosphorylated tau protein).[100] Neuritic plaques, especially Aβ oligomers, are the most neurotoxic plaques, causing structural damage to neurons and loss of function.[104] Research using *in vivo* models has suggested that levels of Aβ plaques correlate with dietary cholesterol. Studies performed in a transgenic mouse model[105] and a rabbit model[106] showed that increased dietary cholesterol increased Aβ plaques within the brain. It was also shown that a decrease in cholesterol intake could attenuate Aβ plaque formation. Inhibition of HMG-CoA reductase, an important enzyme in the mevalonate pathway that leads to cholesterol synthesis, produced similar effects.[100]

TCT has been shown to be effective in blocking HMG-CoA reductase activity by triggering its degradation.[107] Their method of action is similar to statins. TCT and statins share other biological activities such as anticancer and anti-inflammatory properties.[108,109] TCT blocking of HMG-CoA prevents the formation of mevalonate, thereby suppressing the synthesis of cholesterol. This effect was not seen in TCPs, due primarily to their saturated phytyl tail.[100]

Glutamate excitotoxicity also plays an important role in AD. Early in the development of AD, there is chronic neuron depolarization that leads to pathological Ca^{2+} influx. Disruption of glutamate-mediated pathways occurs in the same areas as the buildup of Aβ plaque, suggesting that they are both involved in the pathology of AD.[110] TCTs have been shown to mitigate damage caused by glutamate excitotoxicity, which offers another route of protection against AD.

9.7.3.3 Alcohol-Induced Cognitive Deficits

Exposure of neurons to ethanol causes damage by increasing oxidative stress and activating the inflammatory cascade.[111] This damage can lead to cognitive impairment, especially in the developing brain.[112] Rat models of chronic ethanol exposure displayed deficits in the Morris water maze behavioral test compared to control rats. This deficit was decreased when rats were co-treated with α-TCT and α-TCP. However, TCT was better at preventing the cognitive deficits related to chronic alcohol consumption.[112]

TCT was also shown to attenuate ethanol-induced behavioral deficits in developing brains. Rat pups that were treated with ethanol displayed impaired cognition, which was correlated with increased acetylcholinesterase activity, and increased neuroinflammation. Pups that were treated with ethanol and supplemented with TCT improved their cognitive ability and showed decreased symptoms of neuroinflammation.[111]

9.8 CANCER

Cancer is a leading cause of death worldwide, and brain tumors are a leading cause of cancer deaths in children.[113] For cancer treatment, TCT represents a natural compound to augment current drugs used in cancer therapy. The anticancer properties that TCTs possess are mostly from their ability to suppress cell proliferation and induce apoptosis in various cancer cell lines.[114–116] *In vivo* investigations into the anticancer properties of TCT found that α- and γ-TCT are effective against Ehrlich carcinoma, IMC carcinoma, and sarcoma 180.[117] TCTs are known to be able to cross the blood–brain barrier, making them candidates to aid in the treatment of brain cancers.[118]

The effects of TCTs as anticancer compounds have been tested in the A549 (human lung cancer) and U87MG (human glioblastoma) cell lines. Cancer cell lines treated with 1–100 μM of TCT were shown to be cytotoxic and caused a loss of cell viability. The same treatment of TCT in noncancerous MRC5 cells (human fetal lung fibroblast cells), however, caused no apparent loss of cell viability. In addition, cell proliferation was clearly stunted in the cancerous cells as a result of the treatment when compared to the noncancerous cells. Histological analysis of the cells revealed that the cells' morphology clearly showed signs of apoptosis, such as nuclear fragmentation and cytoplasmic extension. Analysis of DNA damage using the comet assay confirmed the induction of apoptosis. Comparable DNA damage was not seen in the noncancerous cells. TCTs were also found to be activators of caspase-8 activity, which also leads to the induction of apoptosis.[119]

β-TCT was also shown to induce mitochondria-mediated apoptosis in U87MG cells. Treatment of U87MG cells with β-TCT led to a loss of cell viability in a dose- and time-dependent manner. The cells displayed the hallmarks of mitochondrial-mediated apoptosis, including chromatin condensation, nuclear fragmentation, and cytoplasmic extension β-TCT-triggered caspase-8 activity and increased Bid and Bax protein levels, indicating that the pathway for apoptosis is active. The activation of caspase-8 indicates that the intrinsic mitochondrial apoptosis pathway is active. These findings show that β-TCT is able to induce apoptosis in cancer cells, causing the observed loss of cell viability and inhibition of cell proliferation.[120]

A second study conducted by Rahman *et al.* in human glioblastoma cell lines (1321N1, SW1783, and LN18 cells) confirmed these results. γ-TCT was able to potently inhibit cell proliferation in a dose-dependent manner and induce apoptosis in all cell lines. The cells did display differences in sensitivity to γ-TCT, but all of the concentrations required for a physiological effect were lower than reported blood plasma concentrations after supplementation in humans.[121] Additional research investigating the use of TCTs for the brain cancer treatment needs to be carried out; however, current *in vitro* research is promising.

9.9   CONCLUSION

Vitamin E TCTs exhibit powerful biological effects due to their strong antioxidant potential and involvement in regulating biological pathways. The vast majority of research on vitamin E focuses on TCPs, which often becomes generalized to include the entire vitamin E family, including TCTs, as a whole. TCTs mediate unique biological functions that are not shared by TCPs. The newly emerging roles for TCTs in modulating neurological and vascular functions based on animal studies warrant further investigations at the clinical level. The U.S. FDA recognizes them as Generally Recognized as Safe, which allows them to be more easily included in treatment regimens than drugs as they have no negative side effects at physiologically relevant concentrations. In the light of this, and based on published literature, it is plausible that orally supplemented TCTs could be delivered to the brain in high enough concentrations to protect against stroke, cognitive decline, and cancer without any detrimental effects on brain physiology. It is imperative that clinical studies utilizing TCTs be performed in order to advance the potential application of TCTs as therapeutic agents in humans.

REFERENCES

1. Evans HM, Bishop KS. On the existence of a hitherto unrecognized dietary factor essential for reproduction. *Science.* 1922;56:650–651.
2. Litwack G. *Vitamin A.* San Diego, CA: Elsevier Academic Press; 2007.
3. Dormann P. Functional diversity of tocochromanols in plants. *Planta.* 2007;225:269–276.
4. Jiang Q. Natural forms of vitamin E: Metabolism, antioxidant, and anti-inflammatory activities and their role in disease prevention and therapy. *Free Radic Biol Med.* 2014;72:76–90.
5. Cardenas E, Ghosh R. Vitamin E: A dark horse at the crossroad of cancer management. *Biochem Pharmacol.* 2013;86:845–852.
6. Colombo ML. An update on vitamin E, tocopherol and tocotrienol—Perspectives. *Molecules.* 2010;15:2103–2113.
7. Zingg JM. Vitamin E: A role in signal transduction. *Annu Rev Nutr.* 2015;35:135–173.
8. Sen CK, Khanna S, Roy S. Tocotrienols in health and disease: The other half of the natural vitamin E family. *Mol Aspects Med.* 2007;28:692–728.
9. Peh HY, Tan WS, Liao W, Wong WS. Vitamin E therapy beyond cancer: Tocopherol versus tocotrienol. *Pharmacol Ther.* 2016;162:152–169.
10. Burton GW, Joyce A, Ingold KU. First proof that vitamin E is major lipid-soluble, chain-breaking antioxidant in human blood plasma. *Lancet.* 1982;2:327.
11. Kagan VE, Serbinova EA, Packer L. Recycling and antioxidant activity of tocopherol homologs of differing hydrocarbon chain lengths in liver microsomes. *Arch Biochem Biophys.* 1990;282:221–225.
12. Kagan VE, Serbinova EA, Packer L. Generation and recycling of radicals from phenolic antioxidants. *Arch Biochem Biophys.* 1990;280:33–39.
13. Xie W, Merrill JR, Bradshaw WS, Simmons DL. Structural determination and promoter analysis of the chicken mitogen-inducible prostaglandin g/h synthase gene and genetic mapping of the murine homolog. *Arch Biochem Biophys.* 1993;300:247–252.
14. Venkatesh TV, Karunanandaa B, Free DL, Rottnek JM, Baszis SR, Valentin HE. Identification and characterization of an arabidopsis homogentisate phytyltransferase paralog. *Planta.* 2006;223:1134–1144.
15. Porfirova S, Bergmuller E, Tropf S, Lemke R, Dormann P. Isolation of an arabidopsis mutant lacking vitamin E and identification of a cyclase essential for all tocopherol biosynthesis. *Proc Natl Acad Sci U S A.* 2002;99:12495–12500.
16. Sen CK, Khanna S, Rink C, Roy S. Tocotrienols: The emerging face of natural vitamin E. *Vitam Horm.* 2007;76:203–261.
17. Atkinson J. Chemical investigations of tocotrienols: Isotope substitution, fluorophores and a curious curve. *Nesaretnam K.* 2006;6:22.
18. Cahoon EB, Hall SE, Ripp KG, Ganzke TS, Hitz WD, Coughlan SJ. Metabolic redesign of vitamin E biosynthesis in plants for tocotrienol production and increased antioxidant content. *Nat Biotechnol.* 2003;21:1082–1087.
19. Rippert P, Scimemi C, Dubald M, Matringe M. Engineering plant shikimate pathway for production of tocotrienol and improving herbicide resistance. *Plant Physiol.* 2004;134:92–100.

20. Horvath G, Wessjohann L, Bigirimana J, Jansen M, Guisez Y, Caubergs R, Horemans N. Differential distribution of tocopherols and tocotrienols in photosynthetic and non-photosynthetic tissues. *Phytochemistry.* 2006;67:1185–1195.

21. Solomons NW, Orozco M. Alleviation of vitamin A deficiency with palm fruit and its products. *Asia Pac J Clin Nutr.* 2003;12:373–384.

22. Qureshi AA, Mo H, Packer L, Peterson DM. Isolation and identification of novel tocotrienols from rice bran with hypocholesterolemic, antioxidant, and antitumor properties. *J Agric Food Chem.* 2000;48:3130–3140.

23. Sugano M, Koba K, Tsuji E. Health benefits of rice bran oil. *Anticancer Res.* 1999;19:3651–3657.

24. Bryngelsson S, Dimberg LH, Kamal-Eldin A. Effects of commercial processing on levels of antioxidants in oats (*Avena sativa* l.). *J Agric Food Chem.* 2002;50:1890–1896.

25. Tarrago-Trani MT, Phillips KM, Lemar LE, Holden JM. New and existing oils and fats used in products with reduced trans-fatty acid content. *J Am Diet Assoc.* 2006;106:867–880.

26. Blatt DH, Leonard SW, Traber MG. Vitamin E kinetics and the function of tocopherol regulatory proteins. *Nutrition.* 2001;17:799–805.

27. Kaempf-Rotzoll DE, Traber MG, Arai H. Vitamin E and transfer proteins. *Curr Opin Lipidol.* 2003;14:249–254.

28. Traber MG, Arai H. Molecular mechanisms of vitamin E transport. *Annu Rev Nutr.* 1999;19:343–355.

29. Traber MG, Burton GW, Hamilton RL. Vitamin E trafficking. *Ann N Y Acad Sci.* 2004;1031:1–12.

30. Hosomi A, Arita M, Sato Y, Kiyose C, Ueda T, Igarashi O, Arai H, Inoue K. Affinity for alpha-tocopherol transfer protein as a determinant of the biological activities of vitamin E analogs. *FEBS Lett.* 1997;409:105–108.

31. Jishage K, Arita M, Igarashi K, Iwata T, Watanabe M, Ogawa M, Ueda O, Kamada N, Inoue K, Arai H, Suzuki H. Alpha-tocopherol transfer protein is important for the normal development of placental labyrinthine trophoblasts in mice. *J Biol Chem.* 2001;276:1669–1672.

32. Khanna S, Patel V, Rink C, Roy S, Sen CK. Delivery of orally supplemented alpha-tocotrienol to vital organs of rats and tocopherol-transport protein deficient mice. *Free Radic Biol Med.* 2005;39:1310–1319.

33. Khanna S, Roy S, Slivka A, Craft TK, Chaki S, Rink C, Notestine MA, DeVries AC, Parinandi NL, Sen CK. Neuroprotective properties of the natural vitamin E alpha-tocotrienol. *Stroke; a journal of cerebral circulation.* 2005;36:2258–2264.

34. Rink C, Christoforidis G, Khanna S, Peterson L, Patel Y, Khanna S, Abduljalil A, Irfanoglu O, Machiraju R, Bergdall VK, Sen CK. Tocotrienol vitamin E protects against preclinical canine ischemic stroke by inducing arteriogenesis. *J Cereb Blood Flow Metab.* 2011;31:2218–2230.

35. Yap SP, Yuen KH, Wong JW. Pharmacokinetics and bioavailability of alpha-, gamma- and delta-tocotrienols under different food status. *J Pharm Pharmacol.* 2001;53:67–71.

36. Garry PJ, Hunt WC, Bandrofchak JL, VanderJagt D, Goodwin JS. Vitamin A intake and plasma retinol levels in healthy elderly men and women. *Am J Clin Nutr.* 1987;46:989–994.

37. Khosla P, Patel V, Whinter JM, Khanna S, Rakhkovskaya M, Roy S, Sen CK. Postprandial levels of the natural vitamin E tocotrienol in human circulation. *Antioxid Redox Signal.* 2006;8:1059–1068.

38. Park HA, Kubicki N, Gnyawali S, Chan YC, Roy S, Khanna S, Sen CK. Natural vitamin E alpha-tocotrienol protects against ischemic stroke by induction of multidrug resistance-associated protein 1. *Stroke.* 2011;42:2308–2314.

39. Sundram K, Sambanthamurthi R, Tan YA. Palm fruit chemistry and nutrition. *Asia Pac J Clin Nutr.* 2003;12:355–362.

40. Daud ZA, Tubie B, Sheyman M, Osia R, Adams J, Tubie S, Khosla P. Vitamin E tocotrienol supplementation improves lipid profiles in chronic hemodialysis patients. *Vasc Health Risk Manag.* 2013;9:747–761.

41. Gopalan Y, Shuaib IL, Magosso E, Ansari MA, Abu Bakar MR, Wong JW, Khan NA, Liong WC, Sundram K, Ng BH, Karuthan C, Yuen KH. Clinical investigation of the protective effects of palm vitamin E tocotrienols on brain white matter. *Stroke.* 2014;45:1422–1428.

42. Heng EC, Karsani SA, Abdul Rahman M, Abdul Hamid NA, Hamid Z, Wan Ngah WZ. Supplementation with tocotrienol-rich fraction alters the plasma levels of apolipoprotein a-i precursor, apolipoprotein e precursor, and c-reactive protein precursor from young and old individuals. *Eur J Nutr.* 2013;52:1811–1820.

43. Patel V, Rink C, Gordillo GM, Khanna S, Gnyawali U, Roy S, Shneker B, Ganesh K, Phillips G, More JL, Sarkar A, Kirkpatrick R, Elkhammas EA, Klatte E, Miller M, Firstenberg MS, Chiocca EA, Nesaretnam K, Sen CK. Oral tocotrienols are transported to human tissues and delay the progression of the model for end-stage liver disease score in patients. *J Nutr.* 2012;142:513–519.

44. Patel V, Rink C, Khanna S, Sen CK. Tocotrienols: The lesser known form of natural vitamin E. *Indian J Exp Biol.* 2011;49:732–738.

45. Dotan Y, Lichtenberg D, Pinchuk I. No evidence supports vitamin E indiscriminate supplementation. *Biofactors.* 2009;35:469–473.

46. Lee IM, Cook NR, Gaziano JM, Gordon D, Ridker PM, Manson JE, Hennekens CH, Buring JE. Vitamin E in the primary prevention of cardiovascular disease and cancer: The women's health study: A randomized controlled trial. *JAMA.* 2005;294:56–65.

47. Gee PT. Unleashing the untold and misunderstood observations on vitamin E. *Genes Nutr.* 2011;6:5–16.

48. Schneider C. Chemistry and biology of vitamin E. *Mol Nutr Food Res.* 2005;49:7–30.

49. Jandak J, Steiner M, Richardson PD. Reduction of platelet adhesiveness by vitamin E supplementation in humans. *Thromb Res.* 1988;49:393–404.

50. Steiner M. Vitamin E: More than an antioxidant. *Clin Cardiol.* 1993;16:I16–I18.

51. Boscoboinik D, Szewczyk A, Hensey C, Azzi A. Inhibition of cell proliferation by alpha-tocopherol. Role of protein kinase c. *J Biol Chem.* 1991;266:6188–6194.

52. Thakur V, Morley S, Manor D. Hepatic alpha-tocopherol transfer protein: Ligand-induced protection from proteasomal degradation. *Biochemistry.* 2010;49:9339–9344.

53. Zingg JM, Azzi A. Non-antioxidant activities of vitamin E. *Curr Med Chem.* 2004;11:1113–1133.

54. Wagner KH, Kamal-Eldin A, Elmadfa I. Gamma-tocopherol—An underestimated vitamin? *Ann Nutr Metab.* 2004;48:169–188.

55. Suzuki YJ, Tsuchiya M, Wassall SR, Choo YM, Govil G, Kagan VE, Packer L. Structural and dynamic membrane properties of alpha-tocopherol and alpha-tocotrienol: Implication to the molecular mechanism of their antioxidant potency. *Biochemistry.* 1993;32:10692–10699.

56. Khanna S, Roy S, Parinandi NL, Maurer M, Sen CK. Characterization of the potent neuroprotective properties of the natural vitamin E alpha-tocotrienol. *J Neurochem.* 2006;98:1474–1486.

57. Khanna S, Roy S, Ryu H, Bahadduri P, Swaan PW, Ratan RR, Sen CK. Molecular basis of vitamin E action: Tocotrienol modulates 12-lipoxygenase, a key mediator of glutamate-induced neurodegeneration. *J Biol Chem.* 2003;278:43508–43515.

58. Sen CK, Khanna S, Roy S, Packer L. Molecular basis of vitamin E action. Tocotrienol potently inhibits glutamate-induced pp60(c-src) kinase activation and death of ht4 neuronal cells. *J Biol Chem.* 2000;275:13049–13055.

59. Pearce BC, Parker RA, Deason ME, Dischino DD, Gillespie E, Qureshi AA, Volk K, Wright JJ. Inhibitors of cholesterol biosynthesis. 2. Hypocholesterolemic and antioxidant activities of benzopyran and tetrahydronaphthalene analogues of the tocotrienols. *J Med Chem.* 1994;37:526–541.

60. Muller DP, Goss-Sampson MA. Role of vitamin E in neural tissue. *Ann N Y Acad Sci.* 1989;570:146–155.

61. Muller DP, Goss-Sampson MA. Neurochemical, neurophysiological, and neuropathological studies in vitamin E deficiency. *Crit Rev Neurobiol.* 1990;5:239–263.

62. Khanna S, Heigel M, Weist J, Gnyawali S, Teplitsky S, Roy S, Sen CK, Rink C. Excessive alpha-tocopherol exacerbates microglial activation and brain injury caused by acute ischemic stroke. *FASEB J.* 2015;29:828–836.

63. Khanna S, Rink C, Ghoorkhanian R, Gnyawali S, Heigel M, Wijesinghe DS, Chalfant CE, Chan YC, Banerjee J, Huang Y, Roy S, Sen CK. Loss of mir-29b following acute ischemic stroke contributes to neural cell death and infarct size. *J Cereb Blood Flow Metab.* 2013;33:1197–1206.

64. Easton JD, Saver JL, Albers GW, Alberts MJ, Chaturvedi S, Feldmann E, Hatsukami TS, Higashida RT, Johnston SC, Kidwell CS, Lutsep HL, Miller E, Sacco RL, American Heart A, American Stroke Association Stroke C, Council on Cardiovascular S, Anesthesia, Council on Cardiovascular R, Intervention, Council on Cardiovascular N, Interdisciplinary Council on Peripheral Vascular D. Definition and evaluation of transient ischemic attack: A scientific statement for healthcare professionals from the American Heart Association/American Stroke Association Stroke Council; Council on Cardiovascular Surgery and Anesthesia; Council on Cardiovascular Radiology and Intervention; Council on Cardiovascular Nursing; and the Interdisciplinary Council on Peripheral Vascular Disease. The American Academy of Neurology affirms the value of this statement as an educational tool for neurologists. *Stroke.* 2009;40:2276–2293.

65. Moretti A, Ferrari F, Villa RF. Neuroprotection for ischaemic stroke: Current status and challenges. *Pharmacol Ther.* 2015;146:23–34.

66. Rogalewski A, Schneider A, Ringelstein EB, Schabitz WR. Toward a multimodal neuroprotective treatment of stroke. *Stroke.* 2006;37:1129–1136.

67. Contreras MA, Greiner RS, Chang MC, Myers CS, Salem N, Jr., Rapoport SI. Nutritional deprivation of alpha-linolenic acid decreases but does not abolish turnover and availability of unacylated docosahexaenoic acid and docosahexaenoyl-CoA in rat brain. *J Neurochem.* 2000;75:2392–2400.

68. Spector AA. Plasma free fatty acid and lipoproteins as sources of polyunsaturated fatty acid for the brain. *J Mol Neurosci.* 2001;16:159–165; discussion 215–121.

69. Jones CR, Arai T, Rapoport SI. Evidence for the involvement of docosahexaenoic acid in cholinergic stimulated signal transduction at the synapse. *Neurochem Res.* 1997;22:663–670.

70. Sen CK, Rink C, Khanna S. Palm oil-derived natural vitamin E alpha-tocotrienol in brain health and disease. *J Am Coll Nutr.* 2010;29:314S–323S.

71. Beal MF, Howell N, Bodis-Wollner I. *Mitochondria and free radicals in neurodegenerative diseases.* New York: Wiley-Liss; 1997.

72. Kamal-Eldin A, Appelqvist LA. The chemistry and antioxidant properties of tocopherols and tocotrienols. *Lipids.* 1996;31:671–701.

73. Atkinson J, Epand RF, Epand RM. Tocopherols and tocotrienols in membranes: A critical review. *Free Radic Biol Med.* 2008;44:739–764.

74. Packer L, Weber SU, Rimbach G. Molecular aspects of alpha-tocotrienol antioxidant action and cell signalling. *J Nutr.* 2001;131:369S–373S.

75. Kamat JP, Devasagayam TP. Tocotrienols from palm oil as potent inhibitors of lipid peroxidation and protein oxidation in rat brain mitochondria. *Neurosci Lett.* 1995;195:179–182.

76. Khanna S, Roy S, Park HA, Sen CK. Regulation of c-src activity in glutamate-induced neurodegeneration. *J Biol Chem.* 2007;282:23482–23490.

77. Paul R, Zhang ZG, Eliceiri BP, Jiang Q, Boccia AD, Zhang RL, Chopp M, Cheresh DA. Src deficiency or blockade of src activity in mice provides cerebral protection following stroke. *Nat Med.* 2001;7:222–227.

78. Tan S, Schubert D, Maher P. Oxytosis: A novel form of programmed cell death. *Curr Top Med Chem.* 2001;1:497–506.

79. Li Y, Maher P, Schubert D. A role for 12-lipoxygenase in nerve cell death caused by glutathione depletion. *Neuron.* 1997;19:453–463.

80. Adibhatla RM, Hatcher JF. Phospholipase a(2), reactive oxygen species, and lipid peroxidation in CNS pathologies. *BMB Rep.* 2008;41:560–567.

81. Adibhatla RM, Hatcher JF, Dempsey RJ. Effects of citicoline on phospholipid and glutathione levels in transient cerebral ischemia. *Stroke.* 2001;32:2376–2381.

82. Khanna S, Parinandi NL, Kotha SR, Roy S, Rink C, Bibus D, Sen CK. Nanomolar vitamin E alpha-tocotrienol inhibits glutamate-induced activation of phospholipase A2 and causes neuroprotection. *J Neurochem.* 2010;112:1249–1260.

83. Hirrlinger J, Konig J, Keppler D, Lindenau J, Schulz JB, Dringen R. The multidrug resistance protein MRP1 mediates the release of glutathione disulfide from rat astrocytes during oxidative stress. *J Neurochem.* 2001;76:627–636.

84. Park HA, Khanna S, Rink C, Gnyawali S, Roy S, Sen CK. Glutathione disulfide induces neural cell death via a 12-lipoxygenase pathway. *Cell Death Differ.* 2009;16:1167–1179.

85. Hebert SS, Horre K, Nicolai L, Bergmans B, Papadopoulou AS, Delacourte A, De Strooper B. Microrna regulation of Alzheimer's amyloid precursor protein expression. *Neurobiol Dis.* 2009;33:422–428.

86. Kim J, Inoue K, Ishii J, Vanti WB, Voronov SV, Murchison E, Hannon G, Abeliovich A. A microrna feedback circuit in midbrain dopamine neurons. *Science.* 2007;317:1220–1224

87. Sen CK, Ghatak S. MiRNA control of tissue repair and regeneration. *Am J Pathol.* 2015;185:2629–2640.

88. Rink C, Khanna S. Microrna in ischemic stroke etiology and pathology. *Physiol Genomics.* 2011;43:521–528.

89. Kole AJ, Swahari V, Hammond SM, Deshmukh M. Mir-29b is activated during neuronal maturation and targets BH3-only genes to restrict apoptosis. *Genes Dev.* 2011;25:125–130.

90. Rane S, He M, Sayed D, Vashistha H, Malhotra A, Sadoshima J, Vatner DE, Vatner SF, Abdellatif M. Downregulation of mir-199a derepresses hypoxia-inducible factor-1alpha and sirtuin 1 and recapitulates hypoxia preconditioning in cardiac myocytes. *Circ Res.* 2009;104:879–886.

91. Liebeskind DS. Neuroprotection from the collateral perspective. *IDrugs.* 2005;8:222–228.

92. Christoforidis GA, Mohammad Y, Kehagias D, Avutu B, Slivka AP. Angiographic assessment of pial collaterals as a prognostic indicator following intra-arterial thrombolysis for acute ischemic stroke. *AJNR Am J Neuroradiol.* 2005;26:1789–1797.

93. Schierling W, Troidl K, Mueller C, Troidl C, Wustrack H, Bachmann G, Kasprzak PM, Schaper W, Schmitz-Rixen T. Increased intravascular flow rate triggers cerebral arteriogenesis. *J Cereb Blood Flow Metab.* 2009;29:726–737.

94. Hillmeister P, Lehmann KE, Bondke A, Witt H, Duelsner A, Gruber C, Busch HJ, Jankowski J, Ruiz-Noppinger P, Hossmann KA, Buschmann IR. Induction of cerebral arteriogenesis leads to early-phase expression of protease inhibitors in growing collaterals of the brain. *J Cereb Blood Flow Metab.* 2008;28:1811–1823.

95. Courchesne E, Chisum HJ, Townsend J, Cowles A, Covington J, Egaas B, Harwood M, Hinds S, Press GA. Normal brain development and aging: Quantitative analysis at *in vivo* MR imaging in healthy volunteers. *Radiology.* 2000;216:672–682.

96. Liu P, Smith PF, Appleton I, Darlington CL, Bilkey DK. Potential involvement of NOS and arginase in age-related behavioural impairments. *Exp Gerontol.* 2004;39:1207–1222.

97. Morrison JH, Hof PR. Life and death of neurons in the aging brain. *Science.* 1997;278:412–419.

98. Sivonova M, Tatarkova Z, Durackova Z, Dobrota D, Lehotsky J, Matakova T, Kaplan P. Relationship between antioxidant potential and oxidative damage to lipids, proteins and DNA in aged rats. *Physiol Res.* 2007;56:757–764.

99. Talarowska M, Galecki P, Maes M, Gardner A, Chamielec M, Orzechowska A, Bobinska K, Kowalczyk E. Malondialdehyde plasma concentration correlates with declarative and working memory in patients with recurrent depressive disorder. *Mol Biol Rep.* 2012;39:5359–5366.

100. Xia W, Mo H. Potential of tocotrienols in the prevention and therapy of Alzheimer's disease. *J Nutr Biochem.* 2016;31:1–9.

101. Haorah J, Ramirez SH, Floreani N, Gorantla S, Morsey B, Persidsky Y. Mechanism of alcohol-induced oxidative stress and neuronal injury. *Free Radic Biol Med.* 2008;45:1542–1550.

102. Schloesser A, Esatbeyoglu T, Piegholdt S, Dose J, Ikuta N, Okamoto H, Ishida Y, Terao K, Matsugo S, Rimbach G. Dietary tocotrienol/gamma-cyclodextrin complex increases mitochondrial membrane potential and ATP concentrations in the brains of aged mice. *Oxid Med Cell Longev.* 2015;2015:789710.

103. Taridi NM, Abd Rani N, Abd Latiff A, Ngah WZ, Mazlan M. Tocotrienol rich fraction reverses age-related deficits in spatial learning and memory in aged rats. *Lipids.* 2014;49:855–869.

104. Busche MA, Konnerth A. Impairments of neural circuit function in Alzheimer's disease. *Philos Trans R Soc Lond B Biol Sci.* 2016;371.

105. Refolo LM, Malester B, LaFrancois J, Bryant-Thomas T, Wang R, Tint GS, Sambamurti K, Duff K, Pappolla MA. Hypercholesterolemia accelerates the Alzheimer's amyloid pathology in a transgenic mouse model. *Neurobiol Dis.* 2000;7:321–331.

106. Frears ER, Stephens DJ, Walters CE, Davies H, Austen BM. The role of cholesterol in the biosynthesis of beta-amyloid. *Neuroreport.* 1999;10:1699–1705.

107. Song BL, DeBose-Boyd RA. Insig-dependent ubiquitination and degradation of 3-hydroxy-3-methylglutaryl coenzyme A reductase stimulated by delta- and gamma-tocotrienols. *J Biol Chem.* 2006;281:25054–25061.

108. Mo H. The "one-two punch" of isoprenoids to inflammation. *J Nutr Disorders Ther.* 2013;3.

109. McAnally JA, Gupta J, Sodhani S, Bravo L, Mo H. Tocotrienols potentiate lovastatin-mediated growth suppression *in vitro* and *in vivo*. *Exp Biol Med (Maywood).* 2007;232:523–531.

110. Hynd MR, Scott HL, Dodd PR. Glutamate-mediated excitotoxicity and neurodegeneration in Alzheimer's disease. *Neurochem Int.* 2004;45:583–595.

111. Tiwari V, Kuhad A, Chopra K. Suppression of neuro-inflammatory signaling cascade by tocotrienol can prevent chronic alcohol-induced cognitive dysfunction in rats. *Behav Brain Res.* 2009;203:296–303.

112. Tiwari V, Arora V, Chopra K. Attenuation of NF-kappabeta mediated apoptotic signaling by tocotrienol ameliorates cognitive deficits in rats postnatally exposed to ethanol. *Neurochem Int.* 2012;61:310–320.

113. Abubakar IB, Lim KH, Kam TS, Loh HS. Synergistic cytotoxic effects of combined delta-tocotrienol and jerantinine b on human brain and colon cancers. *J Ethnopharmacol.* 2016;184:107–118.

114. Nesaretnam K, Meganathan P. Tocotrienols: Inflammation and cancer. *Ann N Y Acad Sci.* 2011;1229:18–22.

115. Sakai M, Okabe M, Tachibana H, Yamada K. Apoptosis induction by gamma-tocotrienol in human hepatoma hep3b cells. *J Nutr Biochem.* 2006;17:672–676.

116. Yap WN, Chang PN, Han HY, Lee DT, Ling MT, Wong YC, Yap YL. Gamma-tocotrienol suppresses prostate cancer cell proliferation and invasion through multiple-signalling pathways. *Br J Cancer.* 2008;99:1832–1841.

117. Komiyama K, Iizuka K, Yamaoka M, Watanabe H, Tsuchiya N, Umezawa I. Studies on the biological activity of tocotrienols. *Chem Pharm Bull.* 1989;37:1369–1371.

118. Upadhyay J, Misra K. Towards the interaction mechanism of tocopherols and tocotrienols (vitamin E) with selected metabolizing enzymes. *Bioinformation.* 2009;3:326–331.

119. Lim SW, Loh HS, Ting KN, Bradshaw TD, Zeenathul NA. Cytotoxicity and apoptotic activities of alpha-, gamma- and delta-tocotrienol isomers on human cancer cells. *BMC Complement Altern Med.* 2014;14:469.

120. Lim SW, Loh HS, Ting KN, Bradshaw TD, Zeenathul NA. Antiproliferation and induction of caspase-8-dependent mitochondria-mediated apoptosis by beta-tocotrienol in human lung and brain cancer cell lines. *Biomed Pharmacother.* 2014;68:1105–1115.

121. Abdul Rahman A, Jamal AR, Harun R, Mohd Mokhtar N, Wan Ngah WZ. Gamma-tocotrienol and hydroxy-chavicol synergistically inhibits growth and induces apoptosis of human glioma cells. *BMC Complement Altern Med.* 2014;14:213.

10 Emerging Roles of Flavonoids in Brain Health

Arti Parihar and Andrea I. Doseff

CONTENTS

10.1 INTRODUCTION

Dietary compounds, including flavonoids, the most abundant and broadly distributed plant-derived polyphenols, are central to the regulation of brain function. Despite the appreciation of the role of diets in health, the mechanisms by which flavonoids regulate brain function remain unknown. The beneficial effects of flavonoids in brain health have been attributed primarily to their anti-oxidant activities [1]. However, recent findings indicate their ability to regulate neuronal fate, promote synaptic plasticity [2], and ameliorate brain inflammation. Thus, flavonoids may provide alternative opportunities for the prevention of neurodegenerative diseases and loss of cognitive function during normal ageing.

Oxidative stress and neuronal damage have been strongly associated with the pathophysiology of neurodegenerative diseases and ageing. Numerous studies suggest promising effects of flavonoids in the prevention of neurodegenerative diseases [3]. Resveratrol, a flavonoid abundant in red grapes, attenuates free radical production, decreasing oxidative stress [4]. Epicatechin, a polyphenol found in many vegetables and fruits, increases glutathione production in astrocytes, suggesting anti-oxidant activities that facilitate neuronal protection [5]. Since the bioavailability of these polyphenols in the brain is limited, it has been suggested that alternative protective functions are related to their ability to modulate signaling pathways [6] and inflammation in the brain. Thus, the well-known anti-inflammatory activity of flavonoids may reduce neuronal damage, contributing to preserve brain function. Polyphenols such as the theaflavin present in black tea [7], *O*-methylated flavone wogonin [8], and extracts of *Graptopetalum paraguayense* E. Walther leaves reduce microglia-mediated inflammatory activity in cerebral ischemia injury in animal models [9]. The flavanone hesperetin suppresses the activity of nuclear factor kappa-light-chain-enhancer of activated B cells (NF-κB), a key transcription factor in inflammation, in both young and old rats through multiple signal transduction pathways [10]. Moreover, the use of flavonoid-rich plant extracts has been positively correlated with improved cognitive function during normal ageing in both humans and animals. Dietary intervention studies in humans and animals have demonstrated so far promising effects of grape, green tea, cocoa, blueberry, *Ginkgo biloba*, among others, on human brain by improving memory and learning [11–17]. Additional effective regulation of prominent neurotransmitters such as glutamate and gamma-amino butyric acid by flavonoids has also been reported [18–24]. Based on these observations, it is tempting to think that consumption of flavonoids may protect brain performance by enhancing existing neuronal function and/or stimulating neuronal regeneration either directly or by modulation of the immune function. This review focuses on the recent advances in understanding the potential mechanisms responsible of the beneficial effects of flavonoids and their potential role in brain function.

10.2 FLAVONOIDS: AVAILABILITY, ABSORPTION, METABOLISM, AND DISTRIBUTION IN THE BRAIN

Flavonoids are phenolic compounds abundantly present in plants. Structurally, flavonoids possess a 2-phenyl-benzo-γ-pyrane nucleus consisting of two benzene rings A and B linked through a heterocyclic pyran or pyrone ring C (Figure 10.1). Depending on the carbon of the C ring on which B ring is attached, flavonoids are classified into various subgroups, including, among others, flavones (2-phenylchromen-4-one; e.g., apigenin, luteolin, and tangeretin), flavonols (3-hydroxy-2-phenylchromen-4-one; e.g., quercetin, kaempferol, myricetin, and rhamnazin), flavanones (2, 3-dihydro-2-phenylchromen-4-one; e.g., hesperetin, naringenin, and eriodictyol), flavanol (2-phenyl-3,4-dihydro-2H-chromen (flavan)-3-ol, flavan-4-ol, flavan-3,4-diol; e.g., catechins and epicatechins [such as epigallocatechin gallate; EGCG]), flavanonol (3-hydroxy-2,3-dihydro-2-phenylchromen-4-one; e.g., silibinin, taxifolin, and dihydrokaempferol), isoflavones (3-phenylchromen-4-one; e.g., genistein, daidzein, and glycitein), and anthocyanidins (2-phenylchromenylium; e.g., cyanidin, delphinidin, and malvidin) [25,26].

Flavonoids are synthesized by the phenylpropanoid metabolic pathway in plants in response to biotic or abiotic stress conditions. Biosynthetic pathways of flavonoids have been thoroughly studied in several plants including maize, *Arabidopsis thaliana*, petunia, snapdragon, soybean, and barley using biochemical, molecular, and genetic techniques [27,28]. These techniques, along with genetic engineering strategies, have been employed to enhance flavonoid content or to produce flavonoids with unique nutraceutical properties [29]. Paradoxically, in crops with high

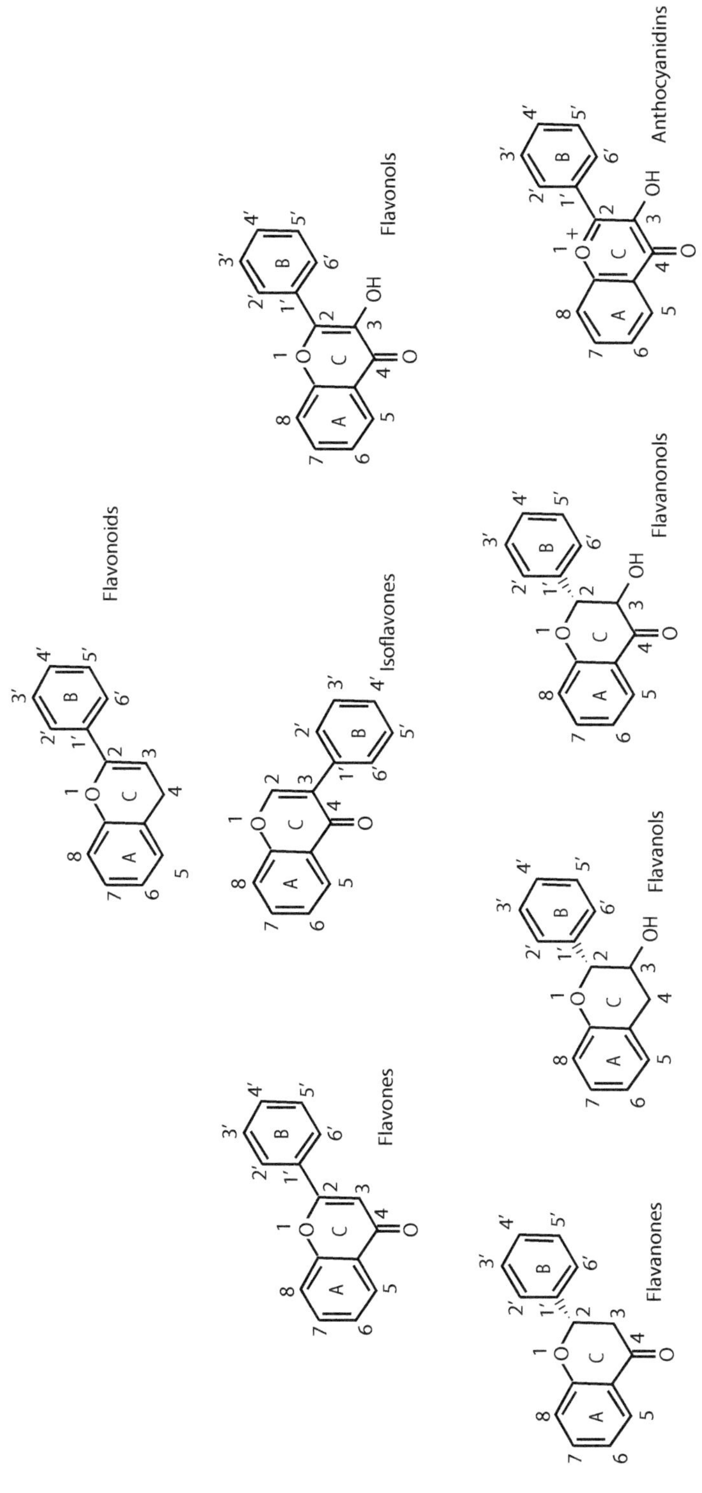

FIGURE 10.1 Chemical structures of flavonoids.

economic value such as maize, genetic lines expressing high levels of flavonoids have been selected against the most marketable varieties containing higher sugar levels. However, recent recognition of flavonoids as health beneficial micronutrients has prompted interest in reintroducing maize lines expressing higher flavonoid levels in our diet [30]. A complex network of transcription factors regulates the synthesis of flavonoids. A better understanding of this network will provide additional opportunities to enhance the flavonoid content in plants, thereby increasing the nutritional value of our diet [31].

The beneficial effects of flavonoids are intimately linked to their availability in foods as well as their levels of consumption, absorption, and bioavailability. For example, consumption of the Mediterranean diet, prominent in southern Europe, includes relatively large amounts of parsley and celery, two vegetables expressing high levels of the flavone apigenin. Apigenin has been shown to have anti-inflammatory activities in several systems [32–34]. Consumption of the Mediterranean diet has been linked to improved cardiac function and prolonged survival of ovarian cancer patients [35,36]. Consumption of flavonoids has been also associated with lower risk of Alzheimer's disease [37]. Thus, adequate nutrition and appropriate bioavailability may help minimize the risk of deteriorating brain function. A limitation in the clinical utilization of flavonoids resides in their limited absorption, which precludes of reaching effective concentrations *in vivo*. The use of functional foods presents promising opportunities to overcome this limitation. The majority of flavonoids are found in plants linked to sugars (glycosides). We showed that flavonoids lacking sugars (aglycones) have increased anti-inflammatory activity in macrophages and *in vivo* [38]. Importantly, foods formulated to increase aglycone content increased the bioavailability and absorption of dietary flavones in mice models [38]. Consumption of aglycone-rich flavonoid containing foods for a week provided immune-regulatory activity in models of acute inflammation, reducing the levels of specific microRNAs [33]. Thus, the development of functional foods that increase aglycone content should help increase flavonoid bioavailability, helping to reach effective concentrations in the brain.

Flavonoids can be metabolized into glucuronidated, sulfonated, and methylated derivatives [39,40]. A better understanding of the formation of these derivatives and their physiological effectiveness is much needed. Studies reported that flavonoids reached the brain, performing anti-oxidant activities and modulating both enzymes and receptors activities [41]. Lipophilicity of flavonoids influences the entry to the brain. Studies showed that flavanones such as hesperetin, naringenin, and some of their metabolites, anthocyanins, cyanidin-3-rutinoside, and pelargonidin-3-glucoside, cross the blood–brain barrier [42]. Rats fed with a blackberry anthocyanin-enriched diet for 15 days showed accumulation of anthocyanins in the brain tissue reaching ~0.25 ± 0.05 nmol/g of tissue [43]. Consumption of 2% w/w blueberry diet for 12 weeks showed levels of anthocyanins ~0.45 nmol/g in hippocampus and cortex [44]. Accumulation of flavanols was found ~2.5 to 2.7 nmol/g tissue in hippocampus and cortex region of brain following the same dietary intervention [44]. Quercetin (50 mg/kg body weight;bw)-fed rats showed decreased levels of oxidative stress in the hippocampus and striatum [45], suggesting its potential in regulating disorders associated with oxidative stress. The flavanol epicatechin, present in cocoa powder, reduced production of reactive oxygen species (ROS), protecting cognitive impairment induced by heat in rat models [46]. Epicatechin induced endothelium nitric oxide synthase (NOS) activity, thus promoting vasodilatation in brain [47–49]. Microscopy studies showed that fisetin reached the brain parenchyma during intraperitoneal or oral administration. Interestingly fisetin was found localized into the nucleoli, suggesting the existence of potential targets in this cellular organelle [50]. Oral administration of quercetin in mice showed significant distribution in the cerebellum [51]. Tangeretin was found in different brain regions, including hypothalamus, striatum, and hippocampus of rat [52]. Oral administration of tangeretin (10 mg/kg bw per day), found in citrus, showed the ability of this flavonoid to reach the brain and provide neuroprotective effects in a Parkinson's disease rat model [52]. These findings suggest that flavonoids can reach the brain, but whether the concentrations reach biological significance will require further investigations.

10.3 FLAVONOIDS: MECHANISMS OF PROTECTION IN THE BRAIN

Neurodegenerative diseases and normal aging are associated with a progressive loss of motor, sensory, and perceptual functions, leading to cognitive and behavioral deficits. The molecular mechanisms responsible for these conditions are just becoming recognized. The beneficial effects of dietary flavonoids in the brain may be due to their neuroprotective activities mediated by their ability to act as anti-oxidants, anti-inflammatory, and anti-protein-aggregates. In addition, plant flavonoids may also modulate signaling pathways involved in neuronal cell fate and inflammation. Limited knowledge of these mechanisms is currently available, but great interest in this area is attracting studies using different models systems. While mounting information is emerging, few clinical studies available seem to indicate that consumption of diets rich in flavonoids improved brain functions such as verbal comprehension, reasoning, decision making, and object recall.

10.3.1 ANTIOXIDANT EFFECTS OF FLAVONOIDS IN NEUROPROTECTION

Some of the beneficial functions of dietary flavonoids are due to their well-accepted anti-oxidant properties. In cellular models, flavonoids showed effectiveness in reducing free radicals in neurons [53]. Similar effects were observed using cocoa extracts rich in the flavonol (-)-epicatechin in neuronal cell cultures [54]. Flavonoid-rich cocoa-derived foods scavenge ROS, attenuating the detrimental effects of superoxide anions such as hydrogen peroxide, hypochlorous acid, and peroxynitrite [55]. Genistein protects neurons from amyloid β (Aβ)-induced cell death by reducing oxidative stress, thereby preventing the activation of the mitogen-activated protein kinase (MAPK) p38 and protecting neurons from cell death [56]. Quercetin and luteolin either as aglycones or C-7-glycosides showed similar scavenged superoxide anions activity, reducing lipid peroxidation and inhibiting H_2O_2-induced cell death in primary rat cortical cells [57]. In animal models, combination of epicatechin and quercetin administered orally reduced neuronal cell death during hypoxic-ischemic brain injury by inhibiting NOS [58]. Flavonoids may provide alternative approaches in regulating mitochondrial function [59], central to brain function. Quercetin, for example, has been found to restore the defects in electron transport system in an experimental model of Parkinson's disease by regulating mitochondrial complex I activity [60]. Genistein showed neuroprotective functions in 6-hydroxydopamine (6-OHDA)-induced neurotoxicity in human neuroblastoma cells [61]. The neuro-protective effects of the apigenin derivative, apigenin-7-O-β-D-(-6″-p-coumaroyl)-glucopyranoside, in cellular models and *in vivo* through the regulation of the transcription factor STAT3 have been reported [62]. Effective concentrations of fla-vonoids range in the order of micromolar in cellular models. In contrast, oral admin-istration of flavonoids in animal models showed nanomolar concentrations in the brain. Thus, the small amount of flavonoids that reaches the brain seems to provide efficient free radical scavenger activity, protecting neurons against oxidative stress. It will be interesting to see whether flavonoids consumed as part of normal diets are also capable to reach the brain and function as anti-oxidants.

10.3.2 ROLE OF FLAVONOIDS IN BRAIN SIGNALING PATHWAYS

The chemical structure of flavonoids plays a critical role in regulating cellular signal-ing pathways involved in neuronal function. Flavonoids alter protein phosphoryla-tion and gene expression, thereby impacting in brain function [63]. In the brain, cells respond to environmental cues by integrating complex regulatory networks that define cellular fate (Figure 10.2). Flavonoids modulate key signaling pathways involved in cell survival including the phosphatidylinositol-3 kinase/protein kinase B (PI3K/Akt) axis, the protein kinase C (PKC), and the MAPK pathways [64,65]. Consumption of blueberry extracts induced the activity of the MAPK, extracellular signal-related kinase (ERK), leading to the activation of the cAMP response element-binding pro-tein (CREB) and the brain-derived neurotrophic factor (BDNF) in the hippocam-pus [44]. The flavonoid baicalein present in *Scutellaria baicalensis* Georgi increased the expression of the transcriptional factor nuclear factor-erythroid 2-related factor 2 (Nrf2)/heme oxygenase 1 protein, promoting anti-oxidant activity, and increased neuronal survival through the PKCα and PI3K/AKT signaling pathways [66].

The flavanone hesperetin activates the PI3K pathway in cortical neurons [65]. Studies have suggested that quercetin, epicatechin, and baicalein inhibit the MAPK c-Jun N-terminal kinase (JNK) in neurons [67,68]. Quercetin, epicatechin, 3′-O-methyl-epicatechin, and hesperetin [67–69] inhibit JNK activity. Genistein, cocoa extracts, and epicatechin [54,56] inhibit the p38 pathway. Flavonoid-rich blueberry extracts activate PI3K/Akt, ERK, and PKC pathways [44].

Additional mechanisms have been proposed to explain the effect of flavonoids in delaying the progression of neurodegenerative diseases such as Alzheimer's and Parkinson's. Gene expression analysis studies showed that green tea polyphenol EGCG decreased the expression of proapoptotic regulators such as Bax, Bad, and Mdm2, while increasing the expression of antiapoptotic Bcl-2, Bcl-w, and Bcl-xL in response to 6-OHDA-induced neuronal stress [64]. These effects were accompanied by inhibition of the proapoptotic proteases caspase-9 and caspase-3 and the apopto-sis signal-regulating kinase 1 [65]. Moreover, flavonoids inhibit brain cell apoptosis initiated by ROS and exacerbated inflammation and their ability to disrupt Aβ aggre-gation [70].

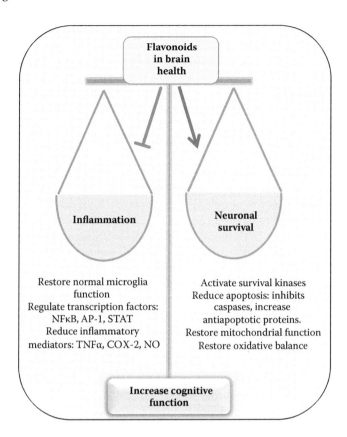

FIGURE 10.2   **(See color insert.)** Schematic model of the role of flavonoids in brain function.

Among flavonoids, epicatechin [71], EGCG [72], quercetin [73], genistein [74,75], daidzein [76], baicalein [77], and apigenin have shown neuroprotection against neurological disorders in preclinical studies. Moreover, apigenin decreases Aβ burden. These effects seemed mediated by the ability of apigenin to inhibit oxidative stress in the cerebral cortex and regulate the signal transduction pathway mediated by ERK-induced phosphorylation of the transcription factor CREB, which modulates the expression of genes involved in synaptic activity [78]. Flavonoid-rich *G. biloba* extracts showed neuroprotective effects in 6-OHDA-induced neurotoxicity in rats, suggesting potential neuromodulatory effects in Alzheimer's and Parkinson's diseases [79,80].

10.3.3   ROLE OF FLAVONOIDS IN APOPTOSIS AND CELL SURVIVAL IN THE BRAIN

In Alzheimer's disease, accumulation of Aβ contributes to neuronal cell death. Treatment of primary neurons with 5-methoxyflavone blocked apoptosis by inhibiting DNA pol-β activity and reducing the characteristic cell cycle arrest induced by Aβ [81]. These findings provide direct evidence that flavonoids can regulate key proteins involved in neuronal cell fate rather than just providing only anti-oxidant protection. Recent studies showed that treatment of rats with quercetin lead to an increase in expression of X-linked inhibitor of apoptosis protein that correlated with an increase of hippocampal neurons [82]. In a traumatic brain injury model of rat, quercetin has been shown to inhibit apoptosis [83]. Extracts of stems and leaves of *S. baicalensis* Georgi inhibited apoptosis and modulated oxidative stress induced by potassium cyanide in primary cortical cell culture in rats [84]. *G. biloba* extract protects the injured neuronal PC12 cells against apoptosis by upregulating Bcl-2 and downregulating Bax and c-myc [85]. In a similar study using PC12 cells, resveratrol has been shown to inhibit apoptosis by downregulating Bax and upregulating Bcl-2 [86]. In Parkinson's disease models, resveratrol prevented neuronal apoptosis by increasing the expression of Bcl-2 [87]. The standardized nonflavone fraction of *G. biloba* extracts (including ginkgolides A, B, C, J and bilobalide) rescued cultured rat hippocampal neurons from apoptosis caused by serum deprivation [88].

Flavonoids, such as tangeretin [52], EGCG [89], genistein [90], and rutin [91], reduced 6-OHDA-induced neurotoxicity in Parkinson's disease rat models. In MPTP (*N*-methyl-4-phenyl-1,2,3,6-tetrahydropyridine) mice models of Parkinson's disease, treatment with EGCG or quercetin showed protective effects in neurons [91,92]. Nanomolar concentrations of flavonoids including pelargonidin, quercetin, hesperetin, caffeic acid, and epicatechin protects nigral neurons induced by endogenous neurotoxin, 5-*S*-cysteinyl-dopamine [93]. *G. biloba* crude extract protects hippocampal neurons from NO and Aβ-induced neurotoxicity [79]. Flavonoids hesperidin and rutin increase neuronal crest cell survival during embryonic development against the chemical toxicity during embryonic development [94]. Fisetin reduces Aβ aggregation, decreasing neuronal apoptosis induced by aluminum chloride *in vivo* [95].

10.4 FLAVONOIDS: REGULATORS OF INFLAMMATION IN THE BRAIN

Inflammation plays a key role in the pathophysiology of brain disease [96]. Chronic inflammatory conditions promote infiltration of inflammatory cells from blood vessels to tissues, inducing dysregulated microenvironments characterized by excessive production of cytokines, changes in cell function, cell damage, and ultimately loss of barrier function. In the brain, increased number of monocytes and activated microglia [97] mediate increased expression of pro-inflammatory cytokines [98]. CD40, a member of the tumor necrosis factor (TNF) receptor family, is a key molecule involved in microglia activation *in vivo* and also highly expressed in cells of the monocytic lineage [99]. Various studies have shown that microglial CD40 expression is significantly induced by interferon-γ and Aβ peptide [100,101]. Increased production of inflammatory cytokines by microglia or monocytes, including TNF-α and other inflammatory mediators such as ROS and NO from these cells, induces neuronal damage. Inflammatory cytokine expression is mediated by transcription factors such as NF-κB, signal transducer and activator or transcription (STAT), and activating protein-1 (AP-1) [102,103]. In addition, flavonoids mediate the induction of the transcription factor Nrf2, responsible for both constitutive and inducible expressions of the anti-oxidant responsive element-regulated genes [104], helping to overcome exacerbated inflammation. Reducing chronic inflammation has been suggested as a beneficial approach to delay and overcome the onset of brain malfunctioning (Figure 10.2). However, so far, the majority of existing drug therapies for neurodegenerative disorders have failed to prevent inflammation-induced neurodegeneration. Therefore, there is an urgent need to develop alternative therapies capable of preventing the progressive neuronal loss resulting from neuroinflammation.

It is tempting to propose that anti-inflammatory flavonoids may help prevent the immune-dysfunction typical of normal ageing or brain disease conditions [36]. Flavones such as apigenin and luteolin showed immune-modulatory activity by reducing STAT-induced expression of CD40 in microglia [105]. Evidence from cellular models showed a reduction of pro-inflammatory mediators such as NO, TNF-α, interleukin (IL)-1β, and inducible NOS (iNOS) expression by flavonoid-rich blueberry extracts [15]. EGCG has been shown to promote attenuations of innate immune cell functions through their direct ROS scavenging [106] properties and also through reduction of TNF-α, IL-1β, and prostaglandin E2 [107]. Luteolin, kaempferol, EGCG, and quercetin [9,108] were also found to exhibit inhibitory property on those pro-inflammatory mediators [109–111]. Administration of luteolin reduced Aβ deposition, inflammatory cytokine expression, and glycogen synthase-3 activation in mice models of Alzheimer's disease [112]. These anti-inflammatory activities of flavonoids seem to be mediated by inhibition of NF-κB and AP-1. Flavonoids also decrease inflammation by inhibiting pro-inflammatory enzymes, such as cyclooxygenase-2 (COX-2), lipoxygenase, and iNOS [113]. Quercetin and kaempferol inhibit COX-2 in rat peritoneal macrophages [114]. Quercetin, apigenin, and luteolin inhibit NO production in lipopolysaccharide (LPS)-treated macrophages through downregulation of inducible NOS [115]. Although selective COX-2 inhibitory flavonoids have not been found with high accuracy, certain flavone derivatives such as apigenin, wogonin, and luteolin are the only COX-2 inhibitory flavonoid reported with high accuracy over COX-1 [116,117]. Flavonoids inhibit p38 and ERK cell signaling cascades and control both iNOS and TNF-α expression in activated glial cells [118]. Luteolin inhibited IL-6 production via the inhibition of the JNK pathway [119]. In LPS-treated mouse macrophage cell lines (RAW), genistein and silybin inhibited TNFα production [120]. Similarly, in LPS-induced RAW cells, quercetin reduced iNOS and TNFα expression by inhibiting MAPK and AP-1 DNA binding [121,122]. Genistein and apigenin inhibit NF-κB in LPS-stimulated macrophages. Our group showed that apigenin reduces the phosphorylation of the p65 subunit of NF-κB required for its transcriptional activity but dispensable for DNA binding [32].

So far, human dietary intervention studies using flavonoids have showed limited effects on markers of inflammation. Cocoa rich in flavanol (900 mg/day) was found to increase cerebral blood flow in grey matter [123]. Similarly, consumption of cocoa tablets increased plasma catechins, and this effect correlated with changes in P-selectin levels in plasma [124]. Green tea polyphenols have been shown to reduce damage in neural tissue via the chelation, or removal, of free iron, a pro-inflammatory mineral [125]. EGCG inhibits TNF-α-induced production of the chemokine monocyte chemoattractant protein 1/chemokine (C-C motif) ligand 2 in bovine coronary artery endothelial cells [126]. Studies conducted by Steptoe et al. [127] found a reduction in C-reactive protein levels, a well-known inflammatory marker, after black tea consumption for six weeks. However, these studies provided no evidence on the absorption of flavonoids after consumption. Mixed results were obtained with supplementation of red wine and grape extracts on inflammatory mediators. While Watzl et al. reported no effects of supplementation of red wine and grape juice on cytokine production [128,129], Estruch and colleagues [130] showed a decrease in plasma fibrinogen, IL-1α, and C-reactive protein accompanied by an increase in EGCG in blood plasma after red wine consumption for four weeks. Similarly nicotinamide adenine dinucleotide phosphate-oxidase activity [131] and TNF-α in postmenopausal women were found to decline [132] after supplementation of red grape juice. Soy consumption for a long period of time caused decline in C-reactive protein and vascular cell adhesion molecule-1 [133,134]. However, other studies found lack of effect of soy supplementation on IL-6 [134,135], C-reactive protein, and TNFα [136,137]. Quercetin supplementation in bikers and runners caused increased quercetin levels in plasma and lower C-reactive protein and IL-8 levels [138,139].

The use of PD-Seq (phage display high-throughput sequencing), a novel approach for small target identification by our team, provided new insights into the potential mechanisms of flavonoids [140]. Using this approach, we have identified several direct targets of apigenin, including RNA binding proteins. The identification of direct targets should provide novel information about how flavonoids regulate brain function. This type of approach in combination with animal and human clinical dietary studies should broaden our understanding on how flavonoids through the regulation of the immune system impact brain function.

10.5   POTENTIAL ROLE OF FLAVONOIDS
IN IMPROVING COGNITIVE FUNCTION

Flavonoids are emerging as potential molecules to improve cognitive function and promoting neuroprotection in ageing or neurodegenerative diseases [141–143]. Diets rich in flavonoids improve brain verbal comprehension, reasoning, decision making, and object recall. Animal behavioral studies suggest that consumption of blueberries and strawberries causes positive effects on age-related deficits in memory [144]. *G. biloba* showed beneficial effects in reversing age-related neuronal and behavioral aging [145]. Naringenin overcomes cognitive impairment in animal models of Alzheimer's disease as suggested by improved performance in Morris water maze paradigm and elevated plus maze tests [146]. Consumption of foods rich in anthocyanins prevented memory deficits [147,148]. Supplementation of blueberries in rats improved long-term and spatial memory [44,149]. Naringenin and rutin alleviate memory deficits in animal models [150].

Encouraging results are emerging from the few dietary human interventions studies using fruits and vegetables rich in flavonoids so far available. Consumption of flavonoid-rich diets showed improvement of several cognitive performance measurements, including memory [151] or visual-oculomotor integration [152]. A double-blind parallel arm study in 90 elderly individuals with mild cognitive impairment showed that consumption of a chocolate drink containing flavanols at high (990 mg/day) and intermediate (520 mg/day) levels increased verbal fluency test scores, visual attention, and task switching [153]. In addition, studies from leaders in the field showed that consumption of fruit flavonoids–rich drinks led to acute improvements in cognitive function in healthy middle-aged males [154]. Despite existing disagreement in the field on the best test to evaluate cognitive function, results with flavonoid diets deserve further investigation. It will be interesting to see future studies to evaluate whether these beneficial effects are transient or persistent.

The mechanisms by which flavonoids improve cognitive functions remain unclear. However, models suggesting the effects of flavonoids in neuroprotection, promoting neurogenesis, and ameliorating inflammation are starting to emerge [155]. The flavonoid oroxylin A attenuates impairment in memory in mice, possibly inducing neuroprotective effects through the reduction of activated microglia as suggested by the decrease in expression of BDNF and CREB phosphorylation [156]. In addition, flavonoids in blueberries were shown to increase the NF-κB levels in brain. These studies so far seem to suggest that flavonoids impact key signaling pathways leading to improved cognitive functions in the brain [155].

10.6   CONCLUDING REMARKS

The value of flavonoids in health is recognized, yet their role in brain function is just emerging. Several mechanisms have been attributed to the beneficial effects of flavonoids, including their anti-oxidant activity and their ability to regulate cellular lifespan and the immune function. Increasing understanding of the brain's molecular networks will help define how these dietary compounds may exert protective effects. The development of foods with increased flavonoid content and improved bioavailability will be imperative to reach effective concentrations *in vivo*, since it is already known that these compounds are likely to be effective at relatively high concentrations. Establishing dietary studies to assess the potential application of these compounds in the prevention or delay of so far devastating and incurable neurological diseases will be invaluable. Future studies to evaluate the effects of flavonoids in cognitive function should provide potential approaches to increase brain function during ageing, thereby contributing to improved quality of life.

ACKNOWLEDGMENTS

This work was supported by Agriculture and Food Research Initiative (AFRI) USDA National Institute of Food and Agriculture grants 2015-67017-23187 (2014-06654) and NSF IOS-1125620 to AID.

REFERENCES

1. Rice-Evans, C.A., Miller, N.J., and Paganga, G. (1996) Structure-antioxidant activity relationships of flavonoids and phenolic acids. *Free Radic. Biol. Med.*, 20, 933–956.
2. Spencer, J.P. (2009) Nutrients and brain health: An overview. *Genes Nutr.*, 4, 225–226.
3. Youdim, K.A., and Joseph, J.A. (2001) A possible emerging role of phytochemicals in improving age-related neurological dysfunctions: A multiplicity of effects. *Free Radic. Biol. Med.*, 30, 583–594.
4. Pandey, K.B., and Rizvi, S.I. (2009) Plant polyphenols as dietary antioxidants in human health and disease. *Oxid. Med. Cell Longev.*, 2, 270–278.

5. Maher, P., Lewerenz, J., Lozano, C., and Torres, J.L. (2008) A novel approach to enhancing cellular glutathione levels. *J. Neurochem.*, 107, 690–700.

6. Singh, M., Arseneault, M., Sanderson, T., Murthy, V., and Ramassamy, C. (2008) Challenges for research on polyphenols from foods in Alzheimer's disease: Bioavailability, metabolism, and cellular and molecular mechanisms. *J. Agric. Food Chem.*, 56, 4855–4873.

7. Cai, F., Li, C.R., Wu, J.L., Chen, J.G., Liu, C., Min, Q., Yu, W., Ouyang, C.H., and Chen, J.H. (2006) Theaflavin ameliorates cerebral ischemia-reperfusion injury in rats through its anti-inflammatory effect and modulation of STAT-1. *Mediators Inflamm.*, 2006, 30490.

8. Cho, J., and Lee, H.K. (2004) Wogonin inhibits ischemic brain injury in a rat model of permanent middle cerebral artery occlusion. *Biol. Pharm. Bull.*, 27, 1561–1564.

9. Kao, T.K., Ou, Y.C., Raung, S.L., Chen, W.Y., Yen, Y.J., Lai, C.Y., Chou, S.T., and Chen, C.J. (2010) Graptopetalum paraguayense E. Walther leaf extracts protect against brain injury in ischemic rats. *Am. J. Chin Med.*, 38, 495–516.

10. Kim, J.Y., Jung, K.J., Choi, J.S., and Chung, H.Y. (2006) Modulation of the age-related nuclear factor-kappaB (NF-κB) pathway by hesperetin. *Aging Cell*, 5, 401–411.

11. Lamport, D.J., Lawton, C.L., Merat, N., Jamson, H., Myrissa, K., Hofman, D., Chadwick, H.K., Quadt, F., Wightman, J.D., and Dye, L. (2016) Concord grape juice, cognitive function, and driving performance: A 12-wk, placebo-controlled, randomized crossover trial in mothers of preteen children. *Am. J. Clin. Nutr.*, 103, 775–783.

12. Haque, A.M., Hashimoto, M., Katakura, M., Tanabe, Y., Hara, Y., and Shido, O. (2006) Long-term administration of green tea catechins improves spatial cognition learning ability in rats. *J. Nutr.*, 136, 1043–1047.

13. Mastroiacovo, D., Kwik-Uribe, C., Grassi, D., Necozione, S., Raffaele, A., Pistacchio, L., Righetti, R., Bocale, R., Lechiara, M.C., Marini, C., Ferri, C., and Desideri, G. (2015) Cocoa flavanol consumption improves cognitive function, blood pressure control, and metabolic profile in elderly subjects: The Cocoa, Cognition, and Aging (CoCoA) Study—A randomized controlled trial. *Am. J. Clin. Nutr.*, 101, 538–548.

14. Galli, R.L., Shukitt-Hale, B., Youdim, K.A., and Joseph, J.A. (2002) Fruit polyphenolics and brain aging: Nutritional interventions targeting age-related neuronal and behavioral deficits. *Ann. N.Y. Acad. Sci.*, 959, 128–132.

15. Lau, F.C., Bielinski, D.F., and Joseph, J.A. (2007) Inhibitory effects of blueberry extract on the production of inflammatory mediators in lipopolysaccharide-activated BV2 microglia. *J. Neurosci. Res.*, 85, 1010–1017.

16. Youdim, K.A., Shukitt-Hale, B., and Joseph, J.A. (2004) Flavonoids and the brain: Interactions at the blood-brain barrier and their physiological effects on the central nervous system. *Free Radic. Biol. Med.*, 37, 1683–1693.

17. Wang, Y., Wang, L., Wu, J., and Cai, J. (2006) The in vivo synaptic plasticity mechanism of EGb 761-induced enhancement of spatial learning and memory in aged rats. *Br. J. Pharmacol.*, 148, 147–153.

18. Zhang, Z., Lian, X.Y., Li, S., and Stringer, J.L. (2009) Characterization of chemical ingredients and anticonvulsant activity of American skullcap (*Scutellaria lateriflora*). *Phytomedicine*, 16, 485–493.

19. Orhan, N., Deliorman, O.D., Aslan, M., Sukuroglu, M., and Orhan, I.E. (2012) UPLC-TOF-MS analysis of *Galium spurium* towards its neuroprotective and anticonvulsant activities. *J. Ethnopharmacol.*, 141, 220–227.

20. Choudhary, N., Bijjem, K.R., and Kalia, A.N. (2011) Antiepileptic potential of flavonoids fraction from the leaves of *Anisomeles malabarica*. *J. Ethnopharmacol.*, 135, 238–242.

21. Medina, J.H., Viola, H., Wolfman, C., Marder, M., Wasowski, C., Calvo, D., and Paladini, A.C. (1998) Neuroactive flavonoids: New ligands for the benzodiazepine receptors. *Phytomedicine.*, 5, 235–243.

22. Abbasi, E., Nassiri-Asl, M., Shafeei, M., and Sheikhi, M. (2012) Neuroprotective effects of vitexin, a flavonoid, on pentylenetetrazole-induced seizure in rats. *Chem. Biol. Drug Des.*, 80, 274–278.

23. Dimpfel, W. (2006) Different anticonvulsive effects of hesperidin and its aglycone hesperetin on electrical activity in the rat hippocampus *in vitro. J. Pharm. Pharmacol.*, 58, 375–379.

24. Nassiri-Asl, M., Naserpour, F.T., Abbasi, E., Sadeghnia, H.R., Sheikhi, M., Lotfizadeh, M., and Bazahang, P. (2013) Effects of rutin on oxidative stress in mice with kainic acid-induced seizure. *J. Integr. Med.*, 11, 337–342.

25. Beecher, G.R. (2003) Overview of dietary flavonoids: Nomenclature, occurrence and intake. *J. Nutr.*, 133, 3248S–3254S.

26. Harborne, J., and Willium, C. (1975) In Harborne JB, M.M.H. (ed), *The Flavonoids.* Chapman and Hall, London, pp. 376–441.

27. Grotewold, E. (2006) The genetics and biochemistry of floral pigments. *Annu. Rev. Plant Biol.*, 57, 761–780.

28. Jiang, N., Doseff, A.I., and Grotewold, E. (2016) Flavonones: From biosynthesis to health benefits. *Plants* (in press).

29. Ververidis, F., Trantas, E., Douglas, C., Vollmer, G., Kretzschmar, G., and Panopoulos, N. (2007) Biotechnology of flavonoids and other phenylpropanoid-derived natural products. Part II: Reconstruction of multienzyme pathways in plants and microbes. *Biotechnol. J.*, 2, 1235–1249.

30. Casas, M.I., Duarte, S., Doseff, A.I., and Grotewold, E. (2014) Flavone-rich maize: An opportunity to improve the nutritional value of an important commodity crop. *Front Plant Sci.*, 5, 440.

31. Yang, F., Ouma, W.Z., Li, W., Doseff, A.I., and Grotewold, E. (2016) *Methods in Enzymology.* Elsevier Inc, New York (in press).

32. Nicholas, C., Batra, S., Vargo, M.A., Voss, O.H., Gavrilin, M.A., Wewers, M.D., Guttridge, D.C., Grotewold, E., and Doseff, A.I. (2007) Apigenin blocks lipopolysaccharide-induced lethality in vivo and proinflammatory cytokines expression by inactivating NF-κB through the suppression of p65 phosphorylation. *J. Immunol.*, 179, 7121–7127.

33. Arango, D., Diosa-Toro, M., Rojas-Hernandez, L.S., Cooperstone, J.L., Schwartz, S.J., Mo, X., Jiang, J., Schmittgen, T.D., and Doseff, A.I. (2015) Dietary apigenin reduces LPS-induced expression of miR-155 restoring immune balance during inflammation. *Mol. Nutr. Food Res.*, 59, 763–772.

34. Zhang, X., Wang, G., Gurley, E.C., and Zhou, H. (2014) Flavonoid apigenin inhibits lipopolysaccharide-induced inflammatory response through multiple mechanisms in macrophages. *PLoS One.*, 9, e107072.

35. Gates, M.A., Vitonis, A.F., Tworoger, S.S., Rosner, B., Titus-Ernstoff, L., Hankinson, S.E., and Cramer, D.W. (2009) Flavonoid intake and ovarian cancer risk in a population-based case-control study. *Int. J. Cancer*, 124, 1918–1925.

36. Parihar, A., Grotewold, E., and Doseff, A. (2015) In Chen, C. (ed), *Pigments in Fruits and Vegetables: Genomics and Dietetics.* Springer Science Ltd., New York, pp. 93–126.

37. Venigalla, M., Gyengesi, E., and Munch, G. (2015) Curcumin and apigenin—Novel and promising therapeutics against chronic neuroinflammation in Alzheimer's disease. *Neural Regen. Res.*, 10, 1181–1185.

38. Hostetler, G., Riedl, K., Cardenas, H., Diosa-Toro, M., Arango, D., Schwartz, S., and Doseff, A.I. (2012) Flavone deglycosylation increases their anti-inflammatory activity and absorption. *Mol. Nutr. Food Res.*, 56, 558–569.

39. Rechner, A.R., Kuhnle, G., Bremner, P., Hubbard, G.P., Moore, K.P., and Rice-Evans, C.A. (2002) The metabolic fate of dietary polyphenols in humans. *Free Radic. Biol. Med.*, 33, 220–235.

40. Chow, H.H., and Hakim, I.A. (2011) Pharmacokinetic and chemoprevention studies on tea in humans. *Pharmacol. Res.*, 64, 105–112.
41. Grosso, C., Valentao, P., Ferreres, F., and Andrade, P.B. (2013) The use of flavonoids in central nervous system disorders. *Curr. Med. Chem.*, 20, 4694–4719.
42. Youdim, K.A., Dobbie, M.S., Kuhnle, G., Proteggente, A.R., Abbott, N.J., and Rice-Evans, C. (2003) Interaction between flavonoids and the blood–brain barrier: *In vitro* studies. *J. Neurochem.*, 85, 180–192.
43. Talavera, S., Felgines, C., Texier, O., Besson, C., Gil-Izquierdo, A., Lamaison, J.L., and Remesy, C. (2005) Anthocyanin metabolism in rats and their distribution to digestive area, kidney, and brain. *J. Agric. Food Chem.*, 53, 3902–3908.
44. Williams, C.M., El Mohsen, M.A., Vauzour, D., Rendeiro, C., Butler, L.T., Ellis, J.A., Whiteman, M., and Spencer, J.P. (2008) Blueberry-induced changes in spatial working memory correlate with changes in hippocampal CREB phosphorylation and brain-derived neurotrophic factor (BDNF) levels. *Free Radic. Biol. Med.*, 45, 295–305.
45. Ishisaka, A., Ichikawa, S., Sakakibara, H., Piskula, M.K., Nakamura, T., Kato, Y., Ito, M., Miyamoto, K., Tsuji, A., Kawai, Y., and Terao, J. (2011) Accumulation of orally administered quercetin in brain tissue and its antioxidative effects in rats. *Free Radic. Biol. Med.*, 51, 1329–1336.
46. Rozan, P., Hidalgo, S., Nejdi, A., Bisson, J.F., Lalonde, R., and Messaoudi, M. (2007) Preventive antioxidant effects of cocoa polyphenolic extract on free radical production and cognitive performances after heat exposure in Wistar rats. *J. Food Sci.*, 72, S203–S206.
47. Patel, A.K., Rogers, J.T., and Huang, X. (2008) Flavanols, mild cognitive impairment, and Alzheimer's dementia. *Int. J. Clin. Exp. Med.*, 1, 181–191.
48. Fisher, N.D., Sorond, F.A., and Hollenberg, N.K. (2006) Cocoa flavanols and brain perfusion. *J. Cardiovasc. Pharmacol.*, 47 Suppl 2, S210–S214.
49. Schroeter, H., Heiss, C., Balzer, J., Kleinbongard, P., Keen, C.L., Hollenberg, N.K., Sies, H., Kwik-Uribe, C., Schmitz, H.H., and Kelm, M. (2006) (-)-Epicatechin mediates beneficial effects of flavanol-rich cocoa on vascular function in humans. *Proc. Natl. Acad. Sci. U.S.A.*, 103, 1024–1029.
50. Krasieva, T.B., Ehren, J., O'Sullivan, T., Tromberg, B.J., and Maher, P. (2015) Cell and brain tissue imaging of the flavonoid fisetin using label-free two-photon microscopy. *Neurochem. Int.*, 89, 243–248.
51. Paulke, A., Eckert, G.P., Schubert-Zsilavecz, M., and Wurglics, M. (2012) Isoquercitrin provides better bioavailability than quercetin: Comparison of quercetin metabolites in body tissue and brain sections after six days administration of isoquercitrin and quercetin. *Pharmazie*, 67, 991–996.
52. Datla, K.P., Christidou, M., Widmer, W.W., Rooprai, H.K., and Dexter, D.T. (2001) Tissue distribution and neuroprotective effects of citrus flavonoid tangeretin in a rat model of Parkinson's disease. *Neuroreport*, 12, 3871–3875.
53. Magalingam, K.B., Radhakrishnan, A.K., and Haleagrahara, N. (2015) Protective mechanisms of flavonoids in Parkinson's disease. *Oxid. Med. Cell Longev.*, 2015, 314560.
54. Ramiro-Puig, E., Casadesus, G., Lee, H.G., Zhu, X., McShea, A., Perry, G., Perez-Cano, F.J., Smith, M.A., and Castell, M. (2009) Neuroprotective effect of cocoa flavonoids on in vitro oxidative stress. *Eur. J. Nutr.*, 48, 54–61.
55. Jonfia-Essien, W.A., West, G., Alderson, P.G., and Tucker, G. (2008) Phenolic content and antioxidant capacity of hybrid variety cocoa beans. *Food Chem.*, 108, 1155–1159.
56. Valles, S.L., Borras, C., Gambini, J., Furriol, J., Ortega, A., Sastre, J., Pallardo, F.V., and Vina, J. (2008) Oestradiol or genistein rescues neurons from amyloid beta-induced cell death by inhibiting activation of p38. *Aging Cell*, 7, 112–118.

57. Kim, S.H., Naveen, K.C., Kim, H.J., Kim, D.H., Cho, J., Jin, C., and Lee, Y.S. (2009) Glucose-containing flavones—Their synthesis and antioxidant and neuroprotective activities. *Bioorg. Med. Chem. Lett.*, 19, 6009–6013.

58. Nichols, M., Zhang, J., Polster, B.M., Elustondo, P.A., Thirumaran, A., Pavlov, E.V., and Robertson, G.S. (2015) Synergistic neuroprotection by epicatechin and quercetin: Activation of convergent mitochondrial signaling pathways. *Neuroscience*, 308, 75–94.

59. Bueler, H. (2009) Impaired mitochondrial dynamics and function in the pathogenesis of Parkinson's disease. *Exp. Neurol.*, 218, 235–246.

60. Karuppagounder, S., Madathil, S.K., Pandey, M., Haobam, R., Rajamma, U., and Mohanakumar, K.P. (2013) Quercetin up-regulates mitochondrial complex-I activity to protect against programmed cell death in rotenone model of Parkinson's disease in rats. *Neuroscience*, 236, 136–148.

61. Gao, Q.G., Xie, J.X., Wong, M.S., and Chen, W.F. (2012) IGF-I receptor signaling pathway is involved in the neuroprotective effect of genistein in the neuroblastoma SK-N-SH cells. *Eur. J. Pharmacol.*, 677, 39–46.

62. Cai, M., Ma, Y., Zhang, W., Wang, S., Wang, Y., Tian, L., Peng, Z., Wang, H., and Qingrong, T. (2016) Apigenin-7-O-β-D-(-6″-p-coumaroyl)-glucopyranoside treatment elicits neuroprotective effect against experimental ischemic stroke. *Int. J. Biol. Sci.*, 12, 42–52.

63. Mansuri, M.L., Parihar, P., Solanki, I., and Parihar, M.S. (2014) Flavonoids in modulation of cell survival signalling pathways. *Genes Nutr.*, 9, 400.

64. Levites, Y., Amit, T., Youdim, M.B., and Mandel, S. (2002) Involvement of protein kinase C activation and cell survival/cell cycle genes in green tea polyphenol (-)-epigallocatechin 3-gallate neuroprotective action. *J. Biol. Chem.*, 277, 30574–30580.

65. Vauzour, D., Vafeiadou, K., Rice-Evans, C., Williams, R.J., and Spencer, J.P. (2007) Activation of pro-survival Akt and ERK1/2 signalling pathways underlie the anti-apoptotic effects of flavanones in cortical neurons. *J. Neurochem.*, 103, 1355–1367.

66. Zhang, Z., Cui, W., Li, G., Yuan, S., Xu, D., Hoi, M.P., Lin, Z., Dou, J., Han, Y., and Lee, S.M. (2012) Baicalein protects against 6-OHDA-induced neurotoxicity through activation of Keap1/Nrf2/HO-1 and involving PKCα and PI3K/AKT signaling pathways. *J. Agric. Food Chem.*, 60, 8171–8182.

67. Ishikawa, Y., and Kitamura, M. (2000) Anti-apoptotic effect of quercetin: Intervention in the JNK- and ERK-mediated apoptotic pathways. *Kidney Int.*, 58, 1078–1087.

68. Schroeter, H., Spencer, J.P., Rice-Evans, C., and Williams, R.J. (2001) Flavonoids protect neurons from oxidized low-density-lipoprotein-induced apoptosis involving c-Jun N-terminal kinase (JNK), c-Jun and caspase-3. *Biochem. J.*, 358, 547–557.

69. Hwang, S.L., and Yen, G.C. (2009) Modulation of Akt, JNK, and p38 activation is involved in citrus flavonoid-mediated cytoprotection of PC12 cells challenged by hydrogen peroxide. *J. Agric. Food Chem.*, 57, 2576–2582.

70. Williams, R.J., and Spencer, J.P. (2012) Flavonoids, cognition, and dementia: Actions, mechanisms, and potential therapeutic utility for Alzheimer disease. *Free Radic. Biol. Med.*, 52, 35–45.

71. Shah, Z.A., Li, R.C., Ahmad, A.S., Kensler, T.W., Yamamoto, M., Biswal, S., and Dore, S. (2010) The flavanol (-)-epicatechin prevents stroke damage through the Nrf2/HO1 pathway. *J. Cereb. Blood Flow Metab*, 30, 1951–1961.

72. Mahler, A., Mandel, S., Lorenz, M., Ruegg, U., Wanker, E.E., Boschmann, M., and Paul, F. (2013) Epigallocatechin-3-gallate: A useful, effective and safe clinical approach for targeted prevention and individualised treatment of neurological diseases? *EPMA. J.*, 4, 5.

73. Dajas, F., Abin-Carriquiry, J.A., Arredondo, F., Blasina, F., Echeverry, C., Martinez, M., Rivera, F., and Vaamonde, L. (2015) Quercetin in brain diseases: Potential and limits. *Neurochem. Int.*, 89, 140–148.

74. Soltani, Z., Khaksari, M., Jafari, E., Iranpour, M., and Shahrokhi, N. (2015) Is genistein neuroprotective in traumatic brain injury? *Physiol Behav.*, 152, 26–31.

75. Wang, S., Wei, H., Cai, M., Lu, Y., Hou, W., Yang, Q., Dong, H., and Xiong, L. (2014) Genistein attenuates brain damage induced by transient cerebral ischemia through up-regulation of ERK activity in ovariectomized mice. *Int. J. Biol. Sci.*, 10, 457–465.

76. Stout, J.M., Knapp, A.N., Banz, W.J., Wallace, D.G., and Cheatwood, J.L. (2013) Subcutaneous daidzein administration enhances recovery of skilled ladder rung walking performance following stroke in rats. *Behav. Brain Res.*, 256, 428–431.

77. Cui, L., Zhang, X., Yang, R., Liu, L., Wang, L., Li, M., and Du, W. (2010) Baicalein is neuroprotective in rat MCAO model: Role of 12/15-lipoxygenase, mitogen-activated protein kinase and cytosolic phospholipase A2. *Pharmacol. Biochem. Behav.*, 96, 469–475.

78. Zhao, L., Wang, J.L., Liu, R., Li, X.X., Li, J.F., and Zhang, L. (2013) Neuroprotective, anti-amyloidogenic and neurotrophic effects of apigenin in an Alzheimer's disease mouse model. *Molecules.*, 18, 9949–9965.

79. Bastianetto, S., Ramassamy, C., Dore, S., Christen, Y., Poirier, J., and Quirion, R. (2000) The *Ginkgo biloba* extract (EGb 761) protects hippocampal neurons against cell death induced by β-amyloid. *Eur. J. Neurosci.*, 12, 1882–1890.

80. Kim, M.S., Lee, J.I., Lee, W.Y., and Kim, S.E. (2004) Neuroprotective effect of *Ginkgo biloba* L. extract in a rat model of Parkinson's disease. *Phytother. Res.*, 18, 663–666.

81. Merlo, S., Basile, L., Giuffrida, M.L., Sortino, M.A., Guccione, S., and Copani, A. (2015) Identification of 5-methoxyflavone as a novel DNA polymerase-β inhibitor and neuroprotective agent against β-amyloid toxicity. *J. Nat. Prod.*, 78, 2704–2711.

82. Hu, K., Li, S.Y., Xiao, B., Bi, F.F., Lu, X.Q., and Wu, X.M. (2011) Protective effects of quercetin against status epilepticus induced hippocampal neuronal injury in rats: Involvement of X-linked inhibitor of apoptosis protein. *Acta Neurol. Belg.*, 111, 205–212.

83. Yang, T., Kong, B., Gu, J.W., Kuang, Y.Q., Cheng, L., Yang, W.T., Xia, X., and Shu, H.F. (2014) Anti-apoptotic and anti-oxidative roles of quercetin after traumatic brain injury. *Cell Mol. Neurobiol.*, 34, 797–804.

84. Miao, G., Zhao, H., Guo, K., Cheng, J., Zhang, S., Zhang, X., Cai, Z., Miao, H., and Shang, Y. (2014) Mechanisms underlying attenuation of apoptosis of cortical neurons in the hypoxic brain by flavonoids from the stems and leaves of *Scutellaria baicalensis* Georgi. *Neural Regen. Res.*, 9, 1592–1598.

85. Serrano-Garcia, N., Pedraza-Chaverri, J., Mares-Samano, J.J., Orozco-Ibarra, M., Cruz-Salgado, A., Jimenez-Anguiano, A., Sotelo, J., and Trejo-Solis, C. (2013) Anti-apoptotic effects of EGb 761. *Evid. Based. Complement Alternat. Med.*, 2013, 495703.

86. Agrawal, M., Kumar, V., Kashyap, M.P., Khanna, V.K., Randhawa, G.S., and Pant, A.B. (2011) Ischemic insult induced apoptotic changes in PC12 cells: Protection by trans resveratrol. *Eur. J. Pharmacol.*, 666, 5–11.

87. Jin, F., Wu, Q., Lu, Y.F., Gong, Q.H., and Shi, J.S. (2008) Neuroprotective effect of resveratrol on 6-OHDA-induced Parkinson's disease in rats. *Eur. J. Pharmacol.*, 600, 78–82.

88. Ahlemeyer, B., and Krieglstein, J. (2003) Pharmacological studies supporting the therapeutic use of *Ginkgo biloba* extract for Alzheimer's disease. *Pharmacopsychiatry*, 36 Suppl 1, S8–S14.

89. Nie, G., Cao, Y., and Zhao, B. (2002) Protective effects of green tea polyphenols and their major component, (-)-epigallocatechin-3-gallate (EGCG), on 6-hydroxydopamine-induced apoptosis in PC12 cells. *Redox. Rep.*, 7, 171–177.

90. Baluchnejadmojarad, T., Roghani, M., Nadoushan, M.R., and Bagheri, M. (2009) Neuroprotective effect of genistein in 6-hydroxydopamine hemi-parkinsonian rat model. *Phytother. Res.*, 23, 132–135.

91. Khan, M.M., Raza, S.S., Javed, H., Ahmad, A., Khan, A., Islam, F., Safhi, M.M., and Islam, F. (2012) Rutin protects dopaminergic neurons from oxidative stress in an animal model of Parkinson's disease. *Neurotox. Res.*, 22, 1–15.

92. Levites, Y., Weinreb, O., Maor, G., Youdim, M.B., and Mandel, S. (2001) Green tea polyphenol (-)-epigallocatechin-3-gallate prevents N-methyl-4-phenyl-1,2,3,6-tetrahydropyridine-induced dopaminergic neurodegeneration. *J. Neurochem.*, 78, 1073–1082.

93. Vauzour, D., Ravaioli, G., Vafeiadou, K., Rodriguez-Mateos, A., Angeloni, C., and Spencer, J.P. (2008) Peroxynitrite induced formation of the neurotoxins 5-S-cysteinyl-dopamine and DHBT-1: Implications for Parkinson's disease and protection by polyphenols. *Arch. Biochem. Biophys.*, 476, 145–151.

94. Nones, J., Costa, A.P., Leal, R.B., Gomes, F.C., and Trentin, A.G. (2012) The flavonoids hesperidin and rutin promote neural crest cell survival. *Cell Tissue Res.*, 350, 305–315.

95. Prakash, D., and Sudhandiran, G. (2015) Dietary flavonoid fisetin regulates aluminium chloride-induced neuronal apoptosis in cortex and hippocampus of mice brain. *J. Nutr. Biochem.*, 26, 1527–1539.

96. McGeer, E.G., and McGeer, P.L. (2003) Inflammatory processes in Alzheimer's disease. *Prog. Neuropsychopharmacol. Biol. Psychiatry*, 27, 741–749.

97. Akiyama, H., Barger, S., Barnum, S., Bradt, B., Bauer, J., Cole, G.M., Cooper, N.R., Eikelenboom, P., Emmerling, M., Fiebich, B.L., Finch, C.E., Frautschy, S., Griffin, W.S., Hampel, H., Hull, M., Landreth, G., Lue, L., Mrak, R., Mackenzie, I.R., McGeer, P.L., O'Banion, M.K., Pachter, J., Pasinetti, G., Plata-Salaman, C., Rogers, J., Rydel, R., Shen, Y., Streit, W., Strohmeyer, R., Tooyoma, I., Van Muiswinkel, F.L., Veerhuis, R., Walker, D., Webster, S., Wegrzyniak, B., Wenk, G., and Wyss-Coray, T. (2000) Inflammation and Alzheimer's disease. *Neurobiol. Aging*, 21, 383–421.

98. Griffin, W.S., and Mrak, R.E. (2002) Interleukin-1 in the genesis and progression of and risk for development of neuronal degeneration in Alzheimer's disease. *J. Leukoc. Biol.*, 72, 233–238.

99. Gerritse, K., Laman, J.D., Noelle, R.J., Aruffo, A., Ledbetter, J.A., Boersma, W.J., and Claassen, E. (1996) CD40–CD40 ligand interactions in experimental allergic encephalomyelitis and multiple sclerosis. *Proc. Natl. Acad. Sci. U.S.A.*, 93, 2499–2504.

100. Townsend, K.P., Shytle, D.R., Bai, Y., San, N., Zeng, J., Freeman, M., Mori, T., Fernandez, F., Morgan, D., Sanberg, P., and Tan, J. (2004) Lovastatin modulation of microglial activation via suppression of functional CD40 expression. *J. Neurosci. Res.*, 78, 167–176.

101. Townsend, K.P., Town, T., Mori, T., Lue, L.F., Shytle, D., Sanberg, P.R., Morgan, D., Fernandez, F., Flavell, R.A., and Tan, J. (2005) CD40 signaling regulates innate and adaptive activation of microglia in response to amyloid beta-peptide. *Eur. J. Immunol.*, 35, 901–910.

102. Santangelo, C., Vari, R., Scazzocchio, B., Di Benedetto, R., Filesi, C., and Masella, R. (2007) Polyphenols, intracellular signalling and inflammation. *Ann. Ist. Super. Sanita*, 43, 394–405.

103. Bode, A.M., and Dong, Z. (2004) Targeting signal transduction pathways by chemopreventive agents. *Mutat. Res.*, 555, 33–51.

104. Gopalakrishnan, A., and Tony Kong, A.N. (2008) Anti-carcinogenesis by dietary phytochemicals: Cytoprotection by Nrf2 in normal cells and cytotoxicity by modulation of transcription factors NF-κB and AP-1 in abnormal cancer cells. *Food Chem. Toxicol.*, 46, 1257–1270.

105. Rezai-Zadeh, K., Ehrhart, J., Bai, Y., Sanberg, P.R., Bickford, P., Tan, J., and Shytle, R.D. (2008) Apigenin and luteolin modulate microglial activation via inhibition of STAT1-induced CD40 expression. *J. Neuroinflammation*, 5, 41.

106. Hashimoto, F., Ono, M., Masuoka, C., Ito, Y., Sakata, Y., Shimizu, K., Nonaka, G., Nishioka, I., and Nohara, T. (2003) Evaluation of the anti-oxidative effect (in vitro) of tea polyphenols. *Biosci. Biotechnol. Biochem.*, 67, 396–401.

107. Zheng, L.T., Ryu, G.M., Kwon, B.M., Lee, W.H., and Suk, K. (2008) Anti-inflammatory effects of catechols in lipopolysaccharide-stimulated microglia cells: Inhibition of microglial neurotoxicity. *Eur. J. Pharmacol.*, 588, 106–113.

108. Chen, J.C., Ho, F.M., Pei-Dawn, L.C., Chen, C.P., Jeng, K.C., Hsu, H.B., Lee, S.T., Wen, T.W., and Lin, W.W. (2005) Inhibition of iNOS gene expression by quercetin is mediated by the inhibition of IkappaB kinase, nuclear factor-κB and STAT1, and depends on heme oxygenase-1 induction in mouse BV-2 microglia. *Eur. J. Pharmacol.*, 521, 9–20.

109. Li, R., Huang, Y.G., Fang, D., and Le, W.D. (2004) (-)-Epigallocatechin gallate inhibits lipopolysaccharide-induced microglial activation and protects against inflammation-mediated dopaminergic neuronal injury. *J. Neurosci. Res.*, 78, 723–731.

110. Park, S.E., Sapkota, K., Kim, S., Kim, H., and Kim, S.J. (2011) Kaempferol acts through mitogen-activated protein kinases and protein kinase B/AKT to elicit protection in a model of neuroinflammation in BV2 microglial cells. *Br. J. Pharmacol.*, 164, 1008–1025.

111. Zhu, L.H., Bi, W., Qi, R.B., Wang, H.D., Wang, Z.G., Zeng, Q., Zhao, Y.R., and Lu, D.X. (2011) Luteolin reduces primary hippocampal neurons death induced by neuroinflammation. *Neurol. Res.*, 33, 927–934.

112. Sawmiller, D., Li, S., Shahaduzzaman, M., Smith, A.J., Obregon, D., Giunta, B., Borlongan, C.V., Sanberg, P.R., and Tan, J. (2014) Luteolin reduces Alzheimer's disease pathologies induced by traumatic brain injury. *Int. J. Mol. Sci.*, 15, 895–904.

113. Kim, H.P., Son, K.H., Chang, H.W., and Kang, S.S. (2004) Anti-inflammatory plant flavonoids and cellular action mechanisms. *J. Pharmacol. Sci.*, 96, 229–245.

114. Welton, A.F., Tobias, L.D., Fiedler-Nagy, C. Anderson, W., Hope, W., Meyers, K., and Coffey, J.W. (1986) Effect of flavonoids on arachidonic acid metabolism. *Prog. Clin. Biol. Res.*, 213, 231–242.

115. Kim, O.K., Murakami, A., Nakamura, Y., and Ohigashi, H. (1998) Screening of edible Japanese plants for nitric oxide generation inhibitory activities in RAW 264.7 cells. *Cancer Lett.*, 125, 199–207.

116. Kim, H.K., Cheon, B.S., Kim, Y.H., Kim, S.Y., and Kim, H.P. (1999) Effects of naturally occurring flavonoids on nitric oxide production in the macrophage cell line RAW 264.7 and their structure–activity relationships. *Biochem. Pharmacol.*, 58, 759–765.

117. Chi, Y.S., Cheon, B.S., and Kim, H.P. (2001) Effect of wogonin, a plant flavone from Scutellaria radix, on the suppression of cyclooxygenase-2 and the induction of inducible nitric oxide synthase in lipopolysaccharide-treated RAW 264.7 cells. *Biochem. Pharmacol.*, 61, 1195–1203.

118. Bhat, N.R., Zhang, P., Lee, J.C., and Hogan, E.L. (1998) Extracellular signal-regulated kinase and p38 subgroups of mitogen-activated protein kinases regulate inducible nitric oxide synthase and tumor necrosis factor-α gene expression in endotoxin-stimulated primary glial cultures. *J. Neurosci.*, 18, 1633–1641.

119. Jang, S., Kelley, K.W., and Johnson, R.W. (2008) Luteolin reduces IL-6 production in microglia by inhibiting JNK phosphorylation and activation of AP-1. *Proc. Natl. Acad. Sci. U.S.A.*, 105, 7534–7539.

120. Cho, J.Y., Kim, P.S., Park, J., Yoo, E.S., Baik, K.U., Kim, Y.K., and Park, M.H. (2000) Inhibitor of tumor necrosis factor-alpha production in lipopolysaccharide-stimulated RAW264.7 cells from *Amorpha fruticosa*. *J. Ethnopharmacol.*, 70, 127–133.

121. Wadsworth, T.L., McDonald, T.L., and Koop, D.R. (2001) Effects of *Ginkgo biloba* extract (EGb 761) and quercetin on lipopolysaccharide-induced signaling pathways involved in the release of tumor necrosis factor-alpha. *Biochem. Pharmacol.*, 62, 963–974.

122. Wadsworth, T.L., and Koop, D.R. (2001) Effects of *Ginkgo biloba* extract (EGb 761) and quercetin on lipopolysaccharide-induced release of nitric oxide. *Chem. Biol. Interact.*, 137, 43–58.

123. Fisher, N.D., and Hollenberg, N.K. (2006) Aging and vascular responses to flavanol-rich cocoa. *J. Hypertens.*, 24, 1575–1580.

124. Murphy, K.J., Chronopoulos, A.K., Singh, I., Francis, M.A., Moriarty, H., Pike, M.J., Turner, A.H., Mann, N.J., and Sinclair, A.J. (2003) Dietary flavanols and procyanidin oligomers from cocoa (*Theobroma cacao*) inhibit platelet function. *Am. J. Clin. Nutr.*, 77, 1466–1473.

125. Mandel, S.A., Amit, T., Kalfon, L., Reznichenko, L., and Youdim, M.B. (2008) Targeting multiple neurodegenerative diseases etiologies with multimodal-acting green tea catechins. *J. Nutr.*, 138, 1578S–1583S.

126. Ahn, H.Y., Xu, Y., and Davidge, S.T. (2008) Epigallocatechin-3-O-gallate inhibits TNF-α induced monocyte chemotactic protein-1 production from vascular endothelial cells. *Life Sci.*, 82, 964–968.

127. Steptoe, A., Gibson, E.L., Vuononvirta, R., Hamer, M., Wardle, J., Rycroft, J.A., Martin, J.F., and Erusalimsky, J.D. (2007) The effects of chronic tea intake on platelet activation and inflammation: A double-blind placebo controlled trial. *Atherosclerosis*, 193, 277–282.

128. Watzl, B., Bub, A., Briviba, K., and Rechkemmer, G. (2002) Acute intake of moderate amounts of red wine or alcohol has no effect on the immune system of healthy men. *Eur. J. Nutr.*, 41, 264–270.

129. Watzl, B., Bub, A., Pretzer, G., Roser, S., Barth, S.W., and Rechkemmer, G. (2004) Daily moderate amounts of red wine or alcohol have no effect on the immune system of healthy men. *Eur. J. Clin. Nutr.*, 58, 40–45.

130. Estruch, R., Sacanella, E., Badia, E., Antunez, E., Nicolas, J.M., Fernandez-Sola, J., Rotilio, D., de Gaetano, G., Rubin, E., and Urbano-Marquez, A. (2004) Different effects of red wine and gin consumption on inflammatory biomarkers of atherosclerosis: A prospective randomized crossover trial. Effects of wine on inflammatory markers. *Atherosclerosis*, 175, 117–123.

131. Castilla, P., Davalos, A., Teruel, J.L., Cerrato, F., Fernandez-Lucas, M., Merino, J.L., Sanchez-Martin, C.C., Ortuno, J., and Lasuncion, M.A. (2008) Comparative effects of dietary supplementation with red grape juice and vitamin E on production of superoxide by circulating neutrophil NADPH oxidase in hemodialysis patients. *Am. J. Clin. Nutr.*, 87, 1053–1061.

132. Zern, T.L., Wood, R.J., Greene, C., West, K.L., Liu, Y., Aggarwal, D., Shachter, N.S., and Fernandez, M.L. (2005) Grape polyphenols exert a cardioprotective effect in pre- and postmenopausal women by lowering plasma lipids and reducing oxidative stress. *J. Nutr.*, 135, 1911–1917.

133. Fanti, P., Asmis, R., Stephenson, T.J., Sawaya, B.P., and Franke, A.A. (2006) Positive effect of dietary soy in ESRD patients with systemic inflammation—Correlation between blood levels of the soy isoflavones and the acute-phase reactants. *Nephrol. Dial. Transplant.*, 21, 2239–2246.

134. Nasca, M.M., Zhou, J.R., and Welty, F.K. (2008) Effect of soy nuts on adhesion molecules and markers of inflammation in hypertensive and normotensive postmenopausal women. *Am. J. Cardiol.*, 102, 84–86.

135. Maskarinec, G., Oum, R., Chaptman, A.K., and Ognjanovic, S. (2009) Inflammatory markers in a randomised soya intervention among men. *Br. J. Nutr.*, 101, 1740–1744.

136. Jenkins, D.J., Kendall, C.W., Connelly, P.W., Jackson, C.J., Parker, T., Faulkner, D., and Vidgen, E. (2002) Effects of high- and low-isoflavone (phytoestrogen) soy foods on inflammatory biomarkers and proinflammatory cytokines in middle-aged men and women. *Metabolism*, 51, 919–924.

137. Ryan-Borchers, T.A., Park, J.S., Chew, B.P., McGuire, M.K., Fournier, L.R., and Beerman, K.A. (2006) Soy isoflavones modulate immune function in healthy postmenopausal women. *Am. J. Clin. Nutr.*, 83, 1118–1125.

138. Nieman, D.C., Henson, D.A., Davis, J.M., Angela, M.E., Jenkins, D.P., Gross, S.J., Carmichael, M.D., Quindry, J.C., Dumke, C.L., Utter, A.C., McAnulty, S.R., McAnulty, L.S., Triplett, N.T., and Mayer, E.P. (2007) Quercetin's influence on exercise-induced changes in plasma cytokines and muscle and leukocyte cytokine mRNA. *J. Appl. Physiol (1985.)*, 103, 1728–1735.

139. Nieman, D.C., Henson, D.A., Davis, J.M., Dumke, C.L., Gross, S.J., Jenkins, D.P., Murphy, E.A., Carmichael, M.D., Quindry, J.C., McAnulty, S.R., McAnulty, L.S., Utter, A.C., and Mayer, E.P. (2007) Quercetin ingestion does not alter cytokine changes in athletes competing in the Western States Endurance Run. *J. Interferon Cytokine Res.*, 27, 1003–1011.

140. Arango, D., Morohashi, K., Yilmaz, A., Kuramochi, K., Parihar, A., Brahimaj, B., Grotewold, E., and Doseff, A.I. (2013) Molecular basis for the action of a dietary flavonoid revealed by the comprehensive identification of apigenin human targets. *Proc. Natl. Acad. Sci. U.S.A.*, 110, E2153–E2162.

141. Vauzour, D., Vafeiadou, K., Rodriguez-Mateos, A., Rendeiro, C., and Spencer, J.P. (2008) The neuroprotective potential of flavonoids: A multiplicity of effects. *Genes Nutr.*, 3, 115–126.

142. Rohdewald, P. (2002) A review of the French maritime pine bark extract (Pycnogenol), a herbal medication with a diverse clinical pharmacology. *Int. J. Clin. Pharmacol. Ther.*, 40, 158–168.

143. Ryan, J., Croft, K., Mori, T., Wesnes, K., Spong, J., Downey, L., Kure, C., Lloyd, J., and Stough, C. (2008) An examination of the effects of the antioxidant pycnogenol on cognitive performance, serum lipid profile, endocrinological and oxidative stress biomarkers in an elderly population. *J. Psychopharmacol.*, 22, 553–562.

144. Joseph, J.A., Shukitt-Hale, B., Denisova, N.A., Prior, R.L., Cao, G., Martin, A., Taglialatela, G., and Bickford, P.C. (1998) Long-term dietary strawberry, spinach, or vitamin E supplementation retards the onset of age-related neuronal signal-transduction and cognitive behavioral deficits. *J. Neurosci.*, 18, 8047–8055.

145. Stoll, S., Scheuer, K., Pohl, O., and Muller, W.E. (1996) *Ginkgo biloba* extract (EGb 761) independently improves changes in passive avoidance learning and brain membrane fluidity in the aging mouse. *Pharmacopsychiatry*, 29, 144–149.

146. Sachdeva, A.K., Kuhad, A., and Chopra, K. (2014) Naringin ameliorates memory deficits in experimental paradigm of Alzheimer's disease by attenuating mitochondrial dysfunction. *Pharmacol. Biochem. Behav.*, 127, 101–110.

147. Shukitt-Hale, B., Carey, A., Simon, L., Mark, D.A., and Joseph, J.A. (2006) Effects of Concord grape juice on cognitive and motor deficits in aging. *Nutrition*, 22, 295–302.

148. Ramirez, M.R., Izquierdo, I., do Carmo Bassols, R.M., Zuanazzi, J.A., Barros, D., and Henriques, A.T. (2005) Effect of lyophilised Vaccinium berries on memory, anxiety and locomotion in adult rats. *Pharmacol. Res.*, 52, 457–462.

149. Lau, F.C., Shukitt-Hale, B., and Joseph, J.A. (2005) The beneficial effects of fruit polyphenols on brain aging. *Neurobiol. Aging*, 26 Suppl 1, 128–132.

150. Ramalingayya, G.V., Nampoothiri, M., Nayak, P.G., Kishore, A., Shenoy, R.R., Mallikarjuna, R.C., and Nandakumar, K. (2016) Naringin and rutin alleviates episodic memory deficits in two differentially challenged object recognition tasks. *Pharmacogn. Mag.*, 12, S63–S70.

151. Macready, A.L., Kennedy, O.B., Ellis, J.A., Williams, C.M., Spencer, J.P., and Butler, L.T. (2009) Flavonoids and cognitive function: A review of human randomized controlled trial studies and recommendations for future studies. *Genes Nutr.*, 4, 227–242.

152. Jagla, F., and Pechanova, O. (2015) Age-Related Cognitive Impairment as a Sign of Geriatric Neurocardiovascular Interactions: May polyphenols play a protective role? *Oxid. Med. Cell Longev.*, 2015, 721514.

153. Desideri, G., Kwik-Uribe, C., Grassi, D., Necozione, S., Ghiadoni, L., Mastroiacovo, D., Raffaele, A., Ferri, L., Bocale, R., Lechiara, M.C., Marini, C., and Ferri, C. (2012) Benefits in cognitive function, blood pressure, and insulin resistance through cocoa flavanol consumption in elderly subjects with mild cognitive impairment: The Cocoa, Cognition, and Aging (CoCoA) study. *Hypertension*, 60, 794–801.

154. Alharbi, M.H., Lamport, D.J., Dodd, G.F., Saunders, C., Harkness, L., Butler, L.T., and Spencer, J.P. (2016) Flavonoid-rich orange juice is associated with acute improvements in cognitive function in healthy middle-aged males. *Eur. J. Nutr.* 55(6):2021–2029.

155. Spencer, J.P. (2008) Flavonoids: Modulators of brain function? *Br. J. Nutr.*, 99 E Suppl 1, ES60–ES77.

156. Kim, D.H., Jeon, S.J., Son, K.H., Jung, J.W., Lee, S., Yoon, B.H., Choi, J.W., Cheong, J.H., Ko, K.H., and Ryu, J.H. (2006) Effect of the flavonoid, oroxylin A, on transient cerebral hypoperfusion-induced memory impairment in mice. *Pharmacol. Biochem. Behav.*, 85, 658–668.

11 Nuts and Brain Health

Sui Kiat Chang and Cesarettin Alasalvar

CONTENTS

11.1 INTRODUCTION

Globally, growing populations and longer mean life spans are leading to a large aging population [1]. This demographic shift comes together with an increased global incidence of non–communicable diseases (NCDs), especially the incidence rate of neurodegenerative diseases in the elderly population [2]. More than 35.6 million people suffer from dementia and nearly 7 million people have been diagnosed with dementia and other neurological disorders in the United States [3]. Neurodegenerative diseases affected the quality of life severely by imposing huge economic burden on both individuals and society. Cures for most degenerative diseases are yet to be discovered, in part due to the irreversible loss of brain cells during the pathogenesis, which often takes more than 10 years to manifest clinically [4].

Neurodegenerative diseases include the chronic exposure to oxidative stress and inflammation, loss of protective signaling, and the accumulation of toxic proteins [5,6]. These factors trigger a cascade of altered molecular events during aging, which will destroy brain cells. Damage to individual brain cells modifies the communications between neurons, which affects memory, cognition, and motor function. All these cellular processes will lead to the pathogenesis of neurodegenerative diseases, such as Alzheimer's disease, Parkinson disease, Huntington disease, amyotrophic lateral sclerosis, prion disease, and dementia [4,7]. Although the central nervous system is especially vulnerable, oxidative stress and inflammation will also affect other organ systems, increasing elderly people's risks of developing other NCDs, such as heart disease, cancer, diabetes, and other age-related disorders. Thus, protection from oxidative stress and inflammation systemically could protect the brain from their direct effects as well as other related pathologies [5–7].

A preventive and sustainable approach is important to dealing with an aging population to reduce the burden of neurodegenerative diseases to maintain brain health [8]. In this context, lifestyle-related factors, particularly diet and nutrition, have demonstrated a great impact in preventing and controlling the morbidity and mortality due to NCDs. Plant-based diet, particularly the Mediterranean diet, demonstrates the most convincing evidence [9–11]. Among the main components of the Mediterranean diet (including olive oils, fruits, vegetables, legumes, nuts, and cereals), only recently has attention focused on nuts. Nuts are nutrient-dense foods and have been a regular constituent of mankind's diet since a long time ago [12]. Nuts are a convenient and tasty snack that contribute to a healthy lifestyle amid today's busy lifestyles. They are usually consumed as whole nuts (either raw or roasted or salted) [12].

Numerous epidemiological studies have demonstrated the benefits of nuts from the Mediterranean diet in reducing the risk of various NCDs, particularly brain-related disorders, such as dementia, cognitive improvement, and neurodegenerative diseases [13–16]. These evidences support the recommendations to include nuts as part of a healthy dietary pattern. Due to this reason, the American Heart Association [17], the U.S. Department of Agriculture (USDA) [18], and the Food and Drug Administration [19] recommend the regular consumption of nuts to the general population. Knowledge on the effect of nut consumption in preventing brain-related disorders, particularly neurodegenerative diseases, has increased rapidly in recent years. Hence, this chapter highlights the contemporary research on the effects of nuts and brain health by discussing the levels of nutritional components and bioactive phytochemicals present in commonly consumed nuts. In addition, the health benefits of nuts contributing to brain health will also be highlighted using various *in vitro*, animal, and human intervention studies.

11.2 OVERVIEW OF NUTS AND THEIR NUTRITIONAL AND PHYTOCHEMICAL COMPOSITIONS

Tree nuts are dry fruits with one seed in which the ovary wall becomes hard at maturity [12]. Common edible tree nuts include almond, Brazil nut, cashew, hazelnut, macadamia, pecan, pine nut, pistachio, and walnut, but the consumer definition also includes peanut, which is botanically legume but has a nutrient profile similar to that of tree nuts and is thus identified as part of the nuts food group [12]. Nuts contain macronutrients, micronutrients, fat-soluble bioactives, and phytochemicals [12,20–25]. Figure 11.1 shows the chemical structures of some fat-soluble bioactives reported in nuts.

Nut proteins contribute about 10%–25% of energy, including individual amino acids, such as L-arginine, which is involved in the production of nitric oxide (NO) an endogenous vasodilator [26]. Table 11.1 shows the compositional and nutritional characteristics of nuts. The fatty acid composition of nuts is saturated fats (4%–15%) and unsaturated fatty acids (30%–60%). Unsaturated fatty acids vary among nuts [27,28]. Nuts are good sources of calcium, magnesium, and potassium, with low amount of sodium, which is important for several pathological conditions, such as bone demineralization, hypertension, and insulin resistance [29]. Nuts are also rich in phytosterols, nonnutritive components of certain plant-foods that exert both

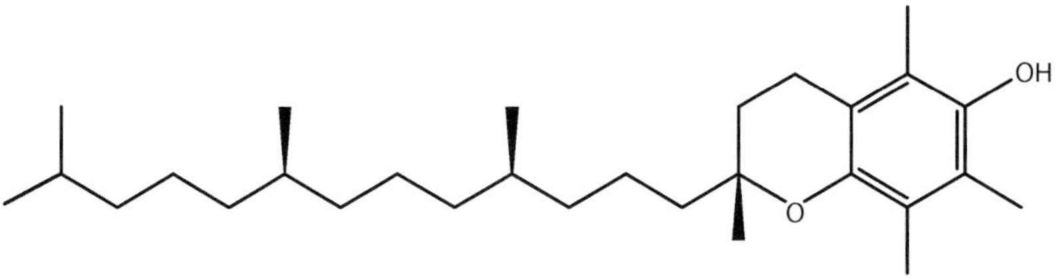

α-Tocopherol

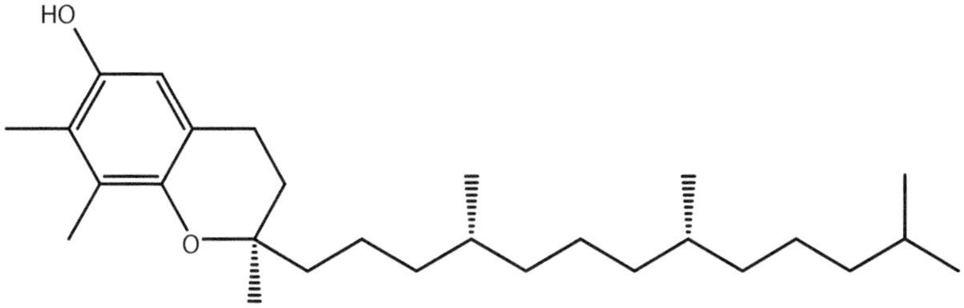

γ-Tocopherol

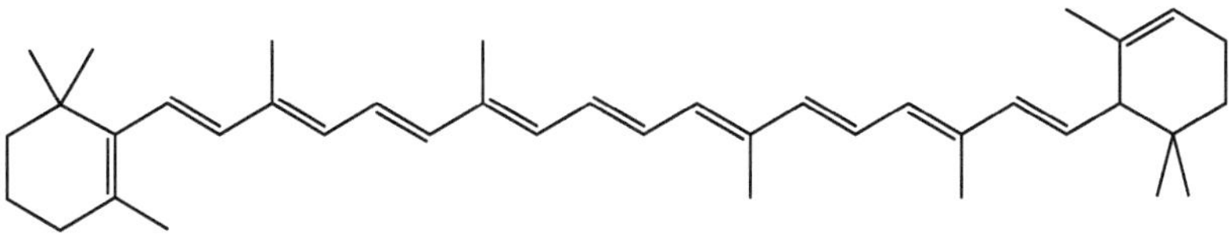

β-Carotene

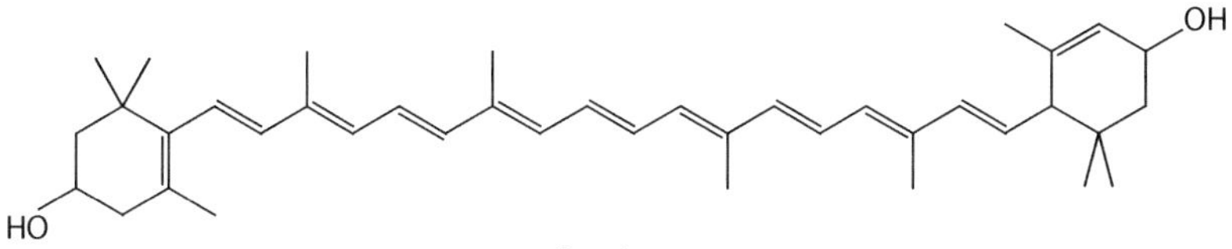

Lutein

FIGURE 11.1 Chemical structures of the major fat-soluble bioactives reported in nuts that contribute to brain health. (*Continued*)

structural (at cellular membrane phospholipids level) and hormonal (estrogen-like) activities [30]. Finally, nuts have been demonstrated to be a rich source of polyphenols, which account for a key role in their antioxidant and anti-inflammatory effects [20–23].

11.3 BRAIN, NUTRIENTS, AND NEUROPROTECTION

Nutrients are bioactive molecules that are essential for human health and functioning [31]. Most cannot be synthesized by the human body (not at all) and need to be obtained from food. The brain is a complex organ with high metabolic rate and high turnover of nutrients. Various nutrient-specific transport systems and physiological mechanisms constantly work to replace the nutrients used by the brain [31].

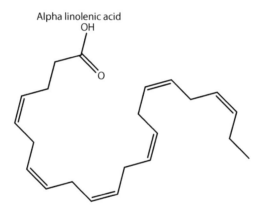

Alpha linolenic acid

Docosahexaenoic acid (DHA)

FIGURE 11.1 (CONTINUED) Chemical structures of the major fat-soluble bioactives reported in nuts that contribute to brain health. (*Continued*)

The brain is metabolically demanding, where it generates lots of reactive oxygen and nitrogen species. However, aging, as well as other factors, disrupts the homeostasis of the endogenous antioxidant defense system, leading to increased oxidative stress *in vivo* [32]. Increased oxidative stress and lipid peroxidation will provoke a series of proinflammatory signals, leading to the death of brain cells. All these are linked directly to the impairment in various cognitive domains, such as, learning, decision making, judgment, problem solving, and memory [33,34].

Fatty acids are the main components of the neuronal membranes, which play significant roles in maintaining structural integrity, modulating enzyme activity, and generating secondary messengers as well as other signaling molecules [35,36].

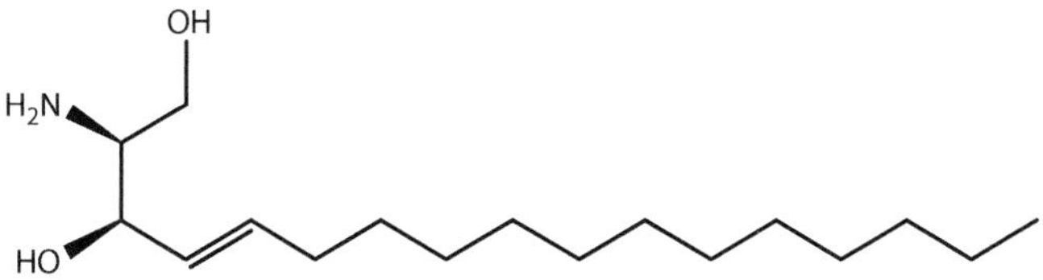

Sphingosine

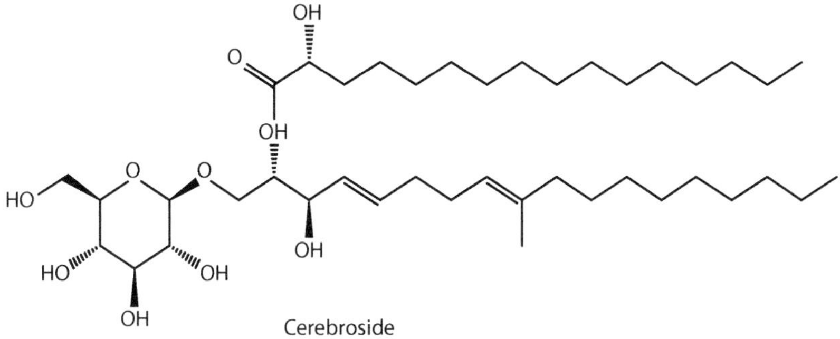

Cerebroside

FIGURE 11.1 (CONTINUED) Chemical structures of the major fat-soluble bioactives reported in nuts that contribute to brain health.

There are two major classes of unsaturated fatty acids: monounsaturated fatty acids (MUFAs) and polyunsaturated fatty acids (PUFAs). PUFA can be classified into two groups: the *n*-6 class (e.g., linoleic and arachidonic acids) and the *n*-3 class (e.g., alpha-linolenic acid [ALA], eicosapentaenoic acid [EPA], and docosahexaenoic acid [DHA]). PUFAs play an important role in the maintenance of cognitive function and have preventive effects against dementia through their antithrombotic and anti-inflammatory properties in addition to their specific effect on neural functions [37]. Specifically, DHA is a key component of membrane phospholipids in the brain, where adequate *n*-3 PUFA status may help maintain neuronal integrity and function. DHA may modify the expression of genes that regulate various biological functions that are important for cognitive health, including neurogenesis and neuronal function [38]. Some studies have found an association between dietary DHA and lower incident rates of dementia [39,40]. Recently, longitudinal studies demonstrated that MUFA may play an important role in the prevention of cognitive decline and dementia due to its anti-inflammatory effects [41,42].

The PUFA composition of neuronal membranes decreases during aging and contributes to the decline of neuronal function observed in aging. This alteration is prevalent in the aged brain, especially the cortex, hippocampus, striatum, and cerebellum, where reduced PUFA concentrations contribute to changes in neuronal morphology and a decrease in membrane fluidity and synaptic plasticity [43–45]. Both EPA and DHA play an important role in brain health by reducing oxidative stress and altering the immune function as well as maintaining synaptic plasticity, neuronal membrane stability, gene expression, and neurogenesis [35,36].

TABLE 11.1

Compositional and Nutritional Characteristics of Nuts (per 100 g nut)

Nutrient	Unit	Almond	Brazil Nut	Cashew	Hazelnut	Macadamia	Pecan	Peanut	Pistachio	Walnut
Energy	kcal	828	876	157	722	962	753	795	689	765
SFA	g	4	15	8	4	12	7	7	6	6
MUFA	g	32	25	24	46	59	40	24	24	9
PUFA	g	12	21	8	8	2	21	16	14	47
Protein	g	21	14	18	17	8	9	25	20	15
Arginine	g	2.5	2.2	2.0	2.2	1.2	3.0	1.2	2.2	2.3
Fiber	g	13	9	3	10	9	10	9	10	7
Vitamin E (ATE)	mg	36.7	7.5	0.26	17.3	0.72	1.53	9.58	3.52	0.82
Vitamin A	µg	0.6	0.0	0.0	12	0.0	33.6	0.0	249	12
Vitamin C	mg	0.0	0.9	0.1	7.2	1.6	1.2	0.0	6.9	1.5
Sodium	mg	1	3	12	0	5	0	18	1	2
Potassium	mg	733	659	660	680	368	410	705	1025	441
Selenium	µg	5.9	25450	5.6	2.8	4.8	4.1	nr	8.6	5.7

Source: Adapted from U.S. Department of Agriculture (USDA), USDA National Nutrient Database for Standard Reference, Release 28, 2016, National Technical Information Service, USDA, Springfield, VA.

Note: ATE, alpha-tocopherol equivalents; MUFA, monounsaturated fatty acids; nr, not reported; PUFA, polyunsaturated fatty acids; SFA, saturated fatty acids.

Other phytochemicals provide direct neuroprotection [35,46–48] as well as indirect protection by improving lipid profiles and endothelial function as well as increased plasma antioxidant capacity together with PUFAs. Polyphenols promote neuronal calcium homeostasis in the striatum and hippocampus, regions of the brain crucial for primary and secondary memory functions [47,48].

11.4 NATURAL ANTIOXIDANTS IN NUTS

Natural antioxidants present in foods include vitamins, minerals, carotenoids, and polyphenols. Every food plant contains numerous types of natural antioxidants with different properties. Antioxidants with different chemical characteristics may act synergistically with one another in an antioxidant network [12,20–22]. The actions of antioxidants have been attributed to their ability to scavenge free radicals, thereby reducing oxidative damage of cellular biomolecules such as lipids, proteins, and DNA. In addition to being free radical scavengers, natural antioxidants function as reducing agents, chelator of pro-oxidant metals, or as quenchers of singlet oxygen [49].

Nuts are good sources of nutrient antioxidants (such as vitamin E and selenium) [28]. Among antioxidant vitamins (A, C, and E), vitamin E is the most abundant in most tree nuts (Table 11.1). Vitamin E is a dominant and the most powerful lipid-soluble antioxidant in the body and serves as the primary defense against lipid peroxidation by protecting the body's cells from free radical damage [50]. The other nuts contain much lower amounts of vitamin E than almond and hazelnut (Table 11.1). In general, nuts are not good sources of vitamins A and C, except pistachio [28]. With respect to selenium, Brazil nut serves as an excellent source of this mineral. One kernel of Brazil nut (approximately 5 g) supplies 174% of selenium for recommended dietary allowance [28].

Nuts are rich in phenolic phytochemicals and have been identified among the richest sources of dietary polyphenols [20–22]. Some of these phytochemicals have been reported to possess strong antioxidant activities (e.g., catechin, quercetin, tannins, ellagic acid, and cyanidin) [12,20–22,49]. Furthermore, it has been demonstrated that phenolic compounds possess much stronger antioxidant activities than nutrient antioxidants [20–22,49]. Phenolics, which are the major group of phytochemicals, can be divided into six groups (phenolic acids, flavonoids, stilbenes, coumarins, lignans, and tannins). Depending on their biochemical structure, flavonoids are classified into anthocyanins, flavones, isoflavones, flavonols, flavanones, and flavan-3-ols [12,21]. Figure 11.2 shows the chemical structures of the representative phenolic compounds reported in nuts. The polyphenol content of various nuts has been documented in nutrient databanks, such as the Phenol-Explorer as well as USDA flavonoid and proanthocyanidin databases. Different classes and levels of phenolic compounds have been reported for different nuts and have also been included in Phenol-Explorer and USDA databases [51–53].

Table 11.2 summarizes the contents of phytochemicals reported in nuts. Total phenolic contents of nuts, expressed as mg of gallic acid equivalents (GAE)/100 g, are in the range of 47–3673. Total phenolic contents of nuts decreased in the order of pecan > walnut > pistachio > hazelnut > peanut > almond > Brazil nut > cashew >

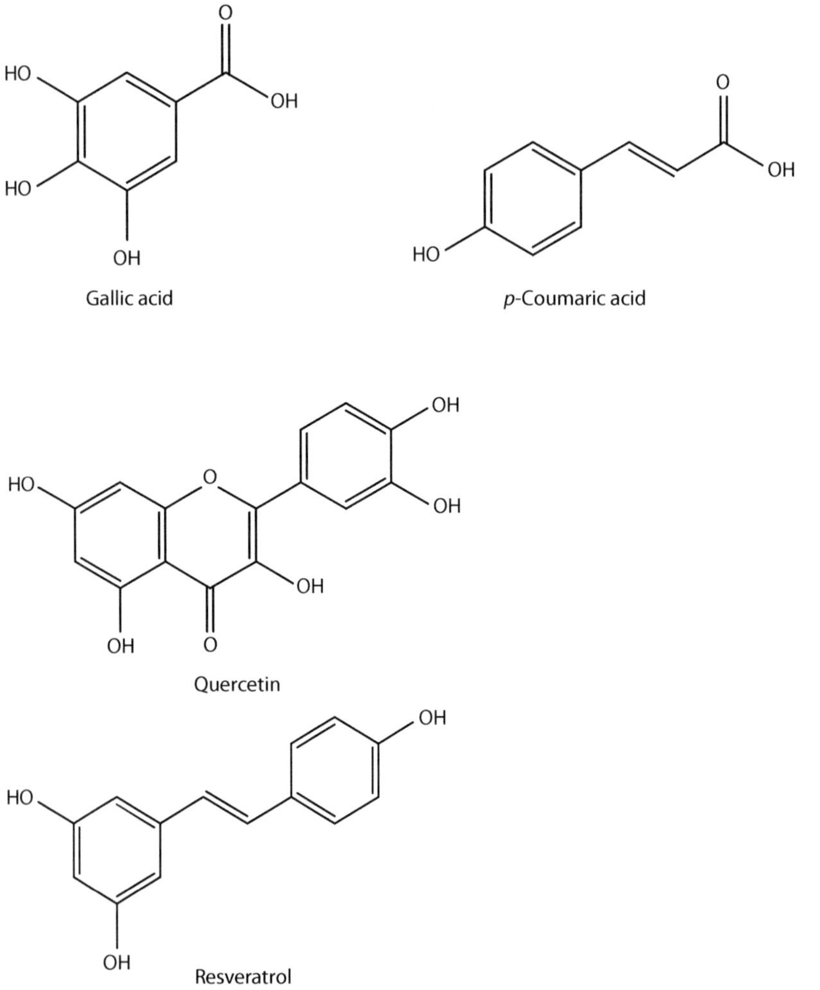

Gallic acid

p-Coumaric acid

Quercetin

Resveratrol

FIGURE 11.2 Structures of the representative phenolic compounds (phenolic acids, flavan-3-ols, flavonols, flavanone, proanthocyanidins, isoflavones) reported in nuts that relate to brain health.   (*Continued*)

macadamia [54,55]. Vinson and Cai [56] determined the levels of free and total polyphenols (μmol catechin equivalents (CE)/g) in both raw and roasted nuts (almond, Brazil nut, cashew, macadamia, peanut, pecan, pistachio, and walnut). Walnut had the highest levels of free and total polyphenols than on both the combined raw and roasted nuts [56]. The antioxidant efficacy in nuts was also determined. The order of decreasing antioxidant efficacy (increasing inhibitory concentration 50%) in nuts is walnut > cashew > hazelnut > pecan > almond > macadamia > pistachio > Brazil nut > peanut. Roasting causes a decline in antioxidant efficacy [56]. A number of

Genistein

Catechin/epicatechin

Secoisolariciresinol

FIGURE 11.2 (CONTINUED)   Structures of the representative phenolic compounds (phenolic acids, flavan-3-ols, flavonols, flavanone, proanthocyanidins, isoflavones) reported in nuts that relate to brain health.   (*Continued*)

methods (*in vitro* chemical and biological assays) have been used to determine the antioxidant activities of nuts (Table 11.3). The antioxidant activities of different nuts vary widely based on the assay type. Walnut demonstrated the highest antioxidant activity in most of the assays determined among nuts [57–59].

In all nuts, most of the antioxidants are located in the skin (pellicle) and less than 10% is retained in most nuts when skin is removed. In other words, the removal of skin from nuts reduces the antioxidant activity significantly [60–62]. For example, almond skin, while representing only 4% of the total weight, contains 70%–100% of the total phenolics present in the nut [62]. Proanthocyanidins (also known as condensed tannins) and hydrolysable tannins are generally the most abundantly found polyphenols in the majority of nuts. Hazelnut and walnut contain the highest total content of proanthocyanidins or condensed tannins among nuts [12,53]. Besides, nuts also have a significant amount of phytates together with flavonoids, phenolic acids, stilbenes, and tyrosols [20–22]. Among nuts, pistachio is the richest source of total isoflavones, total lignans, and total phytoestrogens [57,58]. Walnut, hazelnut, chestnut, pecan, and almond contain the highest phenolic acid contents among nuts [51].

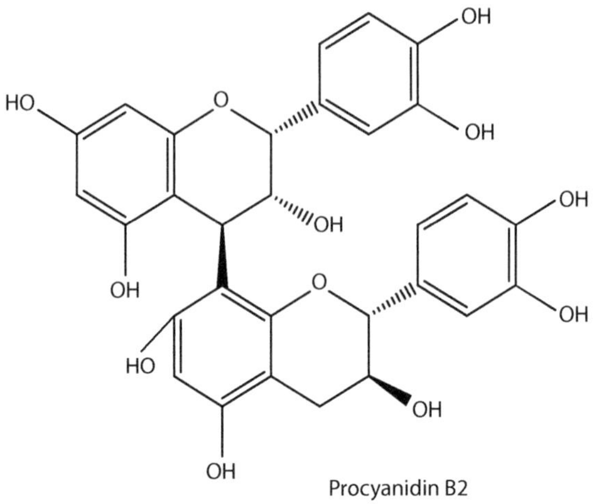

Procyanidin B2

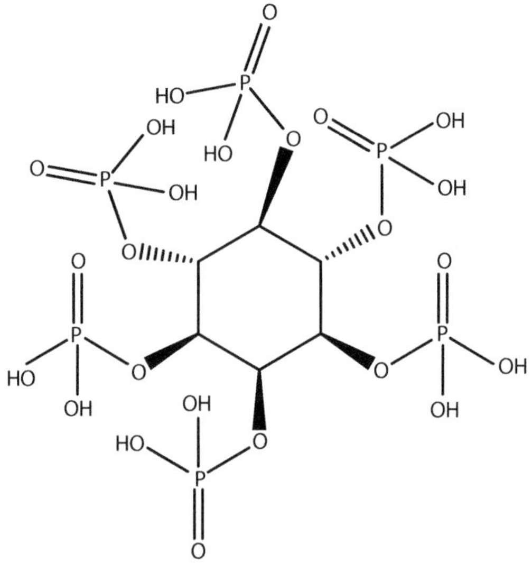

Phytate

FIGURE 11.2 (CONTINUED) Structures of the representative phenolic compounds (phenolic acids, flavan-3-ols, flavonols, flavanone, proanthocyanidins, isoflavones) reported in nuts that relate to brain health.

Using the Spanish cohort of the PREvención con DIeta MEDiterránea (PREDIMED) trial, Tresserra-Rimbau et al. [64] determined the total dietary polyphenol intake of 7200 men and women aged 55–80 years and found total polyphenol intakes at 820 ± 323 mg/day per person with 10 mg/day intake from nuts, 66% and 34% of which are from walnut and other nuts, respectively. Furthermore, 6.4 and 3.6 mg/day per person of flavonoid and phenolic acid intakes were reported [64]. Of the 19.1 mg/day per person of hydroxycinnamic acid intake, 10% was derived from

TABLE 11.2
Reported Phytochemicals in Nuts

Phytochemicals	Almond	Brazil Nut	Cashew	Hazelnut	Macadamia	Pecan	Peanut	Pistachio	Walnut	References
Total phenolics (mg GAE/100 g)	47–418	112–310	137–274	291–835	46–156	1284–2016	0.1–420	867–1657	1558–1625	[54]
Flavonoids (mg/100 g)	15	nr	2	12	nr	34	1.0	16	3	[52]
Total PAC (mg GAE/100 g)	176	nq	2	491	nq	477	11	226	60	[54]
Total isoflavones (µg/100 g)	18	nr	22.1	30.2	nr	3.5	nr	177	53.3	[63]
Total phytosterols (mg/100 g)	165	nr	216	31	155	24	61.6	263	20.2	[63]
Total lignans (µg/100 g)	112	nr	99.4	77.1	nr	25	nr	199	85.7	[63]

Note: GAE, gallic acid equivalents; nq, not quantified; nr, not reported; PAC, proanthocyanidins.

TABLE 11.3
Antioxidant Activities of Nuts Determined Using Various Antioxidant Assays

Antioxidant activities	Almond	Brazil Nut	Cashew	Hazelnut	Macadamia	Pecan	Peanut	Pistachio	Walnut	References
L-ORAC (μmol TE/g)	1.72	5.57	4.74	3.7	2.52	4.16	nr	4.25	4.84	[57]
H-ORAC (μmol TE/g)	42.82	8.62	15.23	92.75	14.43	175	nr	75.57	131	[57]
Total ORAC (μmol TE/g)	44.54	14.19	19.97	96.45	16.95	179	nr	79.83	135	[57]
DPPH (μmol TE/100 g)	1849	1088	994	1214	805	1148	nr	1533	1626	[58]
LDL + VLDL inhibition [μM (IC$_{50}$)]	3.4	6.8	2.2	3.2	4.1	3.4	7.9	5.8	1.8	[56]
FRAP (mmol/100 g)	0.41	0.15	4.67	0.7	0.42	8.33	nr	1.27	23.1	[59]

Note: DPPH, 2,2-diphenyl-1-picrylhydrazyl; FRAP, ferric-reducing ability of plasma; H-ORAC, hydrophilic-oxygen radical absorbance capacity; LDL, low-density lipoprotein; L-ORAC, lipophilic-oxygen radical absorbance capacity; nr, not reported; ORAC, oxygen radical absorbance capacity; TE, trolox equivalents; VLDL, very low-density lipoprotein.

walnut [64]. Based on the data from USDA, the total nut consumption (tree nuts and peanut) in the United States is 162 mg/day. In the EU, 5.0 g of peanut/day is consumed, contributing 71.5 mg of polyphenols. Tree nuts have 76 mg of polyphenols, and thus, nut contribution of polyphenols to the EU diet amounts to 158 mg/day, which is quite similar to the U.S. diet [56].

11.5 EFFECT OF VARIOUS NUTS ON BRAIN HEALTH

Nuts are popular due to their high nutraceutical and pharmaceutical values [12,20–22]. Walnut is the most studied nuts in terms of the health effects on various parameters related to brain health using *in vitro*, animal, and human studies [47,48]. To the best of our knowledge, limited studies have been carried out to determine the effects of Brazil nut, macadamia, chestnut, pine nut, and peanut on various health parameters related to brain health.

11.5.1 RESULTS FROM *IN VITRO* AND ANIMAL STUDIES

Nut extract, particularly walnut extract, has been shown to be neuroprotective against numerous stressors. It has been demonstrated that microglial cells treated with walnut oil protected the cells from increases in inflammation and oxidative stress by inhibiting lipopolysaccharide (LPS)-induced activation of microglial cells [65]. When BV-2 mouse microglial cells were treated with walnut extract prior to LPS stimulation, production of NO and expression of inducible NO synthase were significantly reduced. In the same study, Willis et al. [65] have also shown that walnut extract lowered the production of tumor necrosis factor-alpha, a proinflammatory mediator. Further, calcium buffering in hippocampal cells was substantially altered by LPS and 6-hydroxy dopamine stressors in another study conducted by Joseph et al. [66]. Subsequently, walnut extract was shown to be protective against LPS-induced, but not dopamine-induced, loss of calcium recovery as well as cell death in rat primary hippocampal neurons [67].

Walnut extract was shown to counteract oxidative damage and cell death as observed in patients with Alzheimer's disease in several *in vitro* studies. First, walnut extract alleviated Aβ-induced oxidative stress and cytotoxicity in PC12 cells of rat adrenal medulla. Fibrillar amyloid β (Aβ) protein is the principal component of amyloid plaques in the brains of Alzheimer's patients [68]. Second, Chauhan et al. [69] demonstrated the inhibition of the Aβ formation and defibrillization of Aβ preformed fibrils in walnut treated cells after incubating with 5 mL methanolic walnut extract for two or three days. However, it is still unclear whether this anti-amyloidogenetic activity of walnut is due to the antioxidative properties of the bioactive compounds, such as phenolic compounds and PUFA, or due to the direct/synergistic interaction with Aβ [69].

Carey et al. [67] investigated the cellular mechanisms underlying walnuts' protective effects on neuronal health and functioning in aging brain. Primary hippocampal neurons were pretreated with walnut extract or with the PUFA found in walnut. Subsequently, the cells were treated with dopamine and LPS where cell death and calcium buffering dysregulation were measured [67]. Results demonstrated that walnut oil extract, ALA, and DHA were protective against cell death and calcium dysregulations where the effects were concentration dependent and stressor dependent [67]. Interestingly, linoleic acid and EPA were not effective at protecting hippocampal cells from these insults. The whole-walnut extract was most beneficial, since it does not contribute to cellular toxicity effects [67]. PUFA from walnut was shown to attenuate neuroinflammation by modulating microglial reactivity. Walnut extract had been shown to alter the response stimuli to the chemically induced inflammatory stress in mouse microglial cells through phospholipase D2-mediated internalization of Toll-like receptor 4 [65]. Kim et al. [59] analyzed total fatty acids in commercially available tree nuts and conducted a statistical analysis using a chemometric approach to evaluate the protective potency of nut extracts against oxidative stress in a neuronal cell line. Their results demonstrated that walnut had the highest potency in cell viability and 2′,7′-dichlorofluorescein-dopamine assays. In fact, there was no significant between the potency of walnut and ALA.

Dietary walnut can also improve cognitive function. In one study, rats that ingested 80 mg/day of walnut, besides the standard diet, for 28 days had enhanced learning and memory in the radial arm maze and lowered anxiety on the elevated plus maze [70]. In another study conducted by Willis et al. [46], 19-month-old Fischer (F344) rats were fed diets containing 0%, 2%, 6%, or 9% (wt:wt) ground walnut with skin for eight weeks. Rats fed a diet containing 2% or 6% walnut had improved balance, coordination, and strength. However, rats fed the 9% diet had impaired motor performance and reference memory compared to controls [46]. Working memory in the water maze was also enhanced in rats fed diets containing walnut. The 6% walnut diet gave the best overall results among the aged rats [46]. The 6% diet was shown to contain 5.4 g of ALA and 22.9 g of linoleic acid/kg, which is equivalent to the recommended dietary intake of 1 ounce/day (28 g) of walnut for humans [35,36]. Subsequently, the authors demonstrated that walnut consumption was associated with substantially lower acetylcholinesterase activity in the striatum brain region of aged animals [36].

Almonds have been studied as possible supplemental therapy in cognitive dysfunctions, such as dementia in Alzheimer's disease [71]. Male albino rats were fed 150, 300, or 600 mg almond paste/kg daily for 7–14 days. The dosages were selected by the conversion of conventional human doses into animal doses. The human dose of almonds is five to six nuts daily (6 g) [71]. The almond-fed rats demonstrated significant reversal of scopolamine-induced cognitive impairment using both elevated-plus maze and the passive shock-avoidance paradigm tests. The memory-improving activity of almond could be due to their anticholinesterase, procholinergic, and cholesterol-reducing properties [71].

The brains of aged rats fed walnut were evaluated for polyubiquitinated proteins [72]. An increase in the aggregation of misfolded/damaged polyubiquitinated proteins indicates age-related neurodegenerative diseases. The accumulation of these highly toxic proteins in the brain increases with age, due to the high oxidative stress [72]. Results demonstrated that rats given 6% and 9% walnut diets demonstrated significantly lowered accumulation of polyubiquitinated proteins and activation of autophagy, a neuronal housekeeping function in the striatum and hippocampus [72]. Walnut-fed animals demonstrated upregulation of autophagy by inhibiting phosphorylation of the mechanistic target of rapamycin (mTOR), upregulation of ATG7 and Beclin 1 genes, and turnover of microtubule-associated protein 1 light chain 3 proteins [72]. The clearance of polyubiquitinated protein aggregates, such as sequestosome (p62/SQSTM1), was more profound in the hippocampus, a critical region in the brain involved in memory and cognitive performance [73].

In another scenario, moderate consumption of defatted walnut meal at doses of 1, 2, 3, and 4 kg/day could improve the performance of step-down and Morris water maze tests greatly, enhance the activities of superoxide dismutase (SOD) and acetylcholinesterase in brain tissues of rats, and reduce plasma malondialdehyde level. In short, defatted walnut meal can help to improve the learning and memory abilities by enhancing the plasma antioxidant capacity of rats [74]. Similarly, polyphenol extracts from walnut skin (42%) was demonstrated to improve the learning and memory functions in hypercholesterolemia mice with obesity, hypercholesterolemia, and oxidative stress [75]. These animal studies suggested that walnut may be beneficial in guarding against learning and memory impairment [74,75].

The effects of walnut on neuronal injury and astrocyte reactivity after induction of focal cerebral ischemia in male rats was determined [76]. Results demonstrated that walnut significantly decreased neuronal death induced by cerebral ischemia. Walnut also significantly decreased the mortality rate of the animals due to cerebral ischemia. This shows that walnut may provide protection against the cerebral ischemia-induced injuries in the rat brain through antioxidant and anti-inflammatory mechanisms [76]. In another study conducted by Harandi et al. [77], walnut at doses of 6% and 9% was demonstrated to restore the scopolamine-induced memory-impaired rats significantly and prevented scopolamine neurotoxicity by decreasing the activity of acetylcholinesterase in the whole brain. Similarly, 6% walnut diets were shown to be effective in preventing cisplatin-induced neurotoxicity by attenuating neural damage in the hippocampal region of rats. At the same time, walnut was shown to improve the memory and motor abilities in cisplatin treated rats effectively [78]. In relation, 6% and 9% walnut diet increased the expression of *zif268* (nerve growth factor inducible-A) in the hippocampus of experimental rats. The results suggest that dietary walnut may have protective effects on the aging brain [79]. In another perspective, alcoholic extract of walnut was demonstrated to be effective in reducing anxiety and nervousness in rats based on exploration behavior and anxiety in elevated plus maze, zero maze, and light–dark model [80]. The antianxiety effect of walnut extract warrants further investigations owing to its complex mechanisms.

Dietary intervention of walnut reduced the lipid peroxidation processes and the activities of SOD and catalase induced by 1-methyl-4-phenyl-1,2,3,6-tetrahydropyridine due to its antioxidant function in a Parkinson's disease mouse model [81]. Specifically, the levels of glutathione and activities of glutathione peroxidase were enhanced by walnut diet. The authors suggested that early intervention with walnut-rich diet may help in reducing the risk of developing Parkinson's disease due to its cumulative antioxidant and mitochondrial protective effects of components in walnut [81]. Further, walnut extracts have also been shown to protect neuron cells against oxidative stress induced by epileptic seizure [82]. In another study, walnut has been shown in decreasing mitochondrial dysfunction, B cell lymphoma 2 (Bcl-2)-associated X protein levels and cytochrome *c* release from mitochondria and increased Bcl-2 protein levels, as well as the inhibition of caspase-3 activity. This indicates an inhibition of apoptosis in human epidermal keratinocytes against UVB-induced mitochondria mediated apoptosis [83].

Meanwhile, some recent studies demonstrated the proteins/peptides extracted from walnut may have protective effect on various parameters related to brain health [84,85]. Walnut peptide supplementation ameliorated the cognitive deficits and memory impairment of mice effectively by protecting against the neurotoxicity induced by Aβ25–35 *in vivo* [84]. Meanwhile, walnut peptides (400 or 800 mg/kg) were effective in restoring the levels of antioxidant enzymes as well as inflammatory mediators. In another study conducted by Chen et al. [85], the antioxidant peptides purified from the defatted walnut meal protein hydrolysate was shown to be effective in improving the learning and memory ability in D-galactose memory-impaired mice using the Morris water maze and dark–light avoidance tests. The antioxidant peptides responsible for the ability to combat the D-galactose induced learning and memory impairments were identified as Trp-Ser-Arg-Glu-Glu-Gln-Glu-Arg-Glu-Glu and Ala-Asp-Ile-Tyr-Thr-Glu-Glu-Ala-Gly-Arg, where these novel peptides were not reported and identified in the past [85]. More research should be carried out to understand the mechanisms of these antioxidant peptides in memory improvement activity. All the previous studies [84,85] suggested that walnut proteins/peptides may have a protective effect on brain health by reducing inflammatory responses and modulating the endogenous antioxidant system.

There was only one study determining the effect of pecan on age-related human motor neuron disorder [86]. Mice provided a diet supplemented with 0.05% pecans demonstrated a significant delay in motor neuron function decline, which was accompanied by increased survival of motor neurons and a decrease in reactive gliosis, compared to nonsupplemented mice [86]. This transient delay was accompanied by a preservation of motor neuron number in lumbar spinal cord, along with a corresponding reduction in gliosis and aggregation of SOD-1, both of which are hallmarks of motor neuron decline in these mice and in amyotropic lateral sclerosis [86].

11.5.2 RESULTS FROM HUMAN STUDIES

Evidence restricted to nut consumption alone is scarce; however, several studies have been conducted to determine the dietary patterns, including nuts as a major component. Some epidemiological studies conducted in children and adults have demonstrated a significant association between nut consumption and a higher diet quality score or improved nutrient intakes [87]. O'Neil et al. [87], in a study of 13,292 adults participating in the 1999–2004 National Health and Nutrition Examination Survey (NHANES), observed that tree nut consumers who consume more than 7.09 g/day of nuts or tree nut butters, demonstrated significantly higher intake of fiber, vitamins, and minerals and also a higher total healthy eating index-2005 score. Furthermore, the results of a clinical trial conducted on 124 obese subjects demonstrated that nutritional dietary quality among nut consumers (those eating 42 g hazelnut/day for

12 weeks) was improved compared with the consumption of chocolate, potato crisps, or no additional foods in other groups [88].

Recently, results from several epidemiological studies, such as the Netherlands Cohort study and the Golestan Cohort study, demonstrated an inverse association between nut consumption and mortality [89,90]. A recent comprehensive review and meta-analyses of available articles demonstrated the inverse associations between nut consumption and cardiovascular disease (CVD) outcomes and all-cause mortality [91]. Specifically, peanut and tree nuts were inversely related to mortality (cancer, diabetes, cardiovascular, respiratory, neurodegenerative diseases, and other causes) among the 3202 subjects of both men and women in the Netherlands Cohort study, with evidence for nonlinear dose–response relationships whereas peanut butter was not [89].

Generally, increased nut intake (particularly walnut intake) has been shown to improve cognition among older adults [47,48,92]. Numerous human studies (cross-sectional and prospective studies and randomized controlled trials) measuring the effect of nut consumption on various parameters of cognitive function [93–99] are summarized in Table 11.4. Epidemiological studies, specifically, the prospective Nurses' Health Study demonstrate the association of long-term total nut intake of at least five servings of nut/week and better cognitive function among the 15,467 women in this study [100]. Meanwhile, significant positive associations between walnut consumption and cognitive functions among all adults were demonstrated using NHANES survey. Better outcomes were demonstrated in all cognitive test scores measuring memory, cognitive speed, and flexibility [101].

A large, parallel-group, multicenter randomized, and controlled clinical trial (PREDIMED) provided compelling evidence for the ability of nuts to attenuate depression and age-related cognitive decline [102]. This study, spanning seven years with a minimum follow-up of five years, recruited 7447 persons (55–80 years of age), with three randomized intervention arms [101]. The first group received the Mediterranean diet supplemented with virgin olive oil (1 L/week), the second group received the Mediterranean diet supplemented with 30 g/day of mixed nuts (15 g walnut, 7.5 g almond, and 7.5 g hazelnut), while the third group received a low-fat control diet [102]. After three years, participants who were on the Mediterranean diet supplemented with mixed nuts had substantially improved concentrations of plasma brain-derived neurotrophic factor (BDNF), particularly among the individuals with a history of depression [102]. BDNF is related to actions as synaptic plasticity, neuronal survival and differentiation, axonal elongation, and neurotransmitter release [103]. Low concentrations of BDNF have been associated with neurodegenerative disorders such as epilepsy, Alzheimer's disease, Huntington disease, autism, schizophrenia, and major depression, whereas increased concentrations are associated with the prevention of memory loss and cognitive impairment [104,105]. In relation, beneficial effect of a long-term intervention with a Mediterranean diet supplemented with nuts on depression for diabetic patients was obtained from a subset of subjects in the PREDIMED controlled trial [106].

TABLE 11.4
Selected Human Studies on the Possible Health Effects of Nuts on Brain Health

Subject Characteristics	Study Design and Duration of Study	Amount/Type of Nuts	Outcomes/Mechanism of Actions	References
447 subjects with risk of CVD, male and female, mean age of 69 years	Cross-sectional study, PREDIMED study, and FFQ + cognitive battery	5 g/day (0–60 g) all nuts, 1 g/day walnut (0–30 g)	Walnut (not other nuts) associated with improved working memory. Reduced the risk of stroke by 46%, which indirectly associated with age-related cognitive decline.	[93]
2613 subjects of general population, aged 43–70 years old	The Doetinchem Prospective Cohort Study, FFQ, and cognitive battery	Quintiles of any nut consumption (amount not mentioned)	Higher nut intake associated with better cognitive function (memory, speed, flexibility, and global). No reduction of cognitive decline in nut consumers over five years.	[94]
64 students (male and female), mean age of 21 years	Double-blind, randomized, placebo controlled cross-over (6-week washout) study Healthy and cognitively intact young adults	60 g/day of walnut	Improved inferential verbal reasoning. No significant improvement on mood, nonverbal reasoning, and memory.	[95]
2031 elderly male and female subjects, aged 70–74 years old	Cross-sectional study and FFQ + cognitive battery	Mean intake of nut consumers = 5 g/day	Nut intake associated with better executive function. No significant effect on semantic memory.	[96]

(Continued)

TABLE 11.4 (CONTINUED)
Selected Human Studies on the Possible Health Effects of Nuts on Brain Health

Subject Characteristics	Study Design and Duration of Study	Amount/Type of Nuts	Outcomes/Mechanism of Actions	References
823 elderly healthy male and female subjects, mean age of 62 years	The Sun Prospective Cohort Study, FFQ, and six to eight years follow-up	Quintiles of nut intake (amount not specified) in Mediterranean diet	No significant association between nut intake with better cognitive function over six to eight years.	[97]
552 subjects with CVD risk factors, mean age of 74.6 years	Randomized, multicenter, and primary prevention trial of PREDIMED study Follow up after 6.5 years of intervention	30 g/day of raw mixed nuts (15 g walnut, 7.5 g almond, and 7.5 g hazelnut)	Intervention with Mediterranean diet supplemented with mixed nuts was associated with a better global cognitive performance determined using the MMSE and the CDT.	[98]
334 subjects with high CVD risk, mean age of 66.9 years	Randomized, multicenter, and primary prevention trial of PREDIMED study Follow-up after 4.1 years of intervention	30 g/day of raw mixed nuts (15 g walnut, 7.5 g almond, and 7.5 g hazelnut)	Higher nut intake associated with better cognitive function (memory, speed, flexibility, and global). Mediterranean diet supplemented with nuts is associated with improved cognitive function.	[99]

Note: CDT, clock drawing test; CVD, cardiovascular disease; FFQ, food frequency questionnaire; MMSE, Mini-Mental State Examination; PREDIMED, PREvención con DIeta MEDiterranea.

11.6   PROPOSED MECHANISMS FOR THE EFFECT OF NUTS ON BRAIN HEALTH

The field of nutritional neuroscience has become a well-known discipline with the potential to make significant contributions to our understanding of the relationship between nutrition and cognitive functions [107]. Several nutrients in nuts may be responsible for the observed improvements in cognitive function, Alzheimer's disease, as well as neurodegenerative diseases in relation to brain health [48,73,108,109]. Tree nuts and peanut have similar nutrient profiles, with some variations in micronutrients and macronutrients (Tables 11.1 and 11.2). From the studies reviewed (with the exception of walnut, which has been more extensively researched than other nuts), it is not possible to determine differences in efficacy between different types of nuts. ALA found in nuts is associated with improved endothelial function, inflammation, and neuroprotection in animal models and has been hypothesized to maintain cognitive function in the elderly [108,109]. Recent longitudinal studies support the hypothesis that MUFA may play a protective role toward the prevention of cognitive decline and dementia due to its anti-inflammatory effects on endothelial function [41,42,108]. Nuts contain 2–3 g of L-arginine/100 g (Table 11.1). Various animal and human studies have demonstrated that inflammation can be alleviated by the consumption of arginine-rich foods [110].

Nuts also contain fiber and, when consumed with their skin, contain a significant amount of phenolic compounds [12,22,60,61]. Polyphenols present in nuts, especially walnut, include hydroxycinnamic acids (chlorigenic acid, caffeic acid, *p*-coumaric acid, ferrulic acid, and sinapic acid), hydrobenzoic acids (ellagic acid, gallic acid, protocatechuic acid, syringic acid, and vanillic acid), and other compounds such as glansrin, juglone, and syringaldehyde [12,22,23]. These phenolic compounds from nuts, specifically walnut, contribute to the high antioxidant defense system in healthy humans [111]. Phenolic antioxidants function as free radical terminators and sometimes as metal chelators in oxidation processes. These in turn inhibit cyclooxygenase activity, which is anti-inflammatory activity [49]. A positive association between total and class-specific polyphenol intake and cognitive performance was demonstrated in the Supplémentation en Vitamines et Minéraux Antioxydants (SU. VI.MAX) trial [112]. Results indicate that the intake of catechins, theaflavins, flavonols, and hydroxybenzoic acids was positively associated with language and verbal memory, especially with episodic memory in elderly subjects [112]. On the contrary, negative associations between scores on executive functioning and intake of dihydrochalcones, catechins, proanthocyanidins, and flavonols were reported. The authors conclude that high intake of specific polyphenols, including flavonoids and phenolic acids, may help to preserve verbal memory, which is a vulnerable domain in brain aging process [112].

Numerous evidences indicate that frequent nut consumption benefits cardiovascular risk. Reductions in CVD risk factors may be associated with better brain health since CVD is also associated with the development of cerebrovascular disease, stroke, and mild cognitive impairment [108]. Improved endothelial function had direct benefit on cerebral health and cognitive function [109]. Vitamin E found in nuts may have an essential role in modifying some inflammatory mediators and may be beneficial for cognitive performance and neurodegeneration. Gamma-tocopherol is a powerful antioxidant abundant in walnut, Brazil nut, and pistachio [20,28,50]. Melatonin is another bioactive compound found in walnut. Melatonin deficiency has direct linkage with the degeneration of cholinergic neurons in the basal forebrain and the accumulation of aggregated proteins, such as Aβ peptides, which contributes to cognitive impairment and dementia [113]. A study demonstrated that the consumption of walnut increased the concentrations of blood melatonin, which correlated with an increase in "total antioxidant capacity" *in vivo* [114]. Anti-inflammatory medications offer some protection from Alzheimer's disease, which is in line with the hypothesis that brain cell damage is part of an overall inflammatory reaction [108]. Since inflammation is the major contributing process, nuts, which contain various anti-inflammatory nutrients, such as phenolic compounds, vitamin E, and *n*-3 fatty acids, may be important in maintaining brain health [109].

All these studies give solid support for the hypothesis that the synergistic interaction of all the nutrients and bioactive components in nuts might have a beneficial effect on the brain and all its related functions [115]. Regular nut consumption could be used as a medical nutrition therapy in the attenuation of several neurodegenerative diseases as well as in age-related brain dysfunction [116]. Despite the lack of human intervention studies, observational studies indicate that long-term consumption of even small amounts of nuts may provide benefits for cognitive function and hence overall brain health [96,97,100,108]. More evidence is needed from well-controlled intervention studies before a conclusive benefit can be determined [108,116].

Animal studies indicate that different neural systems in the human body can be affected by short- and long-term supplementation with nuts, as discussed in Section 11.5.1 [70–72,74–86]. The evidence from human trials, although limited, suggests potential positive effects, particularly among older patients, on memory function, global cognition, and depression [47,93–101]. Pribis and Shukitt-Hale [47] recommended that future research on the effects of nuts and brain health should focus on epidemiologic and short- and long-term clinical studies since the evidence from human studies is limited and inconsistent. Specifically, short-term clinical trials could focus on cognitive disorders such as depression, seasonal disorder, and attention-deficit hyperactivity disorder, whereas long-term clinical trials could focus on the prevention and treatment of neurodegenerative diseases and age-related brain dysfunction [47]. Walnut has been extensively studied on its effects on various parameters related to brain health, as discussed in Sections 11.5.1 and 11.5.2 [48,73]. Hence, more research should be conducted using other nuts that were least studied, such as Brazil nut, cashew, macadamia, peanut, and pine nut.

11.7 CONCLUSION

Age-related increases in oxidative stress and inflammation, coupled with metabolic dysfunction, lead to neurodegeneration and cognitive decline, which affect brain health. Nuts contain various macronutrients and micronutrients, as well as bioactive phytochemicals with good health-promoting properties. Evidence suggests that the consumption of nuts as part of a healthy diet could be an effective means of improving diet quality, prolonging life spans, slowing the process of brain aging, and exerting beneficial effects on brain health. More clinical trials and human intervention studies should also be carried out to ascertain the effects of nuts on brain health in relation to their bioaccessibility and bioavailability in the human body.

REFERENCES

1. GBD 2013 Mortality and Causes of Death Collaborators, Global, regional and national age-sex specific all-cause and cause-specific mortality for 240 causes of death, 1990–2013: A systematic analysis for the Global Burden of Disease Study 2013. *Lancet*, 385, 117–171, 2015.
2. Global elderly care in crisis. *Lancet*, 383, 927, 2014.
3. Thies, W. and Bleiler, L., Alzheimer's disease facts and figures. *Alzheimers Dement.*, 7, 208–244, 2011.
4. Marx, J., Neurodegenerative diseases: Picking apart the causes of mysterious dementias. *Science*, 314, 42–43, 2006.
5. Butterfield, D.A., Oxidative stress in neurodegenerative disorders. *Antioxid. Redox Signal*, 8, 1971–1973, 2006.
6. Halliwell, B., Oxidative stress and neurodegeneration: Where are we now? *J. Neurochem.*, 97, 1634–1658, 2006.
7. Floyd, R.A. and Hensley, K., Oxidative stress in brain aging: Implications for therapeutics of neurodegenerative diseases. *Neurobiol. Aging*, 23, 795–807, 2002.
8. Solfrizzi, V., Capurso, C., D'Introno, A., Colacicco, A.M., Santamato, A., Ranieri, M., Fiore, P., Capurso, A., and Panza, F., Lifestyle-related factors in pre-dementia and dementia syndromes. *Expert Rev. Neurother.*, 8, 133–158, 2008.
9. Sofi, F., Macchi, C., Abbate, R., Gensini, G.F., and Casini, A., Mediterranean diet and health status: An updated meta-analysis and a proposal for a literature-based adherence score. *Public Health Nutr.* 17, 2769–2782, 2014.
10. Grosso, G., Mistretta, A., Marventano, S., Purrello, A., Vitaglione, P., Calabrese, G., Drago, F., and Galvano, F., Beneficial effects of the Mediterranean diet on metabolic syndrome. *Curr. Pharm. Des.*, 20, 5039–5044, 2014.
11. Martinez-Gonzalez, M.A., Salas-Salvadó, J., Estruch, R., Corella, D., Fito, M., Ros, E., and PREDIMED Investigators, Benefits of the Mediterranean diet: Insights from the PREDIMED study. *Prog. Cardiovasc. Dis.*, 58, 50–60, 2015.
12. Alasalvar, C. and Shahidi, F., Tree nuts: Composition, phytochemicals, and health effects: An overview, in *Tree Nuts: Composition, Phytochemicals, and Health Effects*, Alasalvar, C. and Shahidi, F. Eds., CRC Press, Taylor & Francis Group, Boca Raton, FL, 2008, pp. 1–10.
13. Bao, Y., Han, J., Hu, F.B., Giovannucci, E.L., Stampfer, M.J., Willett, W.C., and Fuchs, C.S., Association of nut consumption with total and cause-specific mortality. *N. Engl. J. Med.*, 369, 2001–2011, 2013.
14. Grosso, G., Yang, J., Marventano, S., Micek, A., Galvano, F., and Kales, S.N., Nut consumption on all-cause, cardiovascular, and cancer mortality risk: A systematic review and meta-analysis of epidemiologic studies. *Am. J. Clin. Nutr.*, 101, 783–793, 2015.
15. Luo, C., Zhang, Y., Ding, Y., Shan, Z., Chen, S., Yu, M., Hu, F.B., and Liu, L., Nut consumption and risk of type 2 diabetes, cardiovascular disease, and all-cause mortality: A systematic review and meta-analysis. *Am. J. Clin. Nutr.*, 100, 256–269, 2014.
16. Zhou, D., Yu, H., He, F., Reilly, K.H., Zhang, J., Li, S., Wang, B., Ding, Y., and Xi, B., Nut consumption in relation to cardiovascular disease risk and type 2 diabetes: A systematic review and meta-analysis of prospective studies. *Am. J. Clin. Nutr.*, 100, 270–277, 2014.

17. Stone, N.J., Robinson, J.G., Lichtenstein, A.H., Merz, C.N.B., Blum, C.B., Eckel, R.H., Goldberg, A.C., Gordon, D., Levy, D., Lloyd-Jones, D.M., McBride, P., Schwartz, J.S., Shero, S.T., Smith Jr., S.C., Watson, K., and Wilson, P.W.F., 2013 ACC/AHA guideline on the treatment of blood cholesterol to reduce atherosclerotic cardiovascular risk in adults: A report of the American College of Cardiology/American Heart Association Task Force on Practice Guidelines. *Circulation*, 129, S46–S48, 2014.

18. U.S. Food and Drug Administration (FDA), *Qualified Health Claims: Letter of Enforcement Discretion—Nuts and Coronary Heart Disease*, FDA, Rockville, MD, 2003.

19. U.S. Department of Health and Human Services, *2015–2020 Dietary Guidelines for Americans*, 8th Edition, 2015. Published online at: http://health.gov/dietaryguidelines /2015/guidelines/ (accessed June 20, 2016).

20. Alasalvar, C. and Bolling, B.W., Review of nut phytochemicals, fat-soluble bioactives, antioxidant components and health effects. *Br. J. Nutr.*, 113, S68–S78, 2015.

21. Alasalvar, C. and Shahidi, F., Natural antioxidants in tree nuts. *Eur. J. Lipid Sci. Tech.*, 111, 1056–1062, 2009.

22. Bolling, B.W., Chen, C.Y.O., McKay, D.L., and Blumberg, J.B., Tree nut phytochemicals: Composition, antioxidant capacity, bioactivity, impact factors. A systematic review of almonds, Brazils, cashews, hazelnuts, macadamias, pecans, pine nuts, pistachios, and walnuts. *Nutr. Res. Rev.*, 24, 244–275, 2011.

23. Bolling, B.W., McKay, D.L., and Blumberg, J.B., The phytochemical composition and antioxidant actions of tree nuts. *Asia Pac. J. Clin. Nutr.*, 19, 117–123, 2010.

24. Maguire, L.S., O'Sullivan, S.M., Galvin, K., O'Connor, T.P., and O'Brien, N.M., Fatty acid profile, tocopherol, squalene and phytosterol content of walnuts, almonds, peanuts, hazelnuts and the macadamia nut. *Int. J. Food Sci. Nutr.*, 55, 171–178, 2004.

25. Ryan, E., Galvin, K., O'Connor, T.P., Maguire, A.R., and O'Brien, N.M., Fatty acid profile, tocopherol, squalene and phytosterol content of Brazil, pecan, pine, pistachio and cashew nuts. *Int. J. Food Sci. Nutr.*, 57, 219–228, 2006.

26. Brufau, G., Boatella, J., and Rafecas, M., Nuts source of energy and macronutrients. *Br. J. Nutr.*, 96, S24–S28, 2006.

27. Ros, E. and Mataix, J., Fatty acid composition of nuts—Implications for cardiovascular health. *Br. J. Nutr.*, 96, S29–S35, 2006.

28. U.S. Department of Agriculture (USDA), USDA National Nutrient Database for Standard Reference Release 28, 2016. Published online at: http://ndb.nal.usda.gov/ndb/ (accessed June 20, 2016).

29. Segura, R., Javierre, C., Lizarraga, M.A., and Ros, E., Other relevant components of nuts: Phytosterols, folate and minerals. *Br. J. Nutr.*, 96, S36–S44, 2006.

30. Choudhary, S.P. and Tran, L.S., Phytosterols perspectives in human nutrition and clinical therapy. *Curr. Med. Chem.*, 18, 4557–4567, 2011.

31. Morris, M.C., Nutritional determinants of cognitive aging and dementia. *Proc. Nutr. Soc.*, 71, 1–13, 2012.

32. Bishop, N.A., Lu, T., and Yankner, B.A., Neural mechanisms of ageing and cognitive decline. *Nature*, 464, 529–535, 2010.

33. Joseph, J.A., Shukitt-Hale, B., and Casadesus, G., Reversing the deleterious effects of aging on neuronal communication and behavior: Beneficial properties of fruit polyphenolic compounds. *Am. J. Clin. Nutr.*, 81, 313S–316S, 2005.

34. Poulose, S.M., Carey, A.N., and Shukitt-Hale, B., Improving brain signaling in aging: Could berries be the answer? *Expert Rev. Neurother.*, 12, 887–889, 2012.

35. Willis, L.M., Shukitt-Hale, B., and Joseph, J.A., Modulation of cognition and behavior in aged animals: Role for antioxidant- and essential fatty acid-rich plant foods. *Am. J. Clin. Nutr.*, 89, 1602S–1606S, 2009.

36. Willis, L.M., Shukitt-Hale, B., and Joseph, J.A., Dietary polyunsaturated fatty acids improve cholinergic transmission in the aged brain. *Genes Nutr.*, 4, 309–314, 2009.

37. Gillette-Guyonnet, S., Secher, M., and Vellas, B., Nutrition and neurodegeneration: Epidemiological evidence and challenges for future research. *Br. J. Clin. Pharmacol.*, 75, 738–755, 2013.

38. Cederholm, T., Salem, N., and Palmblad, J., ω-3 Fatty acids in the prevention of cognitive decline in humans. *Adv. Nutr.*, 4, 672–676, 2013.

39. Lopez, L.B., Kritz-Silverstein, D., and Barrett-Connor, E., High dietary and plasma levels of the omega-3 fatty acid docosahexaenoic acid are associated with decreased dementia risk: The Rancho Bernardo study. *J. Nutr. Health Aging*, 15, 25–31, 2011.

40. Schaefer, E.J., Bongard, V., Beiser, A.S., Lamon-Fava, S., Robins, S.J., Au, R., Tucker, K.L., Kyle, D.J., Wilson, P.W., and Wolf, P.A., Plasma phosphatidylcholine docosahexaenoic acid content and risk of dementia and Alzheimer disease: The Framingham Heart Study. *Arch. Neurol.*, 63, 1545–1550, 2006.

41. Naqvi, A.Z., Harty, B., Mukamal, K.J., Stoddard, A.M., Vitolins, M., and Dunn, J.E., Monounsaturated, trans, and saturated fatty acids and cognitive decline in women. *J. Am. Geriatr. Soc.*, 59, 837–843, 2011.

42. Vercambre, M.N., Grodstein, F., and Kang, J.H., Dietary fat intake in relation to cognitive change in high-risk women with cardiovascular disease or vascular factors. *Eur. J. Clin. Nutr.*, 64, 1134–1140, 2010.

43. Rabinovitz, S., Carasso, R.L., and Mostofsky, D.I., The role of polyunsaturated fatty acids in restoring the aging neuronal membrane. *Neurobiol. Aging*, 23, 843–853, 2002.

44. Mora, F., Orr, S.K., Trépanier, M.O., and Bazinet, R.P., *n*-3 Polyunsaturated fatty acids in animal models with neuroinflammation. *Prostaglandins Leukot. Essent. Fatty Acids*, 88, 97–103, 2013.

45. Yehuda, S., Segovia, G., and del Arco, A., Aging, plasticity and environmental enrichment: Structural changes and neurotransmitter dynamics in several areas of the brain. *Brain Res. Rev.*, 55, 78–88, 2007.

46. Willis, L.M., Shukitt-Hale, B., Cheng, V., and Joseph, J.A., Dose-dependent effects of walnuts on motor and cognitive function in aged rats. *Br. J. Nutr.*, 101, 1140–1144, 2009.

47. Pribis, P. and Shukitt-Hale, B., Cognition: The new frontier for nuts and berries. *Am. J. Clin. Nutr.*, 100, 347S–352S, 2014.

48. Poulose, S.M., Miller, M.G., and Shukitt-Hale, B., Role of walnuts in maintaining brain health with age. *J. Nutr.*, 144, 561S–566S, 2014.

49. Shahidi, F. and Ambigaipalan, P., Phenolics and polyphenolics in foods, beverages and spices: Antioxidant activity and health effects—A review. *J. Funct. Foods*, 18, 820–897, 2015.

50. Traber, M.G. and Atkinson, J., Vitamin E, antioxidant and nothing more. *Free Radic. Biol. Med.*, 43, 4–15, 2007.

51. Pérez-Jiménez, J., Neveu, V., Vos, F., and Scalbert, A., Identification of the 100 richest dietary sources of polyphenols: An application of the Phenol-Explorer database. *Eur. J. Clin. Nutr.*, 64, S112–S120, 2010.

52. U.S. Department of Agriculture (USDA), USDA Database for the Flavonoid Content of Selected Foods, Release 3.2, 2015. Published online at: http://www.ars.usda.gov/nutrientdata (accessed June 20, 2016).

53. U.S. Department of Agriculture (USDA), USDA Database for the Proanthocyanidin Content of Selected Foods, Release 2, 2015. Published online at: http://www.ars.usda.gov/nutrientdata (accessed June 20, 2016).

54. Rothwell, J.A., Pérez-Jiménez, J., Neveu, V., Medina-Remón, A., M'Hiri, N., García-Lobato, P., Manach, C., Knox, C., Eisner, R., Wishart, D.S., and Scalbert, A., Phenol-Explorer 3.0: A major update of the Phenol-Explorer database to incorporate data on the effects of food processing on polyphenol content. *Database*, 2013. doi: 10.1093/database/bat070, 2013.

55. Rothwell, J.A., Urpi-Sarda, M., Boto-Ordoñez, M., Knox, C., Llorach, R., Eisner, R., Cruz, J., Neveu, V., Wishart, D., Manach, C., Andres-Lacueva, C., and Andres-Lacueva, C., Phenol-Explorer 2.0: A major update of the Phenol-Explorer database integrating data on polyphenol metabolism and pharmacokinetics in humans and experimental animals. *Database*, 2012. doi: 10.1093/database/bas031, 2012.

56. Vinson, J.A. and Cai, Y., Nuts, especially walnuts, have both antioxidant quantity and efficacy and exhibit significant potential health benefits. *Food Funct.*, 3, 134–140, 2012.

57. Thompson, L.U., Boucher, B.A., Liu, Z., Cotterchio, M., and Kreiger, N., Phytoestrogen content of foods consumed in Canada, including isoflavones, lignans, and coumestan. *Nutr. Cancer*, 54, 184–201, 2006.

58. Wu, X., Beecher, G.R., Holden, J.M., Haytowitz, D.B., Gebhardt, S.E., and Prior, R.L., Lipophilic and hydrophilic antioxidant capacities of common foods in the United States. *J. Agric. Food Chem.*, 52, 4026–4037, 2004.

59. Kim, J.K., Shin, E.C., Kim, C.R., Park, G.G., Choi, S.J., Cho, H.Y., and Shin, D.H., Composition of fatty acids in commercially available tree nuts and their relationship with protective effects against oxidative stress-induced neurotoxicity. *Food Sci. Biotechnol.*, 22, 1097–1104, 2013.

60. Blomhoff, R., Carlsen, M.H., Andersen, L.F., and Jacobs, D.R., Health benefits of nuts: Potential role of antioxidants. *Br. J. Nutr.*, 96, S52–S60, 2006.

61. Arcan, I. and Yemenicioğlu, A., Antioxidant activity and phenolic content of fresh and dry nuts with or without the seed coat. *J. Food Comp. Anal.*, 22, 184–188, 2009.

62. Bolling, B.W., Blumberg, J.B., and Chen, C.Y.O. (2010). The influence of roasting, pasteurisation, and storage on the polyphenol content and antioxidant capacity of California almond skins. *Food Chem.*, 123, 1040–1047.

63. Schmitzer, V., Slatnar, A., Veberic, R., Stampar, F., and Solar, A., Roasting affects phenolic composition and antioxidative activity of hazelnuts (*Corylus avellana* L.). *J. Food Sci.*, 76, S14–S19, 2011.

64. Tresserra-Rimbau, A., Medina-Remón, A., Pérez-Jiménez, J., Martínez-González, M.A., Covas, M.I., Corella, D., Salas-Salvadó, J., Gómez-Gracia, E., Lapetra, J., Arós, F., Fiol, M., Ros, E., Serra-Majem, L., Pintó, X., Muñoz, M.A., Saez, G.T., Ruiz-Gutiérrez, V., Warnberg, J., Estruch, R., and Lamuela-Raventós, R.M., Dietary intake and major food sources of polyphenols in a Spanish population at high cardiovascular risk: The PREDIMED study. *Nutr. Metab. Cardiovasc. Dis.*, 23, 953–959, 2013.

65. Willis, L.M., Bielinski, D.F., Fisher, D.R., Matthan, N.R., and Joseph, J.A., Walnut extract inhibits LPS-induced activation of BV-2 microglia *via* internalization of TLR4: Possible involvement of phospholipase D2. *Inflammation*, 33, 325–333, 2010.

66. Joseph, J.A., Shukitt-Hale, B., Brewer, G.J., Weikel, K.A., Kalt, W., and Fisher, D.R., Differential protection among fractionated blueberry polyphenolic families against DA-, Aβ42-and LPS-induced decrements in Ca(2+) buffering in primary hippocampal cells. *J. Agric. Food Chem.*, 58, 8196–8204, 2010.

67. Carey, A.N., Fisher, D.R., Joseph, J.A., and Shukitt-Hale, B., The ability of walnut extract and fatty acids to protect against the deleterious effects of oxidative stress and inflammation in hippocampal cells. *Nutr. Neurosci.*, 16, 13–20, 2013.

68. Muthaiyah, B., Essa, M.M., Chauhan, V., and Chauhan, A., Protective effects of walnut extract against amyloid beta peptide induced cell death and oxidative stress in PC12 cells. *Neurochem. Res.*, 36, 2096–2103, 2011.

69. Chauhan, N., Wang, K.C., Wegiel, J., and Malik, M.N., Walnut extract inhibits the fibrillization of amyloid beta-protein, and also defibrillizes its preformed fibrils. *Curr. Alzheimer Res.*, 1, 183–188, 2004.

70. Haider, S., Batool, Z., Tabassum, S., Perveen, T., Saleem, S., Naqvi, F., Javed, H., and Haleem, D.J., Effects of walnuts (*Juglans regia*) on learning and memory functions. *Plant Foods Hum Nutr.*, 66, 335–340, 2011.

71. Kulkarni, K.S., Kasture, S.B., and Mengi, S.A., Efficacy study of *Prunus amygdalus* (almond) nuts in scopolamine-induced amnesia in rats. *Indian J. Pharmacol.*, 42, 168–173, 2010.

72. Poulose, S.M., Bielinski, D.F., and Shukitt-Hale, B., Walnut diet reduces accumulation of polyubiquitinated proteins and inflammation in the brain of aged rats. *J. Nutr. Biochem.*, 24, 912–919, 2013.

73. Carey, A.N., Poulose, S.M., and Shukitt-Hale, B., The beneficial effects of tree nuts on the aging brain. *Nutr. Aging*, 1, 55–67, 2012.

74. Fan, Y.B., Tao, X.W., Ma, L., Liu, L.J., Liu, Z.G., and Wang, L.M., Effects of defatted walnut meal on learning and memory ability and antioxidant capacity of rats. *Food Sci.*, 17, 069, 2013.

75. Shi, D., Chen, C., Zhao, S., Ge, F., Liu, D., and Song, H., Effects of walnut polyphenol on learning and memory functions in hypercholesterolemia mice. *J. Food Nutr. Res.*, 2, 450–456, 2014.

76. Asadi-Shekaari, M., Basiri, M., and Babaee, A., The protective effects of walnuts (*Juglans regia*) on neuronal death and astrocyte reactivity following middle cerebral artery occlusion in male rats. *Sci. J. Kurdistan Uni. Med. Sci.*, 19, 100–108, 2014.

77. Harandi, S., Golchin, L., Ansari, M., Moradi, A., Shabani, M., and Sheibani, V., Antiamnesic effects of walnuts consumption on scopolamine-induced memory impairments in rats. *Basic Clin. Neurosci.*, 6, 91–100, 2015.

78. Shabani, M., Nazeri, M., Parsania, S., Razavinasab, M., Zangiabadi, N., Esmaeilpour, K., and Abareghi, F., Walnut consumption protects rats against cisplatin-induced neurotoxicity. *Neurotoxicology*, 33, 1314–1321, 2012.

79. Poulose, S., Bielinski, D., Crott, J., Roe, A., Thangthaeng, N., and Shukitt-Hale, B., Effects of aging and walnut-rich diet on DNA methylation and expression of immediate-early genes in critical brain regions. *FASEB J.*, 29, 749–757, 2015.

80. Chandel, H.S., Singh, S., and Pawar, A., Neuropharmacological investigation of *Juglans regia* fruit extract with special reference to anxiety. *Int. J. Drug Res. Technol.*, 2, 461–471, 2012.

81. Essa, M.M., Subash, S., Dhanalakshmi, C., Manivasagam, T., Al-Adawi, S., Guillemin, G.J., and Thenmozhi, A.J., Dietary supplementation of walnut partially reverses 1-methyl-4-phenyl-1, 2, 3, 6-tetrahydropyridine induced neurodegeneration in a mouse model of Parkinson's disease. *Neurochem. Res.*, 40, 1283–1293, 2015.

82. Asadi Shekaari, M., Kalantaripour, T.P., Nejad, F.A., Namazian, E., and Eslami, A., The anticonvulsant and neuroprotective effects of walnuts on the neurons of rat brain cortex. *Avicenna J. Med. Biotechnol.*, 4, 155–158, 2012.

83. Park, G., Kim, H.G., Hong, S.P., Kim, S.Y., and Oh, M.S., Walnuts (seeds of *Juglandis sinensis* L.) protect human epidermal keratinocytes against UVB induced mitochondria mediated apoptosis through upregulation of ROS elimination pathways. *Skin Pharmacol. Physiol.*, 27, 132–140, 2014.

84. Zou, J., Cai, P.S., Xiong, C.M., and Ruan, J.L., Neuroprotective effect of peptides extracted from walnut (*Juglans sigilata* Dode) proteins on Aβ25–35-induced memory impairment in mice. *J. Huazhong Univ. Sci. Technol.*, 36, 21–30, 2016.

85. Chen, H., Zhao, M., Lin, L., Wang, J., Sun-Waterhouse, D., Dong, Y., Zhuang, M., and Su, G., Identification of antioxidative peptides from defatted walnut meal hydrolysate with potential for improving learning and memory. *Food Res. Int.*, 78, 216–223, 2015.

86. Suchy, J., Lee, S., Ahmed, A., and Shea, T.B., Dietary supplementation with pecans delays motor neuron pathology in transgenic mice expressing G93A mutant human superoxide dismutase-1. *Curr. Top. Nutraceutical Res.*, 8, 45–54, 2010.

87. O'Neil, C.E., Keast, D.R., Fulgoni, V.L., and Nicklas, T.A., Tree nut consumption improves nutrient intake and diet quality in US adults: An analysis of National Health and Nutrition Examination Survey (NHANES) 1999–2004. *Asia Pac. J. Clin. Nutr.*, 19, 142, 2010.

88. Tey, S.L., Brown, R., Gray, A., Chisholm, A., and Delahunty, C., Nuts improve diet quality compared to other energy-dense snacks while maintaining body weight. *J. Nutr. Metab.*, 357350, 2011.

89. van den Brandt, P.A. and Schouten, L.J., Relationship of tree nut, peanut and peanut butter intake with total and cause-specific mortality: A cohort study and meta-analysis. *Int. J. Epidemiol.*, 45, 1–10, 2015.

90. Eslamparast, T., Sharafkhah, M., Poustchi, H., Hashemian, M., Dawsey, S.M., Freedman, N.D., Freedman, N.D., Boffetta, P., Alonet, C.C., Etemandi, A., Pourshams, A., Malekshah, A.F., Islami, F., Kamangar, F., Merat, S., Brennan, P., Hekmatdoost, A., and Malekshah, A.F., Nut consumption and total and cause-specific mortality: Results from the Golestan Cohort Study. *Int. J. Epidemiol.*, 45, dyv365, 2016.

91. Mayhew, A.J., de Souza, R.J., Meyre, D., Anand, S.S., and Mente, A., A systematic review and meta-analysis of nut consumption and incident risk of CVD and all-cause mortality. *Br. J. Nutr.*, 115, 212–225, 2016.

92. Huhn, S., Masouleh, S.K., Stumvoll, M., Villringer, A., and Witte, A.V., Components of a Mediterranean diet and their impact on cognitive functions in aging. *Front. Aging Neurosci.*, 7, 2015, 132–142, 2015.

93. Valls-Pedret, C., Lamuela-Raventós, R.M., Medina-Remón, A., Quintana, M., Corella, D., Pintó, X., Martínez-González, M.A., Estruch, R., and Ros, E., Polyphenol-rich foods in the Mediterranean diet are associated with better cognitive function in elderly subjects at high cardiovascular risk. *J. Alzheimer Dis.*, 29, 773–782, 2012.

94. Nooyens, A.C., Bueno-de-Mesquita, H.B., van Boxtel, M.P., van Gelder, B.M., Verhagen, H., and Verschuren, W.M., Fruit and vegetable intake and cognitive decline in middle-aged men and women: The Doetinchem Cohort Study. *Br. J. Nutr.*, 106, 752–761, 2011.

95. Pribis, P., Bailey, R.N., Russell, A.A., Kilsby, M.A., Hernandez, M., Craig, W.J., Grajales, T., Shavlik, D.J., and Sabate, J., Effects of walnut consumption on cognitive performance in young adults. *Br. J. Nutr.*, 107, 1393–1401, 2012.

96. Nurk, E., Refsum, H., Drevon, C.A., Tell, G.S., Nygaard, H.A., Engedal, K., and Smith, A.D., Cognitive performance among the elderly in relation to the intake of plant foods. The Hordaland Health Study. *Br. J. Nutr.*, 104, 1190–1201, 2010.

97. Galbete, C., Toledo, E., Toledo, J.B., Bes-Rastrollo, M., Buil-Cosiales, P., Marti, A., Gullén-Grima, F., and Martínez-González, M.A., Mediterranean diet and cognitive function: The Sun Project. *J. Nutr. Health Aging*, 19, 305–312, 2015.

98. Martínez-Lapiscina, E.H., Clavero, P., Toledo, E., Estruch, R., Salas-Salvadó, J., San Julián, B., Sanchez-Tainta, A., Ros, E., Valls-Pedret, C., and Martinez-Gonzalez, M.Á., Mediterranean diet improves cognition: The PREDIMED-NAVARRA randomised trial. *J. Neurol. Neurosurg. Psychiatry*, 84, 1318–1325, 2013.

99. Valls-Pedret, C., Sala-Vila, A., Serra-Mir, M., Corella, D., de la Torre, R., Martínez-González, M.Á., Martínez-Lapiscina, E.H., Fitó, M., Pérez-Heras, A., Salas-Salvadó, J., Estruch, R., and Ros, E., Mediterranean diet and age-related cognitive decline: A randomized clinical trial. *JAMA Intern. Med.*, 175, 1094–1103, 2015.

100. O'Brien, J., Okereke, O., Devore, E., Rosner, B., Breteler, M., and Grodstein, F., Long-term intake of nuts in relation to cognitive function in older women. *J. Nutr. Health Aging*, 18, 496–502, 2014.

101. Arab, L. and Ang, A., A cross sectional study of the association between walnut consumption and cognitive function among adult us populations represented in NHANES. *J. Nutr. Health Aging*, 19, 284–290, 2015.

102. Sánchez-Villegas, A., Galbete, C., Martinez-González, M.Á., Martinez, J.A., Razquin, C., Salas-Salvadó, J., Estruch, R., Buil-Cosiales, P., and Martí, A., The effect of the Mediterranean diet on plasma brain-derived neurotrophic factor (BDNF) levels: The PREDIMED-NAVARRA randomized trial. *Nutr. Neurosci.*, 14, 195–201, 2011.

103. Karege, F., Schwald, M., and Cisse, M., Postnatal developmental profile of brain-derived neurotrophic factor in rat brain and platelets. *Neurosci. Lett.*, 328, 261–264, 2002.

104. Hashimoto, K., Iwata, Y., Nakamura, K., Tsujii, M., Tsuchiya, K.J., Sekine, Y., Suzuki, K., Minabe, Y., Takei, N., Iyo, M., and Mori, N., Reduced serum levels of brain-derived neurotrophic factor in adult male patients with autism. *Prog. Neuropsychopharmacol. Biol. Psychiatry*, 30, 1529–1531, 2006.

105. Huang, T.L., Effects of antipsychotics on the BDNF in schizophrenia. *Curr. Med. Chem.*, 20, 345–350, 2013.

106. Sánchez-Villegas, A., Martínez-González, M.A., Estruch, R., Salas-Salvadó, J., Corella, D., Covas, M.I., Aros, F., Romaguera, D., Gomez-Gracia, E., Lapetra, J., Pinto, X., Martinez, J.A., Lamuela-Raventos, R.M., Ros, E., Gea, A., Warnberg, J., and Serra-Majem, L., Mediterranean dietary pattern and depression: The PREDIMED randomized trial. *BMC Med.*, 11, 208–219, 2013.

107. Gillette-Guyonnet, S., Van Kan, G.A., Andrieu, S., and Barberger-Gateau, P., IANA Task Force on Nutrition and Cognitive Decline with Aging. *J. Nutr. Health Aging*, 11, 132–152, 2007.

108. Barbour, J.A., Howe, P.R., Buckley, J.D., Bryan, J., and Coates, A.M., Nut consumption for vascular health and cognitive function. *Nutr. Res. Rev.*, 27, 131–158, 2014.

109. Tucker, K.L., Nutrient intake, nutritional status, and cognitive function with aging. *Ann. N. Y. Acad. Sci.*, 1367, 38–49, 2016.

110. Heffernan, K.S., Fahs, C.A., Ranadive, S.M., and Patvardhan, E.A., L-Arginine as a nutritional prophylaxis against vascular endothelial dysfunction with aging. *J. Cardiovasc. Pharmacol. Ther.*, 15, 17–23, 2010.

111. Haddad, E.H., Gaban-Chong, N., Oda, K., and Sabaté, J., Effect of a walnut meal on postprandial oxidative stress and antioxidants in healthy individuals. *Nutr. J.*, 10, 4, 2014.

112. Kesse-Guyot, E., Fezeu, L., Andreeva, V.A., Touvier, M., Scalbert, A., Hercberg, S., and Galan, P., Total and specific polyphenol intakes in midlife are associated with cognitive function measured 13 years later. *J. Nutr.*, 111, 1–8, 2012.

113. Lahiri, D.K., Chen, D.M., Lahiri, P., Bondy, S., and Greig, N.H., Amyloid, cholinesterase, melatonin, and metals and their roles in aging and neurodegenerative diseases. *Ann. N.Y. Acad. Sci.*, 1056, 430–449, 2005.

114. Reiter, R.J., Manchester, L.C., and Tan, D.X., Melatonin in walnuts: Influence on levels of melatonin and total antioxidant capacity of blood. *Nutrition*, 21, 920–924, 2005.

115. Gómez-Pinilla, F., Brain foods: The effects of nutrients on brain function. *Nat. Rev. Neurosci.*, 9, 568–578, 2008.

116. Grosso, G. and Estruch, R., Nut consumption and age-related disease. *Maturitas*, 84, 11–16, 2016.

12 Science behind Usefulness of *Bacopa monnieri* for Memory and Cognition

Chandrasekaran Prasanna Raja, Bethapudi Bharathi, Chinampudur Velusami Chandrasekaran, Nithyanantham Muruganantham, Mundkinajeddu Deepak, and Agarwal Amit

CONTENTS

12.1 INTRODUCTION

12.1.1 COGNITION

Cognition is defined as the process by which an individual organizes information. This includes acquiring information (perception); selecting (attention), representing (understanding) and retaining (memory) information; and using it to guide behavior (reasoning and coordination of motor outputs) (Bostrom and Sandberg, 2009). Memory is considered to be a major component of cognition. Memory functions comprise of three major subprocesses, i.e., encoding, consolidation, and retrieval. During encoding, the perception of a stimulus results in the formation of new memory trace, which is initially highly susceptible to disturbing influences and decay, i.e., forgetting. During consolidation, the labile memory trace is gradually stabilized, possibly involving multiple waves of short- and long-term consolidation processes, which serve to strengthen and integrate the memory into pre-existing knowledge networks. During retrieval, the stored memory is accessed and recalled (Rasch and Born, 2013). The function of cognition is to enable the organism to deal with environmental complexity, enabling it to adapt its behavior to the demands of an ever-changing environment and allowing it to appropriately select and improve the behaviors of a given repertoire. Hence, cognition has been implicated to deal with the day-to-day changing environment (Bosse et al., 2007). Any impairment in the cognition, termed cognitive impairment, affects the quality of life.

12.1.2 COGNITIVE IMPAIRMENT

Cognitive impairment denotes trouble in remembering, learning new things, concentrating, or making decisions that affect everyday life. Cognitive impairment ranges from mild to severe. With mild impairment, people may begin to notice changes in cognitive functions but still be able to do their everyday activities. Severe levels of impairment can lead to losing the ability to understand the meaning or importance of something and the ability to talk or write, resulting in the inability to live independently. Hence, it is very clear that the effects of cognitive impairment range from disturbances in everyday changes to inability to live independently (Centers for Disease Control and Prevention, 2011). Cognitive impairment can be classified into three categories such as age-associated cognitive impairment/decline (AACD), mild cognitive impairment (MCI), and dementia.

12.1.2.1 Age-Associated Cognitive Impairment/Decline (AACD)

AACD is a normal (nonpathological, normative, usual) cognitive ageing that occurs in individuals. Some people however experience a severe deterioration in cognitive skills.

12.1.2.2 Growing Ageing Population and Socioeconomic Impact of Memory Impairment

Population ageing is defined as the increased proportion of older persons in the population and is one of the major problems of this century. Ageing is a dynamic process, determined by the relative size of the younger and older in a population at different moments in time. This demographic transition is a consequence of reductions in mortality followed by reductions in fertility. Together, these reductions eventually lead to smaller proportions of children and larger shares of older people in the population.

Ageing started earlier in developed regions and is beginning to take place in some developing countries. The geriatric population is already a major social and health problem in developed countries that has crossed 524 million, constituting 8% of the world's population. And this number is predicted to increase to about 1.5 billion by 2050, representing 16% of the world's population. In India, the elderly population constitutes 100 million during 2011 and this is expected to rise to 300 million by 2050. So, overall, old age is a period of physical and mental deterioration. The main risk factors are loss of fortune, fall in self-esteem, sense of helplessness, illiteracy, poor health, financial debt, and social discrimination. Of all the problems associated with ageing, health care demands top priority (Ory and Bond, 1989). One of the major health concerns allied with older adults is the risk of memory impairment termed as AACD (Patterson, 1994). The prevalence rate of age-associated cognitive decline is 21% for subjects 60 years and older (Ritchie et al., 2001). It has been reported that there are substantial chances for transition of AACD to MCI (Kidd, 2008).

12.1.2.3   Mild Cognitive Impairment (MCI)

MCI represents a transitional state between the cognitive changes of normal ageing and very early dementia and is becoming increasingly recognized as a risk factor for Alzheimer disease (AD), in which persons experience memory loss to a greater extent than one would expect for age, yet they do not meet currently accepted criteria for clinically probable AD (Petersen et al., 2001). There are a number of forms/practice methods available for screening of MCI, but the most common is "amnestic" MCI, characterized by a set of problems with short-term memory, planning, language, and/or attention that decline more rapidly during the pathological process. MCI is believed to be a high-risk condition for the development of clinically probable AD. The prevalence of MCI due to ageing is between 3% and 19% in older population (Gauthier et al., 2006).

12.1.2.4   Dementia

Dementia in general denotes the decline in mental ability severe enough to interface with daily life. It is a syndrome due to disease of the brain, usually of a chronic or progressive nature, in which there is disturbance of multiple higher cortical functions, including memory, thinking, orientation, comprehension, calculation, learning capability, language, and judgment. Impairments of cognitive function are commonly accompanied, occasionally preceded, by deterioration in emotional control, social behavior, or motivation; however, consciousness is not impaired (British Psychological Society, 2007). A recent estimate of diagnosed and undiagnosed rates finds that the prevalence of late onset dementia is 7.1% among people of 65 or over. The most common form of dementia is AD, which occurs in 50% to 75% of cases, and the next most common is vascular dementia at approximately 20% to 30% of older population (Ray and Davidson, 2014).

12.1.2.5   Transition from AACD to MCI to Dementia

Cognitive decline is an age-associated physiological condition that later gets developed into MCI and dementia, a pathological condition that can be observed in the brain. Cognitive decline occurs along a continuum, and there are no clear set boundaries between normal, MCI, and dementia (Ray and Davidson, 2014). Age-related decline in mental abilities is highly variable between different people, and the prevalence ranges from 3.6% to 38.4%, while the prevalence rates of MCI range from 3% to 42%. There are almost 900 million people aged 60 years and above living worldwide, of which it is estimated that 4.6% to 8.7% of people had dementia (World Alzheimer Report, 2015). Annual conversion rates to dementia for individuals classified according to the Age-Associated Memory Impairment (AAMI) criteria vary between 1% and 3% (Snowdon and Lane, 1994) and 24% (Coria et al., 1992). While the conversion rates to dementia for individuals classified into MCI ranges between 10% and 15% (Bischkopf et al., 2002).

12.1.3   Pharmacological Treatment for Cognitive Enhancement

With the growing ageing population, prevention and treatment of cognitive impairment in the elderly has assumed increasing importance. Interventions for cognitive enhancement range from lifestyle measures to pharmacological treatments (Andrade and Radhakrishnan, 2009). Although there is no cure for cognitive impairments, the current treatments and therapies said to be cognitive enhancer may help in the management of the affected person to think clearly, communicate, remember, and change their mood and personality. Some of the cognitive enhancer drugs that are currently available in the market are as follows:

Donepezil	Cholinesterase inhibitor
Memantine	*N*-Methyl-D-Aspartate (NMDA)-receptor antagonist
Piracetam	Positive allosteric modulator of α-amino-3-hydroxy-5-methly-4-isoxazolepropionic acid (AMPA) receptor
Modafinil	Dopamine transport blocker
Atomoxetine	Inhibits serotonin transporter, nonepinephrine transporter and dopamine transporter
Lisdexamfetamine	Trace amine-associated receptor 1 (TAAR1) agonist
Amphetamine	Trace amine-associated receptor 1 (TAAR1) agonist, G_s and G_q receptor agonist
Rivastigmine	Acetylcholine esterase and butyrylcholinesterase inhibitor
Galantamine	Nicotinic cholinergic receptor agonist

These drugs appear to modulate neurotransmitters to exert their activity. However, the overall effects of drugs seem to be modest and are associated with severe side effects that lead to discontinuation of medications (Husain and Mehta, 2011), thus the penchant toward the usage of herbal medicines is increasing.

12.1.4 HERBAL APPROACH FOR COGNITION AND MEMORY

Herbal approach is an alternative treatment for cognitive and memory impairments. There are several plants like *Hypericum perforatum*, *Bacopa monnieri* (BM), *Ginkgo biloba*, *Panax ginseng*, etc., reported for enhancing cognition. The biological effects of BM are documented in both traditional and modern scientific literature. Traditionally, in Ayurveda it was used as a brain tonic to enhance memory, learning, and concentration (Mukherjee and Dey, 1966). Several studies have been conducted to test for the efficacy of these plants for cognitive and memory related impairments. The systemic reviews have revealed that BM is a potential candidate for cognitive enhancement (Kongkeaw et al., 2014).

12.1.4.1 *Bacopa monnieri*

BM, colloquially called "Brahmi" named after Brahma, the creator of God, is native to India. It is also called water hyssop, Indian pennywort, Jai-Brahmi, *Herpestis*, etc. It is a creeping perennial plant with small oblong leaves and purple flowers that grows as a weed in rice fields, found throughout east Asia and the United States (Figure 12.1). BM was initially described around 6th century AD in texts such as *Charaka Samhita*, *Athar-Ved*, and *Susrutu Samhita* as medhya rasayana, which means the herb that sharpens intellect and attenuates mental deficits. The entire plant was used medicinally and has a long history of use in Ayurveda. Several experiments

FIGURE 12.1 **(See color insert.)** Aerial parts of *Bacopa monnieri.*

have been carried out to establish the medicinal properties of BM. BM has been applied in rodents and cell culture for anti-inflammatory, antineoplastic, hepato-protective, antidepressant, antimicrobial, anticonvulsant, analgesic properties, etc. Although BM has multiple beneficial effects, it is particularly renowned for its nootropic activity, for treating anxiety, intellect and poor memory, and cognitive disabilities. BM is consistently found in many Ayurvedic preparations prescribed for cognitive dysfunction, used to improve memory and cognition (Stough et al., 2008). Different BM alcoholic extracts have been investigated extensively for their neuro-pharmcological and their nootropic activities (Prabhakar et al., 2008, Shinomol et al., 2011, Singh et al., 2008). The present chapter elucidates the science behind the usefulness of BM for memory and cognition.

12.1.4.2 Phytochemistry of BM

Phytochemical investigations carried out on *Bacopa* revealed the presence of a dam-marane type of triterpenoid saponins called bacosides. These saponins were con-sidered responsible for the biological effects of *Bacopa*, of which bacoside A and bacoside B were reported as the major bioactives (Deepak and Amit, 2004). Detailed investigations carried out later revealed that bacoside A is actually a mixture of four saponins viz., bacoside A_3, bacopaside II, bacopasaponin C, and the jujobogenin isomer of bacopasaponin C. These saponins, along with bacopaside I, were known to constitute more than 96% w/w of the total saponins of BM. The identity of baco-side B, one of the so-called bioactives, is reported as ambiguous and was considered as an artefact formed during extraction process. However, bacoside B is still being specified as a marker in some of the marketed *Bacopa* extracts, which needs to be relooked for controlling the quality of *Bacopa* extracts (Figure 12.2; Deepak and Amit, 2013).

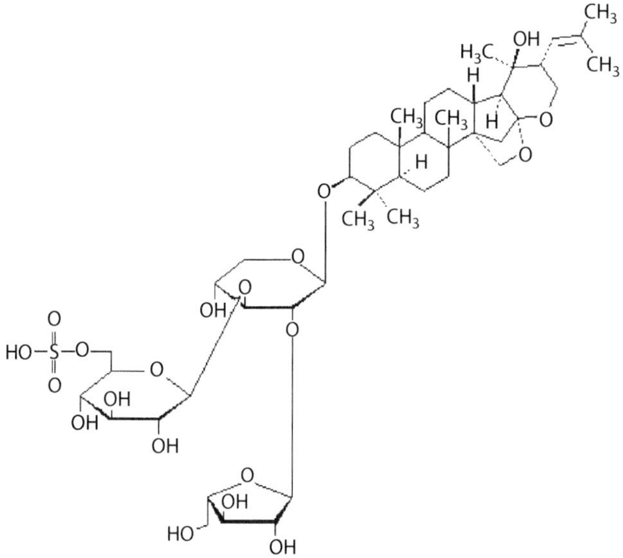

Bacopaside I

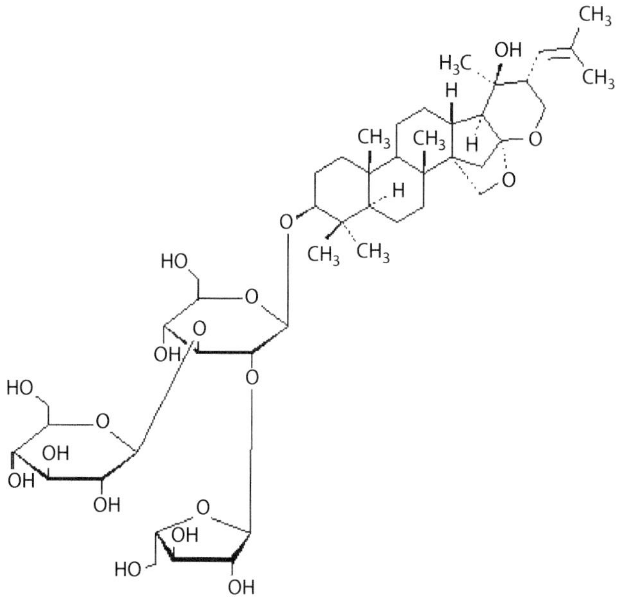

Bacopaside II

FIGURE 12.2   Major bioactive saponins of *Bacopa monnieri*.   (*Continued*)

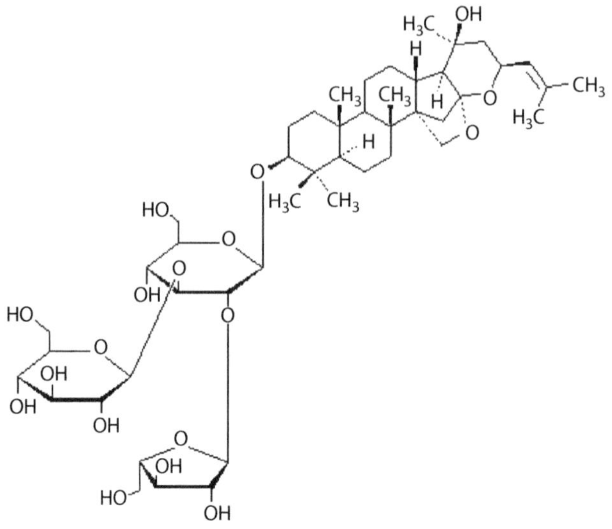

Bacoside A$_3$

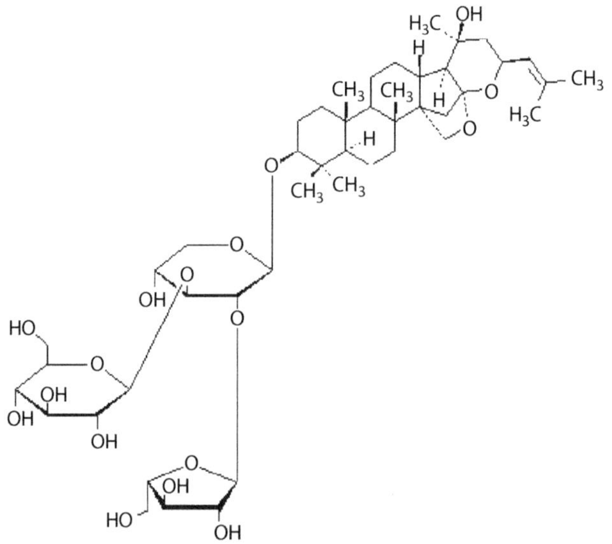

Jujubogenin isomer of Bacopasaponin C

FIGURE 12.2 (CONTINUED)   Major bioactive saponins of *Bacopa monnieri*.   (*Continued*)

12.2   SAFETY STUDIES ON BM

In reference to BM, before understanding the capacity to induce clinical benefits, i.e., efficacy, the likelihood of BM causing any harm has been reviewed. Traditionally, BM has a long history of popular use in Ayurveda for improving learning and memory (Bhanamishra et al., 2010). Commonly, it is perceived that herbs having long

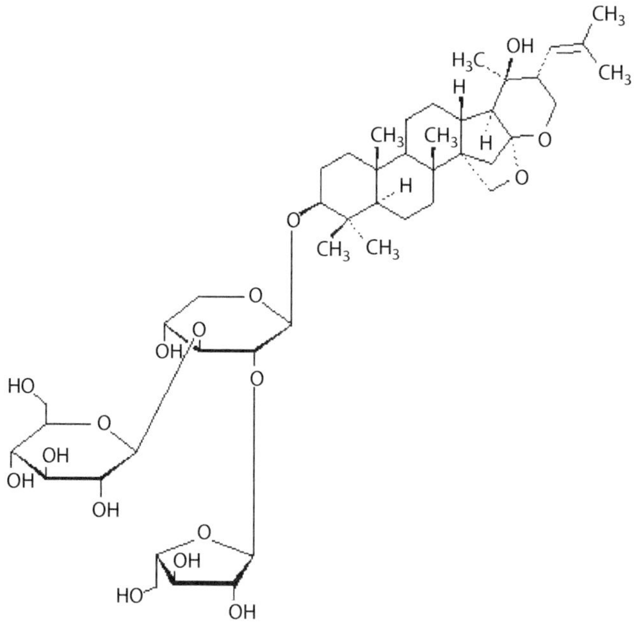

Bacopasaponin C

FIGURE 12.2 (CONTINUED)   Major bioactive saponins of *Bacopa monnieri*.

history of use are recognized as safe at recommended doses (Fong, 2002). In addition to the traditional background, the literature review indicated that broad spectrum of descriptive animal toxicity tests were performed on BM, including acute lethality, repeated-dose studies (subacute and subchronic), reproductive and developmental toxicity, and genotoxicity studies. In addition, a phase I study to evaluate its safety, determine a safe dosage range, and identify the side effects of BM in humans was also performed.

12.2.1 PRECLINICAL SAFETY (*IN VITRO* AND *IN VIVO*) STUDIES

12.2.1.1 Acute Lethality Studies

The median lethal dose (LD_{50}) of orally administered BacoMind, a standardized formulation of BM, in rats is 2400 mg/kg (Allan et al., 2007), 5000 mg/kg for an aqueous extract, and 17,000 mg/kg for an alcoholic extract (Martis and Rao, 1992).

As LD_{50} is greater than 2000 mg/kg, BM can be classified as substance with high safety profile.

12.2.1.2 Repeated Dose Studies

In a subacute toxicity study, BacoMind was administered orally daily for 14 days at a dose of 250, 500, or 1000 mg/kg rat body weight (b.w.). The study findings indicated that except for mild lowering in body weight gain in male rats, it was found to be tolerated well up to a dose of 500 mg/kg (Allan et al., 2007).

In a subchronic oral toxicity study, BacoMind was administered at dose levels of 0, 85, 210, and 500 mg/kg b.w. for 90 days and additional rats were included in the control and high-dose level groups for further period of 28 days to assess for reversibility, persistence, or delayed occurrence of toxic effects. The principal goals of the subchronic study are to identify and characterize the specific organ or organs affected by the test compound after repeated administration and further to establish a no-observed adverse effect level (NOAEL). Repeated dose administration of BacoMind for 90 days did not reveal any evidence of toxicity with respect to clinical signs, food consumption, body weight gain, hematological, and blood biochemistry parameters. Necropsy and histopathological examination did not reveal any remarkable and treatment-related changes. A NOAEL level of 500 mg/kg b.w. was established in rats (Allan et al., 2007).

12.2.1.3 Developmental and Reproductive Toxicity Studies

Reproductive toxicity studies enable us to study the occurrence of adverse effects on the female or male reproductive system that may result from exposure to test substances. A study in male mouse reported that oral administration of BM at a dose of 250 mg/kg b.w. for 28 and 56 days caused reduction in motility, viability, morphology, and number of spermatozoa in cauda epididymis. Histopathological alterations in seminiferous tubules and epididymis were also reported. The study concluded that BM caused reversible suppression of spermatogenesis and fertility (Singh and Singh, 2009).

However, the previously mentioned study revealed reversible male infertility, the phytochemical and analytical specifications were not mentioned. Also, the study observed that the percentage of affected seminiferous tubules in the testes were 96.27% after 28 days of Brahmi treatment vs. 54.13% after 56 days of treatment, and this observation, as a result, leads to contention (Agarwal and Allan, 2010). In addition, a subchronic oral toxicity study on BacoMind for 90 days conducted as per internationally accepted guidelines (OECD test no. 408, 1998) in rats at dose levels of 500 mg/kg did not reveal any evidence of abnormality with respect to absolute and relative organ weights of testes and epididymis. Also, there were no histopathological alterations observed in these organs. Taking these into consideration, the reliability and reproducibility of the results of the study by Singh and Singh (2009) seem to be uncertain. Hence, the antifertility effects reported are dubious in the absence of a study conducted in accordance with internationally accepted guidelines.

Teratology is the study of defects induced during development between conception and birth. The potential of BM to disrupt normal embryonic and/or fetal development (teratogenic effects) is also determined in laboratory animals. The teratogenic studies were conducted in rodent (rat) and nonrodent (rabbits) species. Pregnant rats were administered BM at two dose levels (50 and 100 mg/kg) during organogenesis from 6 to 15 days of gestation, while pregnant rabbits were administered BM at dose levels (26.66 and 53.32 mg/kg) from 6 to 18 days of gestation.

BM administration in pregnant rats and rabbits did not demonstrate any evidence of maternal toxicity and teratogenic effects in the litter. In parents, mortality, abnormal clinical signs, and changes in food and water consumption were not observed. A steady gain in weight was recorded. The number of corpora lutea and implantations were comparable, and no intrauterine death (still birth) was observed. The litter sizes of all the groups were comparable and well within the range of normal limit. Visceral and skeletal examinations of the fetuses showed no significant anomalies. Hence, it can be concluded that BM does not represent a major teratogenic risk (Singh, 2010).

12.2.1.4   Genotoxicity Studies

Genotoxicity tests can be defined as *in vitro* and *in vivo* tests designed to detect compounds that induce genetic damage directly or indirectly by various means. The standard test battery for genotoxicity includes the bacterial reverse mutation test (BRMT), *in vitro* chromosomal damage test with mammalian cells, and *in vivo* micronucleus test (European Medicines Agency, 1998).

The BRMT/Ames test was performed using five test strains of *Salmonella typhimurium*. Strains of salmonella were exposed to various doses of BacoMind, ranging from 61.72 to 5000 μg/plate with and without a metabolic activation. No mutagenic effect was observed with BacoMind up to a dose of 5000 μg/mL in BRMT (Deb et al., 2008).

In vitro chromosomal damage test with mammalian cells was performed as per OECD guideline no. 473 to evaluate the clastogenic effects of BacoMind if any. The clastogenic potential of BacoMind on human lymphocyte cultures was compared with commonly used clastogens mitomycin C (MMC), hydrogen peroxide (H_2O_2), and benzo[a]pyrene (B[a]P). MMC and H_2O_2 effects were studied without metabolic activation, while B[a]P effect was studied with metabolic activation. Treatment of human lymphocytes with BacoMind had no significant effect on the number of chromosomal aberrations observed when compared with controls. Treatment of human lymphocytes with BacoMind along with S-9 mix (with metabolic activation) demonstrated no clastogenic effect. In addition, BacoMind also demonstrated protection of the human lymphocytes against the clastogenic effects of MMC, H_2O_2, and B[a]P (Deb et al., 2008).

Intraperitoneal administration of bacosides A and B, bioactive components of BM, at concentration of 20, 40, and 80 mg/kg was tested for genotoxic potential in mice using *in vivo* micronucleus assay. Bone marrow of treated mice was extracted and evaluated for the proportion of immature erythrocytes. The incidence of micronucleated cell per 1000 polychromatic erythrocytes was scored. The study findings indicated non-genotoxic effects of BM (Giri and Khan, 1996).

In addition to this standard battery of genotoxic tests, bacosides A and B at 20, 40, and 80 mg/kg (*i.p.*) were tested in chromosomal aberration and sister chromatid exchange assays. In both the tests, cytogenic end points such as mitotic and cell replicative indices indicated that bacosides A and B are found to be non-genotoxic (Giri and Khan, 1996).

Overall, several published safety reports are available on BacoMind. The preclinical toxicity studies indicated that BM is safe for consumption. However, it is important that the safety observed in preclinical trials must translate to safety in human clinical trials also.

12.2.2 CLINICAL SAFETY STUDIES

12.2.2.1 Elderly Individuals

In a study on elderly subjects with a mean age of 62.62 years, 40 subjects were administered either 300 or 600 mg of BM once daily for 12 weeks. Laboratory investigations including hematological, biochemical, and electrocardiogram recordings in these elderly individuals did not show any abnormalities, and the results were within the normal limits (Peth-Nui et al., 2012).

12.2.2.2 Young Individuals

BacoMind was evaluated for its safety and tolerability as a phase I dose escalation study in healthy young volunteers aged between 18 and 45 years. Twenty-three volunteers were administered single capsule of BacoMind daily for 30 days, i.e., 300 mg for the first 15 days and 450 mg for the next 15 days. Volunteers were evaluated pre and post treatment. The clinical, hematological, biochemical, and electrocardiographic parameters assessed pre and post treatment did not indicate any untoward effects in any of the treated volunteers. This phase I trial indicated that BacoMind is safe and tolerable in healthy individuals (Pravina et al., 2007).

12.2.2.3 Children

Clinical trials were conducted to evaluate the effects of BacoMind on cognition in children reported that BacoMind did not demonstrate any major adverse effects during the treatment period. Mild gastrointestinal disturbances were reported. Overall, these trials reported that BacoMind was safe and tolerable in children up to six months of treatment period (Dave et al., 2008, 2014).

12.2.3 ADVERSE EFFECTS

BM is safe and well tolerated in elderly, youth, and children. Except for minor gastrointestinal disturbances, i.e., epigastric burning sensation, nausea, abdominal fullness, bloating, abdominal cramping, belching, etc., BM was well tolerated in all age groups from children to elderly population. The gastrointestinal symptoms subsided spontaneously without any need for discontinuation of treatment. A few animal studies indicated adverse effects on fertility and effects of augmenting thyroid function. However, the reports are dubious as adverse effects were not observed in reproductive organs or on the body weight in subchronic toxicity studies (Allan et al., 2007). Probably the effects of BM on fertility and thyroid functioning may be due to contaminants as BM has affinity for preferential accumulation of certain trace metals (Shukla et al., 2007; Srikanth et al., 2013). The effects of BM on fertility and thyroid activity warrants further investigation.

Overall both preclinical and clinical investigations provided conviction towards safety of BM.

12.3   EFFICACY STUDIES OF BM IN ANIMAL MODELS

12.3.1   ANIMAL MODELS ON LEARNING AND MEMORY

Literature review indicated that BM has been extensively researched and proved for its cognition enhancing ability in rodent models. The current section summarizes the preclinical studies peformed on BM with respect to improving cognitive abilities.

Chronic administration of D-galactose has been reported to cause deterioration of cognitive and motor skills that are similar to symptoms of ageing and therefore is regarded as a model of accelerated ageing. KeenMind (CDRI-08) was investigated against D-galactose (D-gal)-induced brain ageing in rats. Experimental groups were subjected to contextual-associative learning task. Administration of KeenMind in the D-gal-treated group attenuated contextual-associative learning deficits; the rats showed more correct responses and retrieved the reward with less latency (Prisila Dulcy et al., 2012).

Elevated plus-maze consisting of two open arms crossed with two closed arms is employed to investigate the effects of BacoMind on learning and memory in mice. The effect of BacoMind was investigated using transfer latency as a parameter for acquisition and retention of memory process. Transfer latency is the time in which animal moves from open arms to enclosed arms and is expressed as inflexion ratio. Decreased transfer latency or increased inflexion ratio is indicative of improvement in learning and memory. BacoMind (60 mg/kg mice b.w.) administered orally for seven days protected the mice from scopolamine-induced learning and memory impairment evident from the significant increase in inflexion ratio as compared to the scopolamine-treated group (Kasture et al., 2007). Neonatal rat pups (10-day-old) administered with BacoMind (40 and 80 mg/kg) for 2, 4, and 6 weeks showed significant improvement in spatial learning and memory in T-maze test. Rat pups administered at a dose of 20 mg/kg of BM showed significant improvement in spatial learning and memory at fourth and sixth week (Vollala et al., 2011).

Passive avoidance behavior is used to examine the effect of drugs/test substances on learning and memory. BacoMind was investigated in passive shock avoidance test. Mice were placed in an apparatus with the electric grid (20 V with AC current of 5 mA) and a shock-free zone (SFZ). The latency to reach SFZ and the mistakes (descents) the animal made in 15 minutes were recorded. BacoMind at 40, 60, and 80 mg/kg showed significant decrease in latency to reach SFZ and number of mistakes in 15 minutes as compared to vehicle control. Further, BacoMind significantly alleviated the scopolamine-induced impairment of retention as compared to the scopolamine-treated group (Kasture et al., 2007). Neonatal rat pups (10-day-old) administered with BacoMind (20, 40 and 80 mg/kg) for 2, 4 and 6 weeks showed significant improvement in passive avoidance test (Vollala et al., 2011).

BacoMind was investigated for its ability to recognize the novel object. Subject to this, a study was conducted with two different kinds of objects to explore the effect of BM on learning, recognition, and memory. Rats were exposed to the familiar arena with two identical objects placed in the first trial. In the second trial, rats were exposed to familiar object and a novel object. The time taken exploring each object and the discrimination index percentage was recorded. BacoMind (27, 40, and 54 mg/kg rat b.w.) showed a significant increase in the discrimination index when compared to the vehicle control. Further, in scopolamine-treated amnesic rats, BacoMind at 40 mg/kg showed a significant increase in the discrimination index as compared to the scopolamine control (Kasture et al., 2007).

Scopolamine is used experimentally for inducing cholinergic-based cognitive deficits, especially spatial learning and memory in animals, and is considered as a reliable tool to study the antiamnesic effects of substances. The Morris water maze model is used to study working memory, reference memory, and task strategy. Rotarod test is conducted to screen the muscle coordination activity of mice. BM at 120 mg/kg orally for six days reversed scopolamine (0.5 mg/kg *i.p.*)-induced both anterograde and retrograde amnesia. BM probably has an effect on the cholinergic system and can be an approach for treating cholinergic-induced cognitive deficits (Saraf et al., 2011). In addition, three different triterpenoid saponins (bacopaside I bacopaside II, and bacopasaponsin C) from BM in scopolamine-induced memory impairment in mice showed nootropic activity when tested by Morris water maze test and step-down test. This illustrates the efficacy of BM saponins in enhancing the cognitive function (Zhou et al., 2009).

Diazepam is well known to cause amnesia by modulating GABAergic system. The effect of BM (120 mg/kg *p.o.*) on diazepam (1.75 mg/kg *i.p.*)-induced amnesia in mice using the Morris water maze showed antiamnesic effects (Prabhakar et al., 2008).

In another study, the effect of bacosides on scopolamine-induced (3 mg/kg, *i.p.*), sodium-nitrite (75 mg/kg, *i.p.*), and BN52021 (15 mg/kg, *i.p.*) induced experimental amnesia in mice was studied using the Morris water maze test. Findings indicated that bacosides attenuated anterograde experimental amnesia induced by scopolamine and sodium nitrite and also reversed BN52021-induced retrograde amnesia (Kishore and Singh, 2005).

Okadaic acid (OKA) is a selective and potent inhibitor of the serine/threonine phosphatases 1 and 2A, and administration of OKA mimics the characteristic features of memory impairment. OKA (200 ng) administered intracerebroventricularly induced memory impairment in rats. BM (40 and 80 mg/kg) administered one hour before OKA injection and continued daily up to 13 days improved memory functions as determined by Morris water maze test on days 13–15. (Dwivedi et al., 2013).

12.3.2 ANIMAL MODELS FOR AD

The key clinical feature of AD is the extracellular deposition of amyloid plaques primarily composed of a 39–43-amino-acid protein called amyloid beta (Aβ). PSAPP is a transgenic mouse that secretes human Aβ peptide, leading to amyloid deposition and serves as a model for AD. BM (40 or 160 mg/kg/day) administered starting at two months of age for either two or eight months lowered Aβ 1–40 and 1–42 levels in the cortex by as much as 60%. Also reversed Y-maze performance and open field hyperlocomotion behavioral changes present in PSAPP mice. The data indicate the prophylactic potential of BM in AD, a serious cognitive disorder (Holcomb et al., 2006).

Ethylcholine aziridinium ion (AF64A) is a cholinergic neurotoxin and is used to induce AD in rodents. Wistar rats administered with alcoholic extract of BM at doses of 20, 40, and 80 mg/kg b.w. orally for a period of two weeks before and one week after the intracerebroventricular administration of AF64A bilaterally demonstrated significant improvement in the escape latency time in the Morris water maze test. Moreover, the reductions of neurons and cholinergic neuron densities were also mitigated, indicating cognitive enhancing effect in AD by BM (Uabundit et al., 2010).

Colchicine-induced cognitive dysfunction is an accepted model for dementia of AD. Intracerebral ventricular colchicine administration causes neuronal death in subventricular zone, resulting in cognitive impairment in rats. The effect of BM administration at the dose of 50 mg/kg rat b.w. for 15 days following colchicine administration reversed cognitive impairment induced by colchicine. In addition, BM attenuated oxidative damage and restored antioxidant enzymes and acetylcholinesterase activities (Saini et al., 2012).

Ibotenic acid, an agonist of the NMDA receptors, injected locally into the nucleus basalis magnocellularis or in the cortex, induces deficits in cholinergic neurotransmission and displays cognitive deficits. Subchronic administration of BM orally at a dose of 10 mg/kg once daily for 14 days demonstrated significant decrease of memory deficits induced by ibotenic acid on day 7 and 14 in comparison to ibotenic acid treated group (Bhattacharya et al., 1999).

Aluminum trichloride ($AlCl_3$) induced cognitive deficit has been widely used for the preclinical screening of promising therapeutic agents against AD. The possible effects of BM alone and synergistic activity of BM with rivastigmine (drug used for AD) in reversing $AlCl_3$-induced learning deficits and memory loss in rats have been investigated. Chronic administration of $AlCl_3$ for 42 days caused significant memory impairment associated with increased retention latency in the Morris water maze task and increased transfer latency in the elevated plus maze test. Administration of BM (100 and 200 mg/kg), rivastigmine (5 mg/kg), and a combination of BM (100 mg/kg) and rivastigmine (5 mg/kg) orally demonstrated significant protection against $AlCl_3$-induced memory impairment in comparison to $AlCl_3$-treated group (Thippeswamy et al., 2013).

Plethora of preclinical *in vitro* and *in vivo* studies indicate that BM has enormous potential to prevent/treat age-related memory impairment and neurotoxin induced mild to severe cognitive impairment including cognitive impairment of AD.

12.4   CLINICAL STUDIES

While the inclination toward herbal medicines as alternative choices of therapy is increasing, the evidence for efficacy in a clinical trial is generally weak. However, a literature review indicates that BM has been investigated in multiple clinical trials in all age groups, from children to geriatric. The current section summarizes the clinical evidence on efficacy of BM in the management of cognitive performance.

12.4.1 CHILDREN

12.4.1.1 Effect of BM in Children Requiring Individual Education Program

Children with borderline intelligent quotient (IQ) between 70 and 90 whose intellectual functioning level affected their ability to keep up with pace in learning and regularly met with failures in schools are considered slow learners. Slow learners require special education services such as Individual Education Program (IEP). BacoMind was investigated for its effect on cognitive enhancement in children requiring IEP. A total of 24 healthy children aged between 4 and 18 years requiring IEP were administered with 225 mg of BacoMind orally for a period of four months. BacoMind treatment showed significant improvement in working memory and short-term memory. In addition, significant improvement was also observed in logical memory, memory related to personal life, visual as well as auditory memory. BacoMind significantly enhanced cognition in slow learners requiring IEP (Dave et al., 2008).

12.4.1.2 Effect of BM in Children and Adolescents with Attention-Deficit Hyperactivity Disorder

Attention-deficit hyperactivity disorder (ADHD) is a clinically heterogeneous disorder of inattention, hyperactivity, and impulsivity or difficulty in controlling behavior. ADHD is a childhood-onset condition and the majority of children continue to manifest symptoms into adulthood, leading to substantial problems affecting relationships, work, and quality of life. The symptoms may lead to functional impairments affecting education, family, and social interactions. ADHD warrants treatment. Sarris et al. (2011) and Pellow et al. (2011) reported that BM is a promising candidate for ADHD.

A clinical trial was conducted to investigate the efficacy of BacoMind in ameliorating ADHD symptoms in children and adolescents. Twenty-seven subjects aged 6 to 12 with an age of onset of ADHD before seven years of age as defined by the Diagnostic and Statistical Manual of Mental Disorders criteria for ADHD were administered with BacoMind (225 mg) of BM extract orally for six months. Following six months of treatment, BacoMind significantly reduced restlessness, impulsivity, attention-deficit, self-control, psychiatric, and learning problems of ADHD children. BacoMind was effective in alleviating ADHD symptoms and was well tolerated by the children (Dave et al., 2014).

Memory Plus containing BM was evaluated for its efficacy in improving memory in children with ADHD. Memory Plus administered at dose of 100 mg per day for 12 weeks significantly improved sentence repetition, logical memory, and paired associate learning tasks. Memory Plus demonstrated significant improvement in learning capabilities of ADHD children and was well tolerated (Negi et al., 2000).

In a randomized double-blind, placebo-controlled trial, 120 children with newly diagnosed ADHD treated with compound herbal preparation (containing BM as one of the ingredients) over a period of four months showed improvement in attention, cognition, and impulse control (Katz et al., 2010).

12.4.1.3 Cognition Enhancing Effect of BM in Normal Children

BM was evaluated for its efficacy in school-going normal children for six weeks. Normal healthy school children administered with BM demonstrated significantly improved mean reaction time (Kaur et al., 1998).

Hence, BM has promising effects on cognitive effects in normal healthy children as well as in children with borderline IQ and attention deficit disorders.

12.4.2 YOUNG

12.4.2.1 Acute Cognition Enhancing Ability of BM in Healthy Individuals

The acute effects of KeenMind (BM extract) were investigated in healthy individuals to test if BM improves cognitive performance during acutely stressful paradigms like multitasking framework and attenuates mood change and cortisol levels. KeenMind administered at two doses, 320 and 640 mg, improved cognitive performance notably at one and two hours post treatment during acutely stressful paradigms. In addition, Keenmind appears to have adaptogenic activity, as evident from reduction in cortisol, and exhibited positive effects on mood (Benson et al., 2014).

The acute effects of KeenMind on mood and performance upon a mentally effortful cognitive task before and after KeenMind administration were investigated. The study findings revealed that KeenMind at 320 mg appears to have acute enhancing effects upon cognitive functioning even in healthy young individuals (i.e., a genuine nootropic effect) (Downey et al., 2013).

Nathan et al. (2001) conducted a study for evaluating the acute cognitive enhancing effect of BM in healthy volunteers with a mean age of 37.4. A single dose of 300 mg (2 × 150 mg) of KeenMind administration did not have any acute effect on cognitive function, as assessed by a battery of neuropsychological tests.

In summary, mixed results are demonstrated by acute/single dose administration of BM in young healthy individuals. Possibly, BM has acute effects in enhancing cognition during stressful conditions rather than normal stress-free situations.

12.4.2.2 Chronic Cognition Enhancing Ability of BM in Healthy Individuals

A double-blind placebo-controlled study examined the chronic effects of KeenMind in 46 healthy human subjects aged between 18 and 60 years with a mean age of 39.4 years. The participants received either 300 mg of KeenMind or placebo for 12 weeks. KeenMind has significantly improved the speed of visual information processing, learning rate, memory consolidation, and state anxiety with maximal effects after 12 weeks of oral intake (Stough et al., 2001).

In another double-blind placebo-controlled study on chronic effects of KeenMind in healthy subjects at an oral dose of 300 mg/day for 90 days on cognition enhancing ability were investigated in cognitive drug research computerized assessment system. KeenMind significantly improved performance on working memory, more specifically spatial memory. In addition, the number of false-positives recorded in the rapid visual information processing task was reduced by KeenMind (Stough et al., 2008).

Roodenrys et al. (2002) replicated the memory-enhancing effects of KeenMind in 76 participants aged between 40 and 65 years. Authors reported a significant decrease in the rate of forgetting newly acquired information after 90 days of treatment with KeenMind.

BacoMind was investigated for its chronic effect on cognition and anxiety in 72 educated healthy urban adults in an age range of 35–60 years. In the randomized double-blind, placebo-controlled parallel study, volunteers were administered 450 mg of placebo or BacoMind for 12 weeks, and cognitive performance and anxiety were assessed post treatment. The authors reported that there was no significant difference in the cognitive performance post treatment, while a trend for lower state anxiety was observed with BacoMind (Sathyanarayanan et al., 2013). The findings of Sathyanarayanan et al. (2013) are not in agreement with the reports by earlier studies on Brahmi (Calabrese et al., 2008; Morgan and Stevens, 2010; Stough et al., 2001). This is probably because the subjects with marginally high cognitive/intellectual functioning at the stipulated age group when compared with the normative data of cognition and with minimal anxiety were included in the study. In such cases, there was little scope of BM showing improvement in cognition.

In conclusion, BM has both acute and chronic effects on enhancing cognitive function while the effects appear to be prominent in chronic conditions such as stress and anxiety.

12.4.3 ELDERLY

12.4.3.1 Effect of BM in Enhancing Cognition in Elderly Individuals

Developed countries have accepted the chronological age of 65 years as the definition of "elderly" or older person, and the United Nations agreed to a cutoff of 60+ years to refer to the older population. As per Britain, the Friendly Societies Act enacted the definition of old age as "any age after 50." Hence, the review considered "elderly" as mean age above 50 years.

A randomized, double-blind, placebo-controlled study evaluated the efficacy and tolerability of BacoMind in elderly subjects. Elderly individuals with a Mini-Mental State Examination score of 24 and above with mean age of 64.9 were included in the study. BacoMind or placebo was given as a oral dose of 450 mg daily for a duration of 12 weeks. The neuropsychology tests revealed that BacoMind significantly improved performance in tests associated with attention and verbal memory and was also found to be well tolerated in elderly participants (Barbhaiya et al., 2008).

Another study in 48 elderly individuals aged above 65 (mean, 73.5) with no complaints of memory impairment compared to others of their age and without signs of dementia were administered either 300 mg BM extract or placebo orally daily for 12 weeks. After treatment, cognitive ability assessed by Rey Auditory Verbal Learning Test (AVLT) and Stroop Task, which indicated a significant increase in cognitive performance in the BM-treated group compared to the placebo group. BM is well tolerated, and this study illustrates the cognitive enhancing effects of BM in elderly individuals (Calabrese et al., 2008).

In an Australian study, 81 healthy participants over 55 years of age (mean, 65.41) received either 300 mg BacoMind or placebo for 12 weeks. The BacoMind-treated group showed significant improvement in verbal learning, memory acquisition, and delayed recall measured by Rey AVLT. BacoMind was found to be effective in enhancing cognitive performance by improving both memory acquisition and retention in healthy elderly individuals (Morgan and Stevens, 2010).

In a randomized, double-blind, placebo-controlled trial, 44 individuals aged 55 and above with Age-Associated Memory Impairment (AAMI) without any evidence of dementia or psychiatric disorder received either 250 mg of standardized BM extract or placebo daily for a period of 12 weeks followed by a placebo period of another four weeks. BM produced significant improvement in mental control, logical memory, and paired associated learning during the 12-week therapy. During the placebo period, the test scores remained the same, indicating sustained effects on cognition until the next four weeks. Thus, BM showed promising effect in managing AAMI (Raghav et al., 2006).

In a randomized, double-blind, placebo-controlled trial, the effect of BM on attention, cognitive processing, working memory, and cholinergic and monoaminergic functions in healthy elders was investigated. Sixty healthy elderly subjects (mean age, 62.62 years) received either a standardized extract of BM (300 and 600 mg) or placebo once daily for 12 weeks. The BM-treated group showed improved attention, cognitive processing, and working memory. Also, suppression of plasma acetyl cholinesterase activity was observed (Peth-Nui et al., 2012).

In summary, the results indicate that BM possesses promising effects on cognitive enhancement in children and young and elderly individuals. In addition, BM was found to be effective in AAMI and could be a promising candidate for MCI and early-phase AD also. Also, further clinical investigation will be needed to ascertain the usefulness of BM in MCI and AD patients.

12.5   MECHANISM OF ACTION

The putative mechanisms of action of BM on cognitive enhancement can be categorized into neuroprotective, neurotrophic activity, and effects of BM on the neurotransmitter release.

12.5.1   Neuroprotective Effects of BM

The damage caused by oxidative stress potentially serves as an early event that initiates the development of cognitive disturbances. BM serves as an excellent free radical scavenging agent and protects the neuronal tissue from oxidative-stress-induced damage. Oxidative stress has been reported to be responsible for AAMI, MCI, AD, and other memory-impairment-related disorders.

12.5.1.1   Neuroprotective Effects of BM via Oxidative Stress Amelioration and Antiapoptotic Pathways

In cell lines, several studies demonstrated neuroprotective effects of BM. Sodium nitroprusside (SNP) is a potent nitric oxide (NO) donor that exerts nitrative stress by up-regulation of inducible NO synthase (iNOS). Pretreatment of rat pheochromocytoma (PC12) cells with BM ameliorated SNP-induced mitochondrial and plasma membrane damage and inhibited NO generation by downregulating iNOS expression. In addition BM pretreatment attenuated SNP-induced apoptotic protein biomarkers such as Bax, Bcl_2, cytochrome-c, and caspase-3 and also attenuated neuronal stress markers like heme oxygenase-1 (HO-1) and iNOS while upregulating brain development neurotrophic factor, indicating its neuroprotective and antiapoptotic function (Ayyathan et al., 2015, Pandareesh and Anand, 2014, Shinomol et al., 2012, Singh et al., 2013).

The neuroprotective effects of BM were investigated in *in vivo* studies also. In rats, standardized extract of BM administered at dose of 5 and 10 mg/kg b.w. for 14 to 21 days demonstrated dose-related increase in superoxide dismutase, catalase, and glutathione peroxidase activities in rat brain frontal cortical, striated, and hippocampal regions, indicating neuroprotective activity via oxidative free radical scavenging activity (Bhattacharya et al., 2000). Such neuroprotective activity via enhancing antioxidant release has also been reported by several authors (Dwivedi et al., 2013, Priya et al., 2013). Dwivedi et al. (2013) demonstrated that BM treatment in rats restored nuclear erythroid related factor 2, Heme oxygenase (HO-1), and glutamate cysteine ligase catalytic subunit expressions led to decreased oxidative stress and neuronal loss in rat brain.

12.5.1.2   Neuroprotective Effects of BM via Attenuation of Neuroinflammation

Neuroinflammation has gained considerable interest in age-associated neurodegeneration and pathologies like senile dementia of the Alzheimer's type due to its slow onset and chronic nature. Upon long-term oral administration of bacosides for a duration of three months, Wistar rats demonstrated significant attenuation of age-dependent elevation of proinflammatory cytokines and iNOS protein expression in middle-aged and aged rat brain cortex (Rastogi et al., 2012). Hence, bacosides are effective against age-related chronic neuroinflammation.

12.5.2   Neurotrophic Effects of BM

BM has been reported for neurotrophic effects. Neurotrophic effects augment proliferation, differentiation, growth, and regeneration of neuronal cells. Brain-derived neurotrophic factor (BDNF) gene expression provides signals to the brain and spinal cord for making BDNF that promotes survival of neurons via playing a role in growth, maturation, and maintenance of neurons. BDNF regulates synaptic plasticity, which is important for learning and memory. KeenMind treatment activated ERK/CREB pathways and, thus, BDNF expression in hippocampus (Preethi et al., 2014). BacoMind treatment in rats showed a significant increase in dendritic intersections and dendritic branching points, indicating dendritic growth stimulating properties (Vollala et al., 2011). Thus, BM was demonstrated to cause neurotrophic effects.

12.5.3 Effects of BM on Neurotransmitter Release

Neurotransmitters such as catecholamines, serotonin, GABA, histamine, and others have been implicated in cognitive processes. BM has been extensively reported to modulate the neurotransmitter profile in the brain. Rats treated with BM (40 mg/kg) by oral gavage from postnatal day 15–29 demonstrated improved learning and memory. The levels of serotonin increased significantly with a concomitant increase in the upregulation of mRNA expression of serotonin synthesizing enzymes tryptophan hydroxylase-2 and serotonin transporter (Prisila et al., 2011).

Blockade of NMDA receptors is known to induce significant deficits in object recognition memory (Malkova et al., 2015). Bacosides are known to upregulate N-methyl-D-aspartale receptor–1 (NMDAR1), glutamate receptor expression and increase CREB phosphorylation and thus augmenting glutaminergic transmission (Hota et al., 2009).

Vesicular glutamate transporter (VGLUT) packs glutamate and is responsible for fast synaptic transmission. BM is reported to increase VGLUT1 density in the frontal cortex, striatum, and CA1 of rat brain (Wetchateng and Piyabhan, 2012).

In aged brain, serotonin, acetylcholine, and dopamine were reduced and bacosides treatment prevented age-related loss of serotonin, acetylcholine, and dopamine (Rastogi et al., 2012).

A comprehensive study was conducted to deduce the molecular mechanism of cognition enhancing activity of BacoMind in a panel of cell-free and receptor-transfected cell assays. BacoMind was found to inhibit Prolyl endopeptidase, Catechol-O-methyl transferase, and Poly (ADP-ribose) polymerase. BacoMind might influence different neurological pathways associated with learning and memory disorders, and age-associated memory impairment due to its antagonistic effect on serotonin 6 and serotonin 2A receptors (Dethe et al., 2016).

Overall, pleiotropic effects of BM such as neuroprotective, neurotrophic, and modulation of neurotransmitter release contributed to the cognitive enhancing effects of BM.

12.6 CONCLUSION

BM has immense potential in preventing and treating cognitive disorders. Children with cognitive disorders encounter challenges in learning, memory, attention, and behavioral changes. Parents and teachers seek interventions to improve their cognitive abilities. In addition, young individuals also seek interventions that enhance cognitive abilities during stressful situations like examinations, multitasking, etc. On the other hand, in elderly, the major risk factor for memory impairment is advanced age, and it has a huge socioeconomic impact on the society. Based on preclinical and clinical evidence, and high safety profile, BM can be considered as an alternative herbal supplement/medicine to all age groups for brain health.

12.7 FUTURE DIRECTIONS

Various clinical trials were performed using subjective parameters and proved the efficacy of BM for cognitive disorders. However, it is important to adopt appropriate objective parameters to confirm the efficacy of BM in double-blind placebo-controlled human clinical trials. Further detailed research is required to understand the active constituent responsible for the beneficial effects in brain. Bioactivity-guided fractionation using *in vivo* and *in vitro* assays might lead to the identification of phytoconstituents that are individually or collectively responsible for neuroprotective or neurotrophic or neurotransmission activity of BM.

REFERENCES

Agarwal A and Allan JJ. Antifertility effects of herbs: Need for responsible reporting. *J Ayurveda Integr Med.* 2010;1:129–131.

Allan JJ, Damodaran A, Desmukh NS, Goudar KS and Amit A. Safety evaluation of a standardized phytochemical composition extracted from *Bacopa monnieri* in Sprague-Dawley rats. *Food Chem Toxicol.* 2007;45:1928–1937.

Andrade C and Radhakrishnan R. The prevention and treatment of cognitive decline and dementia: An overview of recent research on experimental treatments. *Indian J Psychiatry.* 2009;51:12–25.

Ayyathan DM, Chandrasekaran R and Thiagarajan K. Neuroprotective effect of Brahmi, an Ayurvedic drug against oxidative stress induced by methyl mercury toxicity in rat brain mitochondrial-enriched fractions. *Nat Prod Res.* 2015;29:1046–1051.

Barbhaiya HC, Desai RP, Saxena VS, Pravina K, Wasim P, Geetharani P, Joshua JA, Venkateshwarlu K and Amit A. Efficacy and tolerability of BacoMind on memory improvement in elderly participants—A double blind placebo controlled study. *J Pharmacol Toxicol.* 2008;3:425–434.

Benson S, Downey LA, Stough C, Wetherell M, Zangara A and Scholey A. An acute, double-blind, placebo-controlled cross-over study of 320 mg and 640 mg doses of *Bacopa monnieri* (CDRI 08) on multitasking stress reactivity and mood. *Phytother Res.* 2014;28:551–559.

Bhattacharya SK, Bhattacharya A, Kumar A and Ghosal S. Antioxidant activity of *Bacopa monniera* in rat frontal cortex, striatum and hippocampus. *Phytother Res.* 2000;14:174–179.

Bhattacharya SK, Kumar A and Ghosal S. Effect of *Bacopa monniera* on animal models of Alzheimer's disease and pertubed central cholinergic markers of cognition in rats. *Pharmacol Toxicol.* 1999;4:1–12.

Bhanamishra. *Bhavaprakasha Nighantu.* Chunekar KC, Pandey G (eds.). Varanasi: Chaukhambha Bharti Academy. 2010;279–281.

Bischkopf J, Busse A, Angermeyer MC. Mild cognitive impairment—A review of prevalence, incidence and outcome according to current approaches. *Acta Psychiatr Scand.* 2002;106:403–414.

Bosse T, Sharpanskykh A and Treur, J. On the complexity monotonicity thesis for environment, behaviour and cognition. In: Baldoni M, Son TC, van Riemsdijk MB and Winikoff M. (eds.), *Declarative Agent Languages and Technologies,* V. DALT 2007. Lecture Notes in Computer Science, Springer, Berlin, Heidelberg, 2007; 4897: 175–192.

Bostrom N and Sandberg A. Cognitive enhancement: Methods, ethics, regulatory challenges. *Sci Eng Ethics.* 2009;15:311–341.

British Psychological Society. 2007. *A NICE-SCIE Guideline on Supporting People with Dementia and their Carers in Health and Social Care.* NICE Clinical Guidelines, No. 42. National Collaborating Centre for Mental Health (UK), Leicester (UK).

Calabrese C, Gregory WL, Leo M, Kraemer D, Bone K and Oken B. Effects of a standardized *Bacopa monnieri* extract on cognitive performance, anxiety, and depression in the elderly: A randomized double-blind, placebo-controlled trial. *J Altern Complement Med.* 2008;14:707–713.

Centers for Disease Control and Prevention. Cognitive impairment: A call for action. 2011. Retrieved from: http://www.cdc.gov/aging/pdf/cognitive_impairment/cogimp_poilicy _final.pdf.

Coria F, Gomez De Caso JA and Duarte J. Age-associated memory impairment: Nosology and outcome. *J Neurol Suppl.* 1992;2:66.

Dave UP, Dingankar SR, Saxena VS, Joseph JA, Bethapudi B, Agarwal A and Kudiganti V. An open-label study to elucidate the effects of standardized *Bacopa monnieri* extract in the management of symptoms of attention-deficit hyperactivity disorder in children. *Adv Mind Body Med.* 2014;28:10–15.

Dave UP, Wasim P, Joshua AJ, Geetarani P, Murali B, Mayachari AS, Venkateshwarlu K, Saxena VS, Deepak M and Amit A. BacoMind®: A cognitive enhancer in children requiring Individual Education Programme. *J Pharmacol Toxicol.* 2008;3:302–310.

Deb DD, Kapoor P, Dighe RP, Padmaja R, Anand MS, D'Souza P, Deepak M, Murali B and Amit A. *In vitro* safety evaluation and anticlastogenic effect of BacoMind™ on human lymphocytes. *Biomed Environ Sci.* 2008;21:7–23.

Deepak M and Amit A. The need for establishing identities of 'bacoside A and B', the putative major bioactive saponins of Indian medicinal plant *Bacopa monnieri. Phytomedicine.* 2004;11:264–268.

Deepak M and Amit A. 'Bacoside B'—The need remains for establishing identity. *Fitoterapia.* 2013;87:7–10.

Dethe S, Deepak M and Agarwal A. Elucidation of Molecular Mechanism(s) of Cognition Enhancing Activity of Bacomind®: A Standardized Extract of Bacopa Monnieri. Pharmacogn Mag. 2016; 12(Suppl 4):S482-S487.

Downey LA, Kean J, Nemeh F, Lau A, Poll A, Gregory R, Murray M, Rourke J, Patak B, Pase MP, Zangara A, Lomas J, Scholey A and Stough C. An acute, double-blind, placebo-controlled crossover study of 320 mg and 640 mg doses of a special extract of *Bacopa monnieri* (CDRI 08) on sustained cognitive performance. *Phytother Res.* 2013;27:1407–1413.

Dwivedi S, Nagarajan R, Hanif K, Siddiqui HH, Nath C and Shukla R. Standardized Extract of *Bacopa monniera* attenuates okadaic acid induced memory dysfunction in rats: Effect on Nrf2 pathway. *Evid Based Complement Alternat Med.* 2013;294501.

European Medicines Agency (EMEA). *Genotoxicity: A Standard Battery for Genotoxicity Testing of Pharmaceuticals.* 1998. CH Topic S2 B.

Fong HH. 2002. Integration of herbal medicine into modern medical practices: Issues and prospects. *Integr Cancer Ther.* 2002;3:287–293.

Gauthier S, Reisberg B, Zaudig M, Petersen RC, Ritchie K, Broich K, Belleville S, Brodaty H, Bennett D, Chertkow H, Cummings JL, de Leon M, Feldman H, Ganguli M, Hampel H, Scheltens P, Tierney MC, Whitehouse P and Winblad B. International Psychogeriatric Association Expert Conference on mild cognitive impairment. *Lancet.* 2006;367:1262–1270.

Giri AK and Khan KA. Chromosome aberrations, sister chromatid exchange and micronuclei formation analysis in mice after *in vivo* exposure to bacoside A and B. *Cytologia.* 1996;61:99–103.

Holcomb LA, Dhanasekaran M, Hitt AR, Young KA, Riggs M and Manyam BV. Bacopa monniera extract reduces amyloid levels in PSAPP mice. *J Alzheimers Dis.* 2006;9:243–251.

Hota SK, Barhwal K, Baitharu I, Prasad D, Singh SB and Ilavazhagan G. *Bacopa monniera* leaf extract ameliorates hypobaric hypoxia induced spatial memory impairment. *Neurobiol Dis.* 2009;34:23–39.

Husain M and Mehta MA. Cognitive enhancement by drugs in health and disease. *Trends Cogn Sci.* 2011;15:28–36.

Kasture BS, Kasture SV, Joshua JA, Damodaran A and Amit A. Nootropic activity of BacoMind™, an enriched phytochemical composition from *Bacopa monnieri. J Nat Remedies.* 2007;7:166–173.

Katz M, Levine AA, Kol-Degani H and Kav-Venaki L. A compound herbal preparation (CHP) in the treatment of children with ADHD: A randomized controlled trial. *J Atten Disord*. 2010;14:281–291.

Kaur BR, Adhiraj J, Pandit PR, Ajita R, Vijay M, Shanta D, Hemangeeni D, Sudha M and Kamble G. Effect of an Ayurvedic formulation on attention, concentration and memory in normal school going children. *Indian Drugs*. 1998;35:200–203.

Kidd PM. Alzheimer's disease, amnestic mild cognitive impairment, and age-associated memory impairment: Current understanding and progress toward integrative prevention. *Altern Med Rev*. 2008;13:85–115.

Kishore K and Singh M. Effect of bacosides, alcoholic extract of *Bacopa monniera* Linn. (Brahmi), on experimental amnesia in mice. *Indian J Exp Biol*. 2005;43:640–645.

Kongkeaw C, Dilokthornsakul P, Thanarangsarit P, Limpeanchob N and Norman Scholfield C. Meta-analysis of randomized controlled trials on cognitive effects of *Bacopa monnieri* extract. *J Ethnopharmacol*. 2014;151:528–535.

Malkova L, Forcelli PA, Wellman LL, Dybdal D, Dubach MF and Gale K. Blockade of glutamatergic transmission in perirhinal cortex impairs object recognition memory in macaques. *J Neurosci*. 2015;35:5043–5050.

Martis G and Rao A. Neuropharmacological activity of Herpestis monniera. *Fitoterapia*. 1992;63:399–404.

Morgan A and Stevens J. Does *Bacopa monnieri* improve memory performance in older persons? Results of a randomized, placebo-controlled, double-blind trial. *J Altern Complement Med*. 2010;16:753–759.

Mukherjee GD and Dey CD. Clinical trial on Brahmi. I. *J Exp Med Sci*. 1966;10:5–11.

Nathan PJ, Clarke J, Lloyd J, Hutchison CW, Downey L and Stough C. The acute effects of an extract of *Bacopa monniera* (Brahmi) on cognitive function in healthy normal subjects. *Hum Psychopharmacol*. 2001;16:345–351.

Negi KS, Singh YD, Kushwaha KP, Rastogi CK, Rathi AK, Srivastava JS, Asthana OP and Gupta RC. Clinical evaluation of memory enhancing properties of Memory Plus in children with attention deficit hyperactivity disorder. *Indian J Psychiatry*. 2000;42:27.

OECD (Organisation for Economic Co-operation and Development). Guideline for the Testing of Chemicals. Guideline 408: Repeated Dose 90-day Oral Toxicity Study in Rodents; Paris, 1998.

Ory MG and Bond K. Introduction: Health care for an ageing society. In *Ageing and Health Care: Social Science and Policy Perspectives*. Ory G and Bond K (eds). London: Routledge. 1989; 1–24. 1989.

Pandareesh MD and Anand T. Neuroprotective and anti-apoptotic propensity of *Bacopa monniera* extract against sodium nitroprusside induced activation of iNOS, heat shock proteins and apoptotic markers in PC12 cells. *Neurochem Res*. 2014;39:800–814.

Patterson C. Screening for cognitive impairment in the elderly. In: Goldbloom R, editor. The Canadian Task Force on the Periodic Health Examination. Canadian Guide to Clinical Preventive Health Care. Ottawa: Canada Communications Group. 1994:902–911.

Pellow J, Solomon EM and Barnard CN. Complementary and alternative medical therapies for children with attention-deficit/hyperactivity disorder (ADHD). *Altern Med Rev*. 2011;16:323–337.

Petersen RC, Doody R, Kurz A, Mohs RC, Morris JC, Rabins PV, Ritchie K, Rossor M, Thal L and Winblad B. Current concepts in mild cognitive impairment. *Arch Neurol*. 2001;58:1985–1992.

Peth-Nui T, Wattanathorn J, Muchimapura S, Tong-Un T, Piyavhatkul N, Rangseekajee P, Ingkaninan K and Vittaya-Areekul S. Effects of 12-week *Bacopa monnieri* consumption on attention, cognitive processing, working memory, and functions of both cholinergic and monoaminergic systems in healthy elderly volunteers. *Evid Based Complement Alternat Med*. 2012;606424.

Prabhakar S, Saraf MK, Pandhi P and Anand A. *Bacopa monniera* exerts antiamnesic effect on diazepam-induced anterograde amnesia in mice. *Psychopharmacology.* 2008;200:27–37.

Pravina K, Ravindra KR, Goudar KS, Vinod DR, Joshua AJ, Wasim P, Venkateshwarlu K, Saxena VS and Amit A. Safety evaluation of BacoMind in healthy volunteers: A phase I study. *Phytomedicine.* 2007;14:301–308.

Preethi J, Singh HK, Venkataraman JS and Rajan KE. Standardised extract of *Bacopa monniera* (CDRI-08) improves contextual fear memory by differentially regulating the activity of histone acetylation and protein phosphatases (PP1α, PP2A) in hippocampus. *Cell Mol Neurobiol.* 2014;34:577–589.

Prince M, Wimo A, Guerchet M, Ali GC, Wu YT and Prina M. The global impact of dementia: An analysis of prevalence, incidence, cost and trends. World Alzheimer Report. London: Alzheimer's Disease International: 2015;1–82.

Prisila DC, Ganesh A, Geraldine P, Akbarsha MA and Rajan KE. Bacopa monniera leaf extract up-regulates tryptophan hydroxylase (TPH2) and serotonin transporter (SERT) expression: Implications in memory formation. *J Ethnopharmacol.* 2011;134:55–61.

Prisila Dulcy C, Singh HK, Preethi J and Rajan KE. Standardized extract of Bacopa monniera (BESEB CDRI-08) attenuates contextual associative learning deficits in the aging rat's brain induced by D-galactose. *J Neurosci Res.* 2012;90:2053–2064.

Priya V, Poonam S and Behrose SG. Prophylactic efficacy of *Bacopa monnieri* on decabromodiphenyl ether (PBDE-209)-induced alterations in oxidative status and spatial memory in mice. *Asian J Pharm Clin Res.* 2013;6:242–247.

Raghav S, Singh H, Dalai PK, Srivastava JS and Asthana OP. Randomized controlled trial of standardized Bacopa monniera extract in age-associated memory impairment. *Indian J Psychiatry.* 2006;48:238–242.

Rasch B and Born J. About sleep's role in memory. *Physiol Rev.* 2013;93:681–766.

Rastogi M, Ojha RP, Prabu PC, Devi BP, Agrawal A and Dubey GP. Prevention of age-associated neurodegeneration and promotion of healthy brain ageing in female Wistar rats by long term use of bacosides. *Biogerontology.* 2012;13:183–195.

Ray S and Davidson S. Dementia and Cognitive Decline: A Review of the Evidence. London: Age UK Research 2014: 1–38. Age UK Research, 2014.

Ritchie K, Artero S and Touchon J. Classification criteria for mild cognitive impairment: A population-based validation study. *Neurology.* 2001;56:37–42.

Roodenrys S, Booth D, Bulzomi S, Phipps A, Micallef C and Smoker J. Chronic effects of Brahmi (*Bacopa monnieri*) on human memory. *Neuropsychopharmacology.* 2002;27:279–281.

Saini N, Singh D and Sandhir R. Neuroprotective effects of *Bacopa monnieri* in experimental model of dementia. *Neurochem Res.* 2012;37:1928–1937.

Saraf MK, Prabhakar S, Khanduja KL and Anand A. *Bacopa monniera* attenuates scopolamine-induced impairment of spatial memory in mice. *Evid Based Complement Alternat Med.* 2011;2011:236186.

Sarris J, Kean J, Schweitzer I and Lake J. Complementary medicines (herbal and nutritional products) in the treatment of attention deficit hyperactivity disorder (ADHD): Systematic review of the evidence. *Complement Ther Med.* 2011;19:216–227.

Sathyanarayanan V, Thomas T, Einöther SJ, Dobriyal R, Joshi MK and Krishnamachari S. Brahmi for the better? New findings challenging cognition and anti-anxiety effects of Brahmi (*Bacopa monniera*) in healthy adults. Psychopharmacology. 2013;227:299–306.

Shinomol GK, Bharath MM and Muralidhara. Pretreatment with *Bacopa monnieri* extract offsets 3-nitropropionic acid induced mitochondrial oxidative stress and dysfunctions in the striatum of prepubertal mouse brain. *Can J Physiol Pharmacol.* 2012;90:595–606.

Shinomol GK, Muralidhara and Bharath MM. Exploring the role of "Brahmi" (*Bacopa monnieri* and *Centella asiatica*) in brain function and therapy. *Recent Pat Endocr Metab Immune Drug Discov.* 2011;5:33–49.

Shukla OP, Dubey S and Rai UN. 2007. Preferential accumulation of cadmium and chromium: Toxicity in *Bacopa monnieri* L. under mixed metal treatments. *Bull Environ Contam Toxicol.* 2007;78:252–257.

Singh HK. Product Information, Safety Evaluation and Clinical Efficacy of CDRI-08. 2010.

Singh A and Singh SK. Evaluation of antifertility potential of Brahmi in male mouse. *Contraception.* 2009;79:71–79.

Singh M, Murthy V and Ramassamy C. Neuroprotective mechanisms of the standardized extract of *Bacopa monniera* in a paraquat/diquat-mediated acute toxicity. *Neurochem Int.* 2013;62:530–539.

Singh RH, Narsimhamurthy K and Singh G. Neuronutrient impact of Ayurvedic Rasayana therapy in brain aging. *Biogerontology.* 2008;9:369–374.

Snowdon J and Lane F. A longitudinal study of age-associated memory impairment. *Int J Geriatr Psychiatry.* 1994;9:779–787.

Srikanth Lavu RV, Prasad MN, Pratti VL, Meißner R, Rinklebe J, Van De Wiele T, Tack F and Du Laing G. 2013. Trace metals accumulation in *Bacopa monnieri* and their bioaccessibility. *Planta Medica.* 2013;79:1081–1083.

Stough C, Downey LA, Lloyd J, Silber B, Redman S, Hutchison C, Wesnes K and Nathan PJ. Examining the nootropic effects of a special extract of *Bacopa monniera* on human cognitive functioning: 90 day double-blind placebo-controlled randomized trial. *Phytother Res.* 2008;22:1629–1634.

Stough C, Lloyd J, Downey LA, Hutchison CW, Rodgers T and Nathan PJ. The chronic effects of an extract of *Bacopa monniera* (Brahmi) on cognitive function in healthy human subjects. *Psychopharmacology.* 2001;156:481–484.

The global impact of dementia: An analysis of prevalence, incidence, cost and trends. *World Alzheimer Report.* Alzheimer's Disease International: 2015.

Thippeswamy AH, Rafiq M, Viswantha GL, Kavya KJ, Anturlikar SD and Patki PS. Evaluation of Bacopa monniera for its synergistic activity with rivastigmine in reversing aluminum-induced memory loss and learning deficit in rats. *J Acupunct Meridian Stud.* 2013;6:208–213.

Uabundit N, Wattanathorn J, Mucimapura S and Ingkaninan K. Cognitive enhancement and neuroprotective effects of *Bacopa monnieri* in Alzheimer's disease model. *J Ethnopharmacol.* 2010;127:26–31.

Vollala VR, Upadhya S and Nayak S. Enhanced dendritic arborization of hippocampal CA3 neurons by Bacopa monniera extract treatment in adult rats. *Rom J Morphol Embryol.* 2011a;52:879–886.

Vollala VR, Upadhya S and Nayak S. Learning and memory-enhancing effect of Bacopa monniera in neonatal rats. *Bratisl Lek Listy.* 2011b;112:663–669.

Wetchateng T and Piyabhan P. P-1326—Effects of *Bacopa monnieri* on VGLUT1 density in frontal cortex, striatum and hippocampus of schizophrenic rat model. *Eur Psychiatry.* 2012;27:1.

Zhou Y, Peng L, Zhang WD and Kong DY. Effect of triterpenoid saponins from Bacopa monniera on scopolamine-induced memory impairment in mice. *Planta Medica.* 2009;75:568–574.

13 Food-Derived γ-Aminobutyric Acid Improves the Mental State

Hiroshi Shimoda

CONTENTS

13.1 INTRODUCTION

γ-Aminobutyric acid (GABA) is a neurotransmitter in the brain that binds to the GABA receptor and has an inhibitory effect on cerebral neurons.[1] GABA analogs (Figure 13.1) have been prescribed for various neurological diseases. For example, the anticonvulsant pregabalin is used to treat anxiety disorders such as posttraumatic stress disorder,[2] as well as for diabetic peripheral neuropathy, postherpetic neuralgia, and posttraumatic neuropathic pain.[3,4] In addition, gabapentin is effective for alcohol withdrawal,[5] transient insomnia,[6] and headache after spinal anesthesia.[7] Furthermore, both gabapentin and pregabalin were reported to significantly improve pain, sleep quality, and depression in hemodialysis patients with painful peripheral neuropathy.[8] It has been demonstrated that gabapentin increases the cerebral GABA level as a possible mechanism of action.[9] It has also been reported that baclofen, a GABA agonist, improves binge eating,[10] alcohol dependence,[11] persistent hiccups,[12] and complex regional pain syndrome.[13] It has also been found that some neurological medicines, which structurally differ from GABA, alter the cerebral GABA level. For instance, the benzodiazepine receptor agonist zolpidem elevates brain GABA levels in patients with depression.[14]

Since 2000, the development of food-derived GABA for dietary supplements and processed foods has been attracting much attention in Korea and Japan. These products may alleviate stress and have an antihypertensive effect. In this section, I describe recent studies into the influence of food-derived GABA on mental symptoms.

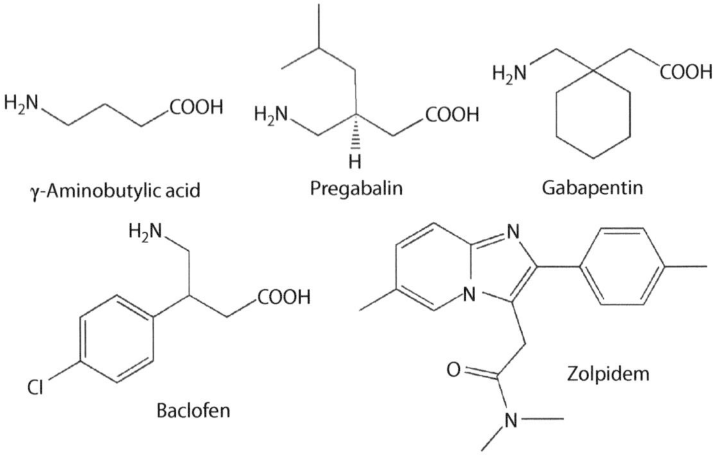

FIGURE 13.1   γ-Aminobutyric acid and its analogues prescribed for neurological diseases.

13.2   GABA IN VARIOUS FOODS

Biosynthesis of GABA from glutamic acid occurs in foods, and fermented foods have a GABA content of several percent. There have been many reports about the GABA content in traditional cuisines such as those of Indonesia, Korea, Japan, and China. Tempeh is a traditional Indonesian soybean paste fermented with *Rizopus* species. It has a GABA content of approximately 1.5%,[15] which varies depending on the species of *Rhizopus* used for fermentation.[16] In Korean cuisine, fermentation is crucial for traditional seasonings, and pickled vegetables. Kimchi (pickled Chinese cabbage) is widely popular among both Koreans and Japanese. During the fermentation of lactate by *Lactobacillus*, GABA gradually accumulates.[17] Recently, various attempts have been made to increase the GABA content in foods by using microorganisms isolated from fermented foods (Table 13.1).[18] Lu *et al.*[19] used *Lactococcus lactis* isolated from Kimchi to ferment sodium glutamate and developed a method that achieved a GABA content of 1.5%. In addition, Lee *et al.*[20] isolated *Lactobacillus brevis* BJ20 from Jot-gal, salty fermented cod gut, and developed a procedure of enriching GABA in sea tangle for dietary supplements.[20] Other Korean

TABLE 13.1
Selected Microorganisms Producing GABA

Name	Source	GABA Production
Lactobacillus paracasei NFRI 7415	Fermented fish (Funa-sushi)	31.1 g/L
Micobacterium purpureus CMU001	Not mentioned	28.3
Lactobacillus brevis NCL 912	Paocai	35.6, 103.7
Lactobacillus brevis GABA 057	Not mentioned	23.3
Lactobacillus buchneri MS	Kimchi	25.8

Source: Dhakal, R., Bajpai, V.K., Baek, K.H., *Braz. J. Microbiol.*, 43, 1230–1241, 2012.

researchers have also made many attempts to increase the GABA content of brown rice by using *Lactobacillus* species. Moo-Chang *et al.*[21] used *Lactobacillus* sakei B2–16 for fermentation of rice bran extract and succeeded in raising the GABA content by 2.4-fold. On the other hand, Seo *et al.*[22] developed new fermented beverages using Kimchi-derived *Lactobacillus brevis* and skim milk.

Developing GABA enrichment methods has also been popular in Japan over the past decade, such as Shochu liquor fermented by *Aspergillus* (0.054%),[23] Japanese citrus (*Satsuma mandarin*, 0.48%),[24] and Funa-zushi, fermented rice with freshwater fish.[25] However, the most extensively studied fermented foods are rice germ and brown rice.[26] At agricultural research centers in several Japanese prefectures, GABA enrichment methods have been developed using locally grown brands of rice.[27,28] Similar attempts have been made by Korean researchers.[29] Although there have only been a few studies of fermented Chinese foodstuffs, the content of GABA in highly fermented tea (Pu-Erh Tea)[30] and pickled vegetables[31] has been reported.

13.3   INFLUENCE OF GABA ON BRAIN FUNCTION IN RODENTS

Dietary intake of GABA has been confirmed to affect cerebral hormones and neurotransmitters in rats. Dietary supplementation (addition of 0.5% GABA to a 20% casein diet) increased the nerve growth factor mRNA level in the brain[32] and growth hormone protein synthesis[33] in ovariectomized female rats. The same researchers also confirmed this effect after a single oral dose of GABA.[34] Thanapreedawat *et al.* fed a diet and water containing 0.5% or 1% GABA to rats for one month and evaluated the changes in memory.[35] Using the novel objective recognition (NOR) test and T-maze test, they showed that object information was retained in the NOR test, and the recognition index was significantly increased after ingestion of GABA (Table 13.2). The accuracy rate was also increased significantly after GABA ingestion (Table 13.3). They concluded that GABA may be involved in long-term object

TABLE 13.2

Effect of GABA Supplementation on Exploring the Object (A1) in the Sample Phase (T1) and Time Spent Exploring the Familiar Object in the Choice Phase (T2)

	Exploration Time	
	Trial 1 (Object A1, Sample Phase)	Trial 2 (Familiar Object A1, Choice Phase)
Control	19.7 ± 1.9	18.6 ± 2.1
0.5% GABA	25.0 ± 1.4	14.9 ± 1.4
1% GABA	24.7 ± 4.0	11.5 ± 2.0*

Source: Thanapreedawat, P., Kobayashi, H., Inui, N., Sakamoto, K., Kim, M., Yoto, A., Yokogoshi, H., *J. Nutr. Sci. Vitaminol.*, 59, 152–157, 2013.

Note: Values are means and SE, $n = 8$.

* Significantly different from the control group at $p < .05$.

TABLE 13.3
Effect of GABA Supplementation on the Recognition Time and Accuracy Rate

	Recognition Time (%)	Accuracy Rate (%)
Control	53.8 ± 1.7	54.2 ± 3.7
0.5% GABA	60.8 ± 1.2*	70.0 ± 3.1*
1% GABA	66.5 ± 2.5*	60.0 ± 2.5*

Source: Thanapreedawat, P., Kobayashi, H., Inui, N., Sakamoto, K., Kim, M., Yoto, A., Yokogoshi, H., *J. Nutr. Sci. Vitaminol.*, 59, 152–157, 2013.

Note: The recognition time (%) after 48 hours of retention and the accuracy rate (%) at a delay time of 900 seconds in rats from the control, 0.5% GABA, and 1% GABA groups. Values are means and SE, *n* = 8.

* Significantly different from the control group at *p* < .05.

recognition memory and working memory. In addition, our research group evaluated the effect of rice-derived GABA on mental fatigue in mice. Reserpine was employed to deplete neurotransmitters and induce depression in mice, which were fed rice bran enriched in GABA and performed forced swimming for 10 minutes while the immobile time was measured during swimming. As a result, reserpine increased the immobile time, while mice given rice bran containing 5% GABA showed dose-dependent shortening of the immobile time (Figure 13.2). The immobile time of mice given 100 mg/kg of GABA was approximately 90 seconds shorter than that of the mice injected with reserpine and the difference was significant (*p* < .05). Based on such evidence, dietary ingestion of GABA seems to affect autonomic function and emotion. However, it is considered that GABA does not cross the blood–brain barrier except under specific conditions.[36,37] Boonstra et al.[38] discussed the

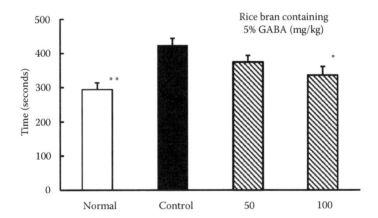

FIGURE 13.2 Effect of GABA-enriched rice bran on the forced swimming test (immobile time in 10 minutes). Data represent the mean and SE of seven to nine mice. Asterisks denote significant differences from the reserpine group at *p* < .05 and **p* < .01.

contradictory effects and influence of GABA on brain function. A recent report also suggested that GABA may indirectly affect brain function via the involvement of growth hormone.[39]

13.4   CLINICAL TRIAL OF DIETARY SUPPLEMENTATION WITH GABA

As described previously, it has not been clarified how orally ingested GABA affects neurotransmitters and receptors in the brain, although several studies have shown that GABA improves neurologic symptoms in humans. Yoto *et al.*[40] administered GABA (100 mg) to 63 subjects and performed both psychological and physiological tests to impose mental stress. As a result, GABA suppressed the decrease of α-waves and β-waves due to the mental stress of the task. The effect of GABA was rapid, being confirmed only 30 minutes after its intake. Similarly, Abdou *et al.*[41] reported that GABA intake significantly increased α-waves and decreased β-waves. These findings suggest that GABA can induce relaxation and improve anxiety. During the previous studies, GABA reduced the salivary immunoglobulin A level, which was increased by stress. Based on such evidence, a confectionary company launched a chocolate containing GABA in Japan several years ago. The psychological stress-reducing effect of this chocolate was confirmed by using the changes in heart rate variability (HRV) and salivary chromogranin A (CgA) during an arithmetic task that was performed to induce stress.[42] Subjects ingested 10 g of chocolate containing 28 mg of GABA and then performed the arithmetic task for 15 minutes. HRV was determined from the electrocardiogram and autonomic nervous activity was estimated on the basis of HRV. The results showed that subjects who ingested the GABA-containing chocolate quickly recovered from the stressful state to normal. Salivary CgA has been recognized as a marker of psychological stress by other researchers.[43] Moreover, the effect of vegetable tablets containing GABA on the autonomic nervous system has been reported in young adults.[44] In a double-blind, randomized controlled trial, a single dose of vegetable tablets containing GABA (31.8 mg) suppressed sympathetic activity, leading to a decrease in blood pressure. On the other hand, Kanehira *et al.*[45] reported the effect of a beverage containing GABA on psychological fatigue. To evaluate the effect of GABA, subjects were assigned an arithmetic task from the Uchida-Kraepelin Psychodiagnostic Test (UKT). Subjects who ingested a beverage containing 50 mg of GABA showed improvement of psychological fatigue as assessed on a visual analogue scale (VAS), and analysis of the Profile of Mood States also indicated that psychological fatigue was significantly improved in the subjects who drank GABA (Table 13.4). Furthermore, the salivary CgA level and cortisol level were significantly lower in the subjects ingesting GABA than in the control group. Moreover, the GABA group had a higher UKT score than the control group. These findings suggest that GABA may reduce both psychological and physical fatigue, as well as causing the improvement of task-solving ability.

In a study employing electroencephalography, oral administration of GABA (100 mg) shortened sleep latency,[46] and answers to questionnaires by the subjects showed that GABA improved their sleep. These results indicate that GABA can help people to fall asleep more quickly.

TABLE 13.4
Profile of Mood State T-Scores Before and After Administration of the UKT

Mood		Before the UKT	After the UKT	Change	P Value vs. Control
Tension-anxiety	Control	56.1 ± 4.7	58.8 ± 5.1	2.7 ± 3.9	
	GABA 25 mg	53.5 ± 4.9	54.8 ± 4.8	1.3 ± 0.8	.9247
	GABA 50 mg	55.2 ± 4.6	54.8 ± 4.7	−0.3 ± 1.5	.7003
Depression-dejection	Control	56.8 ± 4.7	60.1 ± 4.5	3.2 ± 1.9	
	GABA 25 mg	54.2 ± 5.0	57.0 ± 4.5	2.7 ± 1.7	.9855
	GABA 50 mg	56.8 ± 4.5	55.5 ± 5.1	−1.3 ± 1.4	.2463
Anger-hostility	Control	49.8 ± 4.1	50.4 ± 4.3	0.5 ± 0.8	
	GABA 25 mg	50.6 ± 4.8	50.4 ± 4.8	−0.2 ± 0.9	.8611
	GABA 50 mg	52.2 ± 4.9	52.1 ± 5.4	−0.1 ± 0.7	.8164
Vigor	Control	42.5 ± 4.5	45.3 ± 4.4	2.7 ± 0.8	
	GABA 25 mg	41.7 ± 4.5	43.2 ± 4.5	1.4 ± 0.4	.6072
	GABA 50 mg	41.8 ± 4.2	42.4 ± 4.5	0.5 ± 1.2	.2693
Fatigue	Control	56.4 ± 4.9	60.4 ± 4.4	4.0 ± 1.3	
	GABA 25 mg	57.7 ± 4.2	60.3 ± 3.9	2.5 ± 0.7	.4746
	GABA 50 mg	58.6 ± 4.0	53.5 ± 4.3	−5.1 ± 0.8	.0001
Confusion	Control	57.1 ± 5.2	57.3 ± 5.3	0.2 ± 1.3	
	GABA 25 mg	55.6 ± 4.4	54.3 ± 4.6	−1.3 ± 0.9	.6499
	GABA 50 mg	58.1 ± 4.5	59.1 ± 4.0	1 ± 0.9	.8958

Source: Kanehira, T., Nakamura, Y., Nakamura, K., Horie, K., Horie, N., Furugori, K., Sauchi, Y., Yokogoshi, H., *J. Nutr. Sci. Vitaminol.*, 57, 9–15, 2011.
Note: Values are means and SE, *n* = 9.

In our laboratory, we conducted a joint research with the Japan Ministry of Agriculture and the Osaka University School of Medicine to evaluate the effect of rice-germ-derived GABA on the mental state. In a placebo-controlled, double-blind study, GABA-enriched rice germ was given to subjects with insomnia, depression, and autonomic dysfunction. After administration of the GABA-enriched rice germ, all of the symptoms of the subjects showed improvement, including agitation, tiredness, insomnia, depression, and vasomotor instability (Figure 13.3). The effects were statistically significant after eight weeks. In addition, all symptoms became worse again after the administration of a placebo. These results show that the rice-germ-enriched GABA achieved a similar effect to medical GABA on mental symptoms that are considered difficult to treat. Moreover, no side effects of the GABA-enriched rice germ were detected.

We conducted a systematic review of the influence of GABA on the mental state. Seven articles were identified that reported placebo-controlled randomized clinical trials of GABA in healthy subjects. The characteristics of these seven trials are summarized in Table 13.5, while the risk of bias and a summary of the results are displayed in Tables 13.6 and 13.7, respectively. Subjects who took 100 mg of GABA showed significant improvement of α-wave activity (55% or

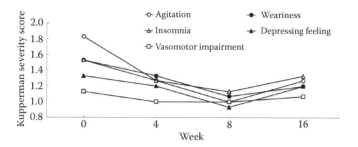

FIGURE 13.3   Effect of GABA-enriched rice germ on the Kupperman severity score.

TABLE 13.5

References Used for a Systematic Review of the Influence of GABA on Mental Symptoms

No.	Reference	Design	Dosage	Primary Outcome	Secondary Outcome
1	Abdou et al., BioFactors, 26, 201–208 (2006)[41]	Single-blind, parallel	100 (mg)	α-wave, β-wave	–
2	Fujibayashi et al., J. Jpn. Soc. Nutr. Food Sci., 61, 129–133 (2008), in Japanese[47]	Double-blind, cross-over	30	Heartbeat	Total autonomic nervous activity
3	Nakamura et al., Int. J. Food Sci. Nutr., Suppl. 5, 106–113 (2009)[42]	Double-blind, cross-over	28	Heart rate	CgA
4	Kanehira et al. J. Nutr. Sci. Vitaminol., 57, 9–15 (2011)[45]	Single-blind, cross-over	50	Fatigue (VAS), Profile of Mood States	CgA, cortisol
5	Yoto et al., Amino Acid, 43, 1331–1337 (2012)[40]	Single-blind, cross-over	100	α-wave	β-wave
6	Yoto et al., Jpn. J. Physiol. Anthropol., 14, 55–59 (2009), in Japanese[48]	Single-blind, cross-over	25	CgA, cortisol	–
7	Yamatsu et al., Jpn. Pharmacol. Ther., 43, 515–519 (2015)[49]	Double-blind, cross-over	28	CgA, fatigue (VAS)	–

15 μV^2, $p < .05$) compared to the placebo group. In another study,[45] a lower dose of GABA (50 mg) significantly reduced CgA (−70%, $p < .05$) and cortisol (−101%, $p < .01$). Parasympathetic nerve activity (PSNA) was significantly maintained (0.12%, $p < .05$) with 28 mg of GABA. However, the effect is questionable because the other report denied the effect of GABA at similar dose. VAS-based evaluation of mental fatigue revealed that 28 and 50 mg of GABA significantly alleviated stress ($p < .05$). In conclusion, a dose of about 30 mg of GABA can improve awareness,

Phytopharmaceuticals for Brain Health

TABLE 13.6

Evaluation of Bias in the Studies

Reference	Selection Bias		Blinding Bias		Risk of Bias Reduced Number Bias		Reported Selective Outcomes	Other Biases	Conclusion
	Randomization Appropriate	Allocation Concealed	Subjects Blinded	Outcome Assessor Blinded	ITT, FAS, PPS	Incomplete Outcome Data			
Abdou et al. (2006)[41]	−1	−1	−1	−1	−1	0	−1	−1	−1
Fujibayashi et al. (2008)[47]	0	0	0	0	−1	0	−1	−1	−1
Nakamura et al. (2009)[42]	−1	−1	0	0	−1	0	−1	−1	−1
Kanehira et al. (2011)[45]	−1	0	0	−1	−2	0	−1	−2	−1
Yoto et al. (2012)[40]	0	0	−1	−1	−1	0	0	−2	−1
Yoto et al. (2009)[48]	0	0	−1	−1	−2	−2	0	−1	−1
Yamatsu et al. (2015)[49]	0	0	0	−1	−2	−1	−1	−1	−1

Note: High: −2, middle: −1, low: 0.

Abbreviations: FAS: full analyst set, ITT: Intention-to-treat, PPS: protocol set.

TABLE 13.7

Summary of Changes in Psychological Parameters After Intake of GABA

	Dose of GABA (mg)	Outcome	Values Before and After Intake								Difference of Change between Active and Placebo	P Value
			Placebo (Before)	Placebo (After)	Difference	P Value	Active (Before)	Active (After)	Difference	P Value		
Abdou et al. (2006)	100	α-wave (%)	100	105	5	–	100	160	60	–	55	<.05
Yoto et al. (2012)	100	α-wave (μV²)	144	94	–50	<.01	145	110	–35	<.05	15	<.05
Fujibayashi et al. (2008)	30	PSNA (%) from heart rate	100	134.9	34.9	–	100	200.3	100.3	<.05	65.4	–
Nakamura et al. (2009)	28	PSNA (%) from heart rate variability	0.46	0.34	–0.12	<.05	0.45	0.45	0	–	0.12	<.05
Nakamura et al. (2009)	28	CgA (pmol/mL)	2.30	4.19	1.89	<.05	2.87	3.17	0.30	–	–1.59	–
Yoto et al. (2009)	25	CgA (ratio)	1	1.6	0.6	–	1	1.2	0.2	–	–0.4	–
Yamatsu et al. (2015)	28	CgA (%)	100	88	–12	–	100	86	–14	–	–2	–
Kanehira et al. (2011)	50	CgA (%)	100	150	50	–	100	80	–20	–	–70	<.05

(Continued)

TABLE 13.7 (CONTINUED)
Summary of Changes in Psychological Parameters After Intake of GABA

	Dose of GABA (mg)	Outcome	Values Before and After Intake								Difference of Change between Active and Placebo	P Value
			Placebo (Before)	Placebo (After)	Difference	P Value	Active (Before)	Active (After)	Difference	P Value		
Kanehira et al. (2011)	50	Cortisol (%)	100	175.6	75.6	—	100	73.9	−26.1	—	−101.7	<.01
Yoto et al. (2009)	25	Cortisol (ratio)	1	0.97	−0.03	—	1	1.02	0.02	—	0.05	—
Yamatsu et al. (2015)	28	Fatigue (VAS)	0	27	27	—	0	12	12	—	−15	<.05
Kanehira et al. (2011)	50	Fatigue (VAS)	0	9.9	9.9	—	0	−10.2	−10.2	—	−20.1	<.05

Note: High: −2, middle: −1, low: 0.

while 50 mg or more is required to alter stress biomarkers such as CgA, cortisol, and brain wave activity.

13.5 CONCLUSION

Whether GABA crosses the blood–brain barrier is still controversial. However, clinical studies mainly conducted in Japan have revealed that oral ingestion of GABA can improve stress biomarkers. In 2015, the Japanese government initiated a new approval system for practical health claims based on clinical results and safety. GABA is one of the dietary ingredients that have been approved for the health claim of alleviating stress. Based on this system, raw materials suppliers and processed food companies have been trying to enhance the development of new processed foods containing GABA. We hope that these attempts result in increased use of GABA again as happened at the beginning of this century.

REFERENCES

1. Luján, R., Shigemoto, R., López-Bendito, G. 2005. Glutamate and GABA receptor signalling in the developing brain. *Neuroscience* 130:567–80.
2. Baniasadi, M., Hosseini, G., Fayyazi, B.M.R., Rezaei, A.A., Mostafavi, T.H. 2014. Effect of pregabalin augmentation in treatment of patients with combat-related chronic posttraumatic stress disorder: A randomized controlled trial. *J. Psychiatr. Pract.* 20:419–27.
3. Moon, D.E., Lee, D.I., Lee, S.C., Song, S.O., Yoon, D.M., Yoon, M.H., Kim, H.K., Lee, Y.W., Kim, C., Lee, P.B. 2010. Efficacy and tolerability of pregabalin using a flexible, optimized dose schedule in Korean patients with peripheral neuropathic pain: A 10-week, randomized, double-blind, placebo-controlled, multicenter study. *Clin. Ther.* 32:2370–85.
4. van Seventer, R., Bach, F.W., Toth, C.C., Serpell, M., Temple, J., Murphy, T.K., Nimour, M. 2010. Pregabalin in the treatment of post-traumatic peripheral neuropathic pain: A randomized double-blind trial. *Eur. J. Neurol.* 17:1082–9.
5. Leung, J.G., Hall-Flavin, D., Nelson, S., Schmidt, K.A., Schak, K.M. 2015. The role of gabapentin in the management of alcohol withdrawal and dependence. *Ann. Pharmacother.* 49:897–906.
6. Rosenberg, R.P., Hull, S.G., Lankford, D.A., Mayleben, D.W., Seiden, D.J., Furey, S.A., Jayawardena, S., Roth, T. 2014. A randomized, double-blind, single-dose, placebo-controlled, multicenter, polysomnographic study of gabapentin in transient insomnia induced by sleep phase advance. *J. Clin. Sleep Med.* 10:1093–100.
7. Vahabi, S., Nadri, S., Izadi, F. 2014. The effects of gabapentin on severity of post spinal anesthesia headache. *Pak. J. Pharm. Sci.* 27:1203–7.
8. Biyik, Z., Solak, Y., Atalay, H., Gaipov, A., Guney, F., Turk, S. 2013. Gabapentin versus pregabalin in improving sleep quality and depression in hemodialysis patients with peripheral neuropathy: A randomized prospective crossover trial. *Int. Urol. Nephrol.* 45:831–7.
9. Cai, K., Nanga, R.P., Lamprou, L., Schinstine, C., Elliott, M., Hariharan, H., Reddy, R., Epperson, C.N. 2012. The impact of gabapentin administration on brain GABA and glutamate concentrations: A 7T ^{1}H-MRS study. *Neuropsychopharmacology.* 37:2764–71.
10. Corwin, R.L., Boan, J., Peters, K.F., Ulbrecht, J.S. 2012. Baclofen reduces binge eating in a double-blind, placebo-controlled, crossover study. *Behav. Pharmacol.* 23: 616–25.

11. Garbutt, J.C., Kampov-Polevoy, A.B., Gallop, R., Kalka-Juhl, L., Flannery, B.A. 2010. Efficacy and safety of baclofen for alcohol dependence: A randomized, double-blind, placebo-controlled trial. *Alcohol. Clin. Exp. Res.* 34:1849–57.

12. Zhang, C., Zhang, R., Zhang, S., Xu, M., Zhang, S. 2014. Baclofen for stroke patients with persistent hiccups: A randomized, double-blind, placebo-controlled trial. *Trials* 15:295.

13. van der Plas, A.A., van Rijn, M.A., Marinus, J., Putter, H., van Hilten, J.J. 2013. Efficacy of intrathecal baclofen on different pain qualities in complex regional pain syndrome. *Anesth. Analg.* 116:211–5.

14. Licata, S.C., Jensen, J.E., Conn, N.A., Winer, J.P., Lukas, S.E. 2014. Zolpidem increases GABA in depressed volunteers maintained on SSRIs. *Psychiatry Res.* 224:28–33.

15. Ishikawa, A., Oka, H., Himemori, M., Yamashita, H., Kimoto, M., Kawasaki, H., Tsuji, H. 2009. Development of a method for the determination of γ-aminobutyric acid in foodstuffs. *J. Nutr. Sci. Vitaminol.* 55:292–5.

16. Aoki, H., Uda, I., Tagami, K., Furuya, Y., Endo, Y., Fujimoto, K. 2003. The production of a new Tempeh-like fermented soybean containing a high level of γ-aminobutyric acid by anaerobic incubation with *Rhizopus. Biosci. Biotechnol. Biochem.* 67:1018–23.

17. Ran, C.Y., Chang, J.Y., Chang, C.C. 2007. Production of γ-aminobutyric acid (GABA) by *Lactobacillus buchneri* isolated from Kimchi and its neuroprotective effect on neuronal cells. *J. Microbiol. Biotechnol.* 17 104–9.

18. Dhakal, R., Bajpai, V.K., Baek, K.H. 2012. Production of GABA (γ-aminobutyric acid) by microorganisms: A review. *Braz. J. Microbiol.* 43:1230–41.

19. Lu, X.X., Xie, C., Gu, Z.X. 2009. Optimization of fermentative parameters for GABA enrichment by *Lactococcus lactis. Czech J. Food Sci.* 27:433–42.

20. Lee, B.J., Kim, J. S., Kang, Y.M., Lim, J.H., Kim, Y.M., Lee, M.S., Jeong, M.H., Ahn, C.B., Je, J.Y. 2010. Antioxidant activity and γ-aminobutyric acid (GABA) content in sea tangle fermented by *Lactobacillus brevis* BJ20 isolated from traditional fermented foods. *Food Chem.* 122:271–6.

21. Moo-Chang, K., Seo, M.J., Cheigh, C.I., Pyun, Y.R., Cho, S.C., Park, H. 2010. Enhanced production of γ-aminobutyric acid using rice bran extracts by *Lactobacillus sakei* B2–16. *J. Microbiol. Biotechnol.* 20:763–6.

22. Seo, M.J., Nam, Y.D., Park, S.L., Lee, S.Y., Yi, S.H., Lim, S.I. 2013. γ-Aminobutyric acid production in skim milk co-fermented with *Lactobacillus brevis* 877G and *Lactobacillus sakei* 795. *Food Sci. Biotechnol.* 22:751–5.

23. Tsuchiya, K., Matsuda, S., Ishida, A., Iwahara, M. 2006. Scale-up test of reactor using *Aspergillus mycelium* and the effect of shochu residue including GABA on blood pressure of SHR rats. *Kumamoto Industrial Research Institute Report* (44):39–42 (in Japanese).

24. Ono, K., Sasayama, S., Hiraki, T. 2011. Studies on the efficient accumulation of γ-aminobutyric acid by the Satsuma mandarin γ-aminobutyric acid wealth method. Ehime *Institute of Industrial Technology Report* (49):1–5 (in Japanese).

25. Komatsuzaki, N., Shima, J., Kawamaoto, S., Momose, H., Kimura, T. 2005. Production of γ-aminobutyric acid (GABA) by *Lactobacillus paracasei* isolated from traditional fermented foods. *Food Microbiol.* 22:497–504.

26. Charoenthaikij, P., Jangchud, K., Jangchud, A., Prinyawiwatkul, W., Tungtrakul, P. 2010. Germination conditions affect selected quality of composite wheat-germinated brown rice flour and bread formulations. *J. Food Sci.* 75:S312–8.

27. Lu, Q., Goto, K., Nishizu, T. 2010. Study on the conditions and quality of GABA enriched brown rice. *J. Japan. Soc. Agric. Machinery* 72:291–6.

28. Yoshida, S., Haramoto, M., Fukuda, T., Mizuno, H., Tanaka, A., Nishimura, M., Nishihara, J. 2015. Optimization of a γ-aminobutyric acid (GABA) enrichment process for Hokkaido white rice and the effects of GABA-enriched white rice on stress relief in humans. *J. Jpn. Soc. Food Sci. Technol.* 62:95–103.

29. Kook, M.C., Cho, S.C. 2013. Production of GABA (gamma amino butyric acid) by lactic acid bacteria. *Korean J. Food Sci.* 33:377–89.
30. Jeng, K.C., Chen, C.S., Fang, Y.P., Hou, R.C.W., Chen, Y.S. 2007. Effect of microbial fermentation on content of statin, GABA, and polyphenols in Pu-Erh tea. *J. Agric. Food Chem.* 55:8787–92.
31. Lin, Q. 2013. Submerged fermentation of *Lactobacillus rhamnosus* YS9 for γ-aminobutyric acid (GABA) production. *Braz. J. Microbiol.* 44:183–7.
32. Tujioka, K., Thanapreedawat, P., Yamada, T., Yokogoshi, H., Horie, K., Kim, M., Tsutsui, K., Hayase, K. 2014. Effect of dietary γ-aminobutyric acid on the nerve growth factor and the choline acetyltransferase in the cerebral cortex and hippocampus of ovariectomized female rats. *J. Nutr. Sci. Vitaminol. (Tokyo).* 60:60–5.
33. Tujioka, K., Ohsumi, M., Horie, K., Kim, M., Hayase, K., Yokogoshi, H. 2009. Dietary γ-aminobutyric acid affects the brain protein synthesis rate in ovariectomized female rats. *J. Nutr. Sci. Vitaminol. (Tokyo).* 55:75–80.
34. Tujioka, K., Okuyama, S., Yokogoshi, H., Fukaya, Y., Hayase, K, Horie, K, Kim, M. 2007. Dietary γ-aminobutyric acid affects the brain protein synthesis rate in young rats. *Amino Acids* 32:255–60.
35. Thanapreedawat, P., Kobayashi, H., Inui, N., Sakamoto, K., Kim, M., Yoto, A., Yokogoshi, H. 2013. GABA affects novel object recognition memory and working memory in rats. *J. Nutr. Sci. Vitaminol.* 59:152–7.
36. Remler, M.P., Marcussen, W.H. 1983. A GABA-EEG test of the blood–brain barrier near epileptic foci. *Appl. Neurophysiol.* 1983 46:276–85.
37. Kakee, A., Takanaga, H., Terasaki, T., Naito, M., Tsuruo, T., Sugiyama, Y. 2001. Efflux of a suppressive neurotransmitter, GABA, across the blood–brain barrier. *J. Neurochem.* 79:110–8.
38. Boonstra, E., de Kleijn, R., Colzato, L.S., Alkemade, A., Forstmann, B.U., Nieuwenhuis, S. 2015. Neurotransmitters as food supplements: The effects of GABA on brain and behavior. *Front Psychol.* doi: 10.3389/fpsyg.2015.01520.
39. Powers, M. 2012. GABA supplementation and growth hormone response. *Med. Sport. Sci.* 59:36–46.
40. Yoto, A., Murao, S., Motoki, M., Yokoyama, Y., Horie, N., Takeshima, K., Masuda, K., Kim, M., Yokogoshi, H. 2012. Oral intake of γ-aminobutyric acid affects mood and activities of central nervous system during stressed condition induced by mental tasks. *Amino Acids* 43:1331–7.
41. Abdou, A.M., Higashiguchi, S., Horie, K., Kim, M., Hatta, H., Yokogoshi, H. 2006. Relaxation and immunity enhancement effects of γ-aminobutyric acid (GABA) administration in humans. *Biofactors* 26:201–8.
42. Nakamura, H., Takishima, T., Kometani, T., Yokogoshi, H. 2009. Psychological stress-reducing effect of chocolate enriched with γ-aminobutyric acid (GABA) in humans: assessment of stress using heart rate variability and salivary chromogranin A. *Int. J. Food Sci. Nutr.* 60 Suppl. 5:106–13.
43. Kanamaru, Y., Kikukawa, A., Shimamura, K. 2006. Salivary chromogranin-A as a marker of psychological stress during a cognitive test battery in humans. *Stress* 9:127–31.
44. Okita, Y., Nakamura, H., Kouda, K., Takahashi, I., Takaoka, T., Kimura, M., Sugiura, T. 2009. Effects of vegetable containing γ-aminobutyric acid on the cardiac autonomic nervous system in healthy young people. *J. Physiol. Anthropol.* 28:101–7.
45. Kanehira, T., Nakamura, Y., Nakamura, K., Horie, K., Horie, N., Furugori, K., Sauchi, Y., Yokogoshi, H. 2011. Relieving occupational fatigue by consumption of a beverage containing γ-amino butyric acid. *J. Nutr. Sci. Vitaminol.* 57:9–15.
46. Yamatsu, A., Yamashita, Y., Maru, I., Yang, J., Tatsuzaki, J., Kim, M. 2015. The improvement of sleep by oral intake of GABA and *Apocynum venetum* leaf extract. *J. Nutr. Sci. Vitaminol.* 61:182–7.

47. Fujibayashi, M., Kamiya, T., Takagaki, K., Moritani, T. 2008. Activation of the autonomic nervous system by the oral ingestion of GABA. *J. Jpn. Soc. Nutr. Food Sci.* (in Japanese) 61:129–33.
48. Yoto, A., Ishihara, S., Li, Y.J., Butterweck, V., Yokogoshi, H. 2009. The stress reducing effect of γ-aminobutyric acid and *Apocynum venetum* leaf extract on changes in the salivary concentration of chromogranin. *Jpn. Soc. Physical Anthropol.* (in Japanese) 14:55–9.
49. Yamatsu, A., Yamashita, Y., Shibata, M., Yoneyama, M., Horie, K., Horie, N., Kim., M., Yokogoshi, H. 2015. The beneficial effects of coffee on stress and fatigue can be enhanced by addition of GABA. *Jpn. Pharmacol. Ther.* 43:515–9.

14 Walnut (*Juglans regia* L) and Its Neuroprotective Action Mechanism on Brain
A Review

Abdul Ilah, Akhtar uz Zaman, Sabia Bano,
Faisal Ismail, and Shahnaz Subhan

CONTENTS

14.1   INTRODUCTION

The Central Asian origin *Juglans regia* L is reported to be distributed throughout the world about 700 years ago [1]. The extraordinary nutritional value and clinical potential perhaps had been the pivotal force of its global distribution since ancient age, and recently, the Food and Agriculture Organization of the United Nations also placed this plant as strategic species for human nutrition [2]. Walnut is a highly nutritive dry fruit of *J. regia* L produced in the month of September through October. It belongs to the family Juglandaceae, and the genus *Juglans* consists of at least 20 sister species [3–5]. The beneficial effect of walnut consumption is to treat wide a variety of human diseases as traditional folk medicine; thus it gradually has drawn special attention of researchers and

Ayurvedic doctors. Researchers and doctors have cited several experimental evidences of remedial potentials of walnut's active constituents by solving various cognitive disorders in human and human brains [6–9]. The diverse range of chemical constituents such as caffeine, polyphenols, numerous ellagitannins, and unsaturated fatty acids (FAs) had been used as antioxidant, anti-inflammatory, and neuroprotective drugs [10,11]. The sources of chemical constituents in *J. regia* are diversely distributed throughout plant parts, such as nuts, green walnuts, shells, barks, pericarp, and leaves. The constituents are scientifically harvested from all these parts of the plant and used as different medical agents such as cosmetics, astringent, digestive, diuretic, for hypertension, for diabetes, antioxidant, brain tonic, etc. [12–14].

The learning and memory function disorder has been cured and maintained with judicial administrations of *J. regia* herbal preparations to the patients [15,16]. There is evidence that the bark of *J. regia* has a high concentration of phenolics compounds with antioxidant, hypoglycemic, and hypocholesterolemic effects in animals [17]. In ancient Chinese medicines, *J. regia* leaves were prescribed to treat skin inflammations, venous insufficiency, hypertension, and ulcers [18,19]. Epidemiological studies suggested that alpha tocopherol, a vitamin E family found in walnut, has antioxidant activity, mainly the prevention of lipid oxidations process. Numerous scientific reports imply that walnut intake is a potential treatment for a variety of neurodegenerative and neurological disorders. Several in vivo and in vitro studies suggested that the active ingredient of *J. regia* could inhibit neurodegenerative disorder and even reverses its normal functions [10,20].

14.2 BIOCHEMISTRY OF *J. REGIA* L

J. regia L contains various chemicals such as FAs, lipids, carbohydrates, proteins, amino acids, phenols, flavones, vitamins, minerals, etc. The biochemically active and potential components of Juglans are reported here.

> *FAs: J. regia* L possesses several monounsaturated FA and polyunsaturated FA (PUFA) in seeds. The available monounsaturated FA are oleic acid (C18:1) 25%, palmitoleic acid (C16:1) 0.77%, and gadoleic acid (C20:1) 0.16%. The PUFAs are alpha linoleic acid (C18:2) 57% and linolenic acid (C18:3) 10%. Besides these unsaturated FAs, there are also found some important saturated FA such as myristic acid (C14:0) 0.2%, palmitic acids (C16:0) 4.2%, stearic acid (C18:0) 1.85%, and arachodonic acid (C20:0) 0.19%. The most important, well-known FA is alpha linoleic acid and is present in large amount (57%). Alpha linoleic acid is believed to play a significant role in brain development and memory enhancement (Figure 14.1) [21].
>
> *Proteins:* The protein content of *J. regia* has been recorded in several scientific reports. The most abundant protein available in walnuts is glutelins (70% of the total seed proteins), and other proteins are present in the seed in lesser quantity, such as globulins (18%), albumins (7%), and prolamins (5%) [22]. Walnut flour is enriched with acidic amino acid residue of aspartate and glutamate and present a relatively high level of arginine. Almost all the essential amino acids are present in large amounts, which are very

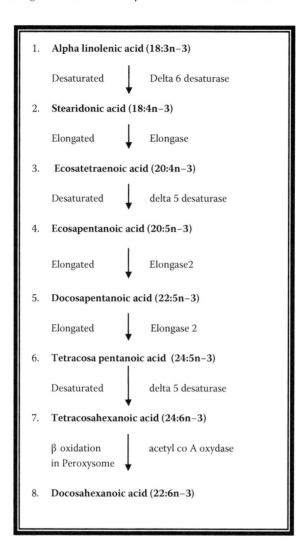

1. **Alpha linolenic acid (18:3n–3)**

    Desaturated   Delta 6 desaturase

2. **Stearidonic acid (18:4n–3)**

    Elongated   Elongase

3.   **Ecosatetraenoic acid (20:4n–3)**

    Desaturated   delta 5 desaturase

4. **Ecosapentanoic acid (20:5n–3)**

    Elongated   Elongase2

5. **Docosapentanoic acid (22:5n–3)**

    Elongated   Elongase 2

6. **Tetracosa pentanoic acid (24:5n–3)**

    Desaturated   delta 5 desaturase

7. **Tetracosahexanoic acid (24:6n–3)**

    β oxidation   acetyl co A oxydase
    in Peroxysome

8. **Docosahexanoic acid (22:6n–3)**

FIGURE 14.1   A glance on metabolic pathways of DHA.

necessary for growth and development of human health. Usually, the lysine and arginine ratio in common vegetable protein is high, but that ratio in *J. regia* is found to be less. This lesser lysine and arginine ratio is considered to be good for health in the reduction of atherosclerosis [23].

Phenolic compounds: Walnut is enriched with polyphenols. The immature fruits, dried leaves, and green husk were used as crude materials for extracting polyphenols through scientific methods for therapeutic purposes. The most frequently available phenolic compounds detected are gallic, protocatechuic, 3-caffeoylquinic, 3-ρ-coumaroylquinic, 4-caffeoylquinic, 4-ρ-coumaroylquinic, and ρ-coumaric acids, as well as quercetin-3-O-deoxyhexoside. Naphthoquinones and flavonoids are also considered to be major phenolic compounds of *J. regia* leaves. 5-hydroxy-1,4-naphthoquinone, also known as juglone, is the most characteristic compound of fresh walnut fruits [24].

Walnut oil and fats: Adequate quantities (approximate 66%) of oils are found to present in walnut of *J. regia*. There are some reports that antioxidant constituents are also found in walnut oil, which have an antioxidant activity. It is thought that walnut oil may be utilized as a source of nutritional antioxidants. The fats are available at about 12.9 g/100 g dry weight in male inflorescence [25].

Carbohydrates: Carbohydrates are also available at about 47.54 g/100 g dry weight in male inflorescence during early flowering stages [25].

Vitamins: Vitamins are one of the most essential organic compounds that are utmost necessary to maintain normal growth and nutrition of the entire body. The walnut is an important source of many such vitamins, for example, niacin (1.12 mg), pantothenic acid (0.570 mg), pyridoxine (0.53 mg), riboflavin (0.150 mg), thiamin (0.541 mg), vitamin A (20 IU), vitamin C (1.3 mg), and vitamin E-y (20.83 mg) [26].

Minerals: For good health, the human body needs a large number of macro-minerals and microminerals; all such minerals have been found in male inflorescence of *J. regia*. These macrominerals are potassium (K, 684.3 mg 100 g^{-1}), phosphorus (P, 356.2 mg 100 g^{-1}), calcium (Ca, 388.2 mg 100 g^{-1}), magnesium (Mg, 330.8 mg 100 g^{-1}), and sodium (Na, 26.1 mg 100 g^{-1}). The microminerals are iron (Fe, 4.3 mg 100 g^{-1}), copper (Cu, 1.8 mg 100 g^{-1}), manganese (Mn, 2.7 mg 100 g^{-1}), and zinc (Zn, 2.7 mg 100 g^{-1}).

Amino acids: The walnut male inflorescences are full of important phytochemicals, and it has been reported that male inflorescence possesses at least 17 different types of amino acids (Table 14.1). This information has been encouraging for both consumers and scientists because of their potential use as nutriceutical. However, as the male inflorescence of walnut flower has a high range phytochemicals with antioxidant capacity, it is suggested that toxicity should be tested to make sure of their safe use as food additives [27].

14.3 PUFA, ABSORPTION, AND METABOLISM

The bioavailability of FAs at the site of action was examined. Radiolabeled 15% linolenic acid (LA) and 3% α-linolenic acid (ALA) were orally administered as a single dose to the experimental model animal [28] and it was found that the brain concentration of LA peaked at eight hours post administration. Although ALA was not detected in brain tissue, ALA metabolites, including ecosapentanoic acid (EPA), docosapentaenoic acid (DPA), and docosahexaenoic acid (DHA), remained elevated up to 25 days post administration [28]. Studies have also shown a decline in FA enzyme activity in the liver of aged animals [29], potentially allowing the presence of ALA and LA in the blood, where it could be taken up in the brain. After consumption of walnut PUFA ALA (18:3n-3; Figure 14.1), it is converted into DHA through a series of sequential desaturation and elongation reactions

TABLE 14.1
Available Amino Acids in Walnut Male Inflorescences

Amino Acid in Walnut	Weight/100 g Dry Weight
Aspartic acid	1.38 g
Essential amino acid threonine	0.68 g
Serine	0.79 g
Glutamic acid	2.01 g
Glycine	0.77 g
Alanine	0.86 g
Cysteine	0.06 g
Essential amino acid, valine	0.71 g
Essential amino acid, methionine	0.21 g
Essential amino acid, isoleucine	0.63 g
Essential amino acid, leucine	0.63 g
Tyrosine	0.45 g
Essential amino acid, phenylalanine	0.92 g
Essential amino acid, histidine	0.45 g
Essential amino acid, lysine	0.65 g
Essential amino acid, arginine	1.33 g
Proline	0.69 g

Source: Wang, C. et al., *J. Food Nutr. Res.* 2 (8), 457-464, 2014.

into essential PUFAs such as EPA (20:5n-3) and DHA (22:6n-3) in the liver. EPA and DHA play an important role in maintaining brain health by reducing oxidative stress and altering the immune function. These PUFAs also play a role in maintaining synaptic plasticity, neuronal membrane stability, gene expression, and neurogenesis [29–31].

14.4   DHA METABOLISM

Step 1. ALA desaturated into stearic acid and subsequently elongated into ecosatetraenoic acid with the help of enzymes delta-6 desaturase, encoded by the FADS2 gene and elongase enzyme.

Step 2. Ecosatetraenoic acid further desaturated into EPA and subsequently elongated into DPA and followed by tetracosapentanoic acid with enzymes delta-5 desaturase encoded by gene FADS 1 and elongase2 enzyme encoded by ELVOL2 gene, respectively. Tetracosapentanoic acid again desaturated into tetracosahexanoic.

All metabolic reactions of step 1 and step 2 take place in the endoplasmic reticulum.

Step 3. Tetracosa hexanoic acid is transported in the peroxysome and catabolise into DHA by single β oxidation reaction with the help of enzymes acyl CoA oxydase encoded by ACOX1 gene and D-bifunctional enzyme encoded by HSD1784 gene and then thiolases (Figure 14.2). A glance on the metabolic pathways of decosahexanoic acid (DHA) is pasted here.

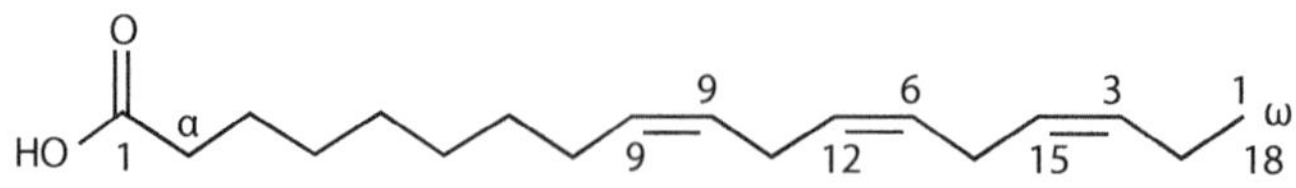

FIGURE 14.2 Chemical structure of ALA, an omega-3 FA (18:3, 9c,12c,15c). A chain of 18 carbons with three double bonds on carbons numbered 9, 12, and 15. Chemists count from the carbonyl carbon , whereas biologists count from the n (ω) carbon. Note that, from the n end (diagram right), the first double bond appears as the third carbon–carbon bond (line segment), hence the name "n-3."

14.5　ABSORPTION AND METABOLISM OF POLYPHENOL

Bioavailability is the quantity of the nutrient that is digested, absorbed, and metabolized in the human body through normal pathways. Most polyphenols are present in food in the form of esters, glycosides, or polymers that cannot be absorbed in native form [17]. Generally, aglycones could be absorbed from the small intestine. Prior to absorption, these compounds are hydrolyzed by intestinal enzymes or by colonic microflora. During the process of the absorption, polyphenols undergo extensive modification; actually, they are conjugated in the intestinal cells, and later, in the liver, they are modified by methylation, sulfation, and/or glucuronidation [18]. The chemical structure of polyphenols determines the rate and extent of absorption circulating in the plasma [32].

14.6　BENEFICIAL EFFECT OF POLYPHENOLS

The polyphenols of walnut have strong antioxidant and anti-inflammatory activity. It also has been shown to improve interneuron signaling, enhance neurogenesis, and have beneficial effects on memory function [33–36]. Previous researches have clearly observed that the antioxidant capability of polyphenols could protect against cognitive impairment [37,38].

Epidemiological studies have constantly shown a contrasting involvement between the risk of chronic human diseases and the eating of polyphenolic-rich diet. The phenolic groups in polyphenols could accept an electron to form relatively stable phenoxyl radicals, in that way disrupting chain oxidation reaction in cellular machinery. It has been clearly proved that polyphenol-rich foods and beverages may increase plasma antioxidant capacity [39,40]. It is well established that consumption of antioxidants has been associated with reduced levels of oxidative damage to lymphocytic DNA. The same observations were noted in polyphenol-rich food and beverages, indicating the protective effects of polyphenols [41]. There are several evidences that have proved that antioxidants and polyphenols may protect cell constituents against oxidative damage and consequently limit the risk of degenerative diseases linked with oxidative stress [42–44].

14.7　BENEFICIAL NEUROPROTECTIVE EFFECT OF FLAVONS

The walnut is full of flavonoids, and its direct effects on impaired neurons by improving their function have been reported several times [45]. The mechanism of neuroprotective actions of dietary flavonoids includes the following: (1) It has great potential to protect neurons against injury induced by neurotoxins and capability to suppress neuroinflammation. (2) It has potential to uphold memory, learning, and cognitive function. These multifaceted actions of flavonoids are accomplished by two processes: (1) by interacting with neuronal signalling cascades in the brain and inhibiting the apoptosis triggered by neurotoxic agents and (2) by upholding the neuronal survival ability and making neuronal differentiation. These included some selective actions on several protein kinase and lipid kinase signalling cascades, especially the phosphoinositide 3-kinase (PI3K)/Akt, also known as protein kinase B, and mitogen-activated protein (MAP) kinase pathways, which are supposed to regulate prosurvival transcription factors and gene expression. It is thought that the concentrations of flavonoids found in the brain might be sufficiently high to make use of such pharmacological activity on receptors, kinases, and transcription factors. And they also are known to induce useful effects on the peripheral and cerebral vascular system, which lead to changes in cerebrovascular blood flow. Such alterations are likely to induce angiogenesis, new nerve cell growth in the hippocampus, and changes in neuronal morphology. The consumption of flavonoid-rich foods, such as berries, nuts, and cocoa, throughout life could hold a potential to limit neurodegeneration and prevent or reverse age-dependent deteriorations in cognitive performance. However, it is unclear about the precise sequential effects of flavonoids on these events. It also is very unclear to say when one needs to start on consuming flavonoids in order to obtain maximum benefits [46].

14.8   ANTI-ALZHEIMER EFFECT OF WALNUT

Walnuts play a pivotal role in preventing Alzheimer disease (AD). Consumption of fruits and vegetable juice enriched with polyphenol was reported to delay the onset of AD, specially with patients who are carriers for the apoliproproteine 4 (APOE4) gene [47]. APOE4 is believed to play a critical role in promoting amyloid accumulation, oxidative stress neurotoxicity, and neurofibrillary tangles [48]. In another similar study, Yasuno et al. [49] reported that daily administration of n-3 PUFAs extracted from fish and *Ginkgo biloba* for three years improved cognitive function in aged persons. Although n-3 PUFAs from animal sources differ from n-3 PUFAs from walnuts, because humans convert ALA from walnuts into EPA and DHA, it could be concluded that walnuts rich in n-3 PUFAs, in combination with other high-antioxidant compounds, can delay the cognitive decline associated with AD [50].

14.9   ANTIDEMENTIA EFFECT OF WALNUT

Melatonin is another active chemical constituent found in walnuts. Melatonin is primarily synthesized by the pineal gland in the brain, and it plays a critical role in regulating circadian rhythms [51]. It is experimentally proven that melatonin deficiency is directly linked to degeneration of cholinergic neurons in the basal forebrain and deposition of amyloid b peptide, which is responsible for cognitive impairment and dementia [52]. Reiter et al. reported that regular consumption of walnuts increases blood melatonin concentrations in association with an increase in "total antioxidant capacity" of the serum, indicating the ability of the blood to detoxify free radicals, leading to improvement in cognitive impairment.

Scientists have observed that 6% walnut supplementation activates two important factors: (1) cyclic adenosine monophosphate (cAMP) response element-binding protein factor and (2) constitutively expressed, nuclear transcription factor that plays an important role in neuronal survival in the hippocampus and striatum [53]. Basically, n-3 FAs are converted to EPA and DHA, which act as precursors for anti-inflammatory eicosanoids and neuroprotection D1 (NPD1) mediator that protect the brain injury, respectively. NPD1 has been shown to reduce the activation of inflammatory signaling mediators such as prostaglandins synthesized from arachidonic acid (long-chain n-6 PUFAs) by cyclooxygenase-2 [54]. Thus, walnut phytochemicals can effectively inhibit prooxidant and proinflammatory mediators, which might be the method of reducing the risk of dementia, seizure, and other neurologic disorders among aged adults.

14.10 WALNUTS AND REDUCED CARDIOVASCULAR RISK FACTORS

Some experimental reports have shown that reductions in cardiovascular disease risk factors are associated with better brain health because cardiovascular disease is also associated with the development of cerebral vascular disease, stroke, and mild cognitive impairment [55,56]. A walnut- and walnut-oil-rich diet reduced inflammatory and cardiovascular risk factors among hypercholesterolemic men and women [57]. In another similar report, it was found that a walnut-enriched diet could also improve endothelium-dependent vasodilatation in type 2 diabetic patients [58] and overweight adults with visceral adiposity [59]. The increased vasculature and better endothelial function have better effect on cerebral health and cognitive function. Generally, frequent walnut intake has been shown to improve cognition among older adults.

14.11 MECHANISM OF DHA ACTIONS ON SPINAL CORD

Spinal cord injury (SCI) occurs very frequently and has a profound impact on both the patient and society. Practical recovery after SCI is often pitiable, and so far, there are no nonsurgical medications. The corticosteroid methyl prednisolone is the only recognized pharmacological treatment that regulates the early inflammatory reaction after SCI in human beings [60]. However, methyl prednisolone is not considered to be the standard care for SCI patients, since its use has severe adverse side effects. Drug administration that is safe and efficacious is vital and ever expected for optimal treatment. In search of potential therapeutic targets, special attention has been given to the molecular events that occur immediately after SCI. Physical damage to neurons, axons, glial cells, and blood vessels induces the discharge of some active biomolecules that lead to glial activation and initiation of secondary inflammation. Various experimental approaches have already been made to obstruct these events, and some have been practiced in human clinical trials. However, some in vivo injury models of SCI were used to repeat some of the typical features of the inflammatory conditions after SCI. Lang-Lazdunski et al. [61] had demonstrated, in an experimental model of transient spinal cord ischemia in rats, that the systematic application of an omega-3 FA, ALA, after injury is found to decrease neuronal loss and improve functional outcome. Moreover, previously, Lim et al. [62] have demonstrated that omega-3 DHA, applied as an intravenous bolus after SCI in rats, was found to induce significant neuroprotection, reduce neuronal cell loss and oligodendrocyte loss, and decrease apoptosis calculated six weeks after injury. Inflammation, oxidative stress, and excitotoxicity are the main factors of the pathogenic mechanism associated with the secondary injury following SCI. It has been shown that DHA can counteract glutamate-induced excitotoxicity. Thus, it is established that DHA has a neuroprotective power and could target the very early events after SCI, within 24 hours. SCI often causes acute loss of motor neurons and degeneration of white matter tracts that lead to irreversible motor dysfunction. The application of DHA has significantly reduced the degree of secondary inflammation and improved motor recovery [63]. SCI triggers the activation of the transcription factor nuclear factor-κB (NF-κB), and this factor is found to play a central role in the regulation of many genes responsible for the synthesis of mediators or proteins involved in secondary inflammation [64]. Now, it has been proved that ω-3 PUFAs inhibit the activation of NF-κB by repressing the phosphorylation of inhibitor kappa B kinase (IK kinase) and preventing the degradation of IκB (inhibitory kappa B), resulting in transactivation of NF-κB target genes [65]. Thus, DHA antagonizes the NF-κB signaling pathway and suppresses the expression of inflammatory genes downstream of NF-κB. Another report confirmed that SCI leads to a substantial increase in the levels of tumor necrosis factor-α (TNF-α). Interestingly, mice with SCI, when treated with DHA, were found to reduce the expression of TNF-α gene. It is well known that NF-κB plays a key role in the regulation of many genes responsible for the production of mediator proteins for inflammation. These genes are particularly for TNF-α, interleukin-1β, cyclooxygenase-2 (COX-2), and inducible nitric acid synthase (iNOS).

Therefore, DHA could be an effective therapy for the inactivation of NF-κB and thereby could control genes like TNF-α, IL-1β, COX-2, and iNOS. Another important mediator of secondary damage after SCI is the apoptosis process [66]. It is experimentally proven that mice with SCI interestingly reduce apoptosis markers when treated with DHA; thus, it is very clear that DHA could be beneficial for the treatment of SCI. Similarly, it has been found that the Bcl-2 gene is expressed in much more higher rate in mice treated with DHA. It can be concluded that DHA is able to slow down apoptotic cell death by upregulating the expression of the Bcl-2 gene family of antiapoptotic proteins. In another kind of experiment regarding oxidative stress, the role of DHA was observed. An oxidative stress model was created by H_2O_2 in culture medium and has been examined in the presence or absence of DHA. This in vitro model demonstrated that the injury induced by H_2O_2-mediated oxidative stress was significantly reduced by supplementation with DHA (1 μM) in the medium [63]. As we know, oxidative stress associated with many central nervous system and peripheral nervous system pathologies, including SCI, could be treated with DHA.

14.12 CONCLUSIONS

Dietary factors play a key role in the development of good health and boosting immunity against several chronic cognitive neuronal disorders [67,68]. Epidemiological studies have shown that diets rich in herbs, fruits, and vegetable are useful for rejuvenation of impaired neurons. SCI has a high occurrence and a profound impact on both the patient and on society. Practical recovery following SCI is often pitiable, and so far, no nonsurgical remedies have been discovered [61]. In an experimental model of transient spinal cord ischemia in rats, the systematic application of an omega-3 FA, ALA, has been found to decrease neuronal loss and improve functional outcome. Moreover, previously, Lim et al. [62] have demonstrated that omega-3 FA (DHA), applied as an intravenous dose after SCI in adult rats, was found to induce significant neuroprotection, reducing neuronal cell loss and oligodendrocyte loss and decreasing the apoptosis. Walnut's active biomolecules seem to have enough potential effects on human health by curing cognitive disorder and enhancing learning and memory functions [69–72]. Several researches have clearly observed that polyphenols could protect against cognitive impairment with their antioxidant capability [69,70]. The polyphenol-enriched walnuts have shown to improve interneuron signaling and enhance neurogenesis and memory function [73–76].

But it is extremely difficult to draw a clear conclusion about the antineurodegenerative efficacy of walnut due to inaccuracy of drug preparation, amounts and frequency of walnut consumption, and possible side effects. It is unclear whether walnut alone or in combination with other nutritional components may have the greatest effects. Well-designed research is necessary to develop standard and active principles that may be therapeutically useful for the treatment of severe neuronal cognitive disorder in coming days.

REFERENCES

1. Thakur, A., and Cahalan, C. (2011). Geographical variation of *Juglans regia* L. in Juglone content: Rapid analysis using micro plate reader. *Curr. Sci.* 100(10): 1483–1485.
2. Taha, N.A., and Al-wadaan. M.A. (2011). Utility and importance of walnut, *Juglans regia* Linn: A review. *Afr. J. Microbiol. Res.* 5(32): 5796–5805. Available online at http://www.academicjournals.org/AJMR.
3. Manning, W.E. (1978). The classification within the Juglandaceae. *Ann. Miss. Bot. Gard.* 65: 1058–1087.
4. Thakur, A. (2011). Juglone: A therapeutic phytochemical from *Juglans regia* L. *J. Med. Plants Res.* 5(22): 5324–5330. Available online at http://www.academicjournals.org /JMPR.
5. Manos, P.S., and Stone, D.S. (2001). Evolution, phylogeny and systematics of the Juglandaceae. *Ann. Miss. Bot. Gard.* 88(2): 231–269.
6. Shi, D., Chen, C., Zhao, S., Ge, F., Liu, D., and Song, H. (2014). Effects of walnut polyphenol on learning and memory functions in hypercholesterolemia mice. *J. Food Nutr. Res.* 2(8): 450–456. Available online at http://pubs.sciepub.com/jfnr/2/8/4.
7. Rea, T.D., Breitner, J.C., Psaty, B.M., Fitzpatrick, A.L., Lopez, O.L., Newman, A.B. and Kuller, L.H. (2005). Statin use and the risk of incident dementia: The Cardiovascular Health Study. *Arch. Neurol.* 62 (7): 1047–1051.
8. Ullrich, C., Pirchl, M., and Humpel, C. (2010). Hypercholesterolemia in rats impairs the cholinergic system and leads to memory deficits. *Mol. Cell. Neurosci.* 45 (4): 408–417.
9. Evola, M., Hall, A., Wall, T., Young, A., and Grammas, P. (2010). Oxidative stress impairs learning and memory in ApoE knockout mice. *Pharmacol. Biochem. Behav.* 96(2): 181–186.

10. Poulose, S.M., Miller, M.G., and Shukitt-Hale, B. (2014). Role of walnuts in maintaining brain health with age. *J. Nutr.* 144(4): 561S–566S.

11. Joseph, J., Cole, G., Head, E., and Ingram, D. (2009). Nutrition, brain aging, and neurodegeneration. *J. Neurosci.* 29(41): 12795–12801.

12. Tahraoui, A., El-Hilaly, J., Israili, Z.H., and Lyoussi, B. (2007). Ethno pharmacological survey of plants used in the traditional treatment of hypertension and diabetes in southeastern Morocco (*Errachidia province*). *J. Ethnopharmacol.* 110(1): 105–117.

13. Stampar, F., Solar, A., Hudina, M., Veberic, R., and Colaric, M. (2006). Traditional walnut liqueur—Cocktail of phenolics. *Food Chem.* 95(4): 627–631.

14. Pereira, J.A., Oliveira, I., Sousa, A., Ferreira, I.C.F.R., Bento, A., and Estevinho, L. (2008). Bioactive properties and chemical composition of six walnut (*Juglans regia* L.) cultivars. *Food Chem. Toxicol.* 46(6): 2103–2111.

15. Haider, S., Batool, Z., Tabassum, S., Perveen, T., Saleem, S., Naqvi, F., and Haleem, D.J. (2011). Effects of walnuts (*Juglans regia*) on learning and memory functions. *Plant Foods Human Nutr.* 66 (4): 335–340.

16. Willis, L.M., Shukitt-Hale, B., Cheng, V., and Joseph, J.A. (2009). Dose-dependent effects of walnuts on motor and cognitive function in aged rats. *Br. J. Nutr.* 101(08): 1140–1144.

17. Almonte-Flores, D.C., Paniagua-Castro, M., Escalona-Cardoso, G., and Rosales-Castro, M. (2015). Pharmacological and genotoxic properties of polyphenolic extracts of *Cedrela odorata* L. and *Juglans regia* L. barks in rodents. *Evid Based Complement. Altern. Med.* 2015. Article ID 187346. http://dx.doi.org/10.1155/2015/187346.

18. Zhao, M.-H., Jiang, Z.-T., Liu, T., and Li, R. (2014). Flavonoids in *Juglans regia* L. leaves and evaluation of in vitro antioxidant activity via intracellular and chemical methods. *Sci. World J.* 2014. Article ID 303878. http://dx.doi.org/10.1155/2014/303878.

19. Cheniany, M., Ebrahimzadeh, H., Vahdati, K., Preece, J.E., Masoudinejad, A., and Mirmasoumi, M. (2013). Content of different groups of phenolic compounds in microshoots of *Juglans regia* cultivars and studies on antioxidant activity. *Acta Physiol. Plant.* 35(2): 443–450.

20. Willis, L.M., Shukitt-Hale, B., and Joseph, J.A. (2009). Dietary polyunsaturated fatty acids improve cholinergic transmission in the aged brain. *Genes Nutr.* 4: 309–314.

21. Martinez, M.L., Labuckas, D.O., Lamarque, A.L., and Maestri, D.M. (2010). Walnut (*Juglans regia* L.): Genetic esources, chemistry, by-products. *J. Sci. Food. Agric.* 90: 1959–1967.

22. Sze-Tao, K.W.C., and Sathe, S.K. (2000). Walnut (*Juglans regia* L): Proximate composition, protein solubility, protein amino acid composition and protein in vitro digestibility. *J. Sci. Food Agric.* 80: 1393–1401.

23. Venkatachakm, M., and Sathe, S.K. (2006). Chemical composition of selected edible nut seeds. *J. Agric. Food. Chem.* 54: 4705–4714.

24. Gawlik-Dziki, U., Durak, A., Pecio, A., and Kowalska, I. (2014). Nutraceutical potential of tinctures from fruits, green husks, and leaves of *Juglans regia* L. *Sci. World J.* 2014. Article ID 501392. http://dx.doi.org/10.1155/2014/501392.

25. Wang, C., Zhang, W., and Pan, X. (2014). Nutritional quality of the walnut male inflorescences at four flowering stages, *J. Food Nutr. Res.* 2(8): 457–464. Available online at http://pubs.sciepub.com/jfnr/2/8/5.

26. USDA. (2010). *National Nutrient Database for Standard Reference. Release 23.*

27. Wang, C., Zhang, W., and Pan, X. (2014). Nutritional quality of the walnut male inflorescences at four flowering stages. *J. Food Nutr. Res.* 2(8): 457–464. doi: 10.12691 /jfnr-2-8-5.

28. Lin, Y.H., and Salem, N. Jr. (2007). Whole body distribution of deuterated linoleic and a-linolenic acids and their metabolites in the rat. *J. Lipid Res.* 48: 2709–2724.

29. Hrelia, S., Bordoni, A., Celadon, M., Turchetto, E., Biagi, P.L., and Rossi, C.A. (1989). Age related changes in linoleate and alpha-linolenate desaturation by rat liver microsomes. *Biochem. Biophys. Res. Commun.* 163: 348–355.

30. Willis, L.M., Shukitt-Hale, B., and Joseph, J.A. (2009). Modulation of cognition and behavior in aged animals: Role for antioxidant- and essential fatty acid rich plant foods. *Am J Clin Nutr.* 89: 1602S–1606S.

31. Willis, L.M., Shukitt-Hale, B., Joseph, J.A. (2009). Dietary polyunsaturated fatty acids improve cholinergic transmission in the aged brain. *Genes Nutr.* 4: 309–314.

32. Pandey, K.B., and Rizvi, S.I. (2009). Plant polyphenols as dietary antioxidants in human health and disease. *Oxid Med Cell Longev.* 2(5): 270–278. Available online at: www.landesbioscience.com/journals/oximed/article/9498.

33. Xu, P.X., Wang, S.W., Yu, X.L., Su, Y.J., Wang, T., Zhou, W.W., and Liu, R.T. (2014). Rutin improves spatial memory in Alzheimer's disease transgenic mice by reducing Aβ oligomer level and attenuating oxidative stress and neuroinflammation. *Behav. Brain Res.* 264: 173–180.

34. Leite, M.R., Wilhelm, E.A., Jesse, C.R., Brandão, R., and Nogueira, C.W. (2011). Protective effect of caffeine and a selective A 2A receptor antagonist on impairment of memory and oxidative stress of aged rats. *Exp. Gerontol.* 46(4): 309–315.

35. Li, L., Tsao, R., Yang, R., Liu, C., Zhu, H., and Young, J.C. (2006). Polyphenolic profiles and antioxidant activities of heartnut (*Juglans ailanthifolia* var. cordiformis) and Persian walnut (*Juglans regia* L.). *J. Agric. Food Chem.* 54(21): 8033–8040.

36. Joseph, J., Cole, G., Head, E., and Ingram, D. (2009). Nutrition, brain aging, and neurodegeneration. *J. Neurosci.* 29(41): 12795–12801.

37. Xu, Y., Zhang, J.J., Xiong, L., Zhang, L., Sun, D., and Liu, H. (2010). Green tea polyphenols inhibit cognitive impairment induced by chronic cerebral hypoperfusion via modulating oxidative stress. *J. Nutr. Biochem.* 21(8): 741–748.

38. Fernández-Fernández, L., Comes, G., Bolea, I., Valente, T., Ruiz, J., Murtra, P. and Unzeta, M., LMN. (2012). Diet, rich in polyphenols and polyunsaturated fatty acids, improves mouse cognitive decline associated with aging and Alzheimer's disease. *Behav. Brain Res.* 228(2): 261–271.

39. Scalbert, A., Manach, C., Morand, C., and Remesy, C. (2005). Dietary polyphenols and the prevention of diseases. *Crit. Rev. Food Sci. Nutr.* 45: 287–306.

40. Arts, I.C.W., and Hollman, P.C.H. (2005). Polyphenols and disease risk in epidemiologic studies. *Am. J. Clin. Nutr.* 81: 317–325.

41. Vitrac, X., Moni, J.P., Vercauteren, J., Deffieux, G., Mérillon, J.M. (2002). Direct liquid chromatography analysis of resveratrol derivatives and flavanonols in wines with absorbance and fluorescence detection. *Anal. Chim Acta.* 458: 103–110.

42. Luqman, S., and Rizvi, S.I. (2006). Protection of lipid peroxidation and carbonyl formation in proteins by capsaicin in human erythrocytes subjected to oxidative stress. *Phytother. Res.* 20: 303–306.

43. García-Lafuente, A., Guillamón, E., Villares, A., Rostagno, M.A., and Martínez, J.A. (2009). Flavonoids as anti-inflammatory agents: Implications in cancer and cardiovascular disease. *Inflamm. Res.* 58: 537–552.

44. Pandey, K.B., and Rizvi, S.I. (2010). Protective effect of resveratrol on markers of oxidative stress in human erythrocytes subjected to in vitro oxidative insult. *Phytother. Res.* 1: S11–S14. doi: 10.1002/ptr.2853.

45. Commenges, D., Scotet, V., Renaud, S., Jacqmin-Gadda, H., Barberger-Gateau, P., and Dartigues, J.F. (2000). Intake of flavonoids and risk of dementia. *Eur J Epidemiol.* 16: 357–363.

46. Spencer, J.P.E. (2009). Flavonoids and brain health: multiple effects underpinned by common mechanisms. *Genes. Nutr.* 4: 243–250. doi: 10.1007/s12263-009-0136-3.

47. Dai, Q., Borenstein, A.R., Wu, Y., Jackson, J.C., and Larson, E.B. (2006). Fruit and vegetable juices and Alzheimer's disease: The Kame Project. *Am. J. Med.* 119: 751–759.
48. Lahiri, D.K. (2004). Apolipoprotein E as a target for developing new therapeutics for Alzheimer's disease based on studies from protein, RNA, and regulatory region of the gene. *J. Mol. Neurosci.* 23: 225–233.
49. Yasuno, F., Tanimukai, S., Sasaki, M., Ikejima, C., Yamashita, F., Kodama, C., Mizukami, K., and Asada, T. (2012). Combination of antioxidant supplements improved cognitive function in the elderly. *J Alzheimer Dis.* 32: 895–903.
50. Pribis, P., Bailey, R.N., Russell, A.A., Kilsby, M.A., Hernandez, M., Craig, W.J., Grajales, T., Shavlik, D.J., and Sabate, J. (2011). Effects of walnut consumption on cognitive performance in young adults. *Br. J. Nutr.* 19: 1–9.
51. Reiter, R.J., Manchester, L.C., and Tan, D.X. (2005). Melatonin in walnuts: Influence on levels of melatonin and total antioxidant capacity of blood. *Nutrition.* 21: 920–924.
52. Lahiri, D.K., Chen, D.M., Lahiri, P., Bondy, S., and Greig, N.H. (2005). Amyloid, cholinesterase, melatonin, and metals and their roles in aging and neurodegenerative diseases. *Ann. NY Acad. Sci.* 1056: 430–449.
53. Poulose, S.M., Bielinski, D.F., and Shukitt-Hale, B. (2013). Walnut diet reduces accumulation of polyubiquitinated proteins and inflammation in the brain of aged rats. *J. Nutr. Biochem.* 24: 912–919.
54. Lukiw, W.J., and Bazan, N.G. (2008). Docosahexaenoic acid and the aging brain. *J. Nutr.* 138: 2510–2514.
55. White, W.B., Wolfson, L., Wakefield, D.B., Hall, C.B., Campbell, P., Moscufo, N., Schmidt, J., Kaplan, R.F., Pearlson, G., and Guttmann, C.R. (2011). Average daily blood pressure, not office blood pressure, is associated with progression of cerebrovascular disease and cognitive decline in older people. *Circulation.* 124: 2312–2319.
56. Roberts, R.O., Knopman, D.S., Geda, Y.E., Cha, R.H., Roger, V.L., and Petersen, R.C. (2010). Coronary heart disease is associated with non-amnestic mild cognitive impairment. *Neurobiol Aging.* 31: 1894–1902.
57. Zhao, G., Etherton, T.D., Martin, K.R., West, S.G., Gillies, P.J., and Kris-Etherton, P.M. (2004). Dietary alpha-linolenic acid reduces inflammatory and lipid cardiovascular risk factors in hypercholesterolemic men and women. *J Nutr.* 134: 2991–2997.
58. Ma, Y., Njike, V.Y., Millet, J., Dutta, S., Doughty, K., Treu, J.A., and Katz, D.L. (2010). Effects of walnut consumption on endothelial function in type 2 diabetic subjects: A randomized controlled crossover trial. *Diabetes Care.* 33: 227–232.
59. Katz, D.L., Davidhi, A., Ma, Y., Kavak, Y., Bifulco, L., and Njike, V.Y. (2012). Effects of walnuts on endothelial function in overweight adults with visceral obesity: A randomized, controlled, crossover trial. *J. Am. Coll. Nutr.* 31: 415–423.
60. Bracken, M.B. (1990). Methyl prednisolone in the management of acute spinal cord injuries. *Med. J.* 153: 368.
61. Lang-Lazdunski, L., Blondeau, N., Jarretou, G., Lazdunski, M., and Heurteaux, C. (2003). Linolenic acid prevents neuronal cell death and paraplegia after transient spinal cord ischemia in rats. *J. Vasc. Surg.* 38: 564–575.
62. Lim, S.N., Huang, W., Hall, J.C., Michael-Titus, A.T., and Priestley, J.V. (2013). Improved outcome after spinal cord compression injury in mice treated with docosahexaenoic acid. *Exp. Neurol.* 239: 13–27.
63. Paterniti, I., Impellizzeri, D., Di Paola, R., Esposito, E., Gladman, S., Yip, P., Priestley, J.V., Michael-Titus, A.T., and Cuzzocrea, S. (2014). Docosahexaenoic acid attenuates the early inflammatory response following spinal cord injury in mice: In-vivo and in-vitro studies. *J. Neuroinflamm.* 11: 6. Available online at http://www.jneuroinflammation .com/content/11/1/6.

64. La Rosa, G., Cardali, S., Genovese, T., Conti, A., Di Paola, R., La Torre, D., Cacciola, F., Cuzzocrea, S. (2004). Inhibition of the nuclear factor-κB activation with pyrrolidine dithiocarbamate attenuating inflammation and oxidative stress after experimental spinal cord trauma in rats. *J. Neurosurg. Spine.* 1: 311–321.

65. Spencer, L., Mann, C., Metcalfe, M., Webb, M., Pollard, C., Spencer, D., Berry, D., Steward, W., and Dennison, A. (2009). The effect of omega-3 FAs on tumour angiogenesis and their therapeutic potential. *Eur J Cancer.* 45: 2077–2086.

66. Paterniti, I., Genovese, T., Mazzon, E., Crisafulli, C., Di Paola, R., Galuppo, M., Bramanti, P., Cuzzocrea, S., and Liver, X. (2010). Receptor agonist treatment regulates inflammatory response after spinal cord trauma. *J. Neurochem.* 112: 611–624.

67. Thies, W., and Bleiler, L. (2011) Alzheimer's disease facts and figures. *Alzheimers Dement.* 7: 208–244.

68. Commenges, D., Scotet, V., Renaud, S., Jacqmin-Gadda, H., Barberger-Gateau, P., and Dartigues, JF. (2000). Intake of flavonoids and risk of dementia. *Eur. J. Epidemiol.* 16: 357–363.

69. Xu, Y., Zhang, J.J., Xiong, L., Zhang, L., Sun, D., and Liu, H. (2010). Green tea polyphenols inhibit cognitive impairment induced by chronic cerebral hypoperfusion via modulating oxidative stress. *J. Nutr. Biochem.* 21(8): 741–748.

70. Fernández-Fernández, L., Comes, G., Bolea, I., Valente, T., Ruiz, J., Murtra, P., and Unzeta, M. (2012). LMN diet, rich in polyphenols and polyunsaturated fatty acids, improves mouse cognitive decline associated with aging and Alzheimer's disease. *Behav. Brain Res.* 228(2): 261–271.

71. Willis, L.M., Shukitt-Hale, B., and Joseph, J.A. (2009). Modulation of cognition and behavior in aged animals: Role for antioxidant- and essential fatty acid rich plant foods. *Am. J. Clin. Nutr.* 89: 1602S–1606S.

72. Willis, L.M., Shukitt-Hale, B., and Joseph, J.A. (2009). Dietary polyunsaturated fatty acids improve cholinergic transmission in the aged brain. *Genes Nutr.* 4: 309–314.

73. Xu, P.X., Wang, S.W., Yu, X.L., Su, Y.J., Wang, T., Zhou, W.W., and Liu, R.T. (2014). Rutin improves spatial memory in Alzheimer's disease transgenic mice by reducing Aβ oligomer level and attenuating oxidative stress and neuroinflammation. *Behav. Brain Res.* 264: 173–180.

74. Leite, M.R., Wilhelm, E.A., Jesse, C.R., Brandão, R., and Nogueira, C.W. (2011). Protective effect of caffeine and a selective A 2A receptor antagonist on impairment of memory and oxidative stress of aged rats. *Exp. Gerontol.* 46(4): 309–315.

75. Poulose, S.M., Miller, M.G., and Shukitt-Hale, B. (2014). Role of walnuts in maintaining brain health with age. *J. Nutr.* 144(4): 561S–566S.

76. Joseph, J., Cole, G., Head, E., and Ingram, D. (2009). Nutrition, brain aging, and neurodegeneration. *J. Neurosci.* 29 (41): 12795–12801.

15 Potential Brain Health Benefits of Royal Jelly

Noriko Hattori, Yukio Narita, Kenji Ichihara, Hiroyoshi Moriyama, and Debasis Bagchi

CONTENTS

15.1 INTRODUCTION

Japanese, on average, live longer according to the demographic data of Ministry of Internal Affairs and Communications, as shown in the population pyramid changes over 100 years (Figure 15.1). The demographic features by ages progressively distort a pyramid shape in 1950, increasing in the number of the aged adults and decreasing in the young adults [1]. The data reveal that the average life expectancy of Japanese has become remarkably higher compared to that a few decades ago. The longevity might be primarily attributed to lifestyle, particularly the Japanese diet, which constitutes fish accessible from the surrounding sea, vegetables and fruits harvested during the four distinct seasons, and also Japanese traditional fermented foods [2,3]. In addition, the availability of the Japanese government medical or health care insurance system plays a pivotal role [1].

For example, aged adults frequently visit clinics or hospitals even for slight illness or as regular health examinations because their medical costs are in part reimbursed by the government. The reimbursement system worked well supporting the aged adults, which potentially prevented them from degenerative diseases. On the other hand, the medical care expenditure significantly hiked with the increase of such frequent health care visitors, thus creating a heavy burden of medical costs on the government. Moreover, caregiver burden has been a serious concern [4]. A possible solution to alleviate the medical

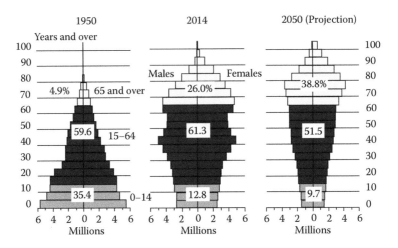

FIGURE 15.1 **(See color insert.)** Changes in the population pyramid in Japan. (From Statistics Bureau, Japan. *Statistical Handbook of Japan 2016*. http://www.stat.go.jp/english /data/handbook/index.htm.)

burden is to go back to the diets that the Japanese used to have. Now, the government provides help on how to take care of health via consultation; the advice is primarily associated with eating habits, physical exercise, and mental health issues such as stress.

As a fact, eating habit are not easily changed in the highly westernized diet society after World War II [2]; it has become common to take supplements within customer choice and to practice self-supplementation. Based on scientific evidence supplements such as FOSHU (Food for Specified Health Use) and/or FFC (Food with Function Claim), enacted in 2015, sellers can make their products with health or function claims to support health benefits [3,5]. For example, if one has a slightly higher blood pressure or is in borderline between healthy and disease states, a number of naturally derived functional food ingredients are now able to help keep at the blood pressure in normal range as FOSHU or FFC. That is, key functional ingredients such as polyphenols (soybean-derived isoflavone and chlorogenic acids in green coffee bean), chitosan, n-3 from fish oils and many others are employed in foods and beverages, capsules and so on [3].

Unlike drug, nutritional supplementation is part of food ingestion, which is not intended to prevent or cure any kinds of diseases by definition [5]. Food with health and/or function claims are scientifically evidenced by randomized clinical trial, and their safety is very much guaranteed with the consumed history of different materials formulated in the supplemental application [2,3,5]. However, such health and function food claim systems direct the consumer to health benefits and well-being to decrease the occurrence of life-related diseases, such as obesity, diabetes mellitus, high blood pressure, cerebrovascular and cardiovascular diseases, and cancer.

Practicing the intake of functional food ingredients, which should be present in FOSHU, is slowly accepted among consumers since the launch of FOSHU products in the marketplace in 2001 [2], with the advocate of the government relaxing regulations of the food labeling. As stated previously, the rapid aging or graying of Japanese, in both genders, causes age-related degenerative diseases (ARDDs) [1].

Dementia, as an example of ARDD, can be defined as the term employed to describe the progressive and profound decline in cognitive function highly attributed to damage or disease in the brain, advanced ages, and so on [6]. Furthermore, a previous study on episodic memory, which memory for unique events and their contexts, demonstrated the remarkable decline of the memory with age [7].

In recent years, various natural functional foods have been studied in vitro, in vivo, and clinical trials to assess the effects of the natural functional foods. Botanical extracts such as *Gingko biloba* extract for the brain health benefit, specifically dementia [8], Bacopa (*Bacopa monniera*) [9], n-3 fatty acids from fish oils [10,11], and polyphenols such as resveratrol [12] have been examined for brain health benefits, including cognition and memory. Among the natural functional food sources, products yielded by bees are very valuable such as beeswax, pollen, propolis, honey, royal jelly (RJ), and so on. In this chapter, RJ is described on the basis of various experimental study results on potential brain health benefits.

15.2 ROYAL JELLY

RJ is a secretion material of the cephalic glands of nurse bees. Its role has attracted academia, because it possesses unique and important biological features and health benefits. In brief, RJ is used as traditional medicine for human beings and yet is widely formulated, for example, in various beverages for health benefits and well-being. A preparation method and some of the functional properties were previously described using water and alkaline RJ [13]. In the industrial scale, a process as described in Figure 15.2 is employed.

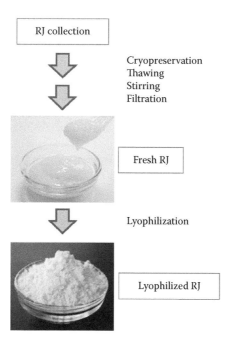

RJ collection

Cryopreservation
Thawing
Stirring
Filtration

Fresh RJ

Lyophilization

Lyophilized RJ

FIGURE 15.2 (**See color insert.**) Processing of RJ for commercial applications.

RJ is enriched with minerals, vitamins, and many other nutritional as well as functional components. Although many functional ingredients are yet to be isolated and identified for their functions, such ingredients for health benefits were previously reported and used as drug and functional food ingredient [14]. In the fresh RJ (moisture content of 62.5%–68.5%), major solid constituents are protein, fructose, glucose, and 10-hydorxy-2-decenoic acid (10-HDA), among which protein constitutes RJ the highest ranging 11.0% to 18.0%. Then, fructose (2.0%–9.0%), glucose (2.0%–9.0%), follows protein and then 10-HDA (>1.4%) [15].

In recent years, much attention has been paid to RJ, which was found to have a wide array of biological effects on human beings. For example, RJ was demonstrated to have multiple biological activities such as anti-inflammatory activity, antioxidant activity, antitumor activity, and others [14]. In the following, some scientific evidence is provided that RJ possesses beneficial brain health effects identifying potential constituents and discussing their mechanism.

15.2.1 EXPERIMENTAL STUDIES

RJ has exhibited marked effects on the nervous system, implying the brain health benefits, based on preclinical studies such as in cellular and animal models and also identifying bioactive compounds that are potentially involved in the positive exertion of the metabolism of the healthy brain functioning.

15.2.2 CELLULAR MODEL STUDY

In the early stage of investigating RJ, we evaluated whether RJ contains bioactive constituents such as nerve growth factor (NGF)-like neurotrophic substances using PC12 cells, which are derived and cloned from a rat pheochromacytoma [16] and also known to be a model system for neuronal differentiation. In the presence of NGF, PC12 cells suppress cellular division and then develop neurites. Furthermore, NGF stimulates PC12 cells to differentiate into neuron-like cells similar to those found in the sympathetic neurons [16]. By contrast, the phosphate-buffered solution of RJ also allowed PC12 cells to induce the development of neurites in serum contained condition [17], although the neurites exhibited the length shorter in RJ-treated cells than in NGF-treated cells. In addition, a synergistic effect was detected on the extension of the neurite length in combination of RJ and NGF as shown in Figure 15.3 [18]. These findings indicate that RJ possibly contains NGF-like neurotrophic substances, stimulating neural differentiation via different mechanism of action from NGF.

We then examined the effects of RJ on differentiation of multipotent neural stem/progenitor cells (NSPCs) obtained from the cerebrums of 15.5-day-old rat embryos [19], for it seems to be that the injured brain has a capacity for self-repair by activated NSPCs. It had been thought that neurogenesis could not occur in the adult brain in the past. It has been, however, revealed that adult neural stem cells and progenitor cells are present in the two areas, a legion around the lateral ventricles and the granular cell layer of the hippocampal dentate gyrus [10–23] and also that NSPCs have self-renewal capacity and multipotent ability to differentiate into neurons, astrocytes, and oligodendrocytes during development [20,24]. Therefore,

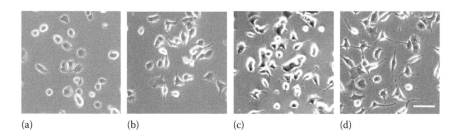

(a)   (b)   (c)   (d)

FIGURE 15.3 Microscopic evaluation of neurite process formation in PC12 cells after various treatments with RJ, NGF, or both in combination. PC12 cells were cultured in a medium supplemented with vehicle (control; a), RJ (diluted 125-fold; b), NGF (25 ng/ml; c), or a combination (d) of RJ (finally diluted 125-fold) and NGF (final concentration of 25 ng/ml). The photographs show morphological changes of PC12 cells at one day after various treatments. Scale bar: 50 μm.

precise control of proliferation and differentiation of multipotent NSPCs is crucial for proper development of the nervous system, and NSPCs in the mature brain are probably a prominent target for the therapy of degenerative neurological disorders such as Alzheimer's, Parkinson's, and depression diseases. NSPCs proliferate vigorously as neurospheres in medium with fibroblast growth factor (FGF)-2 and start to differentiate when FGF-2 is removed from the medium [20,24].

We found that RJ stimulated the proliferation of NSPCs on the first day as shown Figure 15.4 [25] and then markedly reduced the growth of NSPCs on the fifth day in FGF-2 free medium as seen in Figure 15.5 [19], signifying that RJ enhances the proliferation at first and possibly potentiates the differentiation in NSPCs. NGF has also been reported to have the same effect as RJ on PC12 cells, i.e., the induction of proliferation followed by differentiation to neurons [16]. In addition, RJ was found to increase the proportion of the differentiation into neurons, astrocytes, and

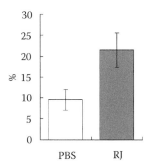

FIGURE 15.4 Effects of RJ on the cell growth of NSPCs. Growing NSPCs as neurospheres were plated on poly-L-ornithine-coated six-well plates and cultured for two days in proliferation medium, and RJ (diluted 250-fold) was added for one day in differentiation medium. Moreover, they were incubated for another four hours with BrdU (10 mM). The cells were fixed and reacted with Alexa Fluor-conjugated anti-BrdU mouse IgG antibody. The number of BrdU-positive cells was counted, and the ratio of them to total cells was calculated. The values represent mean ± SE (n = 4).

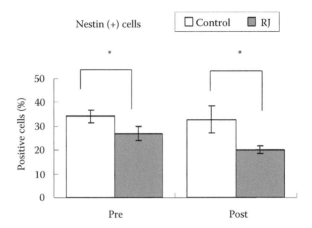

FIGURE 15.5 RJ decreases the ratio of the undifferentiated cells. Dissociated cell suspensions of NSPCs were plated on coverslips coated with poly-L-ornithine. For "pretreatment with RJ," the cells were cultured in the proliferation medium (containing 10 ng/ml FGF-2) with 500-fold diluted RJ or vehicle (control) for two days and then cultured for another five days in the differentiation medium (FGF-2-free) without RJ. And for "posttreatment with RJ," the cells were cultured for two days in the proliferation medium and then cultured for another five days in the differentiation medium (FGF-2-free) with 500-fold diluted RJ or vehicle (control). The values represent mean ± SE (n = 4). A statistical difference between the two values was determined by Student's t-test to be significant at *$p < .05$.

oligodendrocytes in parallel with a decrease in undifferentiated cells [19] by immune-staining assay using antibodies of neuron-specific class III beta-tubulin (Tuj1), glial fibrillary acidic protein, 2′,3′-cyclic-nucleotide 3′-phosphodiesterase (CNPase), and nestin, which are characteristic proteins of neurons [26], astrocytes [27], oligodendrocytes [28], and undifferentiated cells [29], respectively.

These results suggest that RJ facilitates the differentiation of NSPCs into neuronal and glial cells. In cultured NSPCs, the intracellular signal mitogen-activated protein kinase (MAPK) is essential to promote the proliferation of cells [30] as well as PC12 cells, whereas further long-term potentiation (LTP) in the dentate gyrus has been reported to be related to phosphorylation of MAPK and cAMP-response element-binding protein (CREB) [31]. CREB plays a crucial role as a molecular switch to control the conversion of short-term forms of plasticity into long-term forms [32,33]. A recent report also indicated that CREB might importantly contribute to learning and memory as a result of regulating intrinsic neuronal excitability [34]. Interestingly, RJ was also demonstrated to phosphorylate both MAPK and CREB proteins on the cultured NSPCs in the western blotting analysis [19], implying that RJ activates intracellular signals, expected to enhance hippocampal LTP (Figure 15.6).

We lastly isolated and identified two active substances from the RJ solution as a target for the neurite-forming activity in PC12 cells employing ion-exchange and gel-filtration chromatography. Their chemical structures were determined to be adenosine mono phosphate (AMP) and AMP N_1-oxide (Figure 15.7). AMP N_1-oxide

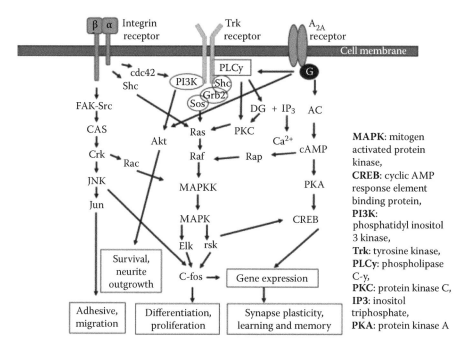

FIGURE 15.6 **(See color insert.)** Intracellular signal transduction pathway.

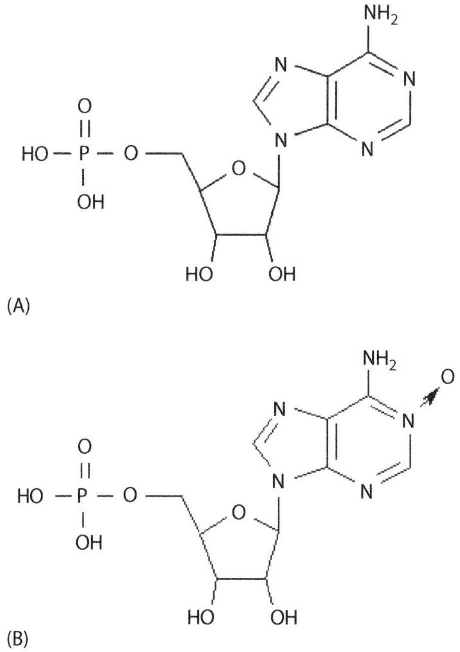

(A)

(B)

FIGURE 15.7 The chemical structures of (A) AMP and (B) AMP N_1-oxide.

induced neurite formation in PC12 cells at over 20-fold lower concentrations than did AMP, and it also inhibited cell growth and enhanced expression of neurofilament M, one of the specific proteins of differentiated cells [18]. AMP N_1-oxide is unique to RJ and identified as the first active substance with the neurogenesis-like effect. This substance is thought to predominantly induce neurite formation through activation of A_{2A} receptors [18]. Previous studies have proposed the functional interactions of A_{2A} receptor-mediated signaling with intracellular NGF signaling, e.g., a supportive effect on NGF-induced neurite outgrowth [35], transactivated TrkA receptor, a high-affinity receptor of NGF [36], and phosphorylation of MAPK (ERK1/2) [37] in PC12 cells. And A_{2A} receptors are known to perform multiple tasks by mediating various signals in nervous systems. It was also found that AMP N_1-oxide-stimulated phosphorylation of not only MAPK but also CREB in a dose-dependent manner [38]. Inhibition of protein kinase A (PKA) but not MAPK activity by the selective inhibitors significantly reduced the AMP N_1-oxide-induced neurite formation [38]. Thus, AMP N_1-oxide is considered to elicit neuronal differentiation of PC12 cells, as evidenced by the inhibition of cell growth followed by the neurite formation through the adenosine A_{2A} receptor-mediated PKA signaling, which may be responsible for characteristic actions of RJ (Figure 15.6).

In immunological staining assay, furthermore, AMP N_1-oxide significantly facilitated the differentiation of NSPCs into astrocytes, but not neurons and oligodendrocytes, in a dose-dependent manner [39]. It was apparent that AMP N_1-oxide phosphorylated signal transducer and activator of transcription 3 (STAT3) [38] and that the phosphorylation of STAT3 facilitated the differentiation of NSPCs into astrocytes [40]. This result supports the promoting effect of AMP N_1-oxide on the differentiation of NSPCs. In addition, 10-HDA, a 10-carbon unsaturated fatty acid that is unique to RJ [41], has been reported to have several pharmacological activities such as extension of the lifespan in *Caenorhabditis elegans* [42]. 10-HDA stimulated the generation of neurons from NSPCs and decreased that of astrocytes [19], suggesting that 10-HDA facilitates the differentiation into neurons and simultaneously suppressed that into glial cells in NSPCs.

These findings thus provide a possibility that RJ facilitates the differentiation of NSPCs into all the types of brain cells, neurons, astrocytes, and oligodendrocytes. Since adenosine derivatives have been shown to play an essential role in modulating neuronal function via adenosine receptors [43], AMP N_1-oxide might be a clue to enhance understanding, at the molecular level, a variety of biological activities of RJ toward the nervous system, and it might provide a novel strategy in nutritional and/or health food (nutraceutical) applications including gradual loss of memory associated with the healthy aged adults.

15.2.3 ANIMAL MODEL STUDY

As a next step, we evaluated the effects of RJ on learning and memory using animal models associated with neurogenesis. Trimethyltin (TMT), an organometallic compound, is well known to induce acute neuronal toxicity that is localized in the hippocampal dentate gyrus and induces cognitive dysfunction [44–48]. At two days after treatment with TMT for mice, the density of dentate granule cells was found to

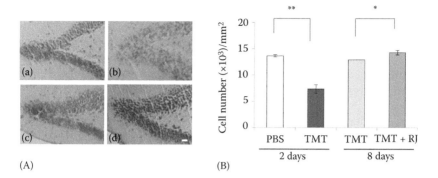

FIGURE 15.8 **(See color insert.)** The ameliorative effect of RJ on the TMT-induced neurodegeneration in the hippocampal dentate gyrus of mice. (A) The Nissl-stained hippocampal dentate gyrus at two days after treatment with PBS (a) or TMT (b) and at eight days after treatment with TMT (c) or TMT + six-day feeding of 5% RJ diet (d). Scale bar: 25 μm. (B) The number of granule cells in the hippocampal dentate gyrus at two days after treatment with PBS or TMT and at eight days after treatment with TMT or TMT + six-day feeding of 5% RJ diet. The values represent mean ± SE (n = 4). A statistical difference between the two values was determined by Student's t-test to be significant at $**p < .001$ and $*p < .05$.

decrease by almost 50% (Figure 15.8Aa and b and B) [49]. Based on the aforementioned evidence, we used TMT-treated mice as an in vivo model to study the effect of RJ on the neurogenesis involved in cognitive ability. The cognitive dysfunction was assessed using a Y-maze test, in which spontaneous alternation behavior is defined as a successive entry into the three arms on overlapping triplet sets [50], representing short-term memory. In the experiment, the mice were treated with TMT and two days later given lyophilized RJ-mixed (1% and 5%) foods for six successive days followed by the Y-maze test. The result was that feeding with the RJ diet dose dependently ameliorated the TMT-induced impairment of the alternation behavior, and specifically, a significant effect was found in the 5% RJ diet group [49]. Since no significant difference was observed in the number of total arm entries among test groups, RJ ameliorated the alternation behavior without affecting locomotor activity in the TMT-treated mice. In addition, RJ feeding significantly attenuated the TMT-induced loss of neuronal cells in the hippocampal dentate gyrus (Figure 15.8Ac and d and B) [49]. These results have proven that a novel, preeminent property of RJ, potentially serving as a functional ingredient, for protecting brain health from some particular neurodegenerative diseases.

As mentioned earlier, NSPCs have self-renewal capacity and multipotent ability of differentiation into neurons, astrocytes, and oligodendrocytes. It is well known that NSPCs are present in some specific areas, especially in the hippocampal dentate gyrus in the adult brain [51]. There are numerous reports linking the relation between stress and hippocampal neurogenesis [52], and further, the majority of them support the idea that acute exposure to stress decreases proliferation of neural progenitor cells in the dentate gyrus [53]. Chronic exposure to stress is also known to reduce hippocampal neurogenesis in both rats and mice as well [53]. It has also been reported that unacceptable stress is harmful to hippocampal progenitor cells, e.g., decreases in proliferation, differentiation, and/or cell survival [54].

Altogether, stress is thought to be a precipitating factor for major depressive disorder and to significantly influence the depressive state. By contrast, the antidepressant fluoxetine induces proliferation and differentiation of neuronal progenitor cells [55]. Thus, stimulatory activation of hippocampal progenitor cells may prevent or alleviate depressive disorders. We preliminarily studied the effect of RJ (4% lyophilized RJ-mixed food) on depression-like behavior, which was assessed as immobility time using a tail suspension test, in mice that had been subjected to unpredictable chronic mild stress (UCMS). In the UCMS rodent model of depression, RJ-fed group showed slightly shorter immobility time compared to non-RJ group, indicating that RJ possibly reduces depressive symptoms (unpublished data).

15.2.4 CLINICAL TRIAL

A randomized placebo-controlled, double-blind trial was conducted in healthy elderly people. Sixty-one subjects of both genders were given beverages containing 3000 mg RJ or a placebo once daily for six months. It is noteworthy that subjects assigned to RJ intake showed better mental health scores, when compared to the placebo group, in the Short-Form 36 Health Survey Questionnaire [56]. The changes in mental health scores thought to be, in part, supported by the antidepressive effects of RJ, which we observed in in vitro and in vivo studies.

15.3 PERSPECTIVES

Although the contribution of foods with health and function claim systems to decrease lifestyle-related diseases (LSRDs) or improve of quality of life is yet to be examined and confirmed with an appropriate methodology to conduct postmarketing surveillance, thus far, the influence of the systems on controlling LSRDs are not apparent. With the new system (FFC), claims associated to memory or brain health will be used on labels of the supplemental or food product in future.

It becomes crucial that the accumulation of data from preclinical studies (in vitro, animal model) in the application of the brain-health-promoting benefits thus far presented encourages additional well-designed randomized clinical trials to substantiate RJ as FOSHU or FFC in Japan or dietary supplements in the different dosage forms such as soft-gel capsules in the United States or some other countries. In addition, a series of safety studies are important to ensure the safety of RJ or any of the natural functional ingredients to meet the requirement.

15.4 CONCLUSIONS

In Japan, along with other aging countries, the hike in the medical care expenditure is a socioeconomic concern [1]. Cognitive dysfunctions associated with age, such as dementias, are prevalent, although robust treatments for cognitive dysfunctions, including dementias, have not yet been developed and are not available. Accepting aging-associated costs and human resource burdens leads into a way to combat aging with self-supplementation, while preclinical and clinical studies showed that RJ is a promising naturally originated functional food for brain health, possibly preventing and/or relieving neurodegenerative disorders in daily life.

REFERENCES

1. Statistics Bureau, Japan. *Statistical Handbook of Japan 2016*. http://www.stat.go.jp/english/data/handbook/index.htm (accessed on March 3, 2016).
2. Bagchi D (Ed). *Nutraceutical and Functional Food Regulations in the United States and Around the World* (2nd.), Academic Press, London, UK, 2014.
3. Ohama H, Ikeda H, Moriyama H. Health foods and foods with health claims in Japan. *Toxicology* 2006;221:95–111.
4. Muraki A, Yamagishi K, Ito Y et al. Caregiver burden for impaired elderly Japanese with prevalent stroke and dementia under long-term care insurance system. *Cerebrovasc Dis.* 2008;25:234–240.
5. http://www.caa.go.jp/rn/index.html (accessed March 3, 2016).
6. Kidd PM. A review of nutrients and botanicals in integrative management of cognitive dysfunction. *Altern Med Rev.* 1999;4:144–161.
7. Nilsson LG. Memory function in normal aging. *Acta Neurol Scand.* 2003;179:7–13.
8. Le Bars PL, Katz MM, Berman N et al. A placebo-controlled, double-blind randomized trial of an extract *Ginkgo biloba* for dementia. North American EGb study group. *JAMA* 1997;278:1327–1332
9. Sharma R, Chaturvedi C, Tewari PV. Efficacy of *Bacopa monniera* in revitalizing intellectual functions in children. *J Res Educ Ind Med.* 1987:1–12.
10. Kidd PM. Omega-3 DHA and EPA for cognitive, behavior, and mood: Clinical findings and structural-functional synergies with cell membrane phospholipids. *Altern Med Rev.* 2007;12:207–227.
11. Sinn N, Milte CM, Street SJ et al. Effects of n-3 fatty acids, EPA v. DHA, on depressive symptoms, quality of life, memory and executive function in older adults with mild cognitive impairment: A 6-month randomised controlled trial. *Br J Nutr.* 2012;107:1682–1693.
12. Ma X, Sun Z, Liu Y et al. Resveratrol improves cognition and reduces oxidative stress in rats with vascular dementia. *Neural Regen Res.* 2013;8:2050–2059.
13. Nagai T, Inoue R. Preparation and the functional properties of water and alkaline extract royal jelly. *Food Chem.* 2004;84:181–186.
14. Ramadan MF, Al-Ghamdi A. Bioactive compounds and health-promoting properties of royal jelly: A review. *J Functional Food.* 2012;4:39–52.
15. Kanelis D, Tananaki C, Liolios V et al. A suggestion for royal jelly specifications. *Arh Hig Rada Tokskol.* 2015;66:275–284.
16. Greene LA, and Tischler AS. Establishment of a noradrenergic clonal line of rat adrenal pheochromocytoma cells which respond to nerve growth factor. *Proc Natl Acad Sci USA*, 1976;73:2424–2428.
17. Hattori N, Nomoto H, Fukumitsu H et al. Royal jelly-induced neurite outgrowth from rat pheochromocytoma PC12 cells requires integrin signal independent of activation of extracellular signal-regulated kinases. *Biomed Res.* 2007;28:139–146.
18. Hattori N, Nomoto H, Mishima S. et al. Identification of AMP N1-oxide in royal jelly as a component neurotrophic toward cultured rat pheochromocytoma PC12 cells. *Biosci Biotechnol Biochem.* 2006;70:897–906.
19. Hattori N, Nomoto H, Fukumitsu H et al. Royal jelly and its unique fatty acid, 10-hydroxy-trans-2-decenoic acid, promote neurogenesis by neural stem/progenitor cells in vitro. *Biomed Res.* 2007;28:261–266.

20. Gage FH, Mammalian neural stem cells. *Science*. 2000;287:1433–1438.
21. Morshead CM, Craig CG, Van der Kooy D. In vivo clonal analyses reveal the properties of endogenous neural stem cell proliferation in the adult mammalian forebrain. *Development*. 1998;125:2251–2261.
22. Lois C, Alvarez-Buylla A. Long-distance neuronal migration in the adult mammalian brain. *Science*. 1994;264:1145–1148.
23. Luskin MB. Neuronal cell lineage in the vertebrate central nervous system. *FASEB J*. 1994;8:722–730.
24. Reynolds BA, Weiss S. Clonal and population analyses demonstrate that an EGF-responsive mammalian embryonic CNS precursor is a stem cell. *Dev Biol*. 1996;175:1–13.
25. Hattori N. Research related to some effects of royal jelly on nervous system. PhD Thesis, Gifu Pharmaceutical University, Japan. 2007.
26. Theocharatos S, Wilkinson DJ, Darling S et al. Regulation of progenitor cell proliferation and neuronal differentiation in enteric nervous system neurospheres. *Plos One*. e54089, 2013;8:1–8.
27. Weinstein DE, Shelanski ML, Liem RK. Suppression by antisense mRNA demonstrates a requirement for the glial fibrillary acidic protein in the formation of stable astrocytic processes in response to neurons. *J Cell Biol*. 1991;112:1205–1213.
28. Sprinkle TJ, Agee JF, Tippins RB et al. Monocional antibody production to human and bovine 2':3'-cylic nucleotide 3'-phosphodiesterase (CNPhase): High-specificity recognation in whole brain acetone powders and powders and conversation of sequence between CNP1 and CNP2. *Brain Res*. 1987;426:349–57.
29. Lendahl U, Zimmerman LB, McKay RD. CNS stem cells express a new class of intermediate filament protein. *Cell*. 1990;60:585–595.
30. Wang B, Gao Y, Xiao Z et al. Erk1/2 promotes proliferation and inhibits neuronal differentiation of neural stem cells. *Neurosci Lett*. 2009;461;252–257.
31. Davis S, Vanhoutte P, Pages C et al. The MAPK/ERK cascade targets both Elk-1 and cAMP response element-binding protein to control long-term potentiation-dependent gene expression in the dentate gyrus in vivo. *J Neurosci*. 2000;20:4563–4572.
32. Lonze BE, Ginty DD. Function and regulation of CREB family transcription factors in the nervous system. *Neuron*. 2002;35:605–623.
33. Barco A, Marie, H. Genetic approaches to investigate the role of CREB in neuronal plasticity and memory. *Mol Neurobiol*. 2011;44:330–349.
34. Benito E, Barco A. CREB's control of intrinsic and synaptic plasticity: Implications for CREB-dependent memory models. *Trends Neurosci*. 2010;33:230–240.
35. Cheng HC, Shih HM, Chern Y. Essential role of cAMP-response element-binding protein activation by A2A adenosine receptors in rescuing the nerve growth factor-induced neurite outgrowth impaired by blockage of the MAPK cascade. *J Biol Chem*. 2002;277:33930–33942.
36. Lee FS, Chao MV. Activation of Trk neurotrophin receptors in the absence of neurotrophins. *Proc Natl Acad Sci USA*. 2001;98:3555–3560.
37. Arslan G, Fredholm BB. Stimulatory and inhibitory effects of adenosine A (2A) receptors on nerve growth factor-induced phosphorylation of extracellular regulated kinases 1/2 in PC12 cells. *Neurosci Lett*. 2000;292:183–186.
38. Hattori N, Nomoto H, Fukumitsu H et al. AMP N1-oxide, a unique compound of royal jelly, induces neurite outgrowth from PC12 cells via signaling by protein kinase A independent of that by mitogen-activated protein kinase. *eCAM*. 2010;7:63–68.
39. Hattori N. Nomoto H, Fukumitsu H et al. AMP N1-oxide potentiates astrogenesis by cultured neural stem/progenitor cells through STAT3 activation. *Biomed Res*. 2007;28:295–299.
40. Fukuda S, Abematsu M, Mori H et al. Potentiation of astrogliogenesis by STAT3-mediated activation of bone morphogenetic protein-Smad signaling in neural stem cells. *Mol Cell Biol*. 2007;27:4931–4937.

41. Takenaka T. Chemical composition of royal jelly. *Honeybee Sci.* 1982;3:69–74.
42. Honda Y, Araki Y, Hata T et al. 10-Hydroxy-2-decenoic acid, the major lipid component of royal jelly, extends the lifespan of *Caenorhabditis elegans* through dietary restriction and target of rapamycin signaling. *J Aging Res.* 2015;2015:425261.
43. Daval JL, Nehlig A, Nicolas F. Physiological and pharmacological properties of adenosine: Therapeutic implications. *Life Sci.* 1991;49:1435–1453.
44. Fiedorowicz A, Figiel I, Kamińska B et al. Dentate granule neuron apoptosis and glia activation in murine hippocampus induced by trimethyltin exposure. *Brain Res.* 2001;912:116–127.
45. Ogita K, Nitta Y, Watanabe M et al. In vivo activation of c-Jun N-terminal kinase signaling cascade prior to granule cell death induced by trimethyltin in the dentate gyrus of mice. *Neuropharmacology.* 2004;47:619–630.
46. Ogita K, Nishiyama N, Sugiyama C et al. Regeneration of granule neurons after lesioning of hippocampal dentate gyrus: Evaluation using adult mice treated with trimethyltin chloride as a model. *J Neurosci Res.* 2005;82:609–621.
47. Niittykoski M, Lappalainen, R, Jolkkonen J et al. Systemic administration of atipamezole, a selective antagonist of alpha-2 adrenoceptors, facilitates behavioural activity but does not influence short-term or long-term memory in trimethyltin-intoxicated and control rats. *Neurosci Biobehav Rev.* 1998;22:735–750.
48. Ishida N, Akaike M, Tsutsumi S et al. Trimethyltin syndrome as a hippocampal degeneration model: Temporal changes and neurochemical features of seizure susceptibility and learning impairment. *Neuroscience.* 1997;81:1183–1191.
49. Hattori N, Ohta S, Sakamoto T et al. Royal jelly facilitates restoration of the cognitive ability in trimethyltin-intoxicated mice. *Evid Based Complement Alternat Med.* 2011;2011:165968.
50. Kornecook TJ, McKinney AP, Ferguson MT et al. Isoform-specific effects of apolipoprotein E on cognitive performance in targeted-replacement miceoverexpressing human APP. *Genes Brain Behav.* 2010;9:182–192.
51. Androutsellis-Theotokis A, Murase S, Boyd JD et al. Generating neurons from stem cells. *Methods Mol Biol.* 2008;438:31–38.
52. Hu P, Wang Y, Liu J et al. Chronic retinoic acid treatment suppresses adult hippocampal neurogenesis, in close correlation with depressive-like behavior. *Hippocampus.* 2016;i26:911–923. doi: 10.1002/hipo.22574
53. Schoenfeld TJ, Gould E. Differential effects of stress and glucocorticoids on adult neurogenesis. *Curr Top Behav Neurosci.* 2013;15:139–164.
54. Kino T. Stress, glucocorticoid hormones, and hippocampal neural progenitor cells: Implications to mood disorders. *Front Physiol.* 2015;6:230. doi: 10.3389/fphys.2015.00230
55. Zusso M, Debetto P, Guidolin D et al. Fluoxetine-induced proliferation and differentiation of neural progenitor cellsisolated from rat postnatal cerebellum. *Biochem Pharmacol.* 2008;76:391–403.
56. Morita H, Ikeda T, Kajita K et al. Effect of royal jelly ingestion for six months on healthy volunteers. *Nutr J.* 2012;11:77. doi: 10.1186/1475-2891-11

16 Quorum Sensing-Disrupting Compounds Derived from Plants
Novel Phytotherapeutics to Control Bacterial Infections

Kartik Baruah, Tom Defoirdt,
Parisa Norouzitallab, and Peter Bossier

CONTENTS

16.1 BACKGROUND

The introduction of antibiotics in the 1930s heralded a new era in the treatment of a large number of fatal diseases, including bacterial infections, which resulted in a marked decrease in the number of deaths from diseases (WHO 2014). However, excessive and nonjudicious use of antibiotics has resulted in the development of multiple drug-resistant bacterial strains. Recently, it has become increasingly difficult to treat many infections since therapeutic options are severely limited, and in some cases nonexistent, due to the emergence and spread of resistant bacteria. Multiple resistant organisms, for instance methicillin-resistant *Staphylococcus aureus*, are an increasing problem in hospital environments, despite the institution of strict control guidelines. Currently, diseases caused by antibiotic-resistant bacteria are the second leading causes of death worldwide (WHO 2014). Apart from rendering treatments ineffective, indiscriminate use of antibiotics, for instance in farmed animals, also constitutes a direct threat to human health and to the environment. The antibiotic resistance determinants that have emerged and/or evolved in the farmed animal environments have been shown to be transmitted by horizontal gene transfer to human pathogens. For example, the *Vibrio cholerae* that caused the 1992 Latin-American epidemic of cholera seemed to have acquired antibiotic resistance as a result of coming into contact with antibiotic-resistant bacteria selected through the heavy use of antibiotics in the Ecuadorian shrimp industry (Angulo 2000). Because of the increasing problems associated with the use of antibiotics, many countries have implemented strict regulations with respect to antibiotic use. One notable example is the ban on the use of antibiotics as growth promoters in animal production in Europe in 2006 (European Parliament and Council Regulation No. 1831/2003). However, there are good indications that this ban could result in a higher occurrence of pathogenic bacteria (such as *Salmonella* spp. *Vibrio* spp.), which in turn could lead to a higher frequency of infections in both animals and final consumers. As a result, the identification of novel drug targets and the development of novel eco-friendly therapeutics constitute an important area of current scientific research. An alternative to killing or inhibiting growth of pathogenic bacteria is the specific attenuation of bacterial virulence, which can be attained by targeting key regulatory systems that mediate the expression of virulence factors. One of the target regulatory systems is quorum sensing (QS), which is defined as a mechanism of gene regulation in which bacteria coordinate the expression of certain genes in response to the presence or absence of small signal molecules. This chapter focuses on the phenomena of QS in commercially important pathogenic bacteria, highlights the signaling molecules and novel regulatory components in QS responses, and furthermore summarizes the various recently learned QS disruptors that can be used as alternatives to conventional antimicrobial agents.

16.2 QS—BRIEF OVERVIEW

QS is a cell-to-cell communication process that enables bacteria to collectively modify behavior in response to changes in cell density, the species composition of the surrounding microbial community, and the characteristics of the environment. QS involves the production, release, and group-wide detection of extracellular signaling molecules. The signaling molecules accumulate in the environment as bacterial population density increases and/or as diffusion rates in the environment decrease. Bacteria monitor changes in the concentration of these molecules to track changes in their cell numbers and to collectively alter global patterns of gene expression. This mechanism was first discovered in the marine bacterium *Vibrio fischeri* and later on was found in many other bacteria. In Gram-negative bacteria, the QS system employs the *N*-acylhomoserine lactone (AHL or autoinducers) family of QS signaling molecules. These consist of five-membered homoserine lactone rings containing varied amide-linked acyl side chains (Figure 16.1) (Baruah et al. 2009). The N-acyl moieties of AHL ranges from 4 to 18 carbons in length and may be saturated or unsaturated, with or without a substituent (usually hydroxy or oxo) on the C3 carbon of the N-linked acyl chain. In addition, other

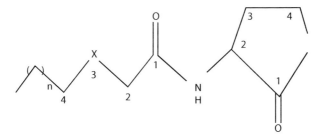

FIGURE 16.1   Basic structure of the AHLs (X = CH2 or CHOH or CO).

types of intercellular signal molecules in Gram-negative bacteria, including cyclic dipeptides and quinolones in *Pseudomonas aeruginosa*, unknown low-molecular-weight substances in *Xanthomonas campestris*, and a volatile fatty acid methyl ester in *Ralstonia solanacearum*, also exist. In contrast, in Gram-positive bacteria, many phenotypes are regulated in a population-density-dependent manner by signaling substances identified as γ-butyrolactones, structurally quite similar to AHLs, and posttranslationally modified peptides. More recently, a third major QS system has been unravelled in the luminescent bacterium *Vibrio campbellii*, especially the genome sequenced strain BB120 (=ATCC BAA-1116), which was previously designated *Vibrio harveyi* (Lin et al. 2010). This species uses three different signals, called Harveyi autoinducer 1 (HAI-1), autoinducer 2 (AI-2), and Cholerae autoinducer 1 (CAI-1).

16.3   QS SYSTEMS

In bacteria, many QS systems have been discovered, and three major systems will be described in more detail in the following:

1. AHL-mediated QS: In Gram-negative bacteria, three different AHL synthase families have been reported: the LuxI, the AinS, and the HdtS family. The LuxI-type proteins are the major class of enzymes responsible for AHL synthesis. These enzymes use S-adenosyl-methionine to synthesize the homoserine lactone ring, whereas the acyl chains come from lipid metabolism, carried by various acyl-carrier proteins. The AHL molecules diffuse freely through the plasma membrane. At low population densities, AHLs are present at a basal concentration level. However, as the population density increases, the AHL concentration increases as well and once a critical concentration is reached, AHL binds to the response regulator (RR) R protein (Figure 16.2) (Xavier & Bassler 2003). The activated LuxR–AHL complex forms dimers or multimers, which in turn act as transcriptional regulator on target genes of the QS systems, either upregulating or down-regulating the expression of target genes.
2. Peptide-mediated QS: The QS system in the Gram-positive bacteria is peptide mediated. A peptide signal (PS) precursor protein is cleaved, releasing

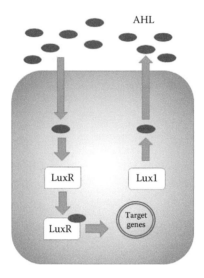

FIGURE 16.2   AHL-mediated QS in Gram-negative bacteria.

the actual signal molecule. The PS is transported out of the cell by an ATP binding cassette transporter. When a critical extracellular PS concentration is reached, a sensor kinase protein is activated to phosphorylate the RR. The phosphorylated RR activates transcription of the target genes (Figure 16.3) (Xavier & Bassler 2003).

3. Multichannel QS in vibrios: One of the signals produced by vibrios is HAI-1 (or AI1), a typical Gram-negative-type AHL identified as

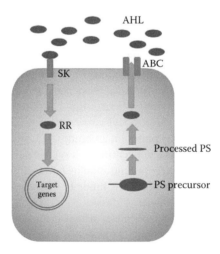

FIGURE 16.3   Peptide-mediated QS in Gram-positive bacteria.

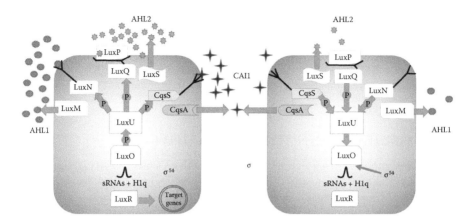

FIGURE 16.4 Multichannel system in *V. harveyi*.

N-(β-hydroxybutanoyl) homoserine lactone, whose biosynthesis is catalyzed by the LuxM enzyme. Second is AI-2, a furanosyl borate diester; and third is CAI-1, whose biosyntheses are catalyzed by LuxS and CqsA enzymes, respectively. All these three autoinducers are detected at the cell surface by the LuxN, LuxP–LuxQ, and CqsS receptor proteins, respectively. At low cell density, the receptors autophosphorylate and transfer phosphate to LuxO via LuxU (Figure 16.4) (Xavier & Bassler 2003). Phosphorylation activates LuxO, which, together with σ^{54}, activates the production of small regulatory RNAs (sRNAs). These sRNAs, together with the chaperone Hfq, destabilize the mRNA encoding the RR LuxR$_{vh}$, and thus, the LuxR$_{vh}$ protein is not produced. At high cell density, the receptor proteins switch from kinase to phosphatases, which results in dephosphorylation of LuxO. Dephosphorylated LuxO is inactive. The sRNAs are therefore not formed and the RR LuxR$_{vh}$ is produced (Henke & Bassler 2004; Figure 16.4).

16.4 AI-2-LIKE MOLECULE PRODUCTION BY OTHER MICROORGANISMS

V. harveyi recognizes AI-1 as a signal produced predominantly by other *V. harveyi* cells. Very few microorganisms have been shown to produce and utilize AI-1 as a QS molecule and so AI-1-based signaling may well be specific to *V. harveyi* cells. In other words, it may be used for intraspecies communication. In contrast, AI-2 represents a new class of bacterial signal molecule and is produced by a wide variety of Gram-negative and Gram-positive bacteria. Unlike for AHL and oligopeptide autoinducers, the AI-2 molecule are identical in all AI-2-producing bacteria studied to date. This finding indicates that AI-2 is a 'universal' signal, which functions in interspecies cell-to-cell communication.

As mentioned earlier, both Gram-positive and Gram-negative bacteria use QS. The biological roles of QS in Gram-positive bacteria have been extensively reviewed elsewhere (Rutherford & Bassler 2012; La Sarre & Federle 2013) and are therefore not discussed in this chapter. Here we focused on QS in Gram-negative bacteria that are responsible for causing diseases in human and farmed (aquatic) animals.

16.5   VIRULENCE PHENOTYPES REGULATED BY QS

QS has been reported to regulate a wide range of activities involved in pathogen–host interaction and microbe–microbe competition. These include luminescence, conjugation, nodulation, symbiosis, swarming, sporulation, biocorrosion, biosynthesis of antibiotic, growth inhibition, plasmid conjugal transfer, surfactants, exopolysaccharide biosynthesis, biofilm formation, and the production of virulence factors, such as lytic enzymes, toxins, siderophores, and adhesion molecules (De Windt et al. 2003). These QS pathogens probably increase their chances to infect their host successfully by maintaining a low profile of virulence factor production to avoid triggering host defenses until a sufficient number of cells is reached to mount an effective attack and overpower the host's immune system. It has been well demonstrated that inactivating the QS system of pathogens, a phenomenon known as quorum quenching, can markedly decrease the pathogenicity of bacteria.

16.6   QS: NOVEL TARGET FOR ANTI-INFECTIVE THERAPY

The discovery that a wide variety of bacteria uses QS to control virulence phenotypes makes it an attractive target for developing novel drugs. By blocking this cell-to-cell communication system, pathogens that use QS to control virulence could potentially be rendered avirulent. There exist several possible ways of interrupting the QS circuitry. For instance, autoinducers and LuxR proteins have a unique specificity for one another. Therefore, analogs that bind to but do not activate LuxR proteins could act as antagonists to prevent autoinducer binding, which in turn would shut down the QS cascade. The ability of autoinducer analogs to inhibit activation of LuxR proteins has already been demonstrated in a number of pathogenic bacteria (Schaefer et al. 1996; Swift et al. 1997; Zhu et al. 1998). Examples of this inhibition have been found to exist in nature. The red marine alga *Delisea pulchra* has developed such a defense mechanism to protect itself from extensive bacterial colonization. The alga

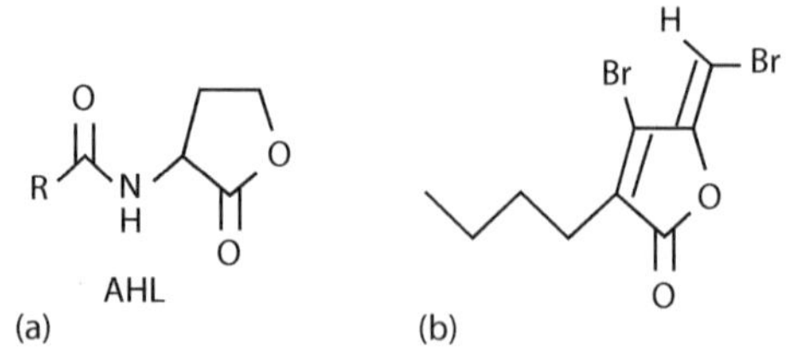

FIGURE 16.5   Structural similarity between (a) AHL and the (b) halogenated furanone.

produces the compound 'halogenated furanones' as antagonists for AHL-mediated QS. Because of their structural similarity with AHLs (Figure 16.5) (Defoirdt et al. 2004), the halogenated furanones bind to LuxR type proteins without activating those (Rasmussen et al. 2000). Givskov et al. (1996) showed that swarming of the pathogen *Serratia liquefaciens* on agar plates could be inhibited completely by adding 100 mg/L of furanone. This furanone could also suppress the expression of bioluminescence genes, located on a reporter plasmid in *S. liquefaciens*, without affecting the growth rate of the bacterium. Furanone was also found to inhibit extracellular toxin production in a pathogenic *V. harveyi* strain (Manefield et al. 2000). Mortality in the invertebrate *Penaeus monodon* was reduced to 50% after intramuscular injection with diluted cell supernatant extracts from *V. harveyi* cultures grown in the presence of the halogenated furanone, compared to extracts from untreated cultures.

In the invertebrate host *Artemia fransciscana*, the compound, at 20 mg/L, increased the survival of animals challenged with several different luminescent vibrios, suggesting that virulence attenuation caused by QS disruption is a general feature for luminescent vibrios (Defoirdt et al. 2006). Consistent with this, Tinh (2007) found that the furanone also neutralized the negative effect of *V. harveyi* BB120 toward gnotobiotic rotifers (an invertebrate). Unfortunately, the furanone seemed to be toxic to both brine shrimp and rotifers. Additionally, there were also reports showing furanones repressing AHL-dependent expression of *V. fischeri* bioluminescence, inhibiting AHL-controlled virulence factor production and pathogenesis in *P. aeruginosa*, inhibiting QS-controlled luminescence and virulence of the black tiger prawn pathogen *V. harveyi*, and finally inhibiting QS-controlled virulence of *Erwina carotovora*. The exact mechanism by which furanones inhibit QS-regulated virulence is unclear; however, a study had suggested that furanones promote rapid turnover of the LuxR protein, reducing the amount of protein available to interact with AHL and act as transcriptional regulator (Manefield et al. 2002). Following the findings that furanones possess QS disruption property, many investigators started investigating the activities of different synthetic AHL and furanone analogues with respect to QS (Hentzer et al. 2002; Smith et al. 2003). A synthetic derivative of the *D. pulchra* halogenated furanones, (5Z)-4-bromo-5-(bromomethylene)-2(5H)-furanone, has been reported the most active AHL antagonist so far. This furanone could almost completely reduce virulence factor expression in pure cultures of *P. aeruginosa* PAO1 (Hentzer et al. 2003). Interestingly, the furanone was equally active on biofilm bacteria compared to planktonic cells. Furthermore, furanone was also active *in vivo* in mouse lungs after infection with a *P. aeruginosa* strain harboring a fluorescent AHL reporter plasmid. The fluorescence signal from the *P. aeruginosa* strain was significantly reduced after subcutaneous injection of the furanone. After eight hours, however, the signal reappeared, indicating that the furanone had cleared from the mouse blood (Hentzer et al. 2003).

Besides *D. pulchra* furanones, antagonists produced by higher plants and microalgae were also shown to interact with bacterial QS. In fact, it was found that exudates from higher plants, such as pea, rice, soybean, tomato, crown vetch, and *Medicago truncatula*, also influence AHL-mediated QS (Teplitski et al. 2000). In contrast to *D. pulchra*, these plants secrete substances that stimulate AHL-dependent QS as well as substances that inhibit such responses (Teplitski et al. 2004). For instance, the microalgae *Chlamydomonas mutablis, Chlorella vulgaris,* and *Chlorella fusca* all stimulated QS-regulated luminescence in wild-type *V. harveyi.* On the contrary, *Chlamydomonas reinhardtii* inhibited AHL-mediated luminescence in several different *Escherichia coli* AHL reporter strains. The inhibition of luminescence was shown not to be due to toxicity or to an inhibition of luminescence as such since the luminescence of a constitutively luminescent derivative of *E. coli* was not inhibited. However, the chemical nature of the QS mimic compounds secreted by higher plants and microalgae still remained to be characterized.

In the search for compounds with a higher QS disruption potential, the synthesis of brominated thiophenones, sulphur analogues of brominated furanones, has recently been reported (Benneche et al. 2011). These compounds caused a similar effect on the QS system of bacteria as brominated furanones; i.e., they decrease the DNA binding activity of LuxR$_{Vh}$. It has been proposed that both types of compounds covalently bind to proteins through an addition–elimination mechanism. Thiophenones are more active than the corresponding brominated furanones, with a concentration of 2.5 μM having a similar effect as approximately 100 μM of the furanones. One thiophenone compound, (Z)-4-((5-(bromomethylene)-2-oxo-2,5-dihydrothiophen-3-yl)methoxy)-4-oxo-butanoic acid, was shown to have an interesting therapeutic potential to treat luminescent vibriosis, with a therapeutic index of about 100 (Defoirdt et al. 2012). Another nontoxic natural compound cinnamaldehyde showed antibacterial properties by mechanism of disrupting bacterial QS. In fact, at subinhibitory concentrations (approximately 100 μM), the compound showed the same QS activity as brominated furanones and thiophenones (Brackman et al. 2008). A major advantage of this compound with respect to practical application is that it is "generally recognized as safe."

In addition to the QS inhibitors described previously, butyrolactone and acetyl-butyrolactone (Figure 16.6) have also been found to repress AHL-mediated QS with no inhibition of *P. aeruginosa* growth (Pan & Ren 2009). Another natural compound, hamamelitannin (2′,5-di-*O*-galloyl-D-hamamelose), a nonpeptide compound isolated from the bark of *Hamamelis virginiana,* was shown to inhibit QS in *S. aureus* and *Staphylococcus epidermidis* at concentrations noninhibitory to growth (Kiran et al. 2008a). This compound was also shown to prevent device-associated infections by methicillin-resistant *S. aureus* and *S. epidermidis* in a rat graft model. Although no effect was observed on *Staphylococcal* growth *in vitro,* hamamelitannin inhibited its infection in a dose-dependent manner *in vivo* in a rat subcutaneous graft model (Kiran et al. 2008b).

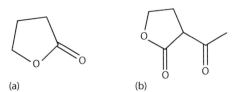

(a)   (b)

FIGURE 16.6 Structure of (a) butyrolactone and (b) acetyl-butyrolactone.

16.7 RELIABLE IDENTIFICATION OF QS INHIBITORS

Many studies claim QS inhibitory activity of different natural and/or synthetic compounds, but many of these studies rely heavily on the inhibition of QS-regulated phenotypes in biosensor strains, which demonstrate a certain phenotype in response to QS molecules (Steindler & Venturi 2007). An important limitation of the use of such reporter strains is that the QS-regulated phenotypes are often codependent on other factors and/or depend on the metabolic activity of the cells. Many reporter strains, for instance, are based on the production of light (bioluminescence) in response to QS molecules. However, bioluminescence is dependent on the metabolic activity of the cells as the reaction requires large amounts of energy. The same is true for reporters based on other phenotypes such as β-galactosidase activity (through *lacZ*) and the production of green fluorescent protein (GFP), which will also be affected by the overall metabolic activity of the cells. Moreover, a test compound might directly interfere with the reporter phenotype rather than with its regulation (e.g., by interfering with bioluminescence biochemistry or folding of GFP or LacZ). Hence, adequate control experiments, in which the compounds are verified to have no effect on the particular phenotype used in the biosensor strain when this phenotype is no longer dependent on QS, should be included (Defoirdt et al. 2013). These additional experiments could include (i) verification that the compound does not affect the biosensor phenotype when it is independent of QS, (ii) assessment of the impact on other phenotypes that are controlled by the QS system of interest, (iii) transcriptomic analyses, (iv) identification of the molecular target of the compound, and (v) sensitive toxicity tests. The inclusion of some of these experiments in studies aiming to discover QS inhibitors should result in more reliable identification of compounds that are true QS inhibitors in the future.

16.8 QS AND NEUROLOGICAL DISEASES

Bacteria communicate with each other by the use of signaling molecules, a process called "quorum sensing." One group of QS molecules includes the oligopeptides, which are mainly produced by Gram-positive bacteria. Recently, these QS peptides were found to biologically influence mammalian cells, promoting *i.a.* metastasis of cancer cells. Moreover, it was found that bacteria can influence different central nervous system related disorders as well, e.g., anxiety, depression, and autism. Research currently focuses on the role of bacterial metabolites in this bacteria–brain interaction, with the role of the QS peptides not yet known. Here, three chemically diverse QS peptides were investigated for their brain influx (multiple time regression technique) and efflux properties in an *in vivo* mouse model (ICR-CD-1) to determine blood–brain transfer properties: PhrCACET1 demonstrated comparatively a very high initial influx into the mouse brain ($K_{in} = 20.87$ µl/(g × minute)), while brain penetrabilities of BIP-2 and PhrANTH2 were found to be low ($K_{in} = 2.68$ µl/ (g × minute)) and very low ($K_{in} = 0.18$ µl/(g × minute)), respectively. All three QS peptides were metabolically stable in plasma (*in vitro*) during the experimental time frame and no significant brain efflux was observed. Initial tissue distribution data showed remarkably high liver accumulation of BIP-2 as well. These results thus support the potential role of some QS peptides in different neurological disorders, thereby enlarging our knowledge about the microbiome–brain axis (Wynendaele et al. 2015).

Some of the neurological diseases are associated with an altered microbiota composition. In rodent models of anxiety, depression, and autism-like behavior, a disrupted microbiome was observed (Grenham et al. 2011; Foster et al. 2013; Kang et al. 2013). Stress hormones (corticosterone and adrenocorticotropic hormone) also rose in germ-free mice compared to normal control mice when exposed to the same stress conditions and pretreatment with *Bifidobacterium infantis* induced more normal hormonal responses, again indicating the influence of the microbiome on stress responses (Sudo 2012; Dinan et al. 2015).

Despite very numerous studies on Alzheimer's disease (AD), especially on amyloid plaques and neurofibrillary tangles, little information has been obtained thus on the causes of the disease. Recently, several evidences have implicated herpes simplex virus type 1 as a strong risk factor for AD when it is present in brain of carriers of the type 4 allele of the gene for apolipoprotein E. A second putative risk factor is the bacterium *Chlamydia pneumoniae*. This pathogen has been identified and localized in AD brain. An infection-based animal model demonstrates that following intranasal inoculation of BALB/c mice with *C. pneumoniae*, amyloid plaques/deposits consistent with those observed in the AD brain develop, thus implicating this infection in the etiology of AD (Itzhaki et al. 2004). The review by Holm and Vikstrom (2014) highlighted these aspects of QS, and *P. aeruginosa* communicates with human cells using the small QS signal molecules AHLs. Both direct and long-range interactions between pathogenic *P. aeruginosa* bacteria and their eukaryotic hosts are important in the outcome of infections. For cell-to-cell communication, these bacteria employ the QS system to pass on information of the density of the bacterial population and collectively switch on virulence factor production, biofilm formation, and resistance development (Holm and Vikstrom, 2014).

16.9 CONCLUDING REMARKS

Here, we have reviewed the diverse role of AHL signaling molecules in bacterial cell-to-cell communication, as well as their potential role in the interaction of bacteria with eucaryotic hosts. As the list of bacteria that use QS systems continues to increase, so does the number of possibilities for exploiting these regulatory mechanisms. As many important human and animal pathogens use QS to regulate virulence, strategies designed to interfere with these signaling systems might have broad applicability for biological control of disease-causing organisms. In the future, it would be interesting to investigate whether more human pathogens utilize QS as part of their pathogenic lifestyle and, if so, whether inhibition of QS can be exploited to control infections.

ACKNOWLEDGMENTS

The authors acknowledge financial support from the Research Foundation Flanders, Fonds Wetenschappelijk Onderzoek-Vlaanderen (FWO), Belgium—postdoc grant to Dr. Kartik Baruah (FWO13/PDO/005) and NEPTUNA project entitled "novel extraction processes for multiple high-value compounds from selected algal source materials."

REFERENCES

Angulo FJ (2000). Antimicrobial agents in aquaculture: Potential impact on health. *APUA Newsletter* 18:1–6.

Baruah K, Norouzitallab P, Bossier P, Sorgeloos P (2009). Quorum sensing disruption: A novel approach to combat disease in aquaculture. *World Aquaculture Magazine* 40 (4):14–16.

Benneche T, Herstad G, Rosenberg M, Assev S, Scheie AA (2011). Facile synthesis of 5-(alkylidene)thiophen-2(5H)-ones. A new class of antimicrobial agents. *RSC Advances* 1:323–332.

Brackman G, Defoirdt T, Miyamoto C et al. (2008). Cinnamaldehyde and cinnamaldehyde derivatives reduce virulence in *Vibrio* spp. by decreasing the DNA-binding activity of the quorum sensing response regulator LuxR. *BMC Microbiology* 8:149.

De Windt W, Boon N, Siciliano SD et al. (2003). Cell density related H-2 consumption in relation to anoxic Fe(0) corrosion and precipitation of corrosion products by *Shewanella oneidensis* MR-1. *Environmental Microbiology* 5:1192–1202.

Defoirdt T, Boon N, Bossier P, Verstraete W (2004). Disruption of bacterial quorum sensing: An unexplored strategy to fight infections in aquaculture. *Aquaculture* 240:69–88.

Defoirdt T, Benneche T, Brackman G et al. (2012). A quorum sensing-disrupting brominated thiophenone with a promising therapeutic potential to treat luminescent vibriosis. *PLoS One* 7:e41788.

Defoirdt T, Brackman G, Coenye T (2013). Quorum sensing inhibitors: How strong is the evidence? *Trends in Microbiology* 21:619–624.

Defoirdt T, Crab R, Wood TK et al. (2006). Quorum sensing-disrupting brominated furanones protect the gnotobiotic brine shrimp *Artemia franciscana* from pathogenic *Vibrio harveyi*, *Vibrio campbellii* and *Vibrio parahaemolyticus* isolates. *Applied Environment and Microbiology* 72:6419–6423.

Dinan TG, Stilling RM, Stanton C et al. (2015). Collective unconscious: How gut microbes shape human behavior. *Journal of Psychiatric Research* 63:1–9. doi: 10.1016/j .jpsychires.2015.02.021. pmid:25772005

Foster JA, McVey Neufeld K-A (2013). Gut–brain axis: How the microbiome influences anxiety and depression. *Trends in Neuroscience* 36:305–312. doi: 10.1016/j.tins.2013.01 .005.pmid:23384445

Givskov M, de Nys R, Manefield M et al. (1996). Eukaryotic interference with homoserine lactone-mediated prokaryotic signalling. *Journal of Bacteriology* 178: 6618–6622.

Grenham S, Clarke G, Cryan JF et al. (2011). Brain–gut–microbiome communication in health and disease. *Frontiers in Physiology* 2:94. doi: 10.3389/fphys.2011.00094 .pmid:22162969

Henke JM, Bassler BL (2004). Three parallel quorum-sensing systems regulate gene expression in *Vibrio harveyi*. *Journal of Bacteriology* 186:6902–6914.

Hentzer M, Riedel K, Rasmussen TB et al. (2002). Inhibition of quorum sensing in Pseudomonas aeruginosa biofilm bacteria by a halogenated furanone compound. *Microbiology* 148:87–102.

Hentzer M, Wu H, Andersen JB et al. (2003). Attenuation of Pseudomonas aeruginosa virulence by quorum sensing inhibitors. *EMBO Journal* 22:3803–3815.

Holm A, Vikström E (2014). Quorum sensing communication between bacteria and human cells: Signals, targets, and functions. *Frontiers in Plant Science* 5:309. http://doi.org /10.3389/fpls.2014.00309

Itzhaki RF1, Wozniak MA, Appelt DM et al. (2004). Infiltration of the brain by pathogens causes Alzheimer's disease. *Neurobiology of Aging* 25(5):619–627.

Kang DW, Park JG, Ilhan ZE et al. (2013). Reduced incidence of Prevotella and other fermenters in intestinal microflora of autistic children. *Plos One* 8(7):e68322

Kiran MD, Adikesavan NV, Cirioni O et al. (2008a). Discovery of a quorum-sensing inhibitor of drug-resistant staphylococcal infections by structure-based virtual screening. *Molecular Pharmacology* 73:1578–1586.

Kiran MD, Giacometti A, Cirioni O et al. (2008b). Suppression of biofilm related, device-associated infections by staphylococcal quorum sensing inhibitors. *International Journal of Artificial Organs* 31:761–770.

La Sarre B, Federle MJ (2013). Exploiting quorum sensing to confuse bacterial pathogens. *Microbiology and Molecular Biology Reviews* 77:73–111.

Lin B, Wang Z, Malanoski AP et al. (2010) Comparative genomic analyses identify the *Vibrio harveyi* genome sequenced strains BAA-1116 and HY01 as *Vibrio campbellii*. *Environmental Microbiology Reports* 2:81–89.

Manefield M, Harris L, Rice SA et al. (2000). Inhibition of luminescence and virulence in the black tiger prawn (*Penaeus monodon*) pathogen *Vibrio harveyi* by intercellular signal antagonists. *Applied and Environmental Microbiology* 66:2079–2084.

Manefield M, Rasmussen TB, Henzter M et al. (2002). Halogenated furanones inhibit quorum sensing through accelerated LuxR turnover. *Microbiology* 148:1119–1127.

Pan J, Ren D (2009). Quorum sensing inhibitors: A patent overview. *Expert Opinion on Therapeutic Patents* 19:1581–1601.

Rasmussen TB, Manefield M, Andersen JB et al. (2000). How *Delisea pulchra* furanones affect quorum sensing and swarming motility in *Serratia liquefaciens* MG1. *Microbiology* 146:3237–3244.

Rutherford ST, Bassler BL (2012). Bacterial quorum sensing: Its role in virulence and possibilities for its control. *Cold Spring Harbor Perspectives in Medicine* 2:a012427.

Schaefer AL, Hanzelka BL, Eberhard A, Greenberg EP (1996). Quorum sensing in *Vibrio fischeri*: Probing autoinducer-LuxR interactions with autoinducer analogs. *Journal of Bacteriology* 178:2897–2901.

Smith KM, Bu Y, Suga H (2003). Induction and inhibition of *Pseudomonas aeruginosa* quorum sensing by synthetic autoinducer analogs. *Chemical & Biology* 10:81–89.

Steindler L, Venturi V (2007). Detection of quorum-sensing N-acyl homoserine lactone signal molecules by bacterial biosensors. *FEMS Microbiology Letters* 266:1–9.

Sudo N (2012). Role of microbiome in regulating the HPA axis and its relevance to allergy. *Chemical Immunology & Allergy* 98:163–175. doi: 10.1159/000336510. pmid:22767063

Swift S, Karlyshev AV, Durant EL et al. (1997). Quorum sensing in *Aeromonas hydrophila* and *Aeromonas salmonicida*: Identification of the LuxRI homologues AhyRI and AsaRI and their cognate signal molecules. *Journal of Bacteriology* 179:5271–5281.

Teplitski M, Chen H, Rajamani S et al. (2004). *Chlamydomonas reinhardtii* secretes compounds that mimic bacterial signals and interfere with quorum sensing regulation in bacteria. *Plant Physiology* 134:137–146.

Teplitski M, Robinson JB, Bauer WD (2000). Plants secrete substances that mimic bacterial *N*-acyl homoserine lactone signal activities and affect population density-dependent behaviors in associated bacteria. *Molecular Plant–Microbe Interaction* 13:637–648.

Tinh NTN (2007). Quorum sensing control for microbial management of the aquaculture food chain rotifers – turbot. PhD Thesis, University of Ghent, Ghent, Belgium.

WHO (2014). *Antimicrobial Resistance: Global Report on Surveillance.* World Health Organization, Geneva, Switzerland.

Wynendaele E, Verbeke F, Stalmans S et al. (2015). Quorum sensing peptides selectively penetrate the blood–brain barrier. *PLoS One* 10(11): e0142071. doi:10.1371/journal .pone.0142071

Xavier KB, Bassler BL (2003). LuxS quorum sensing: More than just a numbers game. *Current Opinion in Microbiology* 6:191–197.

Zhu J, Beaber JW, Moré MI et al. (1998). Analogs of the autoinducer 3-oxooctanoyl-homoserine lactone strongly inhibit activity of the TraR protein of *Agrobacterium tumefaciens. Journal of Bacteriology* 180:5398–5405.

Section III

Molecular Mechanisms

17 Interaction of Phytopharmaceuticals with Neurotransmitters in the Brain

Leah Mitchell-Bush, Jenaye Robinson, Odochi I. Ohia-Nwoko, Emily Kawesa-Bass, Benjamin A. Stancombe, Catherine A. Opere, Ya Fatou Njie-Mbye, and Sunny E. Ohia

CONTENTS

17.1 INTRODUCTION

Chemicals derived from plant sources (phytochemicals) have been used since prehistoric times for the treatment of diseases. Based on the prevalence of brain diseases such as Parkinson's, dementia, amyotrophic lateral sclerosis (ALS), and epilepsy, both preclinical and clinical studies using natural phytochemicals have been vigorously pursued in the search for more effective therapies for these diseases. In this chapter, an attempt has been made to provide the reader with brief descriptions of the pathophysiology of some common brain diseases along with the current therapeutic approaches being used to prevent or ameliorate the symptoms. Furthermore, evidence from both preclinical and clinical studies has been provided to ascertain the efficacies of phytochemicals as potential therapeutic agents for each of these diseases.

17.2 PARKINSON'S DISEASE

Parkinson's disease (PD) is a progressive neurodegenerative brain disorder that affects motor function. Symptoms of PD typically include uncontrolled tremor while at rest, bradykinesia, and rigidity. Nonmotor symptoms are also associated with PD and they include dementia, depression, and psychosis. Since there is no definitive laboratory test that can be used to positively diagnose PD, it is often ruled out among other disease states. The presence of Lewy bodies in various parts of the brain (e.g., substantia nigra, locus coeruleus, and cerebral cortex) and the cardiac sympathetic plexus is used to diagnose PD. The presence of Lewy bodies has also been associated with other neurodegenerative diseases such as Alzheimer disease (AD).

The motor dysfunction observed in PD is primarily due to the depletion of dopamine in the substantia nigra. Thus, the most effective pharmacotherapy for the severe bradykinetic symptoms is Sinemet (carbidopa-levodopa). Carbidopa (a dihydroxyphenylalanine [DOPA] decarboxylase inhibitor) is usually combined with levodopa (L-3,4-DOPA [L-DOPA], a dopamine precursor) to prevent the metabolism of levodopa in the periphery that results in an increase in L-DOPA's absorption, and a subsequent increase in dopamine levels in the central nervous system (CNS). Other approved pharmacotherapies for PD include dopamine agonists, monoamine oxidase B (MAO-B) inhibitors, anticholinergics, amantadine, and catechol-O-methyltransferase inhibitors.

In addition to traditional pharmacotherapies, there is evidence that phytochemicals that are effective in ameliorating the symptoms of PD may have a direct and/or indirect action on dopaminergic neurotransmission. Phytochemicals act on the intermediate products of dopamine biosynthesis (e.g., L-DOPA) or on enzymes that metabolize dopamine (e.g., MAO-B). For example, about 7%–10% of L-DOPA can be extracted from the seed of *Mucuna pruriens*,[1] a legume that is commonly found in tropical climates. HP-200 is a commercially available formulation of *M. pruriens* that has proven to be more effective than synthetic L-DOPA/syndopa in a multicenter clinical trial with 60 patients over a 12-week period. A daily dose of 600 mg/kg of *M. pruriens* is considered nontoxic. Another phytochemical, amur corktree (*Phellodendron amurense*), inhibits human MAO-B enzyme activity and has possible neuroprotective properties that could be useful in PD.[2] Licorice root (*Glycyrrhiza glabra* or *Glycyrrhiza uralensis*) has been reported to have anti-inflammatory activity in microglial cells and can protect nigrostriatal dopaminergic neurons from degeneration in mice.[2] Additionally, zigerone, an alkaloid extracted from ginger, can be used to increase dopamine levels in the brain. The mechanism of action includes increasing the activity of superoxide dismutase and the scavenging activity of superoxide radicals.[3] The U.S. Patent number 6106839 outlines the use of ginger, *M. pruriens*, and *Piper longum* in PD.

The psoralea fruit (*Psoralea corylifolia*, PC) has been reported to inhibit monoamine transport, which affects the uptake of dopamine and norepinephrine.[4] Consequently, PC may be used for the prevention of PD, or in combination with L-DOPA, to improve motor symptoms. Extracts from the Bakuchi seed (*Cyamopsis psoralioides*) can inhibit the human MAO-B enzyme and has been shown to possess neuroprotective properties.[5] It is feasible that extracts from the Bakuchi seed could also be useful in the treatment of PD.

Interestingly, epidemiological studies indicate that there was a 50% reduction in the incidence of PD in people who smoked tobacco.[6] The main psychoactive component isolated from tobacco (*Nicotiana tabacum*) is the pyrrolidine alkaloid, nicotine. Nicotine binds and activates the nicotinic acetylcholine receptor, an essential receptor involved in the regulation of dopamine in the brain. Chronic ingestion of nicotine can increase the number of nicotinic receptors, and this may also have neuroprotective effects. However, more clinical studies are needed to assess the potential therapeutic efficacy of nicotine for PD.

Based on the presumed role of oxidative stress in the pathogenesis of neurodegenerative disorders, the phytochemicals in berry fruits (e.g., anthocyanin and caffeic acid) have been reported to delay the onset of PD.[7] The phytochemicals in berry fruits are effective in PD because of their antioxidative, anti-inflammatory, antiviral, and antiproliferative actions.[7] There is evidence that dietary phytochemicals effective in the prevention or treatment of neurodegenerative diseases such as PD may target neurotrophins (e.g., nerve growth factor [NGF], brain-derived neurotrophic factor, and neurotrophin-3). A decrease in neurotrophins has been implicated in the pathophysiology of many neurodegenerative disorders.

Based on the evidence presented earlier, the use of phytochemicals to treat PD could be a promising alternative to traditional pharmacological therapies. Given that several phytochemicals appear to be beneficial for PD, it is not surprising that there has been scientific and clinical interest in alternative therapeutics for other brain diseases, including dementia, ALS, and epilepsy.

17.3   DEMENTIA

Dementias are brain diseases most notably characterized by a gradual decline in memory. According to the *Diagnostic and Statistical Manual, 5th Edition*, dementias are neurocognitive disorders that can be further classified into major and minor forms. There are several types of dementia that are characterized by varying, but distinct, pathological changes in the brain. According to the World Health Organization, AD is by far the most common, accounting for 60%–70% of dementia cases.[2] Major neuropathological characteristics of AD include the formation of extracellular beta-amyloid plaques and the intraneuronal accumulation of neurofibrillary tangles (comprised of hyperphosphorylated tau protein). The plaques are presumed to disrupt synaptic transmission between neurons while the tangles interfere with the transport of molecules within neurons. The development of plaques and subsequent accumulation of neurofibrillary tangles result in progressive neurodegeneration. A significant reduction in brain volume can be observed in patients with advanced stages of AD.

In addition to AD, there are several other types of dementia. For example, vascular dementia develops after cerebrovascular events. Injury to blood vessels in the brain causes infarcts or bleeding in the brain. The development of dementia depends on the size, location, and number of cerebrovascular events. Dementia with Lewy bodies is caused by the intraneuronal accumulation of the alpha-synuclein protein in the cortex. Frontotemporal dementias are characterized by atrophy of the frontal and temporal lobes of the brain and are a major cause of cognitive dysfunction in patients younger than 65 years of age. Mixed dementia is becoming a more common type of diagnosis in dementia patients. This type of dementia displays the neuropathology and symptomology of several types of co-occurring dementias. For example, the combination of AD and vascular dementia is the most prevalent type of mixed dementia.

Current treatments for dementias are symptomatic, as there are no disease-modifying therapies available. Acetylcholinesterase inhibitors are the most common medications used to treat dementias. These medications prevent the metabolism of acetylcholine, thereby increasing cholinergic neurotransmission and improving cognitive function. The Food and Drug Administration (FDA)-approved cholinesterase inhibitors for AD are tacrine, donepezil, revastigmine, and galantamine. Memantine, an N-methyl-D-aspartate (NMDA) receptor antagonist, is an additional FDA-approved therapeutic for AD. The NMDA receptor is activated by glutamate and is involved in learning and memory. Overstimulation of NMDA receptors can cause ischemia; therefore, agents that block NMDA receptors are especially effective for the treatment of vascular dementia. For patients with moderate to advanced dementia, the combination of a cholinesterase inhibitor and memantine is often used.

Antioxidant therapies that contain vitamin E and selegiline have been used as treatments for AD, with mixed results in randomized trials. Observational studies suggest that there is an association between the use of omega-3 fatty acids and a reduction in the risk of dementia. Randomized control trials have not been able to support this association. One randomized control trial showed that omega-3 fatty acids did not significantly impact cognitive decline when compared to placebo.[8]

Phytochemicals have also been used in the treatment of some types of dementia. Galantamine was first isolated from the Galanthus plants in Eastern Europe and is a selective and competitive inhibitor of acetylcholinesterase. Galantamine is also an allosteric modulator of nicotinic acetylcholine receptors. There is evidence that galantamine administration can be neuroprotective by inhibiting β-amyloid aggregation and modulating amyloid precursor processing. In a recent study, galantamine was shown to delay the progression of amyloid plaque formation and improve behavioral symptoms in a mouse model of AD.[9]

Extracts from sage plants have been used to treat a variety of diseases and ailments in traditional medicine. For instance, extracts from sage plants can inhibit acetylcholinesterase and could have beneficial effects on memory and other AD-associated symptoms. One study demonstrated that essential oil of *Salvia officinalis* at a concentration of 0.5 mg/mL resulted in a 46% inhibition of acetylcholinesterase.[10] Additionally, the neuroprotective characteristic of extracts from sage plant was shown in a study of amyloid-β peptide-induced toxicity in cultured rat pheochromocytoma (PC12) cells. It was proposed that the neuroprotective activity was due to rosmarinic acid, a component of sage.[11]

Extracts from the *Huperzia serrata* plant have been used in traditional Chinese medicine for the treatment of memory impairments. It is composed of quinolizidine-related alkaloids, named huperzines A and B. Huperzine A has been shown to act as a potent acetylcholinesterase inhibitor and also to exhibit neuroprotective properties.[12] A study using $A\beta_{42}$-induced injury in rat neurons showed that huperzine A can reduce intracellular and mitochondrial Aβ.[13] A meta-analysis of randomized controlled trials that administered huperzine A indicated a potential improvement in the cognitive function of Alzheimer patients.[14] However, the authors concluded that the results should be interpreted with caution because of the poor methodologies used in the trials.

Ginseng is another traditional Chinese medicine that is commonly used in many CNS diseases, including AD. The major active compounds in ginseng are ginsenosides. The role of the ginsenosides in treating AD may involve multiple mechanisms of action. There is evidence that ginsenosides can inhibit glutamate and amyloid-β-induced cytotoxicity, as well as tau phosphorylation. Ginsenosides can also increase the uptake of choline and increase the release of acetylcholine in the hippocampus.[15] Randomized control clinical trials indicate that ginsenosides can improve cognitive function. A recent systematic review and meta-analysis of four randomized control trials concluded that the potential benefits of ginseng as a treatment for AD are still inconclusive.[16]

Polyphenols are another class of phytochemicals that may be therapeutically beneficial for patients with AD and other dementias. Resveratrol is a polyphenolic non-flavanoid compound that is present in over 70 different plants. Because resveratrol can produce neuroprotective effects, there are several preclinical studies that have investigated its effects in animal models of dementia. For example, in the senescence accelerated mouse model of AD, 1 mg/kg of dietary resveratrol reduced amyloid-beta accumulation, tau hyperphosphorylation, and cognitive impairments.[17] A reduction in amyloid-beta plaques was also observed after oral resveratrol administration in the APP/PS1 mouse model of AD.[18] Moreover, several studies that utilized a rat model of vascular dementia reported that administration of resveratrol could reduce learning and memory deficits, probably through attenuating oxidative stress and apoptotic cell death in the brain.[19–21]

To date, there has been one clinical trial that assessed the effects of resveratrol in AD patients. The investigators assessed the safety and tolerability of resveratrol, as well as its impact on cerebral spinal fluid (CSF) and plasma levels of amyloid-beta and hyperphosphorylated tau. Daily resveratrol treatment (500-mg increment dose escalation over 13 weeks) was well tolerated and prevented a significant decline in amyloid-beta 40 levels in the plasma and CSF.[22] Low levels of resveratrol and its metabolites were also observed in the plasma and CSF, which the authors suggested might indicate that resveratrol could engage central molecular targets. However, since several other biomarker and cognitive trajectories were unchanged by resveratrol treatment, the authors suggested the results be interpreted with caution and that future large-scale clinical studies are needed to determine the potential benefits of resveratrol in patients with AD.

Epigallocatechin-3-gallate (EGCG) is a polyphenolic flavanoid compound that has potential benefits for dementia patients. EGCG can be extracted from *Camillia sinesis* and is the most abundant catechin in tea. EGCG has powerful antioxidant and anti-inflammatory effects; thus, it has been studied in various animal models of AD and dementia. EGCG reduced amyloid-beta brain levels and attenuated cognitive deficits in APP/PS1 mice and PS-2 mutant mice, potentially through increased neurogenesis (via activation of the NGF–tropomyosin receptor kinase A pathway).[23–25] Although there have been several preclinical studies investigating the potential benefits of EGCG, there is only one clinical trial investigating its affects in early-stage

AD (ClinicalTrails.gov identifier: NCT00951834). The final results of this clinical trial are yet to be reported. It will be interesting to find out if EGCG can be therapeutically effective in human patients.

17.4 AMYOTROPHIC LATERAL SCLEROSIS

ALS (also known as Lou Gehrig's disease) is a progressive neurodegenerative disease characterized by loss of motor neurons in the brain and spinal cord. In Europe, the incidence of ALS is two or three people per 100,000, while in the United States, the incidence rate is 3.9 cases per 100,000 persons. The disease is more common among white, non-Hispanic males and persons 60–69 years of age. Approximately 5%–10% of patients develop an inherited form of familial ALS (FALS), which is associated with more than a dozen genes and usually passed on in an autosomal-dominant manner. About 90%–95% of patients have sporadic ALS (SALS), which occurs without known genetic risk factors and with no family history of the disease.

ALS is a complex, multifactorial disease with a spectrum of phenotypes that could comprise a single disease or result as a manifestation of several closely related disorders with different causes but similar clinical expression. Clinical progression is generally rapid, beginning with a loss of muscle mass, followed by muscle degeneration, paralysis, and respiratory issues. The prognosis for most patients is three to five years after the initial diagnosis before succumbing to the disease due to respiratory or cardiac failure. The etiology underlying ALS is unknown; however, various cellular populations and processes are known to be involved: motor neuron loss and lesions, retraction of motor neurons from neuromuscular junctions, the appearance of inclusion bodies within neurons and astrocytes, ubiquitin-positive protein aggregates in neurons, and blood–brain barrier disruption. Although many gene mutations are associated with the development of FALS, the genetic contribution to SALS is still unclear. Gender, exposure to toxic chemicals, and hazardous environmental exposures during military service are all risk factors for SALS.

Currently, riluzole is the only pharmacological intervention used to attenuate the progression of ALS, which is recommended by the National Institute for Clinical Excellence and approved by the FDA.[26] Riluzole can slightly mitigate respiratory failure and increase the survival time up to three months; however, there are no scientific reports indicating a beneficial role on muscle atrophy, wasting, weakness, muscle spasticity, or quality of life in ALS patients, demonstrating a need for more therapeutics.

Recently, increasing attention has been given to investigating complementary and alternative medicine for ALS. The use of phytopharmaceuticals is based on a theory of encouraging the body to repair itself and maintain its internal balance through the use of herbal remedies. Plant-derived pharmaceutical clinical prescriptions have been conceived and tested clinically over time, thus providing a conceptual framework within which to search for herbal remedies for ALS. A multitude of experimental research information on herbal monomers or compounds associated with neuroprotection or treatment of ALS has been published, exemplifying the use of herbs and herbal extracts and their effectiveness against this disease.

Oxidative stress is a major underlying mechanism associated with several neurodegenerative diseases, including ALS. Excessive reactive oxygen species (ROS) in the brain and impairment of the intrinsic antioxidant system such as that caused by mutations in the SOD1 gene in FALS can lead to the death of neurons.[27] Some herbal remedies have the potential to counteract oxidative stress through their ability to increase the antioxidant activity of the enzymatic/non enzymatic system, reducing the release of intracellular ROS and adjusting gene expression and regulation.

There is a critical correlative relationship between inflammation in the CNS and the occurrence of ALS. Under normal physiological conditions, microglia exist in a resting state; however, under pathological conditions, these cells are rapidly activated and secrete proinflammatory factors through the expression of inflammatory mediators such as tumor necrosis factor-α (TNF-α), inducible nitric oxide synthase (iNOS), or cyclooxygenase, which participate in the pathogenesis of ALS.[27] Evidence suggests that suppression of microglial activation and inhibition of neuroinflammation can attenuate the risk of ALS.[27]

Ginkgo biloba or Maidenhair tree is a well-known medicinal plant native to China with leaves rich in free radical scavenging flavonoids. More than 397 clinical studies have evaluated the beneficial role of *G. biloba* on a multitude of disorders, including dementias, schizophrenia, pulmonary vascular disease, multiple sclerosis, glaucoma, asthma, and acute ischemic stroke, alone and in conjunction with dietary compounds. It has been reported that the *G. biloba* extract possesses a sex-specific neuroprotective role in transgenic (SOD1-G93A) mouse models of ALS. Upon oral administration, *G. biloba* extract was found to significantly mitigate abnormal motor performance and increase survival time. Male transgenic mice displayed a decrease in the loss of spinal cord anterior motor horn neurons and a decrease in overall weight loss after *G. biloba* treatment.[28] Additionally, *G. biloba* extract has been shown to promote healthy mitochondrial function, and in an in vitro study, it was found to protect against glutamate-induced excitotoxicity.[29] Taken together, these evidences suggest the potential role of *G. biloba* as an effective treatment in patients with ALS.

EGCG is the water-soluble main active polyphenolic compound in green tea. It can be used as a radical scavenger and mediate indirect antioxidant effects in neurodegenerative diseases such as ALS due to its strong antioxidative properties. It has been reported that EGCG affects molecular targets such as the mitogen-activated protein kinase (MAPK) pathway, phosphoinositide3-kinase (PI3K), angiogenesis through suppression of vascular endothelial growth factor phosphorylation, DNA methyltransferase, and many others by which it also mediates anticancer activity. The antioxidant effect of EGCG was evaluated in a transgenic SOD1 mouse model of ALS in which EGCG significantly delayed the onset of symptoms and prolonged life span after daily oral administration of the drug. Furthermore, the number of motor neurons was increased and microglial activation was decreased in SOD1-G93A mice treated with EGCG, along with a reduction in inflammatory factors such as nuclear factor-kappa B and caspase 3 cleavage and a reduction in iNOS in the spinal cord.[30] An increase in the expression of the antiapoptotic Bcl-2 protein and a decrease in the expression of the proapoptotic Bax protein were also observed in the spinal cord motor neurons, suggesting that the mechanism of action of EGCG on oxidative stress was associated with the expression of these two genes. Other reports on the neuroprotective actions of EGCG against oxidative stress in wild-type and G93A cells demonstrate an up-regulation of PI3K/Akt and a down-regulation of mitochondrial damage and poly ADP ribose polymerase, resulting in preservation of neurons and neuronal function.[31,32]

Allicin is an organic sulfide present in garlic bulbs of *Liliaceae allium*, which can cross the blood–brain barrier. Diallyl trisulfide (DATS) is the most active monomer in allicin and is the major constituent responsible for its diverse biological activities. DATS has been documented as an inducer of phase II enzymes, which can attenuate oxidative stress, preserve the activity of intrinsic antioxidant enzymes, and play a neuroprotective role in ALS.[33,34] Upon oral administration to SOD1-G93A mice at the clinical onset stage of ALS, DATS was found to significantly extend life span up to one week. DATS treatment activated heme oxygenase-1, a phase II detoxify-ing enzyme implicated in endogenous antioxidation and neuroprotection, especially against oxidative stress and inflammatory insult.[33] SOD1-G93A mice treated with DATS also displayed down-regulation in the expression of glial fibrillary acidic pro-tein, an astrocyte marker, in the lumbar spinal cord.[34] These results indicate that the potential mechanism of action of DATS may be multifactorial and represent favor-able application with regard to the treatment of ALS.

Celastrol is a triterpenoid pigment extracted from *Trupterygium wilfordii* Hook F. or the "thunder duke vine," with anticancer and anti-inflammatory properties. Celestrol was administered daily to SOD-1 G93A mice and was shown to decrease the expression of TNF-α and iNOS and suppress the immunoreactivity of CD40 (a microglia marker) and glial fibrillary acidic protein (an astrocyte maker) in the lumbar spinal cord section of treated mice compared with untreated mice. With celastrol treatment, the onset of disease was markedly delayed, motor function was improved, and the sur-vival time of SOD-1 G93A mice was significantly extended. Celastrol-mediated atten-uation of disease progression was based on inhibition of lipopolysaccharide-elicited activation of the phosphorylation of the MAPK/extracellular signal-related kinases 1/2 signaling pathway and nuclear factor-kappaB.[35] Thus, celastrol can inhibit the activation of microglia and reduce the generation of proinflammatory cytokines such as TNF-α. Human studies are still needed to validate celastrol as an effective candi-date for the treatment and prevention of ALS.

Resveratrol is a polyhydroxy diphenyl ethylene widely found in plants such as grapes, peanuts, muleberry, cassia seeds, *Veratum nigrum*, and *Rhizoma polygoni cuspidate*. Pharmacokinetic analysis has shown that resveratrol has high absorp-tion after oral intake (approximately 75%) but has low bioavailability (approximately <1%) due to extensive metabolism. Despite its low bioavailability, this natural compound possesses a variety of biological and pharmacological actions, includ-ing antioxidant, antiaging, anticancer, anti-Alzheimer, antiviral, anti-inflammatory, antidiabetic, and anti-ischemic properties and neuroprotection and cardioprotective activities. Resveratrol can prevent ALS and cerebrospinal-fluid-induced neurotox-icity and calcium elevation in brain cortical neurons of experimental animals.[36] Resveratrol also upregulates the expression of neuroprotective sirtuin 1 (SIRT1) in mutant hSOD-G93A-bearing motor neurons and improves cell viability while increasing the cellular levels of ATP and preventing apoptosis.[37] Intraperitoneal administration of resveratrol at 20 mg/kg delays the onset of ALS and improves survival time in G93A-SOD1 transgenic mouse model of ALS, via up-regulation of protective heat shock proteins Hsp25 and Hsp70 as well as activation of SIRT1 to deacetylate heat shock factor protein 1.[38] Treatment with resveratrol significantly delays the onset of symptoms as well as improves the lifespan and survival of spinal motor neurons in the SOD1G93A transgenic mouse model of ALS.[39] Resveratrol can serve as a promising therapeutic strategy for ALS; however, further studies are needed to test its effectiveness in human patients.

17.5   EPILEPSY

Epilepsy is a group of neurological disorders defined by epileptic seizures. Causes of epilepsy are classified as genetic, structural/metabolic (e.g., neurodegenerative disease, traumatic brain injury, and stroke), or unknown. Epilepsy is one of the most common neurological disorders, affecting over 20 million people worldwide and with nearly 80% of cases occurring in the developing world. The process in which these changes occur in the brain is known as epileptogenesis, when neurons begin to fire in a hypersynchronous manner. This abnormal hypersynchronous firing of neurons is the pathological hallmark of a seizure.

Treatment of epilepsy is long-term and includes the use of anticonvulsants or anti-epileptic drugs (AEDs), which reduce the frequency or intensity of seizures and, in some cases, surgery. Despite the use of AEDs, 30% of patients continue to have seizures, and serious side effects and toxicity exists with these medications. Therefore, there is a growing interest in the use of natural compounds as a source for a novel anticonvulsant drug.

The fruit of *Terminalia chebula* prepared in an ethanolic extract (200 and 500 mg/kg by mouth) was investigated in mice with seizures induced by maximum electroshock seizure (MES), pentylenetetrazole (PTZ), and picrotoxin test. Both doses significantly reduced the duration of MES-induced seizures, protected mice from PTZ-induced tonic seizures, and delayed the onset of picrotoxin-induced tonic seizures. Taken together, ethanolic extract of *T. chebula* displays anticonvulsant activity, but the mechanism of action has yet to be determined.[40]

The anticonvulsant activity of aqueous extract of the fruit from *P. longum* was investigated in PTZ, strychnine, and 4-aminopyridine-induced seizures in mice. 250 and 500 mg/kg oral doses protected against PTZ-induced convulsions, but not strychnine and 4-aminopyridine. Researchers also reported decreased levels of gamma-aminobutyric acid (GABA) in the brain of extract treated mice compared to control, indicating a potential GABAergic mechanism.[41]

An alkaloid found in the *Piper* genus, piperine, had anticonvulsant activity in a pilocarpine-induced seizure mouse model.[42] Intraperitoneal injections of piperine (2–20 mg/kg) following pilocarpine (350 mg/kg) significantly increased the latency between the first convulsion and also increased survival percentage. This same study was repeated in the presence of the $GABA_A$ modulator diazepam plus piperine and reported an increase in the latency of the first seizure onset, suggesting the GABAergic system is involved in piperine's mechanism of action. Furthermore, when pretreated with the benzodiazepine/GABA receptor antagonist, piperine's pharmacological actions were blocked. Finally, piperine increased striatal levels of GABA, glycine, and taurine and reversed pilocarpine-induced increase in nitrate levels.[42] Taken together, piperine anticonvulsant effects may be due to its effects on the GABAergic system as well as its anti-inflammatory and antioxidant actions. In a clinical study, piperine's ability to enhance the bioavailability of phenytoin in humans was reported by D'Hooge et al. A single dose of piperine in patients with uncontrolled phenytoin-treated epilepsy had a significant increase in area-under-the curve maximum plasma concentration (C_{max}), and absorption rate constant (K_a); thus, piperine enhanced phenytoin bioavailability potentially by increasing absorption.[43]

The genus *Pinellia* (*Araceae*) is mainly found in Eastern Asia. *Pinellia* alkaloids have been reported to increase seizure latency period in penicillin-kindled rats when compared to the untreated groups.[44] Additionally, *Pinellia* alkaloids significantly increased GABA levels and decreased glutamate levels in the hippocampus. These alkaloids also promoted $GABA_A$ receptor expression and upregulated receptor concentration.[44] Thus, the antiepileptic effects caused by *Pinellia* alkaloids may be mediated by the GABAergic and glutamatergic systems.

Huperzine A is a sequiterpene lycopodium alkaloid isolated from the Chinese club moss *Huperzia serrate* and was initially proposed as a noncompetitive NMDA receptor antagonist. When evaluated as an anticonvulsant, oral administration of huperzine A was reported to be active against pentylenetetrazol-induced seizures and somewhat effective against maximal electroshock-induced seizures in mice with peak anticonvulsant activity at one hour. Intraperitoneal huperzine A produced antiepileptic effects in a seizure model induced by long duration, low frequency (6 Hz), a model similar to therapy-resistant partial seizures. In the 6-Hz model, huperzine A ED_{50} values ranged from 0.28 to 0.78 mg/kg, values much lower than that of phenytoyin, carbamazepine, and topiramate, which often lack efficacy in this model without toxicity.[45]

Cannabis contains over a hundred unique compounds known as phytocannabinoids. The primary cannabinoids are Δ-9-tetrahydrocannabinol (THC) and cannabidiol (CBD): THC is responsible for the negative psychoactive effects, whereas CBD lacks psychoactivity.

Early studies on the role of THC in seizures often reported conflicting data, with THC having both proconvulsant and anticonvulsant activity. Literature indicates that THC (2.5–10 mg/kg) decreased the susceptibility of rat hippocampus by seizure discharge.[46] In mice, THC (1–80 mg/kg) offered no protection against pentylenetetrazol-induced seizures. However, at higher doses (160–200 mg/kg) it was effective against electroshock-induced seizures was reduced, but tolerance developed to these effects by the third and fourth day.[47] THC's anticonvulsant and proconvulsant effects may depend on the action of THC at cannabinoid 1 (CB_1) receptors located in glutamatergic or GABAergic neurons. THC may produce proconvulsant effects by inducing a release of excitatory glutamate or inhibit GABA release.

CBD pharmacology is not completely understood but includes multiple mechanisms of actions that do not depend on the $CB_{1/2}$ receptors. At micromolar concentrations, CBD activates 5-hydroxytryptamine $(5\text{-HT})_{1A}$, inhibits serotonin uptake, activates transient receptor potential cation channel subfamily V member RPV channels, and inhibits the uptake of adenosine, norepinephrine, dopamine and GABA. A multitude of preclinical studies looking at CBD and antiepileptic as well as human trials exists. In rats, CBD decreased seizure susceptibility in the hippocampus and protected mice from seizures and mortality induced by leptazol.[48] CBD anticonvulsant activity was also evaluated in the pilocarpine model of temporal lobe seizure and the penicillin model of partial seizures. In the pilocarpine model, CBD (1–100 mg/kg) reduced the incidence of seizures but not mortality.[49] Likewise, in the penicillin model, CBD reduced seizures as well as mortality and the number of animals developing severe seizures.[49] CBD also produced little effect on motor skills, indicating a better safety profile compared to currently available drugs which may have significant motor side effects.[49]

Finally, a double-blind, placebo-controlled clinical trial on the antiepileptic effects of CBD in treatment-resistant epilepsy reports that seven out of eight patients with epilepsy had an improvement with CBD (200–300 mg/day, 8–18 weeks) compared to one out of seven patients who improved with placebo.[50]

17.6   CONCLUSIONS

Although most of the disease states described earlier have distinct phytochemicals that may be therapeutically useful, there are some plant-derived chemicals that have been reported to be effective in several neurodegenerative diseases. For instance, phytochemicals in berry fruits (e.g., caffeic acid and anthocyanin) have been associated with the delayed onset of PD, AD, ischemic disease, and aging.[51,52] The beneficial action of phytochemicals in berry fruits may be due to their antioxidant, anti-inflammatory, antiviral, and antiproliferative actions.[7,53] Venkatesan and coworkers reported that the mechanism of neuroprotective action of phytochemicals in neurodegenerative diseases such as Alzheimer could be linked to an action on neurotrophins, which are important for the survival and maintenance of neuronal population in the brain.[53] Since most of the studies that have described the pharmacological actions of phytochemicals in neurodegenerative diseases are based on animal studies, more clinical studies are needed to corroborate the therapeutic utility of these plant-derived chemicals in these conditions.

REFERENCES

1. Brain KR. Accumulation of L-DOPA in cultures from *Mucuna pruriens. Plant Sci Lett.* 1976;7.3:157–61.
2. Zarmouh NO, Messeha SS, Elshami FM, Soliman KFA. Natural products screening for the identification of selective monoamine oxidase-b inhibitors. *Eur J Med Plants.* 2016;15(1):14802. doi:10.9734/EJMP/2016/26453.
3. Kabuto H, Nishizawa M, Tada M, Higashio C, Shishibori T, Kohno M. Zingerone [4-(4-hydroxy-3-methoxyphenyl)-2-butanone] prevents 6-hydroxydopamine-induced dopamine depression in mouse striatum and increases superoxide scavenging activity in serum. *Neurochem Res Neurochem Res.* 2005;30.3;325–32.
4. Zhao, G, Zheng X-W, Qin G-W, Gai Y, Jiang Z-H, Guo L-H. In vitro dopaminergic neuroprotective and in vivo antiparkinsonian-like effects of δ3,2-hydroxybakuchiol isolated from *Psoralea corylifolia* (L.). *Cell Mol Life Sci Cell Mol Life Sci.* 2009;66.9:1617–29.
5. Mazzio E, Deiab S, Park K, Soliman K. High throughput screening to identify natural human monoamine oxidase B inhibitors. *Phytother Res.* 2013;27(6):818–28. doi:10.1002/ptr.4795.
6. Recent Patents on Endocrine, Metabolic & Immune Drug Discovery, Volume 6, Number 3, September 2012, pp. 181–200(20).

7. Youdim KA, Shukitt-Hale B, Martin A, Wang H, Denisova N, Bickford PC, Joseph JA. Short-term dietary supplementation of blueberry polyphenolics: Beneficial effects on aging brain performance and peripheral tissue function. *Nutr Neurosci*. 2000;3:383–97.

8. Quinn JF, Raman R, Thomas RG et al. Docosahexaenoic acid supplementation and cognitive decline in Alzheimer disease: A randomized trial. *JAMA*. 2010;304(17):1903–11. doi:10.1001/jama.2010.1510.

9. Bhattacharya S, Haertel C, Maelicke A, Montag D. Galantamine slows down plaque formation and behavioral decline in the 5XFAD mouse model of Alzheimer's disease. *PLoS One*. 2014;9(2):e89454. doi:10.1371/journal.pone.0089454.

10. Eidi M, Eidi A, Bahar M. Effects of *Salvia officinalis* L. (sage) Leaves on memory retention and its interaction with cholinergic system. *Nutrition*. 2006;22:321–6.

11. Iuvone T, De Filippis D, Esposito G, D'Amico A, Izzo AA. The spice sage and its active ingredient rosmarinic acid protect PC12 cells from amyloid-β peptide-induced neurotoxicity. *J Pharmacol Exp Ther*. 2006;317:1143–9.

12. Howes MJ, Perry E. The role of phytochemicals in the treatment and prevention of dementia. *Drugs Aging*. 2011;28:439–68.

13. Lei Y, Yang L, Ye CY et al. Involvement of intracellular and mitochondrial Aβ in the ameliorative effects of huperzine A against oligomeric Aβ$_{42}$-induced injury in primary rat neurons. *PLoS One*. 2015;10(5):e0128366. doi:10.1371/journal.pone.0128366.

14. Yang G, Wang Y, Tian J, Liu J-P. Huperzine A for Alzheimer's disease: A systematic review and meta-analysis of randomized clinical trials. *PLoS One*. 2013;8(9):e74916. doi:10.1371/journal.pone.0074916.

15. Kim HJ, Kim P, Shin CY. A comprehensive review of the therapeutic and pharmacological effects of ginseng and ginsenosides in central nervous system. *J Ginseng Res*. 2013;37(1):8–29. doi:10.5142/jgr.2013.37.8.

16. Wang Y, Yang G, Gong J, Lu F, Diao Q, Sun J, Zhang K, Tian J, Liu J. Ginseng for Alzheimer's disease: A systematic review and meta-analysis of randomized controlled trials. *CTMC Curr Top Med Chem*. 2016;16.5:529–36.

17. Porquet D, Casadesus G, Bayod S et al. Dietary resveratrol prevents Alzheimer's markers and increases life span in SAMP8. *Age*. 2013;35:1851–65.

18. Porquet D, Grinan-Ferre C, Ferrer I et al. Neuroprotective role of trans-reveratrol in a murine model of familial Alzheimer's disease. *J Alzheimers Dis*. 2014;42:1209–20.

19. Ma X, Sun Z, Liu Y. Reveratrol improves cognition and reduces oxidative stress in rats with vascular dementia. *Neural Regen Res*. 2013;8:2050–9.

20. Ozacmak VH, Sayan-Ozacmak H, Barut F. Chronic treatment with resveratrol, a natural polyphenol found in grapes, alleviates oxidative stress and apoptotic cell death in ovariectomized female rats subjected to chronic cerebral hypoperfusion. *Nutr Neurosci*. 2016;19(4):178–86.

21. Sun Z, Ma X-R, Jia Y-J et al. Effects of resveratrol on apoptosis in a rat model of vascular dementia. *Exp Ther Med*. 2014;7:843–8.

22. Turner, RS, Thomas, RG, Craft, S et al. A randomized, double-blind, placebo-controlled trial of resveratrol for Alzheimer disease. *Neurology*. 2015;85(16):1383–91.

23. Dragicevic N, Smith A, Lin X. Green tea epigallocatechin-3-gallate (EGCG) and other flavanoids reduce Alzheimer's amyloid-induced mitochondrial dysfunction. *J Alzheimers Dis*. 2011;26:507–21.

24. Lee JW, Lee YK, Ban JO. Green tea(-)-epigallocatechin-3-gallate inhibits beta amyloid-induced cognitive dysfunction through modification of secretase activity via inhibition of ERK and NF-kappaB pathways in mice. *J Nutr*. 2009;139:198701993.

25. Liu M, Chen F, Sha L. (-)-Epigallocatechin-3-gallate ameliorates learning and memory deficits by adjusting the balance of TrkA/p75NTR signaling in APP/PS1 transgenic mice. *Mol Neurobiol*. 2014;49:1350–63.

26. Poppe L, Rué L, Robberecht W, Van Den Bosch L. Translating biological findings into new treatment strategies for amyotrophic lateral sclerosis (ALS). *Exp Neurol.* 2014;262:138–51.

27. Zhang X, Hong Y-L, Xu D-S, Feng Y, Zhao L-J, Ruan K-F, Yang X-J. A review of experimental research on herbal compounds in amyotrophic lateral sclerosis. *Phytother Res Phytother Res.* 2013;28.1:9–21.

28. Ferrante RJ, Klein AM, Dedeoglu A, Beal MF. Therapeutic efficacy of EGb761 (*Gingko biloba* extract) in a transgenic mouse model of amyotrophic lateral sclerosis. *J Mol Neurosci.* 2001;17(1):89–96.

29. Kobayashi MS, Han D, Packer L. Antioxidants and herbal extracts protect ht-4 neuronal cells against glutamate-induced cytotoxicity. *Free Radic Res* 2000;32.2:115–24.

30. Xu Z, Chen S, Li X, Luo G, Li L, Le W. Neuroprotective effects of (-)-epigallocatechin-3-gallate in a transgenic mouse model of amyotrophic lateral sclerosis. *Neurochem Res.* 2006;31(10):1263–9.

31. Koh S-H, Kwon H, Kim KS, Kim J, Kim M-H, Yu H-J, Kim M, Lee K-W, Do BR, Jung HK. Epigallocatechin gallate prevents oxidative-stress-induced death of mutant Cu/Zn superoxide dismutase (G93A) motoneuron cells by alteration of cell survival and death signals. *Toxicology.* 2004;202(3):213–25.

32. Koh S-H, Lee SM, Kim HY, Lee K-Y, Lee YJ, Kim H-T, Kim J, Kim M-H, Hwang MS, Song C. The effect of epigallocatechin gallate on suppressing disease progression of ALS model mice. *Neurosci Lett.* 2006;395(2):103–7.

33. Zhang X, Hong Y-L, Xu D-S, Yi F et al. A review of experimental research on herbal compounds in amyotrophic lateral sclerosis. *Phytother Res.* 2014;28:9–21.

34. Sun MM, Bu H, Li B et al. Neuroprotective potential of phase II enzyme inducer diallyl trisulfide. *Neurol Res.* 2009;31:23–27.

35. Jung HW, Chung YS, Kim YS et al. Celastrol inhibits production of nitric oxide and proinflammatory cytokines through MAPK signal transduction and NF-kappaB in LPS stimulated BV-2 microglial cells. *Exp Mol Med.* 2007;39:715–21.

36. Yáñez M, Galán L, Matías-Guiu J, Vela A, Guerrero A, García AG. CSF from amyotrophic lateral sclerosis patients produces glutamate independent death of rat motor brain cortical neurons: Protection by resveratrol but not riluzole. *Brain Res.* 2011;1423:77–86.

37. Wang J, Zhang Y, Tang L, Zhang N, Fan D. Protective effects of resveratrol through the up-regulation of SIRT1 expression in the mutant hSOD1-G93A-bearing motor neuron-like cell culture model of amyotrophic lateral sclerosis. *Neurosci Lett.* 2011;503(3):250–5.

38. Han S, Choi J-R, Soon Shin K, Kang SJ. Resveratrol upregulated heat shock proteins and extended the survival of G93A-SOD1 mice. *Brain Res.* 2012;1483:112–7.

39. Mancuso R, del Valle J, Modol L, Martinez A, Granado-Serrano AB, Ramirez-Núñez O, Pallás M, Portero-Otin M, Osta R, Navarro X. Resveratrol improves motoneuron function and extends survival in SOD1G93A ALS mice. *Neurotherapeutics.* 2014;11(2):419–32.

40. Debnath J, Sharma UR, Kumar B, Chauhan NS. Anticonvulsant activity of ethanolic extract of fruits of *Terminalia chebula* on experimental animals. *Int J Drug Dev Res.* 2010;2(4):764–8.

41. Juvekar MR, Kulkarni MP, Juvekar AR. Anti-stress, nootropic and anticonvulsant potential of fruit extracts of *Piper longum L. Plant Med.* 2008;74PA244.

42. da Cruz GMP, Felipe CFB, Scorza FA et al. Piperine decreases pilocarpine-induced convulsions by GABAergic mechanisms. *Pharmacol Biochem Behav.* 2013;104:144–53.

43. D' Hooge R, Pei Y-Q, Raes A, Lebrun P, van Bogaert P-P, de Deyn PP. Anticonvulsant activity of piperine on seizures induced by excitatory amino acid receptor agonists. *Arzneimittel-Forschung.* 1996;46(6):557–60.

44. Gu YT, Ma YG, Wang MZ, Wang XF, Xiao Y. The effect of *Pinellia* total alkaloids on the amino acid concentration and the GABAA receptor expression in hippocampus region of epileptic rats. *J Chin Pharm Sci*. 2009;18:252–6.
45. Schachter SC. Botanicals and herbs: A traditional approach to treating epilepsy. *Neurotherapeutics*. 2009;6(2):415–20. doi: 10.1016/j.nurt.2008.12.004.
46. Izquierdo I, Orsingher OA, Berardi AC. Effect of cannabidiol and of other *Cannabis sativa* compounds on hippocampal seizure discharges. *Psychopharmacologia*. 1973;28:95–102.
47. Chesher GB, Jackson DM. Anticonvulsant effects of cannabinoids in mice: Drug interactions within cannabinoids and cannabinoid interactions with phenytoin. *Psychopharmacologia*. 1974;37:255–64.
48. Carlini EA, Leite JR, Tannhauser M, Berardi AC. Cannabidiol and *Cannabis sativa* extract protect mice and rats against convulsive agents. *J Pharm Pharmacol*. 1973;25:664–5.
49. Jones NA, Glyn SE, Akiyama S et al. Cannabidiol exerts anti-convulsant effects in animal models of temporal lobe and partial seizures. *Seizure*. 2012;21:344–52.
50. Cunha JM, Carlini EA, Pereira AE et al. Chronic administration of cannabidiol to healthy volunteers and epileptic patients. *Pharmacology*. 1980;21:175–85.
51. Subash, S, Mustapha ME, Al-Adawi, S et al. Neuroprotective effects of berry fruits on neurodegenerative diseases. *Neural Regen Res*. 2014;9:1557–66.
52. Youdim KA, Joseph JA. A possible emerging role of phytochemicals in improving age-related neurological dysfunctions: A multiplicity of effects. *Free Radic Biol Med*. 2001;30:583–94.
53. Venkatesan R, Ji E, Kim SY. Phytochemicals that regulate neurodegenerative disease by targeting neurotrophins: A comprehensive review. *Biomed Res Int*. 2015:Article ID 814068.

18 Modulation of Neuroprotective Genes by Bioactive Food Components in Senescence-Accelerated Mice

Shigeru Katayama and Soichiro Nakamura

CONTENTS

18.1 INTRODUCTION

Aging is an inevitable part of life for human and is associated with a decline in the physical and functional capacity of tissues and organs. Mild cognitive impairment and dementia, including Alzheimer's disease (AD), in particular, has emerged as a major debilitating illness associated with old age. Physical exercise, dietary interventions, and daily intake of vegetables and fruit containing various phytochemicals contribute to maintaining good health and quality of life during the aging process.[1,2] There has been increasing interest in natural dietary bioactive compounds with potential neuroprotective and brain health benefits, including the prevention and treatment of mild cognitive impairment and dementia. In this chapter, the effects of dietary bioactive compounds on brain function in an animal model of accelerated senescence are reviewed in terms of the mechanisms underlying the modulation of neuroprotective genes.

18.2   SENESCENCE-ACCELERATED MOUSE

The senescence-accelerated mouse (SAM) model was established from the AKR/J strain as a spontaneous murine model of accelerated senescence.[3] A large number of studies have been conducted to elucidate the mechanism underlying senescence acceleration and age-related disorders. Nine major SAM-prone (SAMP) strains and three major SAM-resistant (SAMR) strains have been established. All SAMP strains exhibit accelerated senescence compared to the SAMR strains and each SAMP strain exhibits characteristic disorders.[4] Among these SAM mice, SAMP8 mice spontaneously develop age-related disorders including emotional disorder, impaired immune response, and changes in appearance (lordokyphosis of the spine, cataracts, and loss of hair), when compared with SAMR1 mice (Figure 18.1). SAMP8 furthermore exhibit early-onset learning and memory deficits compared to SAMR1 mice of the same age when challenged with various learning tasks, including passive avoidance, active avoidance, and water maze tests.[5,6] Increased blood–brain barrier (BBB) permeability and astrocyte dysfunction in aged SAMP8 mice diminish the neuroprotective capacity of the brain.[7] The generation of free radical-induced oxidative stress furthermore results in hippocampus damage and consequent cognitive deficits.[8]

The genes affecting learning and memory in SAMP8 have been identified, and the altered expression of these genes is directly or indirectly related to neuronal function in the brain of SAMP8 mice. Morley et al.[9] reported that increases in amyloid precursor protein (APP) play a role in neuronal dysfunction and neurotoxicity, and APP gene expression in the hippocampi of SAMP8 mice is associated with learning and memory deficits. Miyazaki et al.[10] found that hippocampal gene expression levels of glia cell line-derived neurotrophic factor (GDNF) in two-month-old mice were lower in SAMP8 mice than in SAMR1 mice. In the midbrain, the hippocampus, and the forebrain, neurtrophin-3 (NT-3) was furthermore shown to be expressed at lower levels in SAMP8 mice than in SAMR1 mice during early development.[11] These findings suggest that low expression levels of neurotrophic factors may play a role in hippocampal dysfunctions, particularly in age-related learning and memory deficits.

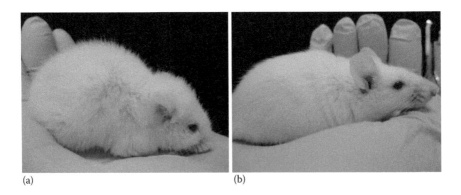

(a)   (b)

FIGURE 18.1   (**See color insert.**) (a) SAMP8 mouse, 42-week-old male. (b) SAMR1 mouse, 42-week-old male.

18.3   EFFECTS OF DIETARY BIOACTIVE COMPONENTS IN SAMP8 MICE

Several studies have indicated that long-term administration of dietary bioactive components prevents learning and memory impairment in SAMP8 mice. Zhao et al.[12] reported that the administration of green tea catechins via drinking water to four-month-old SAMP8 mice for six months prevented spatial learning and memory impairments, demonstrated using the Morris water maze. It was also found that green tea catechin consumption upregulated synaptic plasticity-related proteins including brain-derived neurotrophic factor (BDNF), postsynaptic density protein-95 (PSD-95), and Ca^{2+}/calmodulin-dependent protein kinase II (CaMKII) in the hippocampus. Zhao et al.[13] also demonstrated that four-month-old SAMP8 mice fed ginsenoside in drinking water for seven months exhibited improved learning and memory in the Morris water maze as well as delayed passive avoidance decline in a step-down test. Ginsenoside consumption led to increased expression of PSD-95, phosphor-p-CaMKII, BDNF, phospho-protein kinase A catalyticβ subunit (p-PKA Cβ), protein kinase Cγ subunit (PKCγ), and phospho-cyclic AMP response element binding protein (p-CREB) in the hippocampus, suggesting the involvement of the PKA or PKC/CREB signaling pathway, which plays a crucial role in the consolidation of long-term memory. Armbrecht et al.[14] demonstrated a marked decrease in CAMKII expression in 4- and 12-month-old SAMP8 mice as well as a decrease in CREB protein expression in 12-month-old SAMP8 mice. These bioactive components induce the upregulation of synaptic plasticity-related proteins including BDNF and CREB, both of which are critical for learning and memory formation. Since initial cognition decline is believed to result from dysfunctional synaptic plasticity, the induction of neurotrophic factors such as BDNF and GDNF by dietary bioactive compounds appears to contribute to the prevention of the age-related cognition decline process.

We have investigated the effects of soy peptides composed primarily of dipeptides and tripeptides on age-related cognitive decline in SAMP8. Soy peptides—a product of enzymatic hydrolysis of soy proteins—have been found to demonstrate a number of useful activities including antiobesity, hypocholesterolemic, antihypertensive, antioxidative, and immunomodulatory effects.[15,16] Recent studies have also demonstrated beneficial effects of soy peptides on brain function: Ohinata et al.[17] reported that soymorphins, μ-opioid agonist peptides derived from soy β-conglycinin β-subunit, have anxiolytic-like activity. On the other hand, Maebuchi et al.[18] showed that soy peptide intake improves cognitive dysfunction in subjects with mild cognitive impairment as well as reducing central fatigue and promoting relaxation. It has also been shown that the oral administration of soy peptides results in improvements in immune function, stress conditions, and brain circulation in healthy volunteers.[19]

In our study, soy peptides were orally administered to SAMP8 mice on a diet containing 7% (w/w) soy peptides, which replaces half the dietary casein, from 16 to 42 weeks of age.[20] The suppressive effects of these peptides on the progression of cognitive decline were investigated using the Morris water maze. At 42 weeks of age, the soy peptide-fed SAMP8 mice exhibited superior cognitive ability compared with their control counterparts. Hippocampal gene expression was investigated using DNA microarrays, which revealed that the soy peptide administration promotes synaptic transmission and astrocyte differentiation in SAMP8 mice. Soy peptide administration was also shown to promote neuropeptide signaling in SAMP8 mice. The expression of neurotrophic factors such as BDNF and NT-3 was elevated at both the mRNA and the protein level in the brain of soy peptide-fed mice, and the phosphorylation of the transcription factor CREB in the brain was furthermore markedly up-regulated by soy peptide feeding in SAMP8 mice. These findings suggest that soy peptides have the potential to protect against cognitive impairment via neurotrophic effects.

The BDNF and NT-3 growth factors belong to the neurotrophin family of growth factors and affect the survival and function of neurons in the central nervous system.[21,22] Neurotrophic factors play an important role in the formation of appropriate synaptic connections, both during development and during the learning and memory formation process in adults. It is well known that BDNF exhibits low stability in plasma and low BBB permeability because of its moderately large size and ionic charge.[23] Therapeutic modulation of the levels of neurotrophic factors is thus a promising treatment strategy for neurological and psychiatric disorders in which the levels are dysregulated.

The CREB transcription factor is widely considered essential for the long-term changes in gene expression that positively regulate neuronal plasticity and memory consolidation, and CREB activation can in turn induce BDNF gene expression.[24,25] By binding to its receptor tropomyosin receptor kinase B (TrkB), BDNF is known to activate the phosphatidylinositol-3 (PI3) kinase signaling pathway and the release of neurotransmitters at presynaptic sites.[26,27] A cationic tripeptide sequence (Lys-Lys-Arg) in loop 4 of BDNF reportedly contributes to the binding of BDNF to the p75 neurotrophin receptor and a cyclic pentapeptide (cyclo(-D-Pro-Ala-Lys-Arg-)) was designed as an effective BDNF-like agonist resistant to proteolytic degradation in plasma.[28,29] Massa et al.,[30] on the other hand, demonstrated that the loop 2 subregion b (Ser-Lys-Gly-Gln-Leu) of BDNF is involved in TrkB activation and that BDNF loop 2 mimetics such as the tripeptide Pro-His-Trp exhibit BDNF-like activity. del Carmen Cardenas-Aguayo et al.[31] furthermore demonstrated that tetrapeptides such as Ac-Ile-Lys-Arg-Gly-CONH$_2$ and Ac-Ser-Lys-Lys-Arg-CONH$_2$, which mimic different active regions of BDNF, induce the expression of BDNF via Trk B receptor activation. These findings suggest that soy peptides resulting from the enzymatic hydrolysis of soy proteins contain TrkB agonist-like peptides, which resemble the sequence of the BDNF active site; however, to verify this hypothesis, these BDNF mimetics must be isolated from soy peptides in further studies, thereby allowing for their amino acid sequences to be determined.

A variety of fermented soybean products are widely consumed in Asian countries. Several products such as soybean paste (miso), natto, sufu, and tempeh may contain TrkB agonist-like peptide, the levels of which can be enhanced during fermentation. It would be interesting to investigate the potential relationship between regular intake of fermented soybean products and soundness of brain functions in elderly people.

18.4 SAMP8 MICE AS A MODEL OF EARLY AD

AD is the most common form of dementia worldwide and represents a neurodegenerative disease characterized by a decline in cognitive and memory functions severe enough to interfere with normal activities of daily living.[32,33] The disease is associated with the accumulation of beta-amyloid (Aβ) in extracellular neuritic plaques and tau in intracellular neurofibrillary tangles as well as the loss of neurons from the hippocampus and the cerebral cortex in the brain.[34] Increases in the concentration of Aβ during the course of the disease with subtle effects on synaptic efficacy subsequently lead to a gradual increase in the load of amyloid plaques and a progression in cognitive impairment.

The SAMP8 mouse strain is increasingly being recognized as a model of age-related AD,[35,36] and SAMP8 mice exhibit an AD-related pathology with age: increases in plaques of Aβ, alterations in APP processing by secretase, and increases in aggregates of hyperphosphorylated tau are observed in these mice.[9,37–39] SAMP8 mice have been widely studied, particularly in the context of therapeutic or intervention potency of dietary bioactive compounds in the early stages of AD. Shi et al.[40] found that ginsenoside Rg1 attenuates the cerebral Aβ content by regulating PKA/CREB activity in SAMP8 mice. Sesaminol glucosides reportedly exhibit protective effects against Aβ-induced cytotoxicity in neuronal cells as well as against cognitive deficits in mice via its antioxidant and anti-inflammatory activities.[41–43] Dietary resveratrol has been shown to extend the lifespan of SAMP8 mice and decrease Aβ accumulation. Resveratrol was found to upregulate a disintegrin and metalloprotease domain (ADAM10) expression through deacetylation by Sirtuin 1, thereby inducing the nonamyloidogenic processing of APP.[44] Under nonpathological conditions, a much larger fraction of APP is processed by α-secretase, which yields a soluble, nontoxic APP fragment (Figure 18.2). Specific dietary bioactive compounds such as

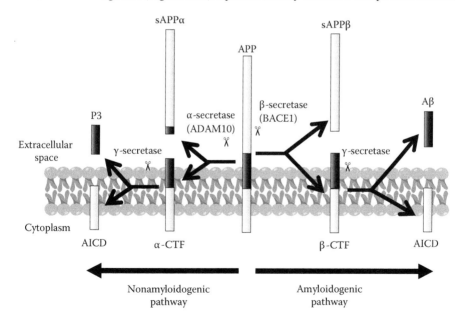

FIGURE 18.2 (See color insert.) The cleavage of amyloid beta-protein precursor (APP) by the secretase enzymes. Aβ, beta-amyloid; AICD, APP intracellular domain; ADAM10, a disintegrin and metalloprotease domain; BACE1, Beta site APP cleaving enxyme I; CTF, C-terminal fragment; sAPP, soluble APP.

resveratrol are therefore expected to contribute to the prevention of Aβ production via the activation of α-secretase.

Some nuclear receptors play a beneficial role in reducing the risk of AD onset. The retinoic acid receptor (RAR)/retinoid X receptor (RXR) transcription factor has been proposed as a potential therapy for disorders of the central nervous system.[45] Recent studies demonstrated an important role for RAR in the activation of the ADAM10 promoter as well as in increasing ADAM10 expression and α-secretase activity.[46] As RAR ligands, retinoids upregulate ADAM10 gene expression and enhance α-secretase activity. Kitaoka et al.[47] reported that the RAR agonist Am80, also known as tambibarotene, increases hippocampal ADAM10 expression in SAMP8 mice. Accordingly, the enhancement of ADAM10 gene expression may represent a promising approach for the prevention of AD. These findings imply that dietary bioactive compounds, like RAR/RXR ligands, suppress the Aβ generation via the upregulation of the nonamyloidogenic pathway of APP cleavage.

18.5 FUTURE DIRECTIONS

The mouse model of the SAM strain P8 (SAMP8) is a useful tool for investigating not only the antiaging effects of bioactive compounds but also the mechanisms underlying aging, including the regulation of gene expression in the brain. In order for a dietary bioactive component to have neuromodulatory effects in the brain, the metabolism and the BBB permeability of the component must be determined. The BBB controls the transport of nutrients and toxins and plays dual roles by (1) preventing circulating toxic agents from reaching the brain (barrier function) and (2) securing a continuous supply of essential nutrients to the brain (carrier function).[48,49] Thus, to be considered potential neuroprotective agents for the brain, it is essential that dietary bioactive compounds and their metabolites can cross through the BBB.

The present work provides useful genetic information on a variety of biochemical processes by which different dietary bioactive compounds can contribute to good health in the elderly.

REFERENCES

1. Fusco, D.; Colloca, G.; Monaco, M. R. L.; Cesari, M. Effects of antioxidant supplementation on the aging process. *Clinical Interventions in Aging.* 2007, *2*, 377.
2. Ferrari, C. K. Functional foods and physical activities in health promotion of aging people. *Maturitas.* 2007, *58*, 327–39.
3. Takeda, T. Senescence-accelerated mouse (SAM): A biogerontological resource in aging research. *Neurobiology of Aging.* 1999, *20*, 105–10.
4. Takeda, T.; Hosokawa, M.; Higuchi, K. Senescence-accelerated mouse (SAM): A novel murine model of senescence. *Experimental Gerontology.* 1997, *32*, 105–9.
5. Miyamoto, M.; Kiyota, Y.; Yamazaki, N.; Nagaoka, A.; Matsuo, T.; Nagawa, Y.; Takeda, T. Age-related changes in learning and memory in the senescence-accelerated mouse (SAM). *Physiology & Behavior.* 1986, *38*, 399–406.
6. Miyamoto, M.; Kiyota, Y.; Nishiyama, M.; Nagaoka, A. Senescence-accelerated mouse (SAM): Age-related reduced anxiety-like behavior in the SAM-P/8 strain. *Physiology & Behavior.* 1992, *51*, 979–85.

7. Pelegrí, C.; Canudas, A. M.; del Valle, J.; Casadesus, G.; Smith, M. A.; Camins, A.; Pallàs, M.; Vilaplana, J. Increased permeability of blood–brain barrier on the hippocampus of a murine model of senescence. *Mechanisms of Ageing and Development*. 2007, *128*, 522–8.

8. Yasui, F.; Ishibashi, M.; Matsugo, S.; Kojo, S.; Oomura, Y.; Sasaki, K. Brain lipid hydroperoxide level increases in senescence-accelerated mice at an early age. *Neuroscience Letters*. 2003, *350*, 66–8.

9. Morley, J. E.; Kumar, V. B.; Bernardo, A. E.; Farr, S. A.; Uezu, K.; Tumosa, N.; Flood, J. F. β-Amyloid precursor polypeptide in SAMP8 mice affects learning and memory. *Peptides*. 2000, *21*, 1761–7.

10. Miyazaki, H.; Okuma, Y.; Nomura, J.; Nagashima, K.; Nomura, Y. Age-related alterations in the expression of glial cell line-derived neurotrophic factor in the senescence-accelerated mouse brain. *Journal of Pharmacological Sciences*. 2003, *92*, 28–34.

11. Kaisho, Y.; Miyamoto, M.; Shiho, O.; Onoue, H.; Kitamura, Y.; Nomura, S. Expression of neurotrophin genes in the brain of senescence-accelerated mouse (SAM) during postnatal development. *Brain Research*. 1994, *647*, 139–44.

12. Li, Q.; Zhao, H.; Zhang, Z.; Liu, Z.; Pei, X.; Wang, J.; Li, Y. Long-term green tea catechin administration prevents spatial learning and memory impairment in senescence-accelerated mouse prone-8 mice by decreasing Aβ 1–42 oligomers and upregulating synaptic plasticity–related proteins in the hippocampus. *Neuroscience*. 2009, *163*, 741–9.

13. Zhao, H.; Li, Q.; Zhang, Z.; Pei, X.; Wang, J.; Li, Y. Long-term ginsenoside consumption prevents memory loss in aged SAMP8 mice by decreasing oxidative stress and upregulating the plasticity-related proteins in hippocampus. *Brain Research*. 2009, *1256*, 111–22.

14. Armbrecht, H. J.; Siddiqui, A. M.; Green, M.; Farr, S. A.; Kumar, V. B.; Banks, W. A.; Patrick, P.; Shah, G. N.; Morley, J. E. SAMP8 mice have altered hippocampal gene expression in long term potentiation, phosphatidylinositol signaling, and endocytosis pathways. *Neurobiology of Aging*. 2014, *35*, 159–68.

15. Singh, B. P.; Vij, S.; Hati, S. Functional significance of bioactive peptides derived from soybean. *Peptides*. 2014, *54*, 171–9.

16. de Oliveira, C. F.; Corrêa, A. P. F.; Coletto, D.; Daroit, D. J.; Cladera-Olivera, F.; Brandelli, A. Soy protein hydrolysis with microbial protease to improve antioxidant and functional properties. *Journal of Food Science and Technology*. 2015, *52*, 2668–78.

17. Ohinata, K.; Agui, S.; Yoshikawa, M. Soymorphins, novel μ opioid peptides derived from soy β-conglycinin β-subunit, have anxiolytic activities. *Bioscience, Biotechnology, and Biochemistry*. 2007, *71*, 2618–21.

18. Maebuchi, M.; Kishi, Y.; Koikeda, T.; Furuya, S. Soy peptide dietary supplementation increases serum dopamine level and improves cognitive dysfunction in subjects with mild cognitive impairment. *Japan Pharmacology & Therapeutics*. 2013, *41*, 67–73.

19. Yimit, D.; Hoxur, P.; Amat, N.; Uchikawa, K.; Yamaguchi, N. Effects of soybean peptide on immune function, brain function, and neurochemistry in healthy volunteers. *Nutrition*. 2012, *28*, 154–9.

20. Katayama, S.; Imai, R.; Sugiyama, H.; Nakamura, S. Oral administration of soy peptides suppresses cognitive decline by induction of neurotrophic factors in SAMP8 mice. *Journal of Agricultural and Food Chemistry*. 2014, *62*, 3563–9.

21. Huang, E. J.; Reichardt, L. F. Neurotrophins: Roles in neuronal development and function. *Annual Review of Neuroscience*. 2001, *24*, 677.

22. Frade, J. M.; Bovolenta, P.; Rodríguez-Tébar, A. Neurotrophins and other growth factors in the generation of retinal neurons. *Microscopy Research and Technique*. 1999, *45*, 243–51.

23. Allen, S. J.; Watson, J. J.; Shoemark, D. K.; Barua, N. U.; Patel, N. K. GDNF, NGF and BDNF as therapeutic options for neurodegeneration. *Pharmacology & Therapeutics.* 2013, *138*, 155–75.

24. Suzuki, A.; Fukushima, H.; Mukawa, T.; Toyoda, H.; Wu, L.-J.; Zhao, M.-G.; Xu, H.; Shang, Y.; Endoh, K.; Iwamoto, T. Upregulation of CREB-mediated transcription enhances both short-and long-term memory. *The Journal of Neuroscience.* 2011, *31*, 8786–802.

25. Shieh, P. B.; Hu, S.-C.; Bobb, K.; Timmusk, T.; Ghosh, A. Identification of a signaling pathway involved in calcium regulation of BDNF expression. *Neuron.* 1998, *20*, 727–40.

26. Yamada, K.; Nabeshima, T. Brain-derived neurotrophic factor/TrkB signaling in memory processes. *Journal of Pharmacological Sciences.* 2003, *91*, 267–70.

27. Schratt, G. M.; Nigh, E. A.; Chen, W. G.; Hu, L.; Greenberg, M. E. BDNF regulates the translation of a select group of mRNAs by a mammalian target of rapamycin-phosphatidylinositol 3-kinase-dependent pathway during neuronal development. *The Journal of Neuroscience.* 2004, *24*, 7366–77.

28. Fletcher, J. M.; Morton, C. J.; Zwar, R. A.; Murray, S. S.; O'Leary, P. D.; Hughes, R. A. Design of a conformationally defined and proteolytically stable circular mimetic of brain-derived neurotrophic factor. *Journal of Biological Chemistry.* 2008, *283*, 33375–83.

29. Fletcher, J. M.; Hughes, R. A. Modified low molecular weight cyclic peptides as mimetics of BDNF with improved potency, proteolytic stability and transmembrane passage in vitro. *Bioorganic & Medicinal Chemistry.* 2009, *17*, 2695–702.

30. Massa, S. M.; Yang, T.; Xie, Y.; Shi, J.; Bilgen, M.; Joyce, J. N.; Nehama, D.; Rajadas, J.; Longo, F. M. Small molecule BDNF mimetics activate TrkB signaling and prevent neuronal degeneration in rodents. *The Journal of Clinical Investigation.* 2010, *120*, 1774–85.

31. del Carmen Cardenas-Aguayo, M.; Kazim, S. F.; Grundke-Iqbal, I.; Iqbal, K. Neurogenic and neurotrophic effects of BDNF peptides in mouse hippocampal primary neuronal cell cultures. *PloS One.* 2013, *8*, e53596.

32. Hardy, J. A.; Higgins, G. A. Alzheimer's disease: The amyloid cascade hypothesis. *Science.* 1992, *256*, 184.

33. Selkoe, D. J. Translating cell biology into therapeutic advances in Alzheimer's disease. *Nature.* 1999, *399*, A23–31.

34. Blennow, K.; de Leon, M. J.; Zetterberg, H. Alzheimer's disease. *Lancet.* 2006, *368*, 387–403.

35. Butterfield, D. A.; Poon, H. F. The senescence-accelerated prone mouse (SAMP8): A model of age-related cognitive decline with relevance to alterations of the gene expression and protein abnormalities in Alzheimer's disease. *Experimental Gerontology.* 2005, *40*, 774–83.

36. Pallas, M.; Camins, A.; Smith, M. A.; Perry, G.; Lee, H.-g.; Casadesus, G. From aging to Alzheimer's disease: Unveiling "the switch" with the senescence-accelerated mouse model (SAMP8). *Journal of Alzheimer's Disease.* 2008, *15*, 615–24.

37. del Valle, J.; Duran-Vilaregut, J.; Manich, G.; Casadesús, G.; Smith, M. A.; Camins, A.; Pallàs, M.; Pelegrí, C.; Vilaplana, J. Early amyloid accumulation in the hippocampus of SAMP8 mice. *Journal of Alzheimer's Disease.* 2010, *19*, 1303–15.

38. Morley, J. E.; Farr, S. A.; Flood, J. F. Antibody to amyloid β protein alleviates impaired acquisition, retention, and memory processing in SAMP8 mice. *Neurobiology of Learning and Memory.* 2002, *78*, 125–38.

39. Canudas, A. M.; Gutierrez-Cuesta, J.; Rodríguez, M. I.; Acuña-Castroviejo, D.; Sureda, F. X.; Camins, A.; Pallàs, M. Hyperphosphorylation of microtubule-associated protein tau in senescence-accelerated mouse (SAM). *Mechanisms of Ageing and Development*. 2005, *126*, 1300–4.

40. Shi, Y.-Q.; Huang, T.-W.; Chen, L.-M.; Pan, X.-D.; Zhang, J.; Zhu, Y.-G.; Chen, X.-C. Ginsenoside Rg1 attenuates amyloid-β content, regulates PKA/CREB activity, and improves cognitive performance in SAMP8 mice. *Journal of Alzheimer's Disease*. 2010, *19*, 977–89.

41. Um, M. Y.; Ahn, J. Y.; Kim, M. K.; Ha, T. Y. Sesaminol glucosides protect β-amyloid induced apoptotic cell death by regulating redox system in SK-N-SH cells. *Neurochemical Research*. 2012, *37*, 689–99.

42. Um, M. Y.; Choi, W. H.; Ahn, J. Y.; Kim, S.; Kim, M. K.; Ha, T. Y. Sesaminol glucosides improve cognitive deficits and oxidative stress in SAMP8 Mice. *Food Science and Biotechnology*. 2009, *18*, 1311–5.

43. Um, M. Y.; Ahn, J. Y.; Kim, S.; Kim, M. K.; Ha, T. Y. Sesaminol glucosides protect beta.-amyloid peptide-induced cognitive deficits in mice. *Biological and Pharmaceutical Bulletin*. 2009, *32*, 1516–20.

44. Porquet, D.; Casadesús, G.; Bayod, S.; Vicente, A.; Canudas, A. M.; Vilaplana, J.; Pelegrí, C.; Sanfeliu, C.; Camins, A.; Pallàs, M. Dietary resveratrol prevents Alzheimer's markers and increases life span in SAMP8. *Age*. 2013, *35*, 1851–65.

45. van Neerven, S.; Kampmann, E.; Mey, J. RAR/RXR and PPAR/RXR signaling in neurological and psychiatric diseases. *Progress in Neurobiology*. 2008, *85*, 433–51.

46. Corbett, G. T.; Gonzalez, F. J.; Pahan, K. Activation of peroxisome proliferator-activated receptor α stimulates ADAM10-mediated proteolysis of APP. *Proceedings of the National Academy of Sciences*. 2015, *112*, 8445–50.

47. Kitaoka, K.; Shimizu, N.; Ono, K.; Chikahisa, S.; Nakagomi, M.; Shudo, K.; Ishimura, K.; Séi, H.; Yoshizaki, K. The retinoic acid receptor agonist Am80 increases hippocampal ADAM10 in aged SAMP8 mice. *Neuropharmacology*. 2013, *72*, 58–65.

48. Reese, T.; Karnovsky, M. J. Fine structural localization of a blood–brain barrier to exogenous peroxidase. *The Journal of Cell Biology*. 1967, *34*, 207–17.

49. Ueno, M. Molecular anatomy of the brain endothelial barrier: An overview of the distributional features. *Current Medicinal Chemistry*. 2007, *14*, 1199–206.

19 Phytochemicals as Antiaggregation Agents in Neurodegenerative Diseases

Eva S.B. Lobbens and Leonid Breydo

CONTENTS

19.1 PROTEIN AGGREGATION IN NEURODEGENERATIVE DISEASES

Misfolding and aggregation of polypeptides and proteins is an important pathological event in many neurodegenerative maladies such as Alzheimer's, Parkinson's, and Huntington's diseases,[1–3] as well as other human diseases and physiological processes.[4,5] Protein aggregation is a complex process, and aggregates can significantly differ in morphology, stability, and self-propagation ability. Protein aggregates are usually classified as amyloid fibrils (structures in which the polypeptides are organized into cross-β-sheets), amorphous aggregates, or amyloid oligomers.[2,6,7] Protein aggregation usually starts from partially folded protein conformations.[8] Over time, aggregates gradually increase in size and gain secondary structure (usually a β-sheet rich one), giving rise to either oligomers or amyloid fibrils (Figure 19.1).

Amyloid fibrils are extended, repetitive, β-sheet rich structures, typically with the morphology of an extended twisted rope. Their high thermodynamic stability comes from association of multiple polypeptide molecules in a cross-β-structure that could be parallel or antiparallel.[9,10] Differences in arrangement of β-sheets can result in populations of fibrils with distinct structure and morphology (aka fibril strains).

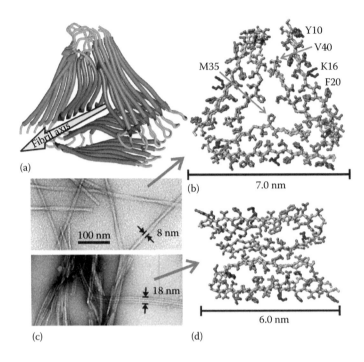

FIGURE 19.1   **(See color insert.)** Amyloid fibrils formed by amyloid β peptide.[18,19] (a) Ribbon representation of fibrils with twisted morphology. (b) Atomic representation of fibrils with twisted morphology viewed down the fibril axis. Hydrophobic, polar negatively charged, and positively charged amino acids are green, magenta, red, and blue, respectively. Unstructured residues 1–8 omitted. (c) Comparison of twisted (upper) and striated ribbon (lower) fibril morphologies by TEM. (d) Atomic representation of fibrils with striated ribbon morphology viewed down the fibril axis.

Structural differences between fibril strains are usually propagated by seeding[11,12] although structural alterations during propagation have been observed.[13–17]

Amyloid oligomers are highly structurally diverse[20] but often belong to one of two broad structural classes (Figure 19.2). Oligomers belonging to the first of these classes resemble fragments of amyloid fibrils,[21–23] while the other structural class encompasses oligomers with antiparallel β-sheet arrangements.[22,24,25] In addition, oligomers with a variety of structures including primarily disordered and primarily α-helical have been examined.[20] The morphology of amyloid oligomers is usually roughly spherical, although other morphologies such as ring-like or 'beads on string'[26,27] have also been observed. Many amyloid oligomers can propagate themselves by incorporating protein monomers and changing their conformation to match the conformation of the seeds.[23,28–30] In some cases, cross-seeding by oligomers has been observed where amyloid oligomers of one protein (e.g., amyloid β) can seed the aggregation of another (e.g., tau), resulting in a potentially dangerous cascade.[31–33]

Given the importance of protein aggregation in a variety of diseases, small molecules that can either inhibit this process or direct it toward less dangerous aggregates are actively sought out. We will briefly examine the examples of these small molecules and their mechanisms of action below.

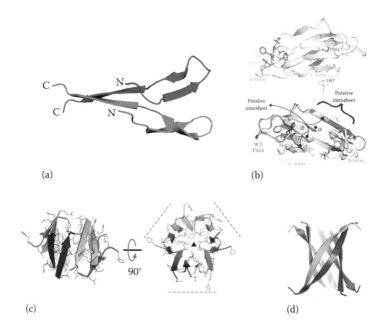

FIGURE 19.2   (See color insert.) Structures of amyloid oligomers.[20] (a) Globulomer, $A\beta_{42}$[34]; (b) hexamer, β2-microglobulin[35]; (c) hexamer, PrP fragment[36]; (d) hexamer, αB-crystallin fragment.[24]

19.2   EFFECTS OF SMALL MOLECULES ON PROTEIN AGGREGATION

19.2.1   NONCOVALENT BINDING

Small molecules can interfere with protein aggregation either via noncovalent binding to the target protein or by modifying it covalently. Hydrophobic (e.g., polyaromatic) small molecules tend to bind to exposed hydrophobic patches of the target protein relatively nonspecifically.[37,38] Since aggregation intermediates tend to have the most exposed hydrophobic surfaces, small molecules usually bind to them preferentially resulting in altered aggregation mechanism. For example, a number of small molecules specifically inhibit formation of either fibrils or oligomers but not both.[39,40] Some hydrophobic small molecules promote conversion of several proteins to highly unstructured oligomers, indicating a common mechanism of action.[41–45] Highly charged small molecules, however, tend to promote fibril formation instead.[46–48] In addition to generic amyloid-binding agents, ligands targeting specific protein aggregates have also been developed.[49–55] By altering the aggregate structure, these aggregation modulators were able to decrease aggregate cytotoxicity of aggregates of a variety of proteins and peptides.[51,52, 56–58]

19.2.2   COVALENT MODIFICATION

Covalent modification of proteins by small molecules can also interfere with their aggregation. For example, reactive aldehydes produced by oxidation of lipids formed lysine adducts with α-synuclein and amyloid β promoting their aggregation into oligomers.[59–62] Sequence-specific covalent aggregation inhibitors have also been designed, and in some cases, they have been shown to promote formation of disordered, less toxic amyloid oligomers.[63] Modification doesn't have to be direct: some small molecules can initiate formation of reactive oxygen species (ROS), leading to modifications at cysteine, methionine, and tyrosine residues. Alternatively, small molecules may serve as ROS scavengers, preventing chemical modification of proteins by ROS. Overall, chemical modifications of proteins by small molecules usually specifically disrupt fibril formation favoring formation of oligomers instead. The likely cause of it is higher conformational variability and generally looser structure of the oligomers that is bound to be more accommodating to modified protein residues.

Plants contain a variety of small molecules belonging to many different structural classes. Here we want to show some examples of plant-derived small molecules shown to interfere with aggregation of proteins implicated in neurodegenerative diseases.

19.3 PHYTOCHEMICALS AS INHIBITORS OF PROTEIN AGGREGATION IN NEURODEGENERATIVE DISEASES

A strong connection exists between diet and human health, and plants have therefore been exploited by man to treat disease symptoms for millennia. Today, the use of plants as medicine has been widely replaced by synthetic pharmaceuticals in the Western world, yet a large part of these are derived from plants. Different approaches can be used to search for biologically active phytochemicals. The most common method is the ethno-directed method, where plants used in the traditional medicine are screened for activity related to their use. Other methods include collecting plants with specific properties or by selecting all available plant material in a specific geographical area, a specific plant family, or simply investigating the compounds of different plant parts of a single species.[64–67] But the question remains whether it is possible to identify and isolate the compounds responsible for the beneficial effects without losing the activity observed when consuming the whole plant.[68–70] Compounds produced by plants are typically divided into two groups, namely, the primary metabolites that regulate vital processes such as growth and the bioactive secondary metabolites that provide protective properties. The term *phytochemical* is primarily used to describe the latter group.[70,71] To date, thousands of phytochemicals have been identified and hundreds of these have been shown to affect human health positively and are thus of great interest to researchers looking for new bioactive medicinal products.[70,71] Several studies have shown that a diet rich in staple foods, such as cereals, vegetables, and fruits, may decrease the risk of attaining age-related neurodegenerative diseases. Although the exact pathogenesis of age-related neurodegenerative diseases remains unresolved, it has long been suggested that oxidative stress might have an important role in the neural decrement. The high content of phytochemicals with antioxidant activity in staple food may therefore be one explanation for the delay in neuronal decrements resulting from oxidative stress in such individuals.[72–74] Another explanation may be the presence of phytochemicals with abilities to interact with the aggregation prone proteins. To limit the scope of this chapter, the phytochemicals have been divided into structural classes (polyphenols, flavonoids, quinones, alkaloids, fatty acids, and carbohydrates), and their activity will be reviewed collectively. Polyphenols are aromatic compounds, which are typically grouped into smaller categories such as vitamins, phenolic acids, and flavonoids. Polyphenols can be found in most staple foods and are known to protect the plants from diseases and ultraviolet light.[37,74–76] The inhibitory activity of the polyphenols against protein fibrillation is dependent on the structure as well as a specific three-dimensional conformation of the polyphenol. It has been shown that two phenolic rings with a minimum of three hydroxyl groups are optimal for efficient inhibition. The phenolic rings inhibit fibrillation by interacting with the aromatic residues of the aggregation prone, partially unfolded state of the proteins and thereby affecting the fibril assembly mechanism and thus improving protein stability.[37,77] Examples of polyphenols with antiamyloidogenic activity are nordihydroguaiaretic acid (NDGA), curcumin,[78–80] rosmarinic acid, resveratrol, purpurogallin, and tannic acid, which inhibited amyloid β fibrillation and dobutamine, which inhibited α-synuclein fibrillation.[37,81,82] Chemical structures are given in Figure 19.3.

The majority of studies investigating the activity of polyphenols have focused on flavonoids. Flavonoids are plant pigments that are known for their antioxidant activity and structure-dependent antiamyloidogenic activity. The inhibitory activity of the flavonoids against alpha-synuclein fibrillation is dependent on the presence of at least one vicinal dihydroxyphenyl moiety, while the number and position of individual hydroxyl groups may have an influence as well. Flavonoids are able to inhibit the progress of α-synuclein fibrillation at any stage by restricting the conformational changes of the protein through covalent modification of the protein by the quinone moiety of the flavonoids.[83,84] Depending on the extent of their activity, incubation of alpha-synuclein with flavonoids results in fewer and shorter fibrils or complete inhibition at the oligomer and protofibril stage.[83] In the case of the amyloid β, flavonoids inhibit fibrillation by binding reversibly to the growing fibril ends competing with the monomer for the binding site.[85] Examples of antiamyloidogenic flavonoids are quercetin, myricetin, morin, kaempferol, gossypetin, catechin, epicatechin, epicatechin gallate, epigallocatechin-3-gallate (EGCG)[86–88] and baicalein,[89] which inhibit amyloid β and α-synuclein fibrillation.[37,83,90] Chemical structures are given in Figure 19.4. When looking more into the formation of Aβ oligomers in the presence of EGCG, it is observed that the C-terminal half of the peptide (residues 22–39) formed a β-sheet, whereas the N-terminal half of the peptide was unstructured.[41] A variety of polyphenols converted α-synuclein fibrils to mostly disordered oligomers via binding to the C-terminal region of α-synuclein. Addition of EGCG to the preformed β-sheet-rich α-synuclein oligomers did not lead to their disaggregation but did inhibit their interaction with the membranes.[88]

Quinones are oxidation products of polyphenols and are common compounds in natural materials. Several quinones have been shown to possess antiamyloidogenic activity.[91] A few examples are 1,4-naphthoquinone and its derivatives phylloquinone and menaquinone, which are able to inhibit α-synuclein fibrillation by interaction with the Lys32 residue. Phylloquinone and menaquinone were suggested to inhibit

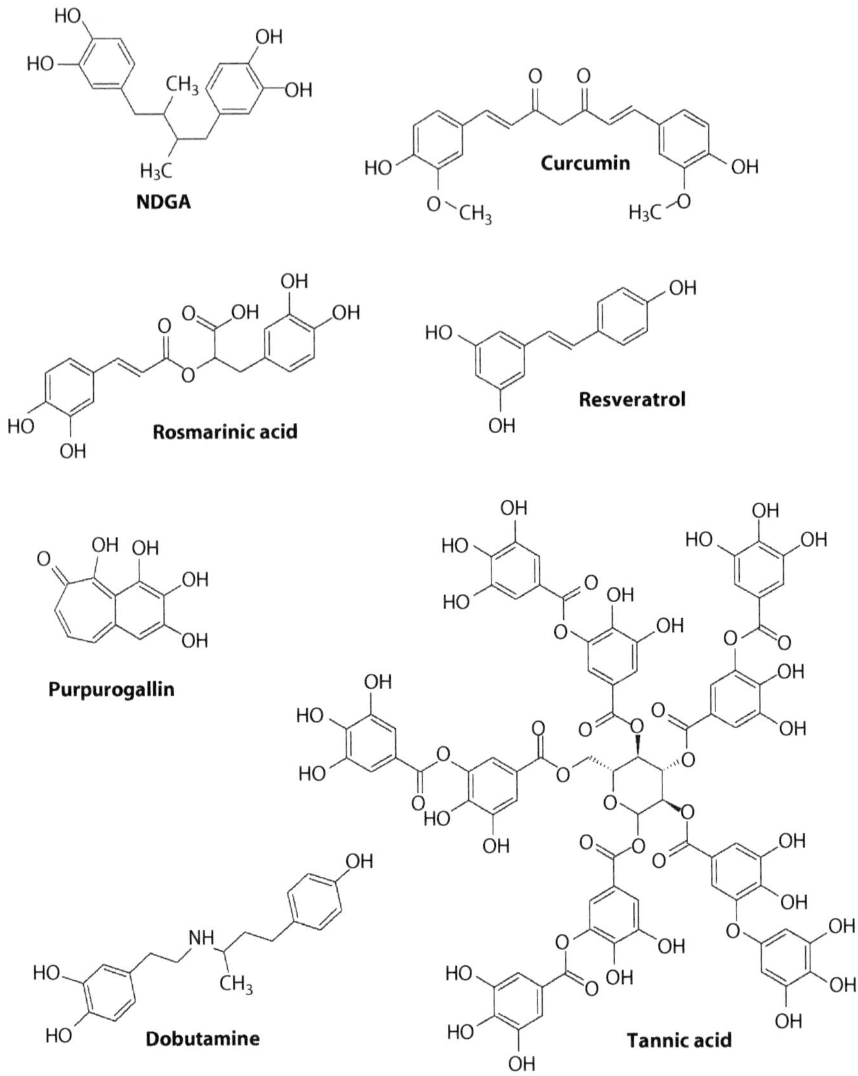

FIGURE 19.3   Chemical structures of polyphenols with antiamyloidogenic activity.[37]

fibrillation by interfering with the nucleus formation, since their activity was only observed when the compounds were added in the beginning of the assay. On the other hand, 1,4-naphthoquinone was able to inhibit α-synuclein fibrillation at any stage of the assay as well as to disaggregate preformed fibrils. It was thus concluded that 1,4-naphthoquinone inhibited fibrillation by interfering with the fibril elongation process or by destabilizing preexisting fibrils.[92] In addition to inhibiting α-synuclein fibrillation, 1,4-naphthoquinone was also shown to inhibit amyloid β fibrillation along with 9,10-anthraquinone, which destabilized the interstrand hydrogen bonds of

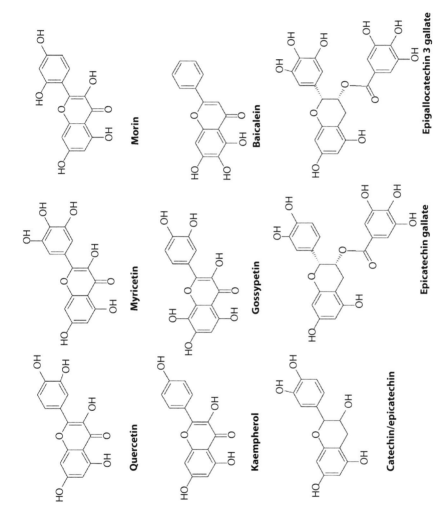

FIGURE 19.4   Chemical structures of flavonoids with antiamyloidogenic activity.[37]

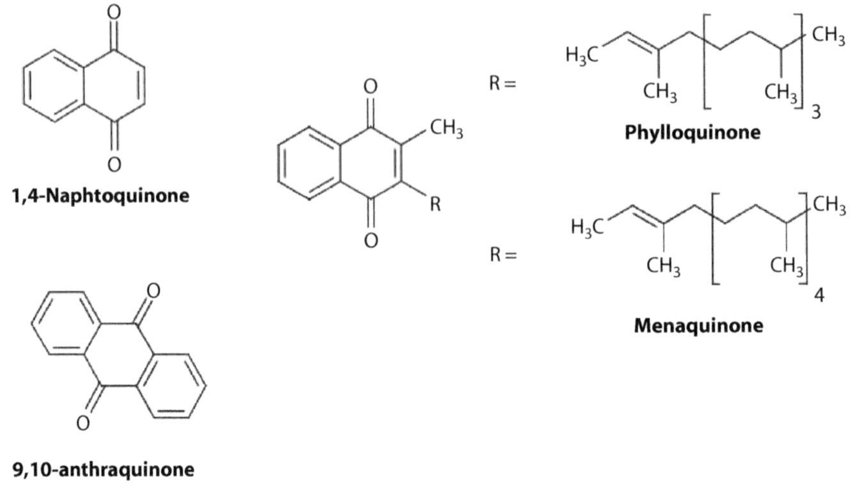

FIGURE 19.5   Chemical structures of quinones with antiamyloidogenic activity.[92,94]

the protein backbone disfavoring the formation of β-sheets.[93,94] Chemical structures are given in Figure 19.5.

Another class of compounds found in a broad range of natural material is the alkaloids. Alkaloids are cyclic, nitrogenous compounds with basic properties and represent almost 50% of all bioactive natural products.[90] Some alkaloids have been shown to inhibit protein aggregation. For example, lobeline, arecoline, and derivatives of isaindigotone were able to reduce aggregation of amyloid β, with 59%, 66%, and 62%, respectively.[90,95,96] Tabersonine was shown to effectively inhibit the formation of Aβ(1–42) fibrils and to convert mature fibrils into largely innocuous amorphous aggregates.[97] Caffeine and its derivatives also inhibited aggregation of amyloid β, α-synuclein, and IAPP.[98–100] Some of these alkaloids are polyaromatic compounds and their mechanism of actions is likely similar to that of polyphenols. Others (e.g., arecoline or pseudopelletierine) are nonaromatic[95,101,102] and thus are likely to operate via a different mechanism. Chemical structures are given in Figure 19.6.

Isoprenoids are compounds derived from isoprene and consist of different types of terpenes. Not much research has been performed on the antiamyloidogenic activity of the group as an entity, but a few compounds have been found to possess this activity. Two examples are the diterpene vitamin A and the terpenoid β-carotene, which are able to inhibit fibrillation of amyloid β.[90,103] Chemical structures are given in Figure 19.7.

Fatty acids are carboxylic acids with a long aliphatic chain and can be found in oils from plants or fish. Fatty acids are essential to human health but are not synthesized in a sufficient quantity by the human body and must therefore be obtained through the diet.[104–106] The most common essential fatty acid in the Western diet is α-linoleic acid. α-Linoleic acid is known to be neuroprotective due to its antioxidant properties, but its antiamyloidogenic activity has yet to be determined. α-Linoleic acid is known to be metabolically converted to another essential fatty acid,

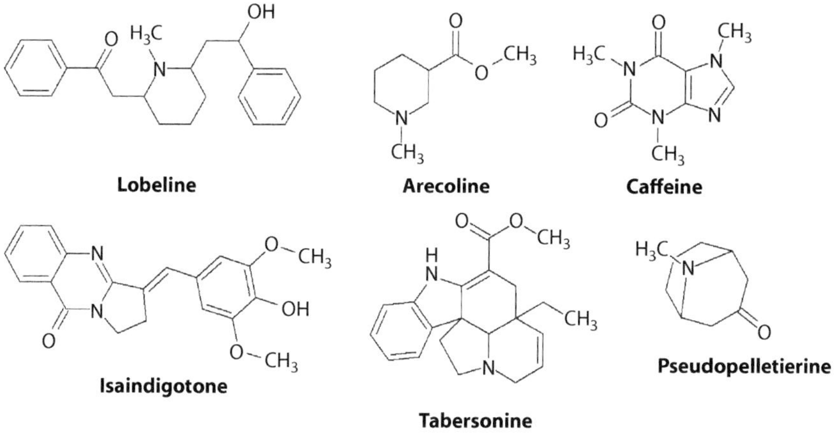

FIGURE 19.6   Chemical structures of alkaloids with antiamyloidogenic activity.[90,95–97]

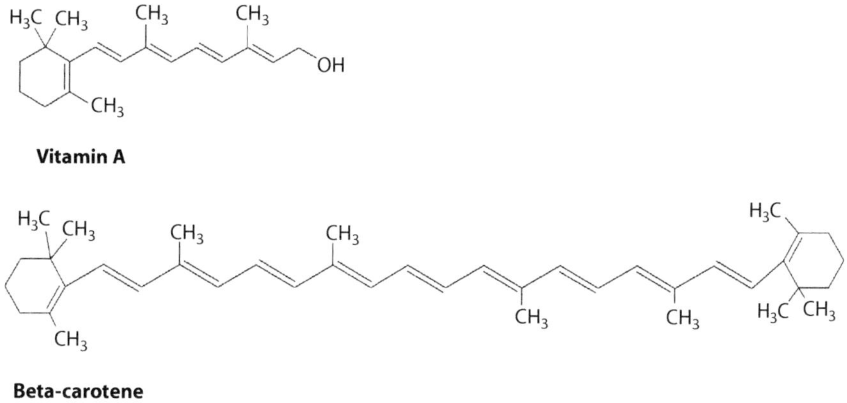

FIGURE 19.7   Chemical structures of terpenes with antiamyloidogenic activity.[90]

docosahexaenoic acid, in the human body.[105,107,108] Chemical structures are given in Figure 19.8. Docosahexaenoic acid is the most abundant n-3 polyunsaturated fatty acid in the human brain and is essential for developing, maintaining, and restoring brain cell function.[109] Docosahexaenoic acid is found in microalgae and is thus also abundant in the marine food chain. Docosahexaenoic acid is known for its antioxidant activity as well as its ability to inhibit amyloid β fibrillation.[104,109–112] Kinetic studies of amyloid β incubated with docosahexaenoic acid have shown that fibrillation is inhibited by inhibition of both nucleus formation and conversion of oligomers into fibrils. In addition to separating the protein molecules by its large hydrophobic volume, docosahexaenoic acid is speculated to inhibit fibrillation by shielding the aromatic residues promoting stacking during fibrillation.[104] Prostaglandins, thromboxanes, and leukotrienes are derivatives of C_{20} polyunsaturated fatty acids.[90] None

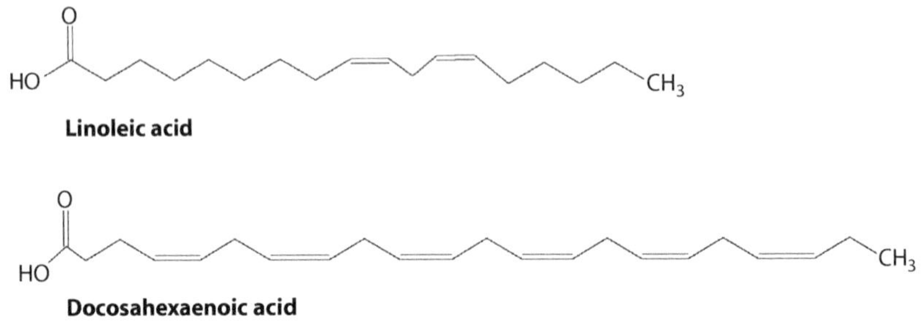

Linoleic acid

Docosahexaenoic acid

FIGURE 19.8   Chemical structures of fatty acids with antiamyloidogenic activity.[90]

of these have to date been shown to inhibit fibrillation of proteins related to neurodegenerative diseases. On the contrary, prostaglandin H2 has been shown to bind covalently to amyloid β accelerating the formation of amyloid β oligomers.[113]

Carbohydrates are formed from water and carbon dioxide as a product of plant photosynthesis. Carbohydrates can be divided into sugars, oligosaccharides, and polysaccharides.[90] To date, no antiamyloidogenic activity has been found in any of the groups, unless coupled to other compounds. An example is the glycosides, which are composed of one or more sugar residues, attached to an aglycone. Many glycosides are known to be bioactive. Two examples of glycosides that are able to inhibit amyloid β fibrillation are the phenylethanoid glycoside acteoside and a flavonol glycoside isolated from the roots of *Panax notoginseng*.[114,115] Chemical structures are given in Figure 19.9.

19.4   EFFECTS OF PHYTOCHEMICALS ON SYMPTOMS OF NEURODEGENERATIVE DISEASES

In addition to interfering with protein aggregation, a variety of phytochemicals were shown to rescue the symptoms of neurodegenerative diseases in cell and animal models by other mechanisms. For example, in the animal models of Alzheimer's disease (AD), the addition of polyphenols (curcumin, resveratrol, and green tea catechins) resulted in lower levels of both oxidized proteins and oxidative metabolites such as malondialdehyde.[116–119] Polyphenols also decreased the levels of both Aβ and tau pathology in animal models and partially rescued cognitive and memory deficits.[118,120–123] They have been tested in AD patients as well, but the results were inconclusive.[124,125] Their mechanisms of action, in addition to interference with protein aggregation, involve direct quenching of ROS and indirect anti-inflammatory effects.[116,126] Flavonoids such as quercetin also counteracted Aβ aggregate-induced oxidative stress in cell cultures,[127] presumably due to their antioxidant activity.

While a number of plant-derived alkaloids can inhibit protein aggregation (see Section 19.3), they were also found to be active against neurodegenerative diseases either by serving as antioxidants[128,129] or by other mechanisms such as acetylcholines-terase (AChE) inhibition.[130–132] Galantamine and rivastigmine act primarily as AChE inhibitors, increasing acetylcholine concentration in the

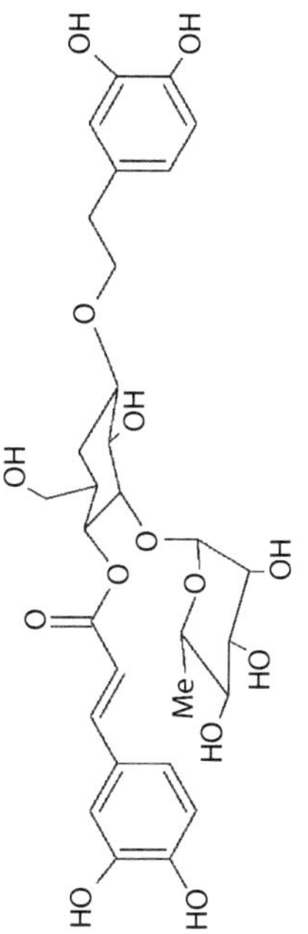

Acteoside

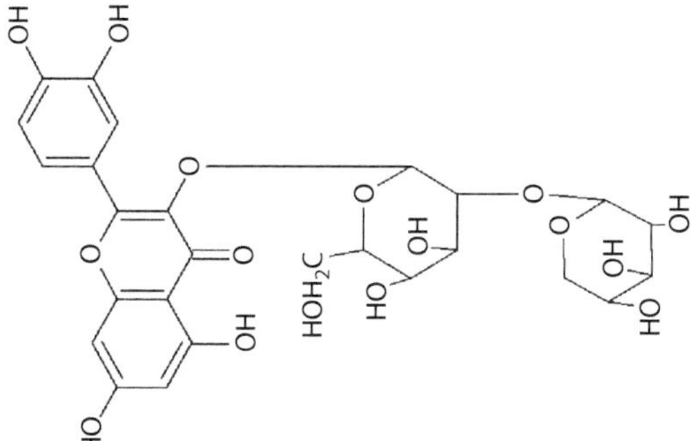

Radix notoginseng flavonol glycoside

FIGURE 19.9 Chemical structures of glycosides with antiamyloidogenic activity.[114,115]

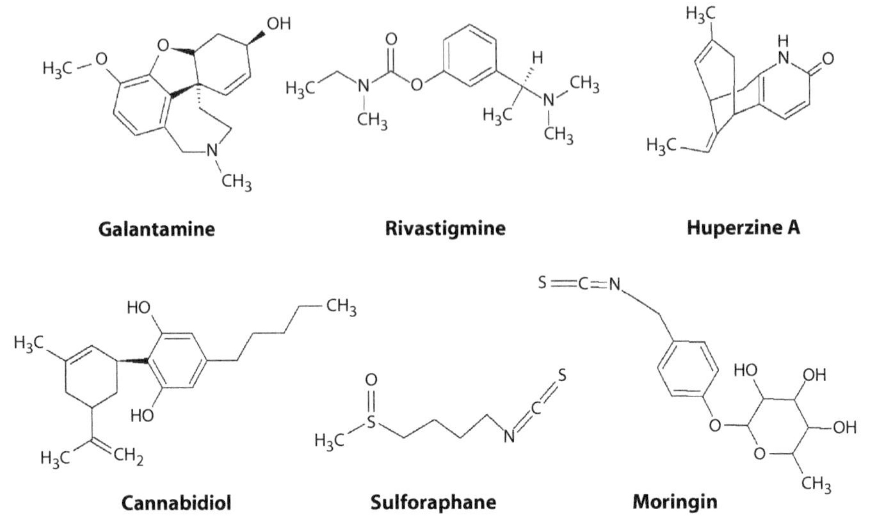

FIGURE 19.10 Chemical structures of compounds shown to rescue the symptoms of neurodegenerative diseases in cell and animal models.[126,132]

synapses.[132,133] Both of these drugs slowed down cognitive decline in mild to moderate AD patients.[134,135] Caffeine, an antagonist of the adenosine A2A receptor, was shown to reduce amyloid burden and memory impairment in animal models of AD.[136,137] Effects of caffeine in human clinical trials were inconclusive.[138] Huperzine A was shown to improve the memory of AD patients in human clinical trials[139] and decrease the amyloid pathology in AD animal models.[140] In addition to acting as an AChE inhibitor, it has also been shown to reduce iron levels in the brain.[140] Cannabidiol and other cannabinoids decreased oxidative stress and improved cognitive functions in AD animal models.[126,141]

Isothiocyanates are sulfur-containing phytochemicals found in cruciferous vegetables such as broccoli and cauliflower (sulforaphane)[126,142] and other plants (moringin).[143,144] They act by activating the anti-inflammatory response and reducing the proinflammatory/proapoptotic responses to stress.[126,145–147] Sulforaphane reduced neuronal loss and plaque load in AD animal models,[148] while moringin improved spatial memory and reduced neurodegeneration.[146] Chemical structures are given in Figure 19.10.

19.5 CONCLUSIONS

Misfolding and aggregation of polypeptides and proteins are important pathological events in many neurodegenerative diseases such as Alzheimer's, Parkinson's, and Huntington's. Plants contain thousands of structurally diverse bioactive secondary metabolites, and their structural diversity makes them excellent lead compounds for drug development. This chapter shows that a variety of phytochemicals can influence both protein aggregation and progression of aggregation-related diseases, making them attractive drug candidates for these diseases.

REFERENCES

1. Chiti, F., and Dobson, C. M. (2006) Protein misfolding, functional amyloid, and human disease, *Annu Rev Biochem 75*, 333–366.
2. Verma, M., Vats, A., and Taneja, V. (2015) Toxic species in amyloid disorders: Oligomers or mature fibrils, *Ann Ind Acad Neurol 18*, 138–145.
3. Uversky, V. N. (2015) Intrinsically disordered proteins and their (disordered) proteomes in neurodegenerative disorders, *Front Aging Neurosci 7*, 18.
4. Maji, S. K., Perrin, M. H., Sawaya, M. R., Jessberger, S., Vadodaria, K., Rissman, R. A., Singru, P. S., Nilsson, K. P., Simon, R., Schubert, D., Eisenberg, D., Rivier, J., Sawchenko, P., Vale, W., and Riek, R. (2009) Functional amyloids as natural storage of peptide hormones in pituitary secretory granules, *Science 325*, 328–332.
5. Si, K., Choi, Y. B., White-Grindley, E., Majumdar, A., and Kandel, E. R. (2010) Aplysia CPEB can form prion-like multimers in sensory neurons that contribute to long-term facilitation, *Cell 140*, 421–435.
6. Jucker, M., and Walker, L. C. (2013) Self-propagation of pathogenic protein aggregates in neurodegenerative diseases, *Nature 501*, 45–51.
7. Lesne, S. E. (2013) Breaking the code of amyloid-beta oligomers, *Int J Cell Biol*, 950783.
8. Uversky, V. N. (2008) Amyloidogenesis of natively unfolded proteins, *Curr Alzheimer Res 5*, 260–287.
9. Sawaya, M. R., Sambashivan, S., Nelson, R., Ivanova, M. I., Sievers, S. A., Apostol, M. I., Thompson, M. J., Balbirnie, M., Wiltzius, J. J., McFarlane, H. T., Madsen, A. O., Riekel, C., and Eisenberg, D. (2007) Atomic structures of amyloid cross-beta spines reveal varied steric zippers, *Nature 447*, 453–457.
10. Nelson, R., Sawaya, M. R., Balbirnie, M., Madsen, A. O., Riekel, C., Grothe, R., and Eisenberg, D. (2005) Structure of the cross-beta spine of amyloid-like fibrils, *Nature 435*, 773–778.
11. Cohen, S. I., Linse, S., Luheshi, L. M., Hellstrand, E., White, D. A., Rajah, L., Otzen, D. E., Vendruscolo, M., Dobson, C. M., and Knowles, T. P. (2013) Proliferation of amyloid-beta42 aggregates occurs through a secondary nucleation mechanism, *Proc Natl Acad Sci U S A 110*, 9758–9763.
12. Arosio, P., Knowles, T. P., and Linse, S. (2015) On the lag phase in amyloid fibril formation, *Phys Chem Chem Phys*.
13. Ghaemmaghami, S., Colby, D. W., Nguyen, H. O., Hayashi, S., Oehler, A., Dearmond, S. J., and Prusiner, S. B. (2013) Convergent replication of mouse synthetic prion strains, *Am J Pathol 182*, 866–874.
14. Weissmann, C., Li, J., Mahal, S. P., and Browning, S. (2011) Prions on the move, *EMBO Rep 12*, 1109–1117.
15. Makarava, N., and Baskakov, I. V. (2013) The evolution of transmissible prions: The role of deformed templating, *PLoS Pathog 9*, e1003759.
16. Gonzalez-Montalban, N., Lee, Y. J., Makarava, N., Savtchenko, R., and Baskakov, I. V. (2013) Changes in prion replication environment cause prion strain mutation, *FASEB J 27*, 3702–3710.
17. Grizel, A. V., Rubel, A. A., and Chernoff, Y. O. (2016) Strain conformation controls the specificity of cross-species prion transmission in the yeast model, *Prion 10*, 269–282.
18. Breydo, L., and Uversky, V. N. (2014) Molecular mechanisms of protein misfolding, In *Bio-nanoimaging: Protein misfolding & aggregation* (Uversky, V. N., and Lyubchenko, Y. L., Eds.), pp 1–16, Elsevier/AP, Academic Press is an imprint of Elsevier, Amsterdam; Boston.
19. Paravastu, A. K., Leapman, R. D., Yau, W. M., and Tycko, R. (2008) Molecular structural basis for polymorphism in Alzheimer's beta-amyloid fibrils, *Proc Natl Acad Sci U S A 105*, 18349–18354.

20. Breydo, L., and Uversky, V. N. (2015) Structural, morphological, and functional diversity of amyloid oligomers, *FEBS Lett 589*, 2640–2648.
21. Chen, S. W., Drakulic, S., Deas, E., Ouberai, M., Aprile, F. A., Arranz, R., Ness, S., Roodveldt, C., Guilliams, T., De-Genst, E. J., Klenerman, D., Wood, N. W., Knowles, T. P., Alfonso, C., Rivas, G., Abramov, A. Y., Valpuesta, J. M., Dobson, C. M., and Cremades, N. (2015) Structural characterization of toxic oligomers that are kinetically trapped during alpha-synuclein fibril formation, *Proc Natl Acad Sci U S A 112*, E1994–E2003.
22. Liu, P., Reed, M. N., Kotilinek, L. A., Grant, M. K., Forster, C. L., Qiang, W., Shapiro, S. L., Reichl, J. H., Chiang, A. C., Jankowsky, J. L., Wilmot, C. M., Cleary, J. P., Zahs, K. R., and Ashe, K. H. (2015) Quaternary structure defines a large class of amyloid-beta oligomers neutralized by sequestration, *Cell Rep 11*, 1760–1771.
23. Wu, J. W., Breydo, L., Isas, J. M., Lee, J., Kuznetsov, Y. G., Langen, R., and Glabe, C. (2010) Fibrillar oligomers nucleate the oligomerization of monomeric amyloid {beta} but do not seed fibril formation, *J Biol Chem 285*, 6071–6079.
24. Laganowsky, A., Liu, C., Sawaya, M. R., Whitelegge, J. P., Park, J., Zhao, M., Pensalfini, A., Soriaga, A. B., Landau, M., Teng, P. K., Cascio, D., Glabe, C., and Eisenberg, D. (2012) Atomic view of a toxic amyloid small oligomer, *Science 335*, 1228–1231.
25. Kayed, R., Canto, I., Breydo, L., Rasool, S., Lukacsovich, T., Wu, J., Albay, R., 3rd, Pensalfini, A., Yeung, S., Head, E., Marsh, J. L., and Glabe, C. (2010) Conformation dependent monoclonal antibodies distinguish different replicating strains or conformers of prefibrillar Abeta oligomers, *Mol Neurodegener 5*, 57.
26. Feng, S., Song, X. H., and Zeng, C. M. (2012) Inhibition of amyloid fibrillation of lysozyme by phenolic compounds involves quinoprotein formation, *FEBS Lett 586*, 3951–3955.
27. Poirier, M. A., Li, H., Macosko, J., Cai, S., Amzel, M., and Ross, C. A. (2002) Huntingtin spheroids and protofibrils as precursors in polyglutamine fibrilization, *J Biol Chem 277*, 41032–41037.
28. Langer, F., Eisele, Y. S., Fritschi, S. K., Staufenbiel, M., Walker, L. C., and Jucker, M. (2011) Soluble Abeta seeds are potent inducers of cerebral beta-amyloid deposition, *J Neurosci 31*, 14488–14495.
29. Walker, L. C., Diamond, M. I., Duff, K. E., and Hyman, B. T. (2013) Mechanisms of protein seeding in neurodegenerative diseases, *JAMA Neurol 70*, 304–310.
30. Lasagna-Reeves, C. A., Castillo-Carranza, D. L., Sengupta, U., Guerrero-Munoz, M. J., Kiritoshi, T., Neugebauer, V., Jackson, G. R., and Kayed, R. (2012) Alzheimer brain-derived tau oligomers propagate pathology from endogenous tau, *Sci Rep 2*, 700.
31. Guerrero-Munoz, M. J., Castillo-Carranza, D. L., Krishnamurthy, S., Paulucci-Holthauzen, A. A., Sengupta, U., Lasagna-Reeves, C. A., Ahmad, Y., Jackson, G. R., and Kayed, R. (2014) Amyloid-beta oligomers as a template for secondary amyloidosis in Alzheimer's disease, *Neurobiol Dis 71*, 14–23.
32. Fang, Y. S., Tsai, K. J., Chang, Y. J., Kao, P., Woods, R., Kuo, P. H., Wu, C. C., Liao, J. Y., Chou, S. C., Lin, V., Jin, L. W., Yuan, H. S., Cheng, I. H., Tu, P. H., and Chen, Y. R. (2014) Full-length TDP-43 forms toxic amyloid oligomers that are present in frontotemporal lobar dementia-TDP patients, *Nat Commun 5*, 4824.
33. Atsmon-Raz, Y., and Miller, Y. (2016) Non-amyloid-beta component of human alpha-synuclein oligomers induces formation of new Abeta oligomers: Insight into the mechanisms that link Parkinson's and Alzheimer's diseases, *ACS Chem Neurosci 7*, 46–55.
34. Yu, L., Edalji, R., Harlan, J. E., Holzman, T. F., Lopez, A. P., Labkovsky, B., Hillen, H., Barghorn, S., Ebert, U., Richardson, P. L., Miesbauer, L., Solomon, L., Bartley, D., Walter, K., Johnson, R. W., Hajduk, P. J., and Olejniczak, E. T. (2009) Structural characterization of a soluble amyloid beta-peptide oligomer, *Biochemistry 48*, 1870–1877.

35. Calabrese, M. F., Eakin, C. M., Wang, J. M., and Miranker, A. D. (2008) A regulatable switch mediates self-association in an immunoglobulin fold, *Nat Struct Mol Biol 15*, 965–971.

36. Apostol, M. I., Perry, K., and Surewicz, W. K. (2013) Crystal structure of a human prion protein fragment reveals a motif for oligomer formation, *J Am Chem Soc 135*, 10202–10205.

37. Porat, Y., Abramowitz, A., and Gazit, E. (2006) Inhibition of amyloid fibril formation by polyphenols: structural similarity and aromatic interactions as a common inhibition mechanism, *Chem Biol Drug Des 67*, 27–37.

38. Andrich, K., and Bieschke, J. (2015) The Effect of (-)-epigallo-catechin-(3)-gallate on amyloidogenic proteins suggests a common mechanism, *Adv Exp Med Biol 863*, 139–161.

39. Necula, M., Kayed, R., Milton, S., and Glabe, C. G. (2007) Small molecule inhibitors of aggregation indicate that amyloid beta oligomerization and fibrillization pathways are independent and distinct, *J Biol Chem 282*, 10311–10324.

40. Necula, M., Breydo, L., Milton, S., Kayed, R., Veer, W. E., Tone, P., and Glabe, C. G. (2007) Methylene blue inhibits amyloid abeta oligomerization by promoting fibrilliza- tion, *Biochemistry 46*, 8850–8860.

41. Lopez del Amo, J. M., Fink, U., Dasari, M., Grelle, G., Wanker, E. E., Bieschke, J., and Reif, B. (2012) Structural properties of EGCG-induced, nontoxic Alzheimer's disease Abeta oligomers, *J Mol Biol 421*, 517–524.

42. Thapa, A., Jett, S. D., and Chi, E. Y. (2015) Curcumin attenuates amyloid-beta aggregate toxicity and modulates amyloid-beta aggregation pathway, *ACS Chem Neurosci 7*, 56–68.

43. Wobst, H. J., Sharma, A., Diamond, M. I., Wanker, E. E., and Bieschke, J. (2015) The green tea polyphenol (-)-epigallocatechin gallate prevents the aggregation of tau pro- tein into toxic oligomers at substoichiometric ratios, *FEBS Lett 589*, 77–83.

44. Bonanomi, M., Natalello, A., Visentin, C., Pastori, V., Penco, A., Cornelli, G., Colombo, G., Malabarba, M. G., Doglia, S. M., Relini, A., Regonesi, M. E., and Tortora, P. (2014) Epigallocatechin-3-gallate and tetracycline differently affect ataxin-3 fibrillogen- esis and reduce toxicity in spinocerebellar ataxia type 3 model, *Hum Mol Genet 23*, 6542–6552.

45. Thapa, A., Jett, S. D., and Chi, E. Y. (2016) Curcumin attenuates amyloid-beta aggre- gate toxicity and modulates amyloid-beta aggregation pathway, *ACS Chem Neurosci 7*, 56–68.

46. Zhang, X., and Li, J. P. (2010) Heparan sulfate proteoglycans in amyloidosis, *Prog Mol Biol Transl Sci 93*, 309–334.

47. Holmes, B. B., and Diamond, M. I. (2014) Prion-like properties of Tau protein: the importance of extracellular Tau as a therapeutic target, *J Biol Chem 289*, 19855–19861.

48. Luo, J., Yu, C. H., Yu, H., Borstnar, R., Kamerlin, S. C., Graslund, A., Abrahams, J. P., and Warmlander, S. K. (2013) Cellular polyamines promote amyloid-beta (Abeta) peptide fibrillation and modulate the aggregation pathways, *ACS Chem Neurosci 4*, 454–462.

49. Kumar, S., Schlamadinger, D. E., Brown, M. A., Dunn, J. M., Mercado, B., Hebda, J. A., Saraogi, I., Rhoades, E., Hamilton, A. D., and Miranker, A. D. (2015) Islet amy- loid-induced cell death and bilayer integrity loss share a molecular origin targetable with oligopyridylamide-based alpha-helical mimetics, *Chem Biol 22*, 369–378.

50. Zheng, X., Liu, D., Klarner, F. G., Schrader, T., Bitan, G., and Bowers, M. T. (2015) Amyloid beta-protein assembly: The effect of molecular tweezers CLR01 and CLR03, *J Phys Chem B 119*, 4831–4841.

51. Cheruvara, H., Allen-Baume, V. L., Kad, N. M., and Mason, J. M. (2015) Intracellular screening of a peptide library to derive a potent peptide inhibitor of alpha-synuclein aggregation, *J Biol Chem 290*, 7426–7435.

52. Wang, Q., Liang, G., Zhang, M., Zhao, J., Patel, K., Yu, X., Zhao, C., Ding, B., Zhang, G., Zhou, F., and Zheng, J. (2014) De novo design of self-assembled hexapeptides as beta-amyloid (Abeta) peptide inhibitors, *ACS Chem Neurosci 5*, 972–981.

53. McKoy, A. F., Chen, J., Schupbach, T., and Hecht, M. H. (2014) Structure-activity relationships for a series of compounds that inhibit aggregation of the Alzheimer's peptide, Abeta42, *Chem Biol Drug Des 84*, 505–512.

54. Saunders, J. C., Young, L. M., Mahood, R. A., Jackson, M. P., Revill, C. H., Foster, R. J., Smith, D. A., Ashcroft, A. E., Brockwell, D. J., and Radford, S. E. (2016) An in vivo platform for identifying inhibitors of protein aggregation, *Nat Chem Biol 12*, 94–101.

55. Nath, A., Schlamadinger, D. E., Rhoades, E., and Miranker, A. D. (2015) Structure-Based Small Molecule Modulation of a pre-amyloid state: Pharmacological enhancement of IAPP membrane-binding and toxicity, *Biochemistry 54*, 3555–3564.

56. Guzior, N., Bajda, M., Skrok, M., Kurpiewska, K., Lewinski, K., Brus, B., Pislar, A., Kos, J., Gobec, S., and Malawska, B. (2015) Development of multifunctional, heterodimeric isoindoline-1,3-dione derivatives as cholinesterase and beta-amyloid aggregation inhibitors with neuroprotective properties, *Eur J Med Chem 92*, 738–749.

57. Ardah, M. T., Paleologou, K. E., Lv, G., Menon, S. A., Abul Khair, S. B., Lu, J. H., Safieh-Garabedian, B., Al-Hayani, A. A., Eliezer, D., Li, M., and El-Agnaf, O. M. (2015) Ginsenoside Rb1 inhibits fibrillation and toxicity of alpha-synuclein and disaggregates preformed fibrils, *Neurobiol Dis 74*, 89–101.

58. Scherzer-Attali, R., Pellarin, R., Convertino, M., Frydman-Marom, A., Egoz-Matia, N., Peled, S., Levy-Sakin, M., Shalev, D. E., Caflisch, A., Gazit, E., and Segal, D. (2010) Complete phenotypic recovery of an Alzheimer's disease model by a quinone-tryptophan hybrid aggregation inhibitor, *PLoS One 5*, e11101.

59. Nasstrom, T., Fagerqvist, T., Barbu, M., Karlsson, M., Nikolajeff, F., Kasrayan, A., Ekberg, M., Lannfelt, L., Ingelsson, M., and Bergstrom, J. (2011) The lipid peroxidation products 4-oxo-2-nonenal and 4-hydroxy-2-nonenal promote the formation of alpha-synuclein oligomers with distinct biochemical, morphological, and functional properties, *Free Radic Biol Med 50*, 428–437.

60. Qin, Z., Hu, D., Han, S., Reaney, S. H., Di Monte, D. A., and Fink, A. L. (2007) Effect of 4-hydroxy-2-nonenal modification on alpha-synuclein aggregation, *J Biol Chem 282*, 5862–5870.

61. Siegel, S. J., Bieschke, J., Powers, E. T., and Kelly, J. W. (2007) The oxidative stress metabolite 4-hydroxynonenal promotes Alzheimer protofibril formation, *Biochemistry 46*, 1503–1510.

62. Bae, E. J., Ho, D. H., Park, E., Jung, J. W., Cho, K., Hong, J. H., Lee, H. J., Kim, K. P., and Lee, S. J. (2013) Lipid peroxidation product 4-hydroxy-2-nonenal promotes seeding-capable oligomer formation and cell-to-cell transfer of alpha-synuclein, *Antioxid Redox Signal 18*, 770–783.

63. Arai, T., Sasaki, D., Araya, T., Sato, T., Sohma, Y., and Kanai, M. (2014) A cyclic KLVFF-derived peptide aggregation inhibitor induces the formation of less-toxic off-pathway amyloid-beta oligomers, *Chembiochem 15*, 2577–2583.

64. Harvey, A. (2000) Strategies for discovering drugs from previously unexplored natural products, *Drug Discov Today 5*, 294–300.

65. Khafagi, I. K., and Dewedar, A. (2000) The efficiency of random versus ethno-directed research in the evaluation of Sinai medicinal plants for bioactive compounds, *J Ethnopharmacol 71*, 365–376.

66. Howes, M. J., and Houghton, P. J. (2012) Ethnobotanical treatment strategies against Alzheimer's disease, *Curr Alzheimer Res 9*, 67–85.

67. Choudhary, I. M., and Atta-ur-Rahman. (1997) Bioactivity-guided isolation of phytochemicals from medicinal plants, In *Phytochemical diversity—A source of new industrial products* (Wrigley, S., Hayes, M., Thomas, R., and Chrystal, E., Eds.), pp 41–52, The Royal Society of Chemistry, Cambridge, UK.
68. Cragg, G. M., and Newman, D. J. (2001) Natural product drug discovery in the next millennium, *Pharm Biol 39 Suppl 1*, 8–17.
69. Patridge, E., Gareiss, P., Kinch, M. S., and Hoyer, D. (2016) An analysis of FDA-approved drugs: natural products and their derivatives, *Drug Discov Today 21*, 204–207.
70. Carkeet, C. (2013) *Phytochemicals: Health promotion and therapeutic potential*, CRC Press, Boca Raton.
71. Wink, M. (2003) Evolution of secondary metabolites from an ecological and molecular phylogenetic perspective, *Phytochemistry 64*, 3–19.
72. Joseph, J. A., Shukitt-Hale, B., and Willis, L. M. (2009) Grape juice, berries, and walnuts affect brain aging and behavior, *J Nutr 139*, 1813S–1817S.
73. Joseph, J. A., Shukitt-Hale, B., Denisova, N. A., Bielinski, D., Martin, A., McEwen, J. J., and Bickford, P. C. (1999) Reversals of age-related declines in neuronal signal transduction, cognitive, and motor behavioral deficits with blueberry, spinach, or strawberry dietary supplementation, *J Neurosci 19*, 8114–8121.
74. Kumar, G. P., and Khanum, F. (2012) Neuroprotective potential of phytochemicals, *Pharmacogn Rev 6*, 81–90.
75. Sgarbossa, A. (2012) Natural biomolecules and protein aggregation: Emerging strategies against amyloidogenesis, *Int J Mol Sci 13*, 17121–17137.
76. Davinelli, S., Sapere, N., Zella, D., Bracale, R., Intrieri, M., and Scapagnini, G. (2012) Pleiotropic protective effects of phytochemicals in Alzheimer's disease, *Oxid Med Cell Longev 2012*, 386527.
77. Howlett, D. R., George, A. R., Owen, D. E., Ward, R. V., and Markwell, R. E. (1999) Common structural features determine the effectiveness of carvedilol, daunomycin and rolitetracycline as inhibitors of Alzheimer beta-amyloid fibril formation, *Biochem J 343 Pt 2*, 419–423.
78. Wang, M. S., Boddapati, S., Emadi, S., and Sierks, M. R. (2010) Curcumin reduces alpha-synuclein induced cytotoxicity in Parkinson's disease cell model, *BMC Neurosci 11*, 57.
79. Ono, K., Hirohata, M., and Yamada, M. (2008) Alpha-synuclein assembly as a therapeutic target of Parkinson's disease and related disorders, *Curr Pharm Des 14*, 3247–3266.
80. Pandey, N., Strider, J., Nolan, W. C., Yan, S. X., and Galvin, J. E. (2008) Curcumin inhibits aggregation of alpha-synuclein, *Acta Neuropathol 115*, 479–489.
81. Ono, K., Hasegawa, K., Naiki, H., and Yamada, M. (2004) Anti-amyloidogenic activity of tannic acid and its activity to destabilize Alzheimer's beta-amyloid fibrils in vitro, *Biochim Biophys Acta 1690*, 193–202.
82. Ono, K., Hasegawa, K., Naiki, H., and Yamada, M. (2004) Curcumin has potent anti-amyloidogenic effects for Alzheimer's beta-amyloid fibrils in vitro, *J Neurosci Res 75*, 742–750.
83. Meng, X., Munishkina, L. A., Fink, A. L., and Uversky, V. N. (2010) Effects of various flavonoids on the alpha-synuclein fibrillation process, *Parkinsons Dis 2010*, 650794.
84. Meng, X., Munishkina, L. A., Fink, A. L., and Uversky, V. N. (2009) Molecular mechanisms underlying the flavonoid-induced inhibition of alpha-synuclein fibrillation, *Biochemistry 48*, 8206–8224.
85. Hirohata, M., Hasegawa, K., Tsutsumi-Yasuhara, S., Ohhashi, Y., Ookoshi, T., Ono, K., Yamada, M., and Naiki, H. (2007) The anti-amyloidogenic effect is exerted against Alzheimer's beta-amyloid fibrils in vitro by preferential and reversible binding of flavonoids to the amyloid fibril structure, *Biochemistry 46*, 1888–1899.

86. Ehrnhoefer, D. E., Bieschke, J., Boeddrich, A., Herbst, M., Masino, L., Lurz, R., Engemann, S., Pastore, A., and Wanker, E. E. (2008) EGCG redirects amyloidogenic polypeptides into unstructured, off-pathway oligomers, *Nat Struct Mol Biol 15*, 558–566.

87. Bieschke, J., Russ, J., Friedrich, R. P., Ehrnhoefer, D. E., Wobst, H., Neugebauer, K., and Wanker, E. E. (2010) EGCG remodels mature alpha-synuclein and amyloid-beta fibrils and reduces cellular toxicity, *Proc Natl Acad Sci U S A 107*, 7710–7715.

88. Lorenzen, N., Nielsen, S. B., Yoshimura, Y., Vad, B. S., Andersen, C. B., Betzer, C., Kaspersen, J. D., Christiansen, G., Pedersen, J. S., Jensen, P. H., Mulder, F. A., and Otzen, D. E. (2014) How epigallocatechin gallate can inhibit alpha-synuclein oligomer toxicity in vitro, *J Biol Chem 289*, 21299–21310.

89. Hong, D. P., Fink, A. L., and Uversky, V. N. (2008) Structural characteristics of alpha-synuclein oligomers stabilized by the flavonoid baicalein, *J Mol Biol 383*, 214–223.

90. Samuelsson, G., and Bohlin, L. (2009) *Drugs of natural origin. A treatise of pharmacognosy, 6th edition*, Swedish Pharmaceutical Press, Stockholm, Sweden.

91. Gong, H., He, Z., Peng, A., Zhang, X., Cheng, B., Sun, Y., Zheng, L., and Huang, K. (2014) Effects of several quinones on insulin aggregation, *Sci Rep 4*, 5648.

92. da Silva, F. L., Coelho Cerqueira, E., de Freitas, M. S., Goncalves, D. L., Costa, L. T., and Follmer, C. (2013) Vitamins K interact with N-terminus alpha-synuclein and modulate the protein fibrillization in vitro. Exploring the interaction between quinones and alpha-synuclein, *Neurochem Int 62*, 103–112.

93. Bermejo-Bescos, P., Martin-Aragon, S., Jimenez-Aliaga, K. L., Ortega, A., Molina, M. T., Buxaderas, E., Orellana, G., and Csaky, A. G. (2010) In vitro antiamyloidogenic properties of 1,4-naphthoquinones, *Biochem Biophys Res Commun 400*, 169–174.

94. Convertino, M., Pellarin, R., Catto, M., Carotti, A., and Caflisch, A. (2009) 9,10-Anthraquinone hinders beta-aggregation: How does a small molecule interfere with Abeta-peptide amyloid fibrillation?, *Protein Sci 18*, 792–800.

95. Krazinski, B. E., Radecki, J., and Radecka, H. (2011) Surface plasmon resonance based biosensors for exploring the influence of alkaloids on aggregation of amyloid-beta peptide, *Sensors (Basel) 11*, 4030–4042.

96. Yan, J. W., Li, Y. P., Ye, W. J., Chen, S. B., Hou, J. Q., Tan, J. H., Ou, T. M., Li, D., Gu, L. Q., and Huang, Z. S. (2012) Design, synthesis and evaluation of isaindigotone derivatives as dual inhibitors for acetylcholinesterase and amyloid beta aggregation, *Bioorg Med Chem 20*, 2527–2534.

97. Kai, T., Zhang, L., Wang, X., Jing, A., Zhao, B., Yu, X., Zheng, J., and Zhou, F. (2015) Tabersonine inhibits amyloid fibril formation and cytotoxicity of Abeta(1–42), *ACS Chem Neurosci 6*, 879–888.

98. Kardani, J., and Roy, I. (2015) Understanding caffeine's role in attenuating the toxicity of alpha-synuclein aggregates: Implications for risk of Parkinson's disease, *ACS Chem Neurosci 6*, 1613–1625.

99. Mohan, A., Roberto, A. J., Mohan, A., Liogier-Weyback, L., Guha, R., Ravishankar, N., Rebello, C., Kumar, A., and Mohan, R. (2015) Caffeine as treatment for Alzheimer's disease: A review, *J Caffeine Res 5*, 61–64.

100. Cheng, B., Liu, X., Gong, H., Huang, L., Chen, H., Zhang, X., Li, C., Yang, M., Ma, B., Jiao, L., Zheng, L., and Huang, K. (2011) Coffee components inhibit amyloid formation of human islet amyloid polypeptide in vitro: Possible link between coffee consumption and diabetes mellitus, *J Agric Food Chem 59*, 13147–13155.

101. Echeverria, V., and Zeitlin, R. (2012) Cotinine: A potential new therapeutic agent against Alzheimer's disease, *CNS Neurosci Ther 18*, 517–523.

102. Ono, K., Hirohata, M., and Yamada, M. (2007) Anti-fibrillogenic and fibril-destabilizing activity of nicotine in vitro: Implications for the prevention and therapeutics of Lewy body diseases, *Exp Neurol 205*, 414–424.

103. Takasaki, J., Ono, K., Yoshiike, Y., Hirohata, M., Ikeda, T., Morinaga, A., Takashima, A., and Yamada, M. (2011) Vitamin A has anti-oligomerization effects on amyloid-beta in vitro, *J Alzheimers Dis 27*, 271–280.

104. Hossain, S., Hashimoto, M., Katakura, M., Miwa, K., Shimada, T., and Shido, O. (2009) Mechanism of docosahexaenoic acid-induced inhibition of in vitro Abeta1–42 fibrillation and Abeta1–42-induced toxicity in SH-S5Y5 cells, *J Neurochem 111*, 568–579.

105. Burdge, G. C., and Wootton, S. A. (2002) Conversion of alpha-linolenic acid to eicosapentaenoic, docosapentaenoic and docosahexaenoic acids in young women, *Br J Nutr 88*, 411–420.

106. Hitchcock, C., and Nichols, B. W. (1971) *Plant lipid biochemistry: The biochemistry of fatty acids and acyl lipids with particular reference to higher plants and algae*, Academic Press, London, New York.

107. Packer, L., Witt, E. H., and Tritschler, H. J. (1995) alpha-lipoic acid as a biological antioxidant, *Free Radic Biol Med 19*, 227–250.

108. Packer, L., Tritschler, H. J., and Wessel, K. (1997) Neuroprotection by the metabolic antioxidant alpha-lipoic acid, *Free Radic Biol Med 22*, 359–378.

109. Hashimoto, M., and Hossain, S. (2011) Neuroprotective and ameliorative actions of polyunsaturated fatty acids against neuronal diseases: Beneficial effect of docosahexaenoic acid on cognitive decline in Alzheimer's disease, *J Pharmacol Sci 116*, 150–162.

110. Gamoh, S., Hashimoto, M., Sugioka, K., Shahdat Hossain, M., Hata, N., Misawa, Y., and Masumura, S. (1999) Chronic administration of docosahexaenoic acid improves reference memory-related learning ability in young rats, *Neuroscience 93*, 237–241.

111. Hashimoto, M., Hossain, S., Shimada, T., Sugioka, K., Yamasaki, H., Fujii, Y., Ishibashi, Y., Oka, J., and Shido, O. (2002) Docosahexaenoic acid provides protection from impairment of learning ability in Alzheimer's disease model rats, *J Neurochem 81*, 1084–1091.

112. Hashimoto, M., Shahdat, H. M., Yamashita, S., Katakura, M., Tanabe, Y., Fujiwara, H., Gamoh, S., Miyazawa, T., Arai, H., Shimada, T., and Shido, O. (2008) Docosahexaenoic acid disrupts in vitro amyloid beta(1–40) fibrillation and concomitantly inhibits amyloid levels in cerebral cortex of Alzheimer's disease model rats, *J Neurochem 107*, 1634–1646.

113. Boutaud, O., Ou, J. J., Chaurand, P., Caprioli, R. M., Montine, T. J., and Oates, J. A. (2002) Prostaglandin H2 (PGH2) accelerates formation of amyloid beta1–42 oligomers, *J Neurochem 82*, 1003–1006.

114. Choi, R. C., Zhu, J. T., Leung, K. W., Chu, G. K., Xie, H. Q., Chen, V. P., Zheng, K. Y., Lau, D. T., Dong, T. T., Chow, P. C., Han, Y. F., Wang, Z. T., and Tsim, K. W. (2010) A flavonol glycoside, isolated from roots of *Panax notoginseng*, reduces amyloid-beta-induced neurotoxicity in cultured neurons: Signaling transduction and drug development for Alzheimer's disease, *J Alzheimers Dis 19*, 795–811.

115. Kurisu, M., Miyamae, Y., Murakami, K., Han, J., Isoda, H., Irie, K., and Shigemori, H. (2013) Inhibition of amyloid beta aggregation by acteoside, a phenylethanoid glycoside, *Biosci Biotechnol Biochem 77*, 1329–1332.

116. Kim, J., Lee, H. J., and Lee, K. W. (2010) Naturally occurring phytochemicals for the prevention of Alzheimer's disease, *J Neurochem 112*, 1415–1430.

117. Lim, G. P., Chu, T., Yang, F., Beech, W., Frautschy, S. A., and Cole, G. M. (2001) The curry spice curcumin reduces oxidative damage and amyloid pathology in an Alzheimer transgenic mouse, *J Neurosci 21*, 8370–8377.

118. Kumar, A., Naidu, P. S., Seghal, N., and Padi, S. S. (2007) Neuroprotective effects of resveratrol against intracerebroventricular colchicine-induced cognitive impairment and oxidative stress in rats, *Pharmacology 79*, 17–26.

119. Choi, Y. T., Jung, C. H., Lee, S. R., Bae, J. H., Baek, W. K., Suh, M. H., Park, J., Park, C. W., and Suh, S. I. (2001) The green tea polyphenol (-)-epigallocatechin gallate attenuates beta-amyloid-induced neurotoxicity in cultured hippocampal neurons, *Life Sci 70*, 603–614.

120. Ishrat, T., Hoda, M. N., Khan, M. B., Yousuf, S., Ahmad, M., Khan, M. M., Ahmad, A., and Islam, F. (2009) Amelioration of cognitive deficits and neurodegeneration by curcumin in rat model of sporadic dementia of Alzheimer's type (SDAT), *Eur Neuropsychopharmacol 19*, 636–647.

121. Rezai-Zadeh, K., Arendash, G. W., Hou, H., Fernandez, F., Jensen, M., Runfeldt, M., Shytle, R. D., and Tan, J. (2008) Green tea epigallocatechin-3-gallate (EGCG) reduces beta-amyloid mediated cognitive impairment and modulates tau pathology in Alzheimer transgenic mice, *Brain Res 1214*, 177–187.

122. Haque, A. M., Hashimoto, M., Katakura, M., Hara, Y., and Shido, O. (2008) Green tea catechins prevent cognitive deficits caused by Abeta1–40 in rats, *J Nutr Biochem 19*, 619–626.

123. Wang, P., Su, C., Li, R., Wang, H., Ren, Y., Sun, H., Yang, J., Sun, J., Shi, J., Tian, J., and Jiang, S. (2014) Mechanisms and effects of curcumin on spatial learning and memory improvement in APPswe/PS1dE9 mice, *J Neurosci Res 92*, 218–231.

124. Brondino, N., Re, S., Boldrini, A., Cuccomarino, A., Lanati, N., Barale, F., and Politi, P. (2014) Curcumin as a therapeutic agent in dementia: A mini systematic review of human studies, *ScientificWorldJournal 2014*, 174282.

125. Turner, R. S., Thomas, R. G., Craft, S., van Dyck, C. H., Mintzer, J., Reynolds, B. A., Brewer, J. B., Rissman, R. A., Raman, R., Aisen, P. S., and Alzheimer's Disease Cooperative, S. (2015) A randomized, double-blind, placebo-controlled trial of resveratrol for Alzheimer disease, *Neurology 85*, 1383–1391.

126. Libro, R., Giacoppo, S., Soundara Rajan, T., Bramanti, P., and Mazzon, E. (2016) Natural phytochemicals in the treatment and prevention of dementia: An overview, *Molecules 21*, 518.

127. Mancuso, C., Siciliano, R., Barone, E., and Preziosi, P. (2012) Natural substances and Alzheimer's disease: From preclinical studies to evidence based medicine, *Biochim Biophys Acta 1822*, 616–624.

128. Huang, M., Chen, S., Liang, Y., and Guo, Y. (2016) The role of berberine in the multi-target treatment of senile dementia, *Curr Top Med Chem 16*, 867–873.

129. Imenshahidi, M., and Hosseinzadeh, H. (2016) *Berberis vulgaris* and berberine: An update review, *Phytother Res*.

130. Fu, R. H., Wang, Y. C., Chen, C. S., Tsai, R. T., Liu, S. P., Chang, W. L., Lin, H. L., Lu, C. H., Wei, J. R., Wang, Z. W., Shyu, W. C., and Lin, S. Z. (2014) Acetylcorynoline attenuates dopaminergic neuron degeneration and alpha-synuclein aggregation in animal models of Parkinson's disease, *Neuropharmacology 82*, 108–120.

131. Dewapriya, P., Li, Y. X., Himaya, S. W., Pangestuti, R., and Kim, S. K. (2013) Neoechinulin A suppresses amyloid-beta oligomer-induced microglia activation and thereby protects PC-12 cells from inflammation-mediated toxicity, *Neurotoxicology 35*, 30–40.

132. Konrath, E. L., Passos Cdos, S., Klein, L. C., Jr., and Henriques, A. T. (2013) Alkaloids as a source of potential anticholinesterase inhibitors for the treatment of Alzheimer's disease, *J Pharm Pharmacol 65*, 1701–1725.

133. Mehta, M., Adem, A., and Sabbagh, M. (2012) New acetylcholinesterase inhibitors for Alzheimer's disease, *Int J Alzheimers Dis 2012*, 728983.

134. Birks, J. S., and Grimley Evans, J. (2015) Rivastigmine for Alzheimer's disease, *Cochrane Database Syst Rev*, CD001191.

135. Schneider, L. S., Mangialasche, F., Andreasen, N., Feldman, H., Giacobini, E., Jones, R., Mantua, V., Mecocci, P., Pani, L., Winblad, B., and Kivipelto, M. (2014) Clinical trials and late-stage drug development for Alzheimer's disease: An appraisal from 1984 to 2014, *J Intern Med 275*, 251–283.

136. Laurent, C., Eddarkaoui, S., Derisbourg, M., Leboucher, A., Demeyer, D., Carrier, S., Schneider, M., Hamdane, M., Muller, C. E., Buee, L., and Blum, D. (2014) Beneficial effects of caffeine in a transgenic model of Alzheimer's disease-like tau pathology, *Neurobiol Aging 35*, 2079–2090.

137. Han, K., Jia, N., Li, J., Yang, L., and Min, L. Q. (2013) Chronic caffeine treatment reverses memory impairment and the expression of brain BNDF and TrkB in the PS1/APP double transgenic mouse model of Alzheimer's disease, *Mol Med Rep 8*, 737–740.

138. Kim, Y. S., Kwak, S. M., and Myung, S. K. (2015) Caffeine intake from coffee or tea and cognitive disorders: a meta-analysis of observational studies, *Neuroepidemiology 44*, 51–63.

139. Xing, S. H., Zhu, C. X., Zhang, R., and An, L. (2014) Huperzine A in the treatment of Alzheimer's disease and vascular dementia: A meta-analysis, *Evid Based Complement Alternat Med 2014*, 363985.

140. Huang, X. T., Qian, Z. M., He, X., Gong, Q., Wu, K. C., Jiang, L. R., Lu, L. N., Zhu, Z. J., Zhang, H. Y., Yung, W. H., and Ke, Y. (2014) Reducing iron in the brain: A novel pharmacologic mechanism of huperzine A in the treatment of Alzheimer's disease, *Neurobiol Aging 35*, 1045–1054.

141. Aso, E., Sanchez-Pla, A., Vegas-Lozano, E., Maldonado, R., and Ferrer, I. (2015) Cannabis-based medicine reduces multiple pathological processes in AbetaPP/PS1 mice, *J Alzheimers Dis 43*, 977–991.

142. Zhang, Y., Talalay, P., Cho, C. G., and Posner, G. H. (1992) A major inducer of anti-carcinogenic protective enzymes from broccoli: Isolation and elucidation of structure, *Proc Natl Acad Sci U S A 89*, 2399–2403.

143. Galuppo, M., Giacoppo, S., Iori, R., De Nicola, G. R., Milardi, D., Bramanti, P., and Mazzon, E. (2015) 4(alpha-L-rhamnosyloxy)-benzyl isothiocyanate, a bioactive phytochemical that defends cerebral tissue and prevents severe damage induced by focal ischemia/reperfusion, *J Biol Regul Homeost Agents 29*, 343–356.

144. Giacoppo, S., Galuppo, M., De Nicola, G. R., Iori, R., Bramanti, P., and Mazzon, E. (2015) 4(alpha-l-Rhamnosyloxy)-benzyl isothiocyanate, a bioactive phytochemical that attenuates secondary damage in an experimental model of spinal cord injury, *Bioorg Med Chem 23*, 80–88.

145. Jazwa, A., Rojo, A. I., Innamorato, N. G., Hesse, M., Fernandez-Ruiz, J., and Cuadrado, A. (2011) Pharmacological targeting of the transcription factor Nrf2 at the basal ganglia provides disease modifying therapy for experimental parkinsonism, *Antioxid Redox Signal 14*, 2347–2360.

146. Sutalangka, C., Wattanathorn, J., Muchimapura, S., and Thukham-mee, W. (2013) Moringa oleifera mitigates memory impairment and neurodegeneration in animal model of age-related dementia, *Oxid Med Cell Longev 2013*, 695936.

147. Giacoppo, S., Galuppo, M., Montaut, S., Iori, R., Rollin, P., Bramanti, P., and Mazzon, E. (2015) An overview on neuroprotective effects of isothiocyanates for the treatment of neurodegenerative diseases, *Fitoterapia 106*, 12–21.

148. Zhang, R., Miao, Q. W., Zhu, C. X., Zhao, Y., Liu, L., Yang, J., and An, L. (2015) Sulforaphane ameliorates neurobehavioral deficits and protects the brain from amyloid beta deposits and peroxidation in mice with Alzheimer-like lesions, *Am J Alzheimers Dis Other Demen 30*, 183–191.

20 Synergism among Natural Products in Neuroprotection and Prevention of Brain Cancer

*Darakhshanda Neelam, Syed Akhtar Husain,
Tanveera Tabasum, and Mohd Maqbool Lone*

CONTENTS

20.1 INTRODUCTION

Over the past decades, herbal medicine has become a topic of global importance, making an impact on both world health and international trade. Medicinal plants continue to play a central role in the healthcare system of a large proportion of the world's population. This is particularly true in developing countries, where herbal medicine has a long and uninterrupted history of use. Recognition and development of the medicinal and economic benefits of these plants are on the increase in both developing and industrialized nations. Continued usage of herbal medicine by a large proportion of the population in the developing countries is largely due to the high cost of Western pharmaceuticals and healthcare.

Among the human diseases, herbal medicine has long been used to treat neural symptoms. Structural diversity of medicinal herbs makes them a valuable source of novel lead compounds against therapeutic targets that are newly discovered by genomics, proteomics, and high-throughput screening. Various combinations of the active components of the plants are assessed for their synergetic effects in the remedy of neuroprotective potential and brain cancers. This article reviews many such structures and their related chemistry along with the recent advances in understanding the mechanism of action or with additive effects for their potential neuroprotective properties [1].

20.2 NOOTROPISM

Nootropics is a term used by proponents of smart drugs to describe medical drugs and nutritional supplements that have a positive effect on brain function; "nootropic" is derived from Greek and means acting on the mind [2]. Drugs to improve neurofunction generally work by altering the balance of particular chemicals (neurotransmitters) in the brain. Some act by selective enhancement of cerebral blood flow, cerebral oxygen usage metabolic rate, and cerebral glucose metabolic rate in chronic impaired human brain function, i.e., multi-infarct (stroke) dementia, senile dementia of the Alzheimer type and pseudo-dementia, and ischemic cerebral (poor brain blood flow) infarcts. One of the mechanisms suggested to dementia is decreased cholinergic activity in the brain. Therefore, cholinergic drugs (of plant origin) like muscarinic agonists (e.g., arecoline, pilocarpine, etc.), nicotinic agonists (e.g., nicotine), and cholinesterase inhibitors (e.g., huperzine) can be employed for improving memory [2].

There are more than 120 traditional medicines that are being used for the therapy of central nervous system (CNS) disorders in Asian countries [3]. In the Indian system of medicine, the following medicinal plants have shown promising activity in neuropsychopharmacology: *Allium sativum, Bacopa monniera, Centella asiatica, Celastrus paniculatus, Nicotiana tabaccum, Withania somnifera, Ricinus communis, Salvia officinalis, Ginkgo biloba, Huperiza serrata, Angelica sinensis, Uncaria tomentosa, Hypericum perforatum, Physostigma venosum, Acorus calmus, Curcuma longa, Terminalia chebula, Crocus sativus, Enhydra fluctuans, Valeriana wallichii, Glycyrrhiza glabra,* etc. In Chinese medicine, numerous plants have been used to treat stroke, and some of them are *Ledebouriella divaricata, Scutellaria baicalensis, Angelica pubescens, Morus alba, Salvia miltiorrhiza, Uncaria rhynchophylla,* and *Ligusticum chuanxiong* [4].

20.3 SYNERGETIC EFFECTS BY DIETARY APPROACH

Herbal products contain complicated mixtures of organic chemicals, which may include fatty acids, sterols, alkaloids, flavonoids, glycosides, saponins, tannins, terpenes, and so forth. Proponents of herbal medicines describe a plant's therapeutic value as coming from the synergistic effects of the various components of the plants, in contrast to the individual chemicals of conventional medicines isolated by pharmacologists; therefore, it is believed that traditional medicines are effective, with few or no side effects. Synergetic effects of bilobalide and ginkgolides present in *G. biloba* have classified it as nootropic agent [4]. *G. biloba* (Ginkgoaceae) is also known as maiden hair tree, kew tree, ginkyo, yinhsing, and is indigenous to East Asia. Phytoconstituents include terpenoids bilobolide, ginkgolides, flavanoids, steroids (sitosterol and stigmasterol), and organic acids (ascorbic, benzoic shikimic, and vanillic acid). Leaf extract contains 24% of flavonoids and 6% of terpenic lactones, giving this extract its unique polyvalent pharmacological action. The flavonoid fraction is mainly composed of three flavonols, quercetin, keampferol, and isorhamnetin, whereas terpenic derivatives are represented by diterpenic lactones, the ginkgolides, and a sesquiterpenic trilactone, the bilobalide [5]. It has been used in traditional Chinese medicine for the improvement of memory loss associated with abnormalities in blood circulation. The herb shows memory-enhancing action by increasing the supply of oxygen and helps the body to eliminate free radicals thereby improving memory [5]. *B. monniera*, "Brahmi," has been used in the Ayurvedic system of medicine for centuries. Traditionally, it was used as a brain tonic to enhance memory development and learning and to provide relief to patients with anxiety or epileptic disorders [6]. Bacopa include many active constituents including the alkaloids brahmine and herpestine, saponins D-mannitol and hersaponin, and monnierin. Other active constituents have since been identified, including betulic acid, stigmastarol, beta-sitosterol, as well as numerous bacosides and bacopasaponins. The constituents responsible for Bacopa's cognitive effects are bacosides A and B [7]. The bacosides aid in the repair of damaged neurons by enhancing kinase activity, neuronal synthesis, and restoration of synaptic activity and, ultimately, nerve impulse transmission [8]. *C. asiatica* is a psychoactive medicinal plant used for centuries in Ayurvedic system of medicine as a medhya rasayna due to synergetic effect of major bioactive compounds of highly variable triterpenoid saponins, including asiaticoside (AS), oxyasiaticoside, centelloside, brahmoside, brahminoside, thankunoside, isothankunoside, and related sapogenins [9,10]. It also contains triterpenoid acids viz. asiatic acid, madecassic acid, brahmic acid, isobrahmic acid, betulic acid, etc.

CNS cells are able to combat oxidative stress using some limited resources, like vitamins, bioactive molecules, lipoic acid, antioxidant enzymes, and redox sensitive protein transcriptional factors. However, this defense system can be activated/modulated by nutritional antioxidants such as polyphenols, flavonids, terpenoids, fatty acids etc. Plant-derived alternative antioxidants (AOXs) are regarded as effective in controlling the effects of oxidative damage. In nature, AOXs are grouped as endogenous or exogenous. The endogenous group includes enzymes (and trace elements part-of) like superoxidase dismutase (zinc, manganese, and copper), glutathione peroxide (selenium), and catalase and proteins like albumin, transferrin, ceruloplasmin, metallothionein, and haptoglobin. The most important exogenous AOXs are dietary phytochemicals (such as polyphenols, quinones, flavonoids, catechins, coumarins, terpenoids) and the smaller molecules like ascorbic acid (vitamin C), alpha-tocopherol (vitamin E), beta-carotene, vitamin E, and supplements [11,12]. AOXs offer a promising approach in the control of or slowing down the progression of neurodegenerative disorders such as Alzheimer's disease, Parkinson's disease, Huntington's disease, amyotrophic lateral sclerosis, and ischemic and hemorrhagic stroke [12,13].

20.4   RATIONALE FOR FLAVONOID EFFECTS

Epidemiological studies of human populations and experiments in animal models of neurodegenerative disorders have provided evidence that phytochemicals in fruits and vegetables can protect the nervous system against disease [14,15]. Catechins are polyphenols exhibiting neuroprotective activities that are mediated, in part, by activation of protein kinase C and transcription factors that induce the expression of cell-survival genes [16]. Catechins have been suggested to suppress the pathogenesis of Alzheimer's disease and to protect neurons against the processes of Alzheimer's disease. Studies have shown that catechins and their metabolites activate multiple signaling pathways that can exert cell survival and anti-inflammatory actions, including altering the expression of proapoptotic and antiapoptotic proteins and upregulating antioxidant defenses [17]. Flavanones such as hesperetin, naringenin, and their in vivo metabolites, along with some dietary anthocyanins, cyanidin-3-rutinoside, and pelargonidin-3-glucoside, have been shown to traverse the blood–brain barrier (BBB) in relevant in vitro and in situ models [18,19]. Anthocyanins can possibly cross the monolayer in BBB models in vitro. Individual flavonoids such as the citrus flavanone tangeretin have been observed to maintain nigro-striatal integrity and functionality following lesioning with 6-hydroxydopamine, suggesting that it may serve as a potential neuroprotective agent against the underlying pathology associated with Parkinson's disease [18]. Alkaloids may affect the CNS, including nerve cells of the brain and spinal cord, which control many direct body functions and the behavior. Ergot alkaloids have marked effects on blood flow, which was originally thought to be the main mechanism of action. Galantamine is a tertiary alkaloid, belonging to the phenanthrene chemical class, which occurs naturally in the daffodil (*Narcissus tazetta*), snowdrop (*Galanthus nivalis*), and the snowflake (*Leucojum aestivum*). The drug also belongs to a class of drugs called cholinesterase inhibitors and is capable of stimulating nicotinic receptors that further enhance cognition and memory [20]. Tropane alkaloids like atropine, hyoscyamine, and scopolamine from *Datura* affect the spinal cord and CNS. Iso-quinoline alkaloid such as morphine, which is isolated from *Papaver somniferum* (opium poppy), is a highly potent analgesic and a narcotic drug. Indole alkaloids contain the indole carbon–nitrogen ring, which is also found in the fungal alkaloids ergine and psilocybin and the mind-altering drug lysergic acid diethylamide. These alkaloids may interfere or compete with the action of serotonin in the brain [20]. Morphin's mechanism of action is strong binding to the μ-opioid receptor in the CNS, resulting in an increase in gamma aminobutyric acid (GABA) in the synapses of the brain [38]. Vinpocetine is an alkaloid obtained from *Vinca minor* and is a highly potent vasodilator. Clinical studies of vinpocetine report selective enhancement of cerebral blood flow and metabolism, including enhanced glucose uptake, which may protect against the effects of hypoxia and ischemia [21]. Huperzine A, a sesquiterpene alkaloid purified from the Chinese medicinal herb *Huperia serrata*, exhibits a broad range of neuroprotective actions. Huperzine A ameliorated learning and memory impairments and improved spatial working memory.

Dietary monounsaturated fatty acids and polyunsaturated fatty acids (PUFAs) were shown to slow cognitive decline in animals and in humans. n23 and n26 (omega-3 and omega-6) PUFAs found in nuts, such as walnuts, contain the monounsaturated fatty acid oleic acid (8:1) and the n26 and n23 PUFAs linoleic acid and a-linolenic acid [22]. Numerous studies have shown that consuming diets deficient in n23 fatty acids will impair cognitive functioning [23]. The structure of neurons is critical to their function as the cells must maintain appropriate electrical gradients across the membrane, with normal anchor receptors and the ion channels in proper position to communicate with other cells, and be able to release and reabsorb unmetabolized neurotransmitters. These properties depend on the fatty acid composition of the neuronal membrane [24]. The fatty acid composition of neuronal membranes declines during aging, but dietary supplementation with essential fatty acids was shown to improve membrane fluidity and PUFA content. In addition to affecting membrane biophysical properties, PUFAs in the form of phospholipids in neuronal membranes can also directly participate in signaling cascades to promote neuronal function, synaptic plasticity, and neuroprotection [24].

Sulforaphane is an isothiocyanate present in high amounts in broccoli, brussels sprouts, and other cruciferous vegetables. Several studies have reported neuroprotective effects of sulforaphane in animal models of both acute and chronic neurodegenerative conditions. In a rodent model of stroke, sulforaphane administration reduced the amount of brain damage, brain edema, and protected retinal pigment epithethelial [25]. Naphthodianthrones such as hypericin and pseudohypericin are predominant components of *H. perforatum*. There is strong evidence that hypericin and pseudohypericin contribute to the antidepressant action. Inhibition of monoamine oxidase is one mechanism by which some antidepressants operate to increase levels of neurotransmitters such as serotonin, norepinephrine, or dopamine [26]. This chemical appears to block synaptic reuptake of serotonin, dopamine, and norepinephrine [27]. Blocking neurotransmitter reuptake elevates their synaptic concentration. This represents another mechanism by which synthetic antidepressants may operate [27]. Accumulating cell culture and animal model data show that dietary curcumin is a strong candidate for use in the prevention or treatment of major disabling age-related neurodegenerative diseases like Alzheimer's, Parkinson's, and stroke. Curcumin has been shown to reverse chronic stress-induced impairment of hippocampal neurogenesis and increase expression of brain-derived neurotrophic factor (BDNF) in an animal model of depression [28].

Resveratrol, a phytophenol present in high amounts in red grapes, exhibits antioxidant activity. However, more recent findings showed that resveratrol enters the CNS rapidly following peripheral administration and can protect neurons in the brain and spinal cord against ischemic injury. Administration of resveratrol to rats reduced ischemic damage to the brain in a model of stroke and also protected spinal cord neurons against ischemic injury [29]. Resveratrol can protect cultured neurons against nitric oxide-mediated oxidative stress-induced death. Similarly, resveratrol protected dopaminergic neurons in midbrain slice cultures against metabolic and oxidative insults, a model relevant to Parkinson's disease [29]. Resveratrol protected cells against the toxicity of mutant huntingtin in worm and cell culture models [30]. In models relevant to Alzheimer's disease, resveratrol protected neuronal cells from being killed by amyloid β-peptide and promoted the clearance of amyloid β-peptide from cultured cells [31]. Organosulfur compounds, such as allium and allicin, are present in high amounts in garlic and onions and have been shown to be neuroprotective. In addition to their antioxidant activities, allyl-containing sulfides might activate stress-response pathways, resulting in the upregulation of neuroprotective proteins such as mitochondrial uncoupling proteins [32]. Moreover, allicin activates transient receptor potential (TRP) ion channels in the plasma membrane of neurons. Numerous other phytochemicals also activate TRP channels in neurons, including isothiocyanates, garlic alliums, and cannabinoids, resulting in adaptive cellular stress responses [32].

Numerous phytochemicals modify neuronal excitability by activating or inhibiting specific receptors or ion channels. Approximately 50 neurotransmitters belonging to diverse chemical groups have been identified in the brain. Acetylcholine (Ach), the first neurotransmitter to be characterized, has a very significant presence in the brain; recently, it was determined that Ach is essential for learning and memory [33]. Ach has been a special target for investigations for almost two decades because its deficit, among other factors, has been held responsible for senile dementia and other degenerative cognitive disorders, including Alzheimer's disease. A well-characterized example is capsaicin, the phytochemical responsible for the striking noxious physiological effects of hot peppers; capsaicin activates specific Ca2+ channels called vanilloid receptors. There is now an impressive array of natural products that are known to influence the function of ionotropic receptors for GABA, the major inhibitory neurotransmitter in the brain. A wide range of plant-derived flavonoids, terpenes, and related substances modulate the function of ionotropic GABA receptors [34]. Such GABA modulators have been found in fruits (e.g., grapefruit), vegetables (e.g., onions), various beverages (including tea, red wine, and whiskey), and in herbal preparations (such as *G. biloba* and ginseng). Many of the chemicals first used to study ionotropic GABA receptors are of plant origin including the antagonists bicuculline (from *Dicentra cucullaria*) [34] and picrotoxin (from *Anamirta cocculus*) [35] and the agonist muscimol (from *Amanita muscaria*) [36]. These substances are known to cross the BBB and are thus able to influence brain function. Low levels of activation of glutamate receptors can enhance synaptic plasticity and protect neurons against dysfunction and degeneration, in part by inducing the expression of neurotrophic factor. In addition, several phytochemicals are well known for their actions on the nervous system by activating TRP calcium channels in the nerve cell membrane. Capsaicin (a noxious chemical in hot peppers), allicin (a pungent agent in garlic), and tetrahydrocannabinol all activate TRP channels [36].

20.5 THERAPEUTIC APPROACH WITH SYNERGETIC EFFECTS FOR BRAIN CANCER

Noncytotoxic natural products possess pleiotropic properties and represent a possible therapeutic approach for cancers, including brain cancer. To date, most mechanistic studies on dietary chemopreventive agents have utilized single dietary agents at high concentrations, which are unlikely to be achieved by food intake [37]. Bioactive compounds with similar effects will sometimes result in exaggerated or diminished effects when used simultaneously. Synergistic interaction can be achieved if the constituents of compound mixtures affect distinct targets or interact with one another to improve the solubility and, in turn, enhance the bioavailability of one or several substances of the multicompound combination. Hypothetically, a combination of compounds can affect several targets, such as enzymes, substrates, metabolites and proteins, receptors, ion channels, DNA/RNA, monoclonal antibodies, signal cascades, and physicochemical mechanisms [38]. Thus, the use of compounds in combination may target complementary sites of action, resulting in the inhibition of the proliferation of cancer cells. Synergistic interaction of modified plant bioactives, [40]-gingerol, epigallocatechingallate (EGCG), and AS, which are frequently found in a traditional Asian diet, and a vitamin E isomer mixture, tocotrienol-rich fraction (TRF), has been proved effective against glioma cancer cell lines. Each chosen compound showed anticancer activities, with overlapping and different molecular actions and targets. TRF exerts its antitumor effects by enhancing immune response [39], whereas [40]-gingerol induces apoptosis by affecting the mitochondrial signaling pathway and modulating p53 [40]. EGCG exerts epigenetic control by inhibiting DNA methyltransferases and histone acetyltransferase to obstruct tumor cell proliferation [40], whereas AS significantly inhibits azoxymethane-induced tumorigenesis in the intestines of F344 rats and HepG2 human hepatoma cells. Uncariatomentosa (cat's claw) is a rain forest climber that has been traditionally used in Peru to treat tumors. Uncariatomentosa contains many antioxidant compounds, including polyphenols, triterpines, campesterol, stigmasterol, and beta-sitosterol, used to treat brain tumors [41]. Vincristine is usually given in combination with other anticancer agents to treat acute lymphocytic leukaemia, Wilm's tumor, neuroblastoma, and rhabdomyosarcoma [42].

20.6 CONCLUSION

As the demand for phytotherapeutic agents is growing, there is need for their scientific validation before plant-derived extracts gain wider acceptance and use. Natural products may provide a new source of beneficial neuropsychotropic drugs. These studies provide a phytochemical basis for some of the effects that these herbal preparations have on brain function, neuroprotection, and cytostatic or cytotoxic effects on cancer cells. The health-promoting effects of a medicinal herb depict that different phytochemicals produce in vivo additive and/or synergistic effects, thus amplifying (or reducing/inhibiting) their activities. For polyphenols or other phytochemicals, the presence of receptors or transporters in brain tissues remains to be ascertained; compounds with multiple targets appear as a potential and promising class of therapeutics for the treatment of neurodisorders and diseases with a multifactorial etiology.

REFERENCES

1. Pueyo IU, Calvo MI. 2009. Phytochemical study and evaluation of antioxidant, neuroprotective and acetylcholinesterase inhibitor activities of *Galeopsis ladanum* L. extracts. *Pharmacognosy Mag.* (5):287–90.
2. Houghton PJ, Raman A. 1998. *Laboratory handbook for fractionation of natural extracts.* London: Chapman and Hall. p. 199.
3. Kumar V. 2006. Potential medicinal plants for CNS disorders: An overview. *Phytother Res.* (20):1023–35.
4. Gong X, Sucher NJ. 1999. Stroke therapy in traditional Chinese medicine (TCM): Prospects for drug discovery and development. *Trends Pharmacol Sci.* (20):191–6.
5. Chandrasekaran K, Mehrabian Z, Spinnewyan B, Drieu Kand Kiskum G. 2001. Neuroprotective effects of bilobalide, a component of *Ginkgo biloba* extract (EGb 761), in gerbil global brain ischemia. *Brain Res.* (922):282–92.
6. Anand T, Naika M, Swamy MS and Khanum F. 2011. Antioxidant and DNA damage preventive properties of *Bacopa monniera* (L) Wettst. *Free Rad Antioxidants.* (1):89–95.

7. Singh HK, Dhawan BN. 1997. Neuropsychopharmacological effects of the Ayurvedic nootropic *Bacopa monniera* Linn (Brahmi). *Indian J Pharmacol.* (29):359–65.

8. Russo A, Borrelli F. 2005. *Bacopa monniera.* A reputed nootropic plant: An overview. *Phytomedicine.* (12):305–17.

9. Nalini K, Aroor A, Karanth K, Rao A. 1992. *Centella asiatica* fresh leaf aqueous extract on learning and memory and biogenic amine turnover in albino rats. *Fitoterapia.* (63):232–7.

10. Anand T, Naika M, Kumar PG, Khanum F. 2011. Antioxidant and DNA damage: Preventive properties of *Centella asiatica* (L). Urb. *Pharmacognosy Mag.* (2):53–8.

11. Larson RA. 1988. The antioxidants of higher plants. *Phytochemistry.* (27):969–78.

12. Berger MM. 2005. Can oxidative damage be treated nutritionally? *Clin Nutr.* (24):172–83.

13. Floyd RA. 1999. Antioxidants, oxidative stress and degenerative neurological disorders. *Proc Soc Exp Biol Med.* (222):236–45.

14. Liu RH. 2003. Health benefits of fruit and vegetables are from additive and synergistic combinations of phytochemicals. *Am J Clin Nutr.* (78):517S–20S.

15. Joseph JA, Bartus RT, Clody DE. 2005. Reversing the deleterious effects of aging on neuronal communication and behavior: Beneficial properties of fruit polyphenolic compounds. *Am J Clin Nutr.* (81):313S–6S.

16. Mandel SA, Avramovich-Tirosh Y, Reznichenko L, Zheng H, Weinreb O, Amit T. 2005. Multifunctional activities of green tea catechins in neuroprotection. Modulation of cell survival genes, iron-dependent oxidative stress and PKC signaling pathway. *Neurosignals.* (14):46–60.

17. Sutherland BA, Rahman RM, Appleton I. 2005. Mechanisms of action of green tea catechins, with a focus on ischemia-induced neurodegeneration. *J Nutr Biochem.* (17):291–306.

18. Youdim KA, Qaiser MZ, Begley DJ. 2004. Flavonoid permeability across an in situ model of the blood–brain barrier. *Free Radic Biol Med.* (36):592–604.

19. Youdim KA, Spencer JP, Schroeter H, Rice-Evans C. 2002. Dietary flavonoids as potential neuroprotectants. *Biol Chem.* (383):503–19.

20. Pearson VE. 2001. Galantamine. A new Alzheimer drug with a past life. *Ann Pharmacother.* (35):1406–13.

21. Halliwell B. 2007. Oxidative stress and cancer: Have we moved forward? *Biochem J.* (401):1–11.

22. Crews C, Hough P, Godward J. 2005. Study of the main constituents of some authentic walnut oils. *J Agric Food Chem.* (53):4853–60.

23. McCann JC, Ames BN. 2005. Is docosahexaenoic acid, an n23 longchain polyunsaturated fatty acid, required for development of normal brain function? An overview of evidence from cognitive and behavioral tests in humans and animals. *Am J Clin Nutr.* (82):281–95.

24. Yehuda S, Rabinovitz S, Carasso RL. 2002. The role of polyunsaturated fatty acids in restoring the aging neuronal membrane. *Neurobiol Aging.* (23):843–53.

25. Zhao J, Kobori N, Aronowski J, Dash PK. 2006. Sulforaphane reduces infarct volume following focal cerebral ischemia in rodents. *Neurosci Lett.* (393):108–12.

26. Schulz V, Haänsel R, Tyler V. 1998. *Rational phytotherapy: A physician's guide to herbal medicine.* Berlin: Springer-Verlag; pp. 1–155.

27. Chatterjee SS, Bhattacharya SK, Wonneman M, Singer A, Muller WE. 1998. Hyperforin as a possible antidepressant component of Hypericum extracts. *Life Sci.* (63):499–510.

28. Xu Y, Ku B, Cui L, Li X, Barish PA, Foster TC et al. 2007. Curcumin reverses impaired hippocampal neurogenesis and increases serotonin receptor 1A mRNA and brain-derived neurotrophic factor expression in chronically stressed rats. *Brain Res.* (1162):9–18.

29. Huang SS, Tsai MC, Chih CL, Hung LM, Tsai SK. 2001. Resveratrol reduction of infarct size in Long-Evans rats subjected to focal cerebral ischemia. *Life Sci.* (69):1057–65.
30. Parker JA, Arango M, Abderrahmane S et al. 2005. Resveratrol rescues mutant polyglutamine cytotoxicity in nematode and mammalian neurons. *Nat Genet.* (37):349–50.
31. Marambaud P, Zhao H, Davies P. 2005. Resveratrol promotes clearance of Alzheimer's disease amyloid-beta peptides. *J Biol Chem.* (283):7377–82.
32. Oi YM, Imafuku C, Shishido Y, Kominato S, Nishimura K. 1999. Allyl-containing sulfides in garlic increase uncoupling protein content in brown adipose tissue, and noradrenaline and adrenaline secretion in rats. *J Nutr.* (129):336–42.
33. Winkler J, Suhr ST, Gage FH, Thal LJ, Fischer LJ. 1995. Essential role of neocorticat acetylcholine in spatial memory. *Nature.* (375):484–7.
34. Curtis DR, Duggan AW, Felix D, Johnston GA. 1970. Bicuculline and central GABA receptors. *Nature.* (228):676–7.
35. Jarboe CH, Porter LA, Buckler RT. 1968. Structural aspects of picrotoxinin action. *J Med Chem.* (2):729–31.
36. Johnston GA. 2005. GABAA receptor channel pharmacology. *Curr Pharm Des.* (11):1867–85.
37. Straetemans R, O'Brien T, Wouters L, van Dun J, Janicot M, Bijnens L, Burzykowski T, Aerts M. 2005. Design and analysis of drug combination experiments. *Biom J.* (47):299–308.
38. Ulrich-Merzenich G, Panek D, Zeitler H, Vetter H, Wagner H. 2010. Drug development from natural products: Exploiting synergistic effects. *Indian J Exp Biol.* (48):208–219.
39. Nigam N, George J, Srivastava S, Roy P, Bhui K, Singh M, Shukla Y. 2010. Induction of apoptosis by-gingerol associated with the modulation of p53 and involvement of mitochondrial signaling pathway in B[a]P-induced mouse skin tumorigenesis. *Cancer Chemother Pharmacol.* (65):687–696.
40. Meeran SM, Ahmed A, Tollefsbol TO. 2010. Epigenetic targets of bioactive dietary components for cancer prevention and therapy. *Clin Epigenet.* (1):101–16.
41. Vermeil C, Morin O. 1976. Experimental role of the unicellular algae Prototheca and Chlorella (Chlorellaceae) in anti-cancer immunogenesis (murine BP8 sarcoma). *C R Seances Soc Biol Fil.* 170(3):6469.
42. Shibata S, Nishikawa Y, Tanaka M, Fukuoka F, Nakanishi M. 1968. Antitumouractivities of lichen polysaccharides. *J Cancer Res Clin Oncol.* 71(1):102–4.

21 Deconstructing the Mechanisms of Bioactives from Food Plants in the Management of Alzheimer's Disease

Zaina Bibi Ruhomally, Shameem Fawdar,
Vidushi Shradha Neergheen-Bhujun,
and Theeshan Bahorun

CONTENTS

21.1 INTRODUCTION

With over 46 million people living with dementia globally and its anticipated rise to 131.5 million by 2050, this clinical condition represents a major public health issue and a socioeconomic burden. Substantial efforts are directed toward reducing risk and improving treatment, care, and cure to mitigate the deleterious effects of dementia on the individual. Alzheimer's disease (AD) is the most common cause of dementia, accounting for an estimated 60%–80% of cases (World Alzheimer Report, 2015). The initial changes in hippocampal volume and the accumulation of the protein fragment β-amyloid (Aβ) and the protein tau are detected 15 and 5 years, respectively, prior to symptom onset, providing a wide time frame for therapeutic action to target disease progression and limit cognitive impairment primarily due to neuronal damage and death (Alzheimer Disease Facts and Figures, 2015). Given the scale of this clinical condition potentiated by a global ageing population, it is vital to reflect on how natural resources such as bioactive foods, medicinal herbs, and plant-derived biofactors can be used to reduce the risks as well as manage the disease. With the current pharmacological therapy providing only short-term improvement for AD, prophylactic intervention poses as judicious course of action to alleviate AD burden on the population.

AD is the most common neurodegenerative disorder to date, with no known cure or preventive therapy on the horizon. This disease progressively compromises both memory and cognition, culminating in a state of full dependence and dementia (Ansari and Khodagholi, 2013; Casey, 2012). Currently, AD is the main cause of dementia in the elderly and the emergence of new dementia cases each year is 4.9 million (49% of the total) in Asia, 2.5 million (25%) in Europe, 1.7 million (18%) in the Americas, and 0.8 million (8%) in Africa (World Alzheimer Report, 2015). The scenario is exacerbated by the fast ageing population since the incidence of dementia increases exponentially with increasing age, and it is projected that by the year 2050, the world's population aged 60 years and higher is expected to reach a total of 2 billion, 55% more than in 2015 (World Health Organization, 2015). Between 2015 and 2050, the number of elderly people living in high-income countries is forecast to increase by 56%, compared with 138% in upper middle income countries, 185% in lower middle income countries, and 239% in low-income countries, thereby contributing to a significantly high proportion of new cases arising in the Asian and African continents (World Alzheimer Report, 2015). This situation will negatively impact on the economic burden of AD the country's health care system (Kim et al., 2009).

Classic drugs, such as acetylcholinesterase (AChE) inhibitors, fail to slow disease progression and display several side effects that reduce the patient's adherence to pharmacotherapy (Silva et al., 2014). Despite a high investment in research, only five approved drugs are available for the management of AD. Hence, there is a need to consider alternative therapies besides the pharma-directed solutions. Use of natural products (NPs) from food plants and culinary herbs has gained popularity in recent years in neuroprotection. Since the interplay between oxidative stress and inflammation has been well established, compounds with remarkable antioxidant and antineuroinflammatory properties are of interest in addition to their ability to modulate cellular stress response pathways (Kris-Etherton et al., 2002). Likewise, AChE and Aβ cleaving enzyme (BACE-1) inhibitors, as well as modulators of receptor excitotoxicity, are of relevance. Accumulated evidence suggests that naturally occurring phytocompounds, such as polyphenolic antioxidants found in fruits, vegetables, herbs, and nuts, may potentially hinder neurodegeneration and improve memory and cognitive function. Increasing epidemiological studies advocate that a healthy diet and nutrition might be a vital modifiable risk factor for AD (Hu et al., 2013). Moreover, a higher consumption of fruits and vegetables was found to be associated with a declined risk of AD and dementia (Hughes et al., 2010). Fruits and vegetables contain a variety of bioactive compounds with antioxidant and anti-inflammatory properties (carotenoids, vitamin E, vitamin C, and polyphenols), thus aiding in the protective role of these compounds against neurodegeneration (Dai et al., 2006; Morris et al., 2006). Additionally, Barberger-Gateau et al. (2007) investigated the frequency of fruit and vegetable intake in subjects aged 65 years and higher in Bordeaux(France). After 3.6 years of follow-up, it was concluded that the risk of dementia decreased by about 30% with daily consumption of fruits and vegetables compared to their rare consumption. Thus, the mechanism of action of bioactive compounds in parallel with the use of novel pharmaceutical drug design and delivery techniques certainly provide the opportunity to explore these molecules as complementary to conventional medicine (Drever et al., 2009).

This book chapter therefore focuses on the role of phytochemicals from food plants which act on multiple and novel target sites as potential alternatives to preventing and improving AD outcome. The chapter also describes the possible molecular mechanisms by which bioactive compounds from functional foods exert their protective effects with the scope for developing novel therapeutic targets to manage and eventually mitigate neurodegeneration.

21.2 PATHOLOGY OF THE DISEASE

The observed symptoms for AD are related to cognitive decline; memory loss; confusion; problems with reading, writing, and speaking; along with changes in mood and personality. As the disease progresses, AD patients gradually withdraw from work and social activities to total dependence on caregivers (Freeman, 2005). Alois Alzheimer first described the disease in 1906, and to date, its cause remains elusive apart from the fewer than 5% of cases that show genetic predisposition with mutations found in one of three genes: amyloid precursor protein (APP) and presenilin-1 and -2 (Calcul et al., 2012). Alteration of any of these genes can lead to the production of excessive amounts of amyloid beta peptides in the brain forming the amyloid plaques (National Library of Medicine). Extensive studies on the pathology of the disease have made important advancement to unravel the mechanism of disease progression and delineate potential therapeutic targets. Clinically, there are two major neuropathological features for the diagnosis of AD: the extracellular beta-amyloid plaque formation and the intracellular neurofibrillary tangle (NFT) formation (Nelson et al., 2009). The former comprises amyloid-β protein (Aβ) while the latter involves NFTs consisting of paired helical filaments of hyperphosphorylated tau protein. These histopathological lesions are mainly confined to the hippocampus region of the brain and in the cerebral cortex, the two large forebrain domains related to memory and other higher cognitive functions. The characteristic pathology progressively leads to the typical clinical symptoms, for example, memory impairment, general cognitive decline, and personality changes associated with AD. The causes of AD are still rather poorly known, with different observed etiologies such as Aβ overproduction, genetics, Aβ impaired clearance, and NFT formation, leading to senile plaques and extensive neuronal death. However, several studies and evidence point to Aβ as critical in the pathogenesis of AD. According to the amyloid cascade hypothesis, Aβ peptides form aggregates and toxic assemblies, which initiate several processes leading to neuronal dysfunction and ultimately large-scale neuronal cell death (Haass and Selkoe, 2007). The prevalence of AD varies among several different factors, including age, genetics, and comorbidities. There is no definitive cure for AD; however, promising research and development for prophylactic measures and therapeutic interventions are underway.

Excitatory amino acids are the main excitatory neurotransmitters in the cerebral cortex and hippocampus of the brain (Dong et al., 2009). Glutamatergic neurons have been identified to form the main excitatory system in the brain, playing an essential role in regulating neurophysiological functions. Under normal conditions, glutamate is the major excitatory neurotransmitter in the mammalian central nervous system (CNS), producing an excitatory response that is generated following an interaction of glutamate with receptors comprising numerous channels. However, excessive activation of these glutamate receptors can lead to neuronal dysfunction and cell death, a process called excitotoxicity (Sims and Zaidan, 1995).

Glutamate exerts its effect on neurons by interacting with excitatory amino acid receptors on the postsynaptic cell membrane, including N-methyl-D-aspartate (NMDA), α-amino-3-hydroxy-5-methyl-4-isoxazole propionate, and kainate receptors. These ionotropic receptors are ligand-gated ion channels permeable to various cations (Danysz and Parsons, 2003). NMDA receptors are Ca^{2+} favoring glutamate-gated ion channels that are expressed in most central neurons and responsible for neuronal injury in brain areas vulnerable to glutamate toxicity (Rothstein, 1996). Prolonged activation of huge amounts of NMDA receptors (especially the NR1/NR2 B-subtype) leads to a rise in intracellular calcium loads and catabolic enzyme actions, which can activate a cascade of events, ultimately leading to apoptosis or necrosis. These downstream effects include mitochondrial membrane depolarization, caspase activation, production of toxic oxygen, and nitrogen-free radicals and hence cellular toxicity (Dong et al., 2009).

Research has shown that neuronal cell death in AD seems to be dependent mostly on NMDA receptor activation (Greenamyre and Young, 1989). Lately, it was found that NMDA receptor activation can stimulate APP processing which in turn produces amyloid β-peptide (Gordon-Krajcer et al., 2002). Consequently, excess glutamate can lead to overstimulation of the NMDA receptor, leading to excessive amyloid β-peptide production with subsequent oxidative stress-induced neurotoxicity (Butterfield et al., 2001). However, glutamate can be converted to glutamine by glutamine synthetase in the glia. Thus, increasing the glutamine synthetase enzyme action can contribute toward the management of the physiological and behavioral features of AD (Varadarajan et al., 2000). Figure 21.1 highlights the main stages leading to the pathology of AD.

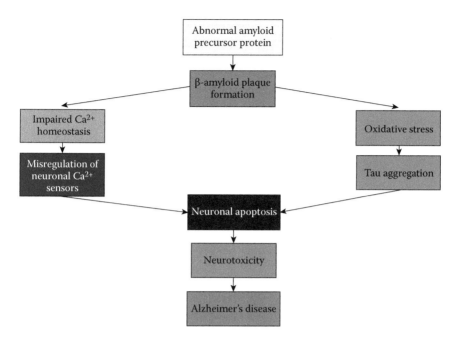

FIGURE 21.1   **(See color insert.)** Pathology of Alzheimer's disease.

21.3 PLANT BIOACTIVE COMPOUNDS AS PROPHYLACTIC AGENTS

Nature harbors a wide variety of flora and fauna, which provides an excellent source of chemically diverse compounds with great therapeutic potential (Bhatnagar and Kim, 2010). NPs are usually referred to as small organic molecules (molecular weight below 3000 Da) or secondary metabolites and are produced by plants, animals, and microorganisms, but their occurrence can be limited to a particular taxonomic family, genus, species, or even organism (Martins et al., 2014). These NPs and their derivatives have historically been used as therapeutic agents (Koehn and Carter, 2005). During the last few years, much research has been carried out to identify novel compounds from the NP pool to be putative candidates for drug discovery. As such, more than 60% of drugs that are on the market have been derived from natural sources (Molinari, 2009; Katiyar et al., 2012). Thus, NPs are still considered as a formidable supply of potential drug leads, with more than 1 million novel chemical entities discovered (Carter, 2011; Dias et al., 2012). Besides, there is a variety of natural sources that are still unexploited for the discovery of novel bioactive compounds.

Reflecting a recent trend of research in NPs, an increasing number of phenolic compounds have been reported with fascinating prophylactic properties. Polyphenols are one of the most common classes of NPs ubiquitously present in our diet and comprise a wide variety of molecules that have several hydroxyl groups on aromatic rings (D'Archivio et al., 2007). This class of plant metabolites contains more than 8000 known compounds, distributed in 10 classes based on their structure, the main groups being flavonoids, phenolic acids, phenolic alcohols, stilbenes, and lignans (D'Archivio et al., 2007; Tsao, 2010). Among them, flavonoids are known to have the most diverse structures, with at least 4000 of them referenced in the literature to date. Depending on the oxidation state of the central pyran ring, flavonoids can themselves be subdivided into many subclasses; flavonols, flavones, flavanones, anthocyanidins, flavanols, and isoflavones, as shown in Figure 21.2 (Bahorun et al., 2003).

Phenolic compounds possess a wide spectrum of biological activities, including antioxidant, antimutagenic, anticarcinogenic, antimicrobial, as well as anti-inflammatory properties (Soobrattee et al., 2005). These phenolics exhibit their antioxidative activity via several mechanisms of action, inter alia, as reducing agents, singlet oxygen quenchers, hydrogen donating antioxidants, free radical scavengers, and metal ion chelators (Rice-Evans et al., 1996; Soobrattee et al., 2005). In addition, in view of their pluripharmacological properties, they can exert modulatory actions in cells by interacting with multiple cellular molecular targets (Williams et al., 2004; Soobrattee et al., 2005).

Traditionally, a variety of plants have been used for the management and treatment of neurodegenerative diseases and cognitive disorders. For example, curcumin, commonly known as diferuloylmethane, derived from turmeric plant (*Curcuma longa*), possesses pluripharmacological properties such as antioxidant, anti-inflammatory, antiproliferative, and neuroprotective effects (Wang et al., 2010). Extensive preclinical studies have revealed that curcumin has therapeutic potential against a wide array of human diseases (Aggarwal and Harikumar, 2009) and affects several pathways involved in AD like oxidative stress, tau phosphorylation, or neuroinflammation (Zhu et al., 2004). Moreover, resveratrol, present in red grape skin, nuts, and red wine, is an eminent polyphenolic flavonoid with powerful antioxidant activity. Besides the anti-inflammatory, antioxidant, and anticancer activities of resveratrol, it

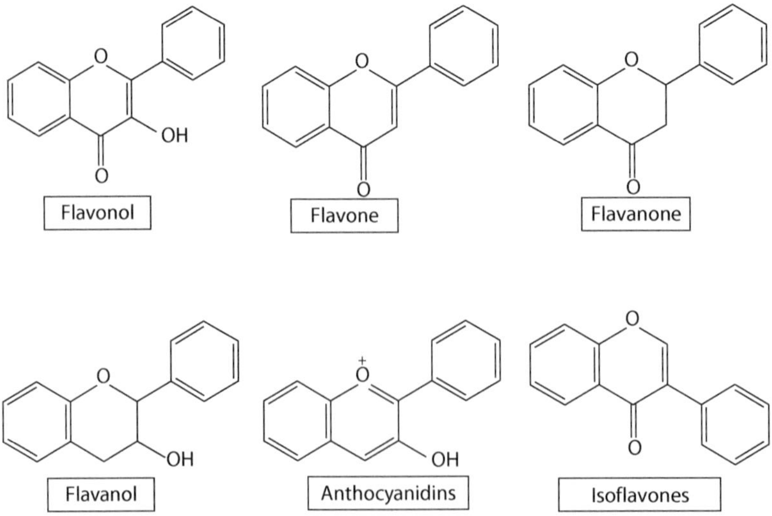

FIGURE 21.2 Flavonoid subclasses.

also acts as a prophylactic agent in the management of cardiovascular disease, type 2 diabetes, and neurologic diseases (Shakibaei et al., 2009).

21.4   BIOACTIVE COMPOUNDS FOR BRAIN HEALTH AND FOR MANAGEMENT OF AD

21.4.1   AChE INHIBITORS

AD is associated with a deficiency of the neurotransmitter acethylcholine in the synaptic cleft of the cerebral cortex (Afshari et al., 2016). AChE, also known as acetyl-hydrolases, catalyzes the hydrolysis of acetylcholine to choline and acetic acid, leading to cognitive impairment (Frank and Gupta, 2005). Inhibition of AChE has been considered as a promising strategy for the treatment of several neurological disorders including AD (Borisovskaya et al., 2014). There are a few synthetic medicines on the market for the treatment of memory loss, namely, Tacrine and Donepezil. Donepezil is a selective AChE inhibitor (AChEi) that binds to the peripheral anionic site and also delays amyloid plaque formation (Arce et al., 2009). However, there is compelling evidence that these compounds have limited effectiveness and exert numerous adverse effects, including gastrointestinal disturbances, in a subgroup of patients. Thus, in this scenario, there is growing interest for search of novel molecules with anti-AChE activity from plant sources (Melzer, 1998; Schulz, 2003).

Moreover, several *in vitro* and *in vivo* studies have revealed that numerous plant species and phytoconstituents can been considered as a potential source of inhibitors with the ability to inhibit AChE, henceforth increasing the level of acetylcholine in the brain and improving neurological disorders (Mukherjee et al., 2007). Contributing to the global search for novel AChEi, marine organisms and fungus are increasingly being investigated. For example, anticholinesterase compounds such as sargaquinoic acid was found in marine seaweed *Sargassum sagamianum* (Choi et al., 2007). Moreover, bioactives like arisugacin, sporothrin, and curvularin, having anticholinesterase activity, have been discovered from fungi (Otoguro et al., 1997; Wen et al., 2009; Kumar et al., 2013). Concomitantly, most of the AChEis belong to the alkaloid groups, such as indole and steroidal alkaloids. Conversely, other nonalkaloidal potent AChEis were discovered in phenolic and terpenoid groups (Mukherjee et al., 2007; Murray et al., 2013).

Galantamine, an alkaloid isolated from the plant *Galanthus woronowii*, has been observed to elicit mild to moderate AChE inhibitory activity by interacting with the anionic sites and also acts as an allosteric ligand. Since it has an allosteric potentiating effect at the cholinergic receptors, it affects other neurotransmitters, including monoamines and glutamate, resulting in an improvement in cognitive dysfunction, depression and other psychiatric disorders (Ago et al., 2011). Likewise, *Ginkgo biloba*, used in traditional Indian medicine for cognition and memory improvement, has been tested for its AChE inhibitory activity. *G. biloba* extracts showed a dose-dependent inhibitory activity on AChE activity (Das et al., 2002). The essential oils and the monoterpenes, namely, geraniol and limonene, of various plants have been reported to inhibit AChE (Mukherjee et al., 2007).

21.4.2 Aβ Cleaving Enzyme (BACE-1)

The human Aβ cleaving enzyme (BACE-1) is considered as a primary inhibitory target for preventing and treating AD. Reverse transcription polymerase chain reaction analysis revealed that the flavonoid acacetin reduced both the human *APP* and *BACE-1* mRNA levels in the transgenic *Drosophila melanogaster* AD models, suggesting an important role in the transcriptional regulation of human *BACE-1* and *APP* (Wang et al., 2015). In addition, a number of naturally occurring secondary metabolites like the alkaloids epiberberi and groenlandicine (Jung et al., 2009), flavonoids (e.g., epigallocatechin gallate, neocorylin) (Jeon et al., 2003; Choi et al., 2008), benzopyranoids (e.g., aloeresin D and C-2′-decoumaroyl-aloeresin G) (Lv et al., 2008), phenylpropanoids (e.g., *p*-coumaric acid) (Youn and Jun, 2012), stilbenoids (e.g., resveratrol) (Choi et al., 2011), and tannins (e.g., geraniin and corilagin) (Youn and Jun, 2013) demonstrated significant inhibition of BACE-1 in a dose-dependent manner.

21.4.3 Curbing Oxidative Stress with Antioxidants

Mitochondrial dysfunction is an important intracellular lesion linked with several diseases, including neurodegenerative disorders. Mitochondrial dysfunction encompasses modifications of mitochondrial respiratory chain enzymes and induces mitochondrial permeability transition pore, formation of reactive oxygen species (ROS), structural abnormalities of mitochondria, oxidative stress, and apoptosis. These aberrations are known to occur early in AD (Kumar and Singh, 2015). In addition, a role for the mitochondrial build-up of proteins such as plasma-membrane-associated APP and cytosolic alpha synuclein in the pathogenesis of mitochondrial dysfunction in AD is known (Devi and Anandatheerthavarada, 2010).

Highly reactive free radicals and oxygen species, including singlet oxygen, hydrogen peroxide, superoxide, nitric oxide (NO) anion radical, and hydroxyl radical, are present in biological systems from a wide variety of reactions. These free radicals are known to oxidize nucleic acids, proteins, lipids, or DNA resulting in oxidative stress, a condition implicated in the pathogenesis of various diseases, such as atherosclerosis, ischemic heart disease, cancer, AD, Parkinson's disease, and other degenerative diseases (Valko et al., 2006; Shahbudin et al., 2011). Thus, the balance between the production and neutralization of ROS and reactive nitrogen species is vital since an overproduction of these reactive species can lead to the deleterious effects of oxidative stress (Wiernsperger, 2003).

The brain has inadequate ability to thwart oxidative stress due to its high lipid content, high aerobic metabolic activity, and low catalase (CAT) activity. Moreover, the high oxygen demand of the brain, about 20% of the oxygen available through respiration, makes it more susceptible to oxidative damage (Weiss and Fintelmann, 2000; Gilgun-Sherki et al., 2001). Consequently, phytoantioxidants have garnered enormous attention in modern medicine as well as in the traditional system due to their beneficial effects (Kumar and Khanum, 2012). Natural extracts rich in antioxidants are generally recognized as safe, and the confirmed therapeutic potentials of these NPs include actions such as protection from cellular and molecular damage by free radicals and reduction of radical-induced tissue injury to finally delay the progress of chronic diseases (Balboa et al., 2013).

Antioxidants play an important role in the protection of the human body against damage by ROS (Govindarajan et al., 2005). Antioxidant system includes enzymatic and nonenzymatic components. Superoxide dismutase, glutathione peroxidase, and CAT are the major antioxidant enzymes, whereas nonenzymatic antioxidants consist of endogenous components (uric acid, reduced glutathione, and albumin) in addition to dietary antioxidants such as carotenoids, flavonoids, ascorbic acid, and α tocopherol (Rietveld and Wiseman, 2003). A panoply of simple and reliable *in vitro* antioxidant assays are generally used for determining the antioxidant activity of various extracts based on mechanisms such as ability to quench free radical by hydrogen donation or electron transfer (Balboa et al., 2013).

Over the last two decades, compelling and multifarious evidences arising from both *in vitro* and *in vivo* studies strongly demonstrate the therapeutic beneficial effects of phytoconstituents in neuroprotection and against neurodegenerative diseases (Davinelli et al., 2012; Kumar and Khanum, 2012). Many of these have yielded numerous lead compounds for the pharmaceutical industry. Concurrently, a number of the phytochemical-based antioxidants have demonstrated neuroprotective actions in both animal and culture models of neurological problems by reducing or reversing cellular damage and by slowing progression of neuronal cell loss (Kumar and Khanum, 2012; Venkatesan et al., 2015). Polyphenolic catechins have been reported to exert a protective role in neurodegeneration. In addition to the known antioxidant activities of catechins, they can also act as iron chelators and radical scavengers, and these neuroprotective effects can be due to the activation of protein kinase C and modulation of transcriptional factors that induce the expression of cell-survival genes (Mandel et al., 2005).

Similarly, a number of natural antioxidants can hinder the oxidation of biomolecules like protein and lipid and thus prevent the formation of ROS, acting as upstream regulator of oxidative stress (Uttara et al., 2009). For example, *G. biloba* (EGb 761), a prominent Chinese medicinal plant, grew in interest owing to its antioxidant properties that curbs Aβ toxicity after plaque formation (Yao et al., 2001). Moreover, 17-estradiol, a chemical moiety that resembles vitamin E in its structure, can inhibit neuronal cell death either by impeding H_2O_2 formation or thwarting Aβ toxicity. The antioxidant effects of these phenolic compounds are due to interaction of their functional groups with the redox metals (Green and Simpkins, 2000). Resveratrol, a phytophenol present in great quantity in red grapes, has demonstrated potent antioxidant activity. Recent findings have shown that in models pertinent in AD, resveratrol can help protect neuronal cell death by amyloid β-peptide and reduce the levels of amyloid β-peptide from cultured cells (Marambaud et al., 2005).

Curcumin at micromolar levels improves mitochondrial membrane potential and cytochrome c release, prevents the activation of caspase-3, and increases the expression of Bcl-2 (Zhu et al., 2004). Omega-3 polyunsaturated fatty acids (ω-3 PUFAs) are groups of essential fatty acids, functioning as energy substrates (Cole et al., 2009). Consequently, ω-3 PUFAs play a major role in the management of neurological disorders, as further evidenced by the study of Eckert et al. (2011). Furthermore, it has been demonstrated that reduced consumption of ω-3 PUFAs increases the risk of AD (Cole et al., 2009).

Increased intracellular calcium levels due to disturbance of homeostatic metal metabolic pathway can cause neuronal cell death owing to dysregulated microtubules association and axonal transport. Several *in vitro* studies have revealed the competencies of natural oxidants such as Taxol in averting neuronal cell death by inhibiting microtubule disaggregation (Burke et al., 1994). Capsaicin, a well-known phytochemical from hot peppers, has been shown to activate specific Ca^{2+} channels called vanilloid receptors. These can influence the function of gamma-aminobutyric acid receptors, the principal inhibitory neurotransmitter in the brain. Additionally, low levels of activation of glutamate receptors can improve synaptic flexibility and safeguard neurons against dysfunction and deterioration, in part by inducing the expression of neurotrophic factor (Kumar and Khanum, 2012). Besides, a wide variety of phytochemicals are also eminent for their beneficial actions on the nervous system owing to their activation of transient receptor potential (TRP) calcium channels in the nerve cell membrane. For instance, capsaicin from hot pepper and allicin from garlic can both activate the TRP channels (Johnston, 2005).

21.4.4 NEUROTROPHINS AND AD

In addition to being important for the correct development and growth of the vertebrate nervous system, neurotrophins also play an essential role in adult neuron survival, maintenance, and regeneration of particular neuronal populations (Allen et al., 2011). Several neurotrophins were identified as neuronal survival-promoting proteins in animal models, including nerve growth factor (NGF), brain-derived neurotrophic factor (BDNF), neurotrophin-3 (NT-3), and NT-4/5 (Venkatesan et al., 2015). These neurotrophins can bind with nanomolar affinity to a pan-neurotrophin receptor p75NTR, and each binds separately, with picomolar affinity, to specific tyrosine kinase receptors (Trks): NGF binds TrkA; BDNF and NT-4 bind TrkB; and NT-3 binds TrkC. A decrease in neurotrophins has been linked to the pathology of a number of neurodegenerative diseases (Cho et al., 2013).

Thus, phytochemicals from natural sources could be valuable candidates as a means of controlling neurotrophin levels and its receptor. The administration of a mixture of olive polyphenols (hydroxytyrosol, phenolic acids, secoiridoid acids, and other polyphenols) may increase the levels of NGF and BDNF and the expression of their receptors, TrkA and TrkB, respectively, in the rodent brain, playing a vital role in learning and memory processes and in the proliferation and migration of endogenous progenitor cells present in the brain (De Nicoló et al., 2013). In addition, flavonoid-rich foods are capable of inducing enhancements in memory and cognition in animals and humans. Pure flavanols and anthocyanins, at levels similar to those in blueberry, were found to induce significant improvements in spatial working memory, and these changes were paralleled by increases in hippocampal BDNF. Conversely, the regional increase of BDNF mRNA expression in the hippocampus seemed to be primarily boosted by anthocyanins (Rendeiro et al., 2013).

Besides the pluripharmacological activities of polyphenols, the latter have been coupled with increased expression of BDNF, assisting in the reversal of neuronal degeneration and behavior deficits (Wollen, 2010). Polyphenols are also shown to exhibit their neuroprotective properties through modulation of particular cellular signaling pathways associated with cognitive processes such as synaptic plasticity—particularly, pathways that signal cAMP-response element-binding protein (CREB), a transcription factor regulating genes coding for BDNF (Gomez-Pinilla and Nguyen, 2012). Likewise, studies on rats have proved that polyphenols can improve cognitive processes by modulation of extracellular signal-regulated kinase/CREB signaling pathways and protection from ROS-specific damage such as lipid peroxidation and neuroinflammation (Spencer, 2009). Additionally, curcumin supplementation was given to rats exposed to a mild fluid striking injury. It was demonstrated that curcumin thwarted cognitive impairment in these animal models by increasing levels of CREB, BDNF, and synapsin I, a protein that modulates synaptic actions. Furthermore, owing to its lipophilicity, curcumin can readily cross the blood–brain barrier to exert neuroprotective effects (Wu et al., 2006).

21.4.5 ANTINEUROINFLAMMATORY ACTIVITY

Promising studies indicate that the interplay between oxidative stress and inflammation is one of the main determinants of brain ageing and cognitive decline. Several histological studies have exposed the presence of activated microglia and reactive astrocytes around Aβ plaques in brains of AD patients. This chronic activation of microglia stimulates production of cytokines and specific reactive substances that worsen Aβ pathology (Mishra and Palanivelu et al., 2008). *In vitro* and *in vivo* work authenticated that several inflammatory alterations, such as microgliosis, astrocytosis, and the existence of proinflammatory substances, accompany the deposition of amyloid-β (Aβ) peptide. Curcumin inhibits Aβ-induced expression of Egr-1 protein and Egr-1 DNA-binding activity in THP-1 monocytic cells. The role of Egr-1 in amyloid peptide-induced cytochemokine gene expression in monocytes has also been studied. Curcumin, by inhibition of Egr-1 DNA-binding activity, has shown to reduce inflammation. Moreover, chemotaxis of monocytes, which can arise in response to chemokines from activated microglia and astrocytes in the brain, can be dwindled by curcumin (Pendurthi and Rao, 2000; Giri et al., 2004).

Concomitantly, several bioactive elements from food plants also exhibit compelling anti-inflammatory activities by inhibiting cyclooxygenase-1 (COX-1) that surrounds amyloid plaque in microglia, which can lead to oxidative stress. *Zingiber officinalis* has been reported to inhibit COX-1 enzyme activity (Ali et al., 2013). Likewise, isothiocyanates, found especially in Brassica vegetables, are active in the brain, through mechanisms that involve the modulation of the inflammatory pathways along with the decline in the activation of cell death by apoptosis. Numerous experimental models of AD disorders showed that isothiocyanates are capable of significantly decreasing nuclear factor (NF)-kB translocation and subsequent proinflammatory cytokines including interleukin (IL)-1β and tumor necrosis factor-alpha (TNF-α) as well as the activation of oxidative species generation (inducible NO synthase and nitrotyrosine) and neuronal apoptotic death pathway (caspase3 and Bax/Bcl-2 unbalance) (Giacoppo et al., 2015) (Figure 21.3).

Furthermore, curcumin can inhibit COX-2, phospholipases, transcription factors, and enzymes involved in metabolizing the membrane phospholipids into

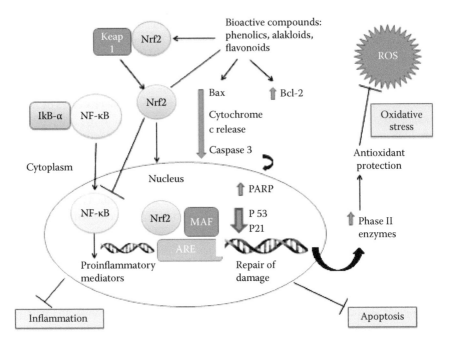

FIGURE 21.3 **(See color insert.)** Summary of the mechanisms underlying neuroprotective effects of bioactive compounds. (Adapted from Giacoppo, S., Galuppo, M., Montaut, S., Iori, R., Rollin, P., Bramanti, P., Mazzon, E., *Fitoterapia*, 106, 12–21, 2015.)

prostaglandins. Inhibition of AP-1 and NF-κB owing to a decline in ROS level has shown to inhibit the transcription of proinflammatory cytokines TNF-α and IL-1β (Park and Kim, 2002; Kim et al., 2005). Triphala, an Indian Ayurvedic herbal preparation containing *Terminalia chebula*, at a dosage of 1 g/kg, demonstrated an anti-inflammatory effect on mice. This study presented that the mechanisms of anti-inflammatory properties of this plant include inhibition of lysosomal enzyme release, a significant decrease in inflammatory mediator TNF-α, and finally, a decrease in beta-glucuronidase and lactate dehydrogenase enzymes release. The presence of phenolic constituents and flavonoids may possibly be responsible for the anti-inflammatory effects of Triphala (Sabina and Rasool, 2008).

Phlorotannins are the group of tannin derivatives, which belong to the polyphenolic substances and consist of polymers of phloroglucinol (1, 3, 5-tryhyroxybenzene) units and are synthesized via the acetate-malonate pathway in marine seaweeds. Among marine algae, brown algae have been identified to accumulate a variety of phloroglucinol-based polyphenol (Wijesekara and Kim, 2010). Phlorotannins, such as dieckol, were capable of significantly diminishing the expression and the release of proinflammatory mediators and cytokines, such as NO, prostaglandin E2 (PGE2), IL-1β, and TNF-α, through down-regulation of NF-κB, p38 kinase activation, and/or inhibition of ROS signal in microglial cells (Jung et al., 2009; Barbosa et al., 2014). Therefore, dieckol can exhibit potential effects in the management of AD. Likewise, hymenialdisine, an alkaloid isolated from marine sponges, such as *Acanthella aurantianca* and *Stylissa massa*, has been demonstrated to inhibit phosphorylation of the protein tau, with promising prospects against human neurodegenerative diseases (Meijer et al., 2000; Tasdemir et al., 2002). Similarly, 11-dehydrosinulariolide, obtained from formosan soft coral, *Sinularia flexibilis*, show neuroprotective properties as a promising candidate for the treatment of Parkinson's disease (Chen et al., 2012).

21.5 CONCLUSION

As indicated by the Dominantly Inherited Alzheimer Network study, the pathological changes associated with AD begin decades before the manifestation of clinical symptoms. Moreover, the pathogenesis of AD is a multifaceted process comprising both genetic and environmental factors; consequently, development of effective disease-modifying drugs is proving to be a difficult task. Dietary antioxidants/herbal extracts/plant biofactors can significantly contribute to modulate complex pathways of degenerative diseases and their progression.

However, effective uses of phytochemicals rely on their ability to cross the blood–brain barrier, in order to reach the target sites (receptors and transporters) of the CNS receptors. These bioactives represent enormous promise as an alternative approach to preventing and ameliorating AD because of their interaction with multiple targets involved in neurodegeneration. Understanding the molecular mechanisms of neuroprotection, oxidative stress, and most importantly, immune function will certainly facilitate definition of prophylactic potentials for use. Further epidemiological and follow-up studies are thus warranted in order to explore their maximum therapeutic potential, for novel and effective utilization as pharmaceuticals and nutraceuticals for the treatment and/or prevention of neurodegenerative diseases.

REFERENCES

Afshari, A.R., Sadeghnia, H.R., Mollazadeh, H. 2016. A review on potential mechanisms of *Terminalia chebula* in Alzheimer's disease. *Adv Pharmacol Sci* Article ID 8964849:1–14.

Aggarwal, B.B., Harikumar, K.B. 2009. Potential therapeutic effects of curcumin, the anti-inflammatory agent, against neurodegenerative, cardiovascular, pulmonary, metabolic, autoimmune and neoplastic diseases. *Int J Biochem Cell Biol* 41:40–59.

Ago, Y., Koda, K., Takuma, K., Matsuda, T. 2011. Pharmacological aspects of the acetylcholinesterase inhibitor galantamine. *J Pharmacol Sci* 11:6–17.

Ali, S.K., Hamed, A.R., Soltan, M.M. et al. 2013. *In-vitro* evaluation of selected Egyptian traditional herbal medicines for treatment of Alzheimer disease. *BMC Complement Altern Med* 13:121, http://www.biomedcentral.com/1472-6882/13/121.

Allen, S.J., Watson, J.J., Dawbarn, D. 2011. The neurotrophins and their role in Alzheimer's disease. *Curr Neuropharmacol* 9:559–573.

Alzheimer disease facts and figures. https://www.alz.org/facts/downloads/facts_figures_2015.pdf (accessed 29 April 2014).

Ansari, N., Khodagholi, F. 2013. Natural products as promising drug candidates for the treatment of Alzheimer's disease: Molecular mechanism aspect. *Curr Neuropharmacol* 11:414–429.

Arce, M.P., Rodriguez-Franco, M.I., Gonzalez-Munoz, G.C., Perez, C., Lopez, B., Villarroya, M., Lopez, M.G., Garcia, A.G., Conde, S. 2009. Neuroprotective and cholinergic properties of multifunctional glutamic acid derivatives for the treatment of Alzheimer's disease. *J Med Chem* 52:7249–7257.

Bahorun, T., Gurib-Fakim, A., Aruoma, O.I. 2003. Plant bioactive components as antioxidant prophylactic agents. In *Molecular and Therapeutic Aspects of Redox Biochemistry*, eds. T. Bahorun and A. Gurib-Fakim, 171–189. OICA International (UK) Limited, London, England.

Balboa, E.M., Conde, E., Moure, A., Falque, E., Dominguez, H. 2013. *In vitro* antioxidant properties of crude extracts and compounds from brown algae. *Food Chem* 138:1764–1785.

Barberger-Gateau, P., Raffaitin, C., Letenneur, L., Berr, C., Tzourio, C., Dartigues, J.F., Alpérovitch, A. 2007. Dietary patterns and risk of dementia: The Three-City cohort study. *Neurology* 69:1921–30.

Barbosa, M., Valentão, P., Andrade, P.B. 2014. Bioactive compounds from macroalgae in the new millennium: Implications for neurodegenerative diseases. *Mar Drugs* 12:4934–4972.

Bhatnagar, I., Kim, S.K., 2010. Marine antitumour Drugs: Status, shortfalls and strategies. *Mar Drugs* 8:2702–2720.

Borisovskaya, A., Pascualy, M., Borson, S. 2014. Cognitive and neuropsychiatric impairments in Alzheimer's disease: Current treatment strategies. *Curr Psychiatry* 16:1–9.

Burke, W.J., Raghu, G., Strong, R. 1994. Taxol protects against calcium-mediated death of differentiated rat pheochromocytoma cells. *Life Sci* 55:313–9.

Butterfield, D.A., Drake, J., Pocernich, C. 2001. Evidence of oxidative damage in Alzheimer's disease brain: Central role for amyloid beta-peptide. *Trends Mol Med* 7:548–554.

Calcul, L., Zhang, B., Jinwal, U.K., Dickey, C.A., Baker, B.J. 2012. Natural products as a rich source of tau-targeting drugs for Alzheimer's disease. *Future Med Chem* 4:1751–1761.

Carter, G.T. 2011. Natural products and pharma: Strategic changes spur new opportunities. *Nat Prod Rep* 28:1783–1789.

Casey, G. 2012. Alzheimer's and other dementias. *Kai Tiaki Nurs N Z* 18:20–24.

Chen, W.F., Chakraborty, C., Sung, C.S., Feng, C.W., Jean, Y.H., Lin, Y.Y., Hung, H.C., Huang, T.Y., Huang, S.Y., Su, T.M., Sung, P.J., Sheu, J.H., Wen, Z.H. 2012. Neuroprotection by marine-derived compound, 11-dehydrosinulariolide, in an *in vitro* Parkinson's model: A promising candidate for the treatment of Parkinson's disease. *Naunyn-Schmiedeberg Arch Pharmacol* 385:265–275.

Cho, T., Ryu, J.K., Taghibiglou, C., Ge Y., Chan, A.W., Liu, L., Lu, J., McLarnon, J.G., Wang, Y.T. 2013. Long-term potentiation promotes proliferation/survival and neuronal differentiation of neural stem/progenitor cells. *PLoS ONE*, 8, Article ID e76860.

Choi, B.W., Ryu, G., Park, S.H., Kim, E.S., Shin, J., Roh, S.S., Shin, H.C., Lee, B.H. 2007. Anticholinesterase activity of plastoquinones from *Sargassum sagamianum*: Lead compounds for Alzheimer's disease therapy. *Phytother Res* 21:423–426.

Choi, C.W., Choi, Y.H., Cha, M.R., Kim, Y.S., Yon, G.H., Hong K.S., Park, W.K., Kim, Y.H., Ryu, S.Y. 2011. *In vitro* BACE1 inhibitory activity of resveratrol oligomers from the seed extract of *Paeonia lactiflora*. *Planta Med* 77:374–376.

Choi, Y.H., Yon, G.H., Hong, K.S., Yoo, D.S., Choi, C.W., Park, W.K., Kong, J.Y., Kim, Y.S., Ryu, S.Y. 2008. *In vitro* BACE-1 inhibitory phenolic components from the seeds of *Psoralea corylifolia*. *Planta Med* 74:1405–1408.

Cole, G.M., Ma, Q.-L., Frautschy, S.A. 2009. Omega-3 fatty acids and dementia. *Prostaglandins Leukot Essent Fatty Acids* 81:213–221.

D'archivio, M., Filesi, C., Benedetto, R.D., Gargiulo, R., Giovannini, C., Masella, R. 2007. Polyphenols, dietary sources and bioavailability. *Ann Ist Super Sanita* 43:348–361.

Dai, Q., Borenstein, A.R., Wu, Y., Jackson, J.C., Larson, E.B. 2006. Fruit and vegetable juices and Alzheimer's disease: The Kame Project. *Am J Med.* 119:751–759.

Danysz, W., Parsons, C.G. 2003. The NMDA receptor antagonist memantine as a symptomatological and neuroprotective treatment for Alzheimer's disease: Preclinical evidence. *Int J Geriatr Psychiatry* 18:23–32.

Das, A., Shanker, G., Nath, C., Pal, R., Singh, S., Singh, H.K. 2002. A comparative study in rodents of standardized extracts of *Bacopa monniera* and *Ginkgo biloba* anticholinesterase and cognitive enhancing activities. *Pharmacol Biochem Behav* 73:893–900.

Davinelli, S., Sapere, N., Zella, D., Bracale, R., Intrieri, M., Scapagnini, G. 2012. Pleiotropic protective effects of phytochemicals in Alzheimer's disease. *Oxid Med Cell Longev.* doi: 10.1155/2012/386527. Epub May 28, 2012.

De Nicoló, S., Tarani, L., Ceccanti, M., Maldini, M., Natella, F., Vania, A., Chaldakov, G.N., Fiore, M. 2013. Effects of olive polyphenols administration on nerve growth factor and brain-derived neurotrophic factor in the mouse brain. *Nutrition* 29:681–687.

Devi, L., Anandatheerthavarada, H.K. 2010. Mitochondrial trafficking of APP and alpha synuclein: Relevance to mitochondrial dysfunction in Alzheimer's and Parkinson's diseases. *Biochim Biophys Acta* 1802:11–19.

Dias, D.A., Urban, S., Roessner, U. 2012. A historical overview of natural products in drug discovery. *Metabolites* 2:303–336.

Dong, X.X., Wang, Y., Qin, Z.H. 2009. Molecular mechanisms of excitotoxicity and their relevance to pathogenesis of neurodegenerative diseases. *Acta Pharmacol Sin* 30:379–387.

Drever, B.D., Anderson, W.G., Riedel, G., Kim, D.H., Ryu, J.H., Choi, D.Y., Platt, B. 2009. The seed extract of *Cassia obtusifolia* offers neuroprotection to mouse hippocampal cultures. *J Pharmacol Sci* 107:380–392.

Eckert, G.P., Chang, S., Eckmann, J. et al. 2011. Liposome-incorporated DHA increases neuronal survival by enhancing non-amyloidogenic APP processing. *Biochim Biophys Acta* 1808:236–243.

Frank, B., Gupta, S. 2005. A review of antioxidants and Alzheimer's disease. *Ann Clin Psychiatry* 17:269–286.

Freeman, W.J. 2005. Alzheimer: The life of a physician and the career of a disease. *J Am Med Assoc* 293:745–746.

Giacoppo, S., Galuppo, M., Montaut, S., Iori, R., Rollin, P., Bramanti, P., Mazzon, E. 2015. An overview on neuroprotective effects of isothiocyanates for the treatment of neurodegenerative diseases. *Fitoterapia* 106:12–21.

Gilgun-Sherki, Y., Melamed, E., Offen, D. 2001. Oxidative stress induced-neurodegenerative diseases: The need for antioxidants that penetrate the blood brain barrier. *Neuropharmacol* 40:959–975.

Giri, R.K., Rajagopal, V., Kalra, V.K. 2004. Curcumin, the active constituent of turmeric, inhibits amyloid peptide-induced cytochemokine gene expression and CCR5-mediated chemotaxis of THP-1 monocytes by modulating early growth response-1 transcription factor. *J Neurochem* 91:1199–210.

Gomez-Pinilla, F., Nguyen, T.T. 2012. Natural mood foods: The actions of polyphenols against psychiatric and cognitive disorders. *Nutr Neurosci* 15:127–133.

Gordon-Krajcer, W., Salińska, E., Lazarewicz, J.W. 2002. *N*-methyl-D-aspartate receptor-mediated processing of beta-amyloid precursor protein in rat hippocampal slices: *In vitro*—Superfusion study. *Folia Neuropathol* 40:13–17.

Govindarajan, R., Vijayakumar, M., Pushpangadan, P. 2005. Antioxidant approach to disease management and the role of Rasayana herbs of Ayurveda. *J Ethnopharmacol* 99:165–178.

Green, P.S., Simpkins, J.W. 2000. Neuroprotective effects of estrogens: Potential mechanisms of action. *Int J Dev Neurosci* 18:347–358.

Greenamyre, J.T., Young, A.B. 1989. Excitatory amino acids and Alzheimer's disease. *Neurobiol Aging* 10:593–602.

Haass, C., Selkoe D.J. 2007. Soluble protein oligomers in neurodegeneration: Lessons from the Alzheimer's amyloid β-peptide. *Nat Rev Mol Cell Biol* 8:101–112.

Hu, N., Yu, J., Tan, L., Wang, Y., Sun, L., Tan, L. 2013. Nutrition and the risk of Alzheimer's disease. *BioMed Res Int* 13: Article ID 524820, 1–12.

Hughes, T.F., Andel, R., Small, B.J., Borenstein, A.R., Mortimer, J.A., Wolk, A., Johansson, B., Fratiglioni, L., Pedersen, N.L., Gatz, M. 2010. Midlife fruit and vegetable consumption and risk of dementia in later life in Swedish twins. *Am J Geriatr Psychiatry* 18:413–420.

Jeon, S.Y., Bae, K., Seong, Y.H., Song, K.S. 2003. Green tea catechins as a BACE1 (β-secretase) inhibitor. *Bioorg Med Chem Lett* 13:3905–3908.

Johnston, G.A. 2005. GABAA receptor channel pharmacology. *Curr Pharm Des* 11:1867–1885.

Jung, H.A., Min, B.S., Yokozawa, T., Lee, J.h., Kim, Y.S., Choi, J.S. 2009. Anti-Alzheimer and antioxidant activities of *Coptidis rhizoma* alkaloids. *Biol Pharm Bull* 32:1433–1438.

Katiyar, C., Gupta, A., Kanjilal, S., Katiyar, S. 2012. Drug discovery from plant sources: An integrated approach. *Ayu* 33:10–19.

Kim, G.Y., Kim, K.H., Lee, S.H., Yoon, M.S., Lee, H.J., Moon, D.O. 2005. Curcumin inhibits immunostimulatory function of dendritic cells: MAPKs and translocation of NF-B as potential targets. *J Immunol* 174:8116–8124.

Kim, J., Basak, J.M., Holtzman, D.M. 2009. The role of apolipoprotein E in Alzheimer's disease. *Neuron* 63:287–303.

Koehn, F.E., Carter, G.T. 2005. The evolving role of natural products in drug discovery. *Nat Rev Drug Discov* 4:206–220.

Kris-Etherton, P.M., Harris, W.S., Appel, L.J. 2002. Fish consumption, fish oil, omega-3 fatty acids, and cardiovascular disease. *Circulation* 106:2747–2757.

Kumar, A., Singh, A. 2015. A review on mitochondrial restorative mechanism of antioxidants in Alzheimer's disease and other neurological conditions. *Front Pharmacol* 6:206.

Kumar, C.G., Mongolla, P., Sujitha, P., Joseph, J., Babu, K.S., Suresh, G., Ramakrishna, K.V., Purushotham, U., Sastry, G.N., Kamal, A. 2013. Metabolite profiling and biological activities of bioactive compounds produced by *Chrysosporium* lobatum strain BK-3 isolated from Kaziranga National Park, Assam India. *Springerplus* 2:122.

Kumar, G.P., Khanum, F. 2012. Neuroprotective potential of phytochemicals. *Pharmacogn Rev* 6:81–90.

Lv, L., Yang, Q.Y., Zhao, Y., Yao, C.S., Sun, Y., Yang, E.J., Song, K.S., Mook-Jung, I., Fang, W.S. 2008. BACE1 (beta-secretase) inhibitory chromone glycosides from *Aloe vera* and *Aloe nobilis*. *Planta Med* 74:540–545.

Mandel, S.A., Avramovich-Tirosh, Y., Reznichenko, L., Zheng, H., Weinreb, O., Amit, T. 2005. Multifunctional activities of green tea catechins in neuroprotection. Modulation of cell survival genes, iron dependent oxidative stress and PKC signaling pathway. *Neurosignals* 14:46–60.

Marambaud, P., Zhao, H., Davies, P. 2005. Resveratrol promotes clearance of Alzheimer's disease amyloid-beta peptides. *J Biol Chem* 280:37377–37382.

Martins, A., Vieira, H., Gaspar, H., Santos, S. 2014. Marketed marine natural products in the pharmaceutical and cosmeceutical industries: Tips for success. *Mar Drugs* 12:1066–1101.

Meijer, L., Thunnissen, A.M., White, A.W., Garnier, M., Nikolic, M., Tsai, L.H., Walter, J., Cleverley, K.E., Salinas, P.C., Wu, Y.Z. 2000. Inhibition of cyclin-dependent kinases, GSK-3[beta] and CK1 by hymenialdisine, a marine sponge constituent. *Chem Biol* 7:51–63.

Melzer, D. 1998. New drug treatment for Alzheimer's diseases: Lessons for healthcare policy *BMJ* 316:762–764.

Mishra, S., Palanivelu, K. 2008. The effect of curcumin (turmeric) on Alzheimer's disease: An overview. *Ann Indian Acad Neurol* 11:13–19.

Molinari, G. 2009. Natural products in drug discovery: Present status and perspectives. *Adv Exp Med Biol* 655:13–27.

Morris, M.C., Evans, D.A., Tangney, C.C., Bienias, J.L., Wilson, R.S. 2006. Associations of vegetable and fruit consumption with age-related cognitive change. *Neurology* 67:1370–1376.

Mukherjee, P.K., Kumar, V., Mal, M., Houghton, P.J. 2007. Acetylcholinesterase inhibitors from plants. *Phytomedicine* 14:289–300.

Murray, A.P., Faraoni, M.B., Castro, M.J., Alza, N.P., Cavallaro, V. 2013. Natural AChE inhibitors from plants and their contribution to Alzheimer's disease therapy. *Curr Neuropharmacol* 11:388–413.

Nelson, P.T., Braak, H., Markesbery, W.R. 2009. Neuropathology and cognitive impairment in Alzheimer disease: A complex but coherent relationship. *J Neuropathol Exp Neurol* 68:1–14.

Otoguro, K., Kuno, F., Ōmura, S. 1997. Arisugacins, selective acetylcholinesterase inhibitors of microbial origin. *Pharmacol Ther* 76:45–54.

Park, S.Y., Kim, D.S. 2002. Discovery of natural products from *Curcuma longa* that protect cells from beta-amyloid insult: A drug discovery effort against Alzheimers disease. *J Nat Prod* 65:1227–1231.

Pendurthi, U.R., Rao, L.V. 2000. Suppression of transcription factor Egr-1 by curcumin. *Thromb Res* 97:179–189.

Rendeiro, C., Vauzour, D., Rattray M., Waffo-Téguo, P., Mérillon, J.M., Butler, L.T., Williams, C.M., Spencer, J.P. 2013. Dietary levels of pure flavonoids improve spatial memory performance and increase hippocampal brain-derived neurotrophic factor. *PLoS One* 8:63535.

Rice-Evans, C.A., Miller, N.J., Paganga, G. 1996. Structure-antioxidant activity relationships of flavonoids and phenolic acids. *Free Radic Biol Med* 20:933–938.

Rietveld, A., Wiseman, S. 2003. Antioxidant effects of tea: Evidence from human clinical trials. *J Nutr* 133:3285S–3292S.

Rothstein, J.D. 1996. Excitatory hypothesis. *Neurology* 47(4 Suppl 2):S19–S25.

Sabina, E.P., Rasool, M. 2008. An *in vivo* and *in vitro* potential of Indian Ayurvedic herbal formulation Triphala on experimental gouty arthritis in mice. *Vasc Pharmacol* 48:14–20.

Schulz, V. 2003. Ginkgo extract or cholinesterase inhibitors in patients with dementia: What clinical trials and guidelines fail to consider. *Phytomedicine* 10:74–79.

Shahbudin, S., Deny, S., Zakirun, A.M.T., Haziyamin, T.A.H., Akbar, J.B., Taher, M. 2011. Antioxidant properties of soft coral *Dendronephthya* sp. *Int J Pharmacol* 7:263–267.

Shakibaei, M., Harikumar, K.B., Aggarwal, B.B. 2009. Resveratrol addiction: To die or not to die. *Mol Nutr Food Res* 53:115–128.

Silva, S.L., Vellas, B., Elemans, S., Luchsinger, J., Kamphuis, P., Yaffe, K., Sijben, J., Groenendijk, M., Stijnen, T. 2014. Plasma nutrient status of patients with Alzheimer's disease: Systematic review and meta-analysis. *Alzheimer Dement* 10:485–502.

Sims, N.R., Zaidan, E. 1995. Biochemical changes associated with selective neuronal death following short-term cerebral ischaemia. *Int J Biochem Cell Biol* 27:531–550.

Soobrattee, M.A., Neergheen, V.S., Luximon-Rama, A., Aruoma, O.I., Bahorun, T. 2005. Phenolics as potential antioxidant therapeutics agents: Mechanism and actions. *Mutat Res* 579:200–213.

Spencer, J.P. 2009. The impact of flavonoids on memory: Physiological and molecular considerations. *Chem Soc Rev* 38:1152–1161.

Tasdemir, D., Mallon, R., Greenstein, M., Feldberg, L., Kim, S., Collins, K., Wojciechowicz, D., Mangalindan, G., Concepcion, G., Harper, M.K., Ireland, C.M. 2002. Aldisine alkaloids from the Philippine sponge *Stylissa massa* are potent Inhibitors of mitogen-activated protein kinase kinase-1 (MEK-1) *J Med Chem* 45:529–532.

Tsao, R. 2010. Chemistry and biochemistry of dietary polyphenols. *Nutrients* 2:1231–1246.

Uttara, B., Singh, A.V., Zamboni, P., Mahajan, R.T. 2009. Oxidative stress and neurodegenerative diseases: A review of upstream and downstream antioxidant therapeutic options. *Curr Neuropharmacol* 7:65–74.

Valko, M., Rhodes, C.J., Moncol, J., Izakovic, M., Mazur, M. 2006. Free radicals, metals and antioxidants in oxidative stress-induced cancer. *Chemico-biol Interact* 160:1–40.

Varadarajan, S., Yatin, S., Aksenova, M., Butterfield, D.A. 2000. Review: Alzheimer's amyloid b-amyloid peptide-associated free radical oxidative stress and neurotoxicity. *J Struct Biol* 130:184–208.

Venkatesan, R., Ji, E., Kim, S.Y. 2015. Phytochemicals that regulate neurodegenerative disease by targeting neurotrophins: A comprehensive review. *BioMed Res Int* 2015, Article ID 814068.

Wang, R., Li, Y., H., Xu, Y., Li, Y.B., Wu, H.L., Guo, H., Zhang, J.Z., Zhang, J.J., Pan, X.Y., Li, Z.J. 2010. Curcumin produces neuroprotective effects via activating brain-derived neurotrophic factor/TrkB-dependent MAPK and PI-3K cascades in rodent cortical neurons. *Prog Neuro-Psychopharmacol Biol Psychiatry* 34:147–153.

Wang, X., Perumalsamy, H., Kwon, H.W., Na, Y.E., Ahn, Y.J. 2015. Effects and possible mechanisms of action of acacetin on the behavior and eye morphology of Drosophila models of Alzheimer's disease. *Sci Rep* 5:16127.

Weiss, R.F., Fintelmann, V. 2000. Herbal medicine. *Stuttgart* 3–20.

Wen, L., Cai, X., Xu, F., She, Z., Chan, W.L., Vrijmoed, L.L., Jones, E.B., Lin, Y. 2009. Three metabolites from the mangrove endophytic fungus *Sporothrix sp.* (#4335) from the South China Sea. *J Org Chem* 74:1093–1098.

Wiernsperger, N.F. 2003. Oxidative stress as a therapeutic target in diabetes: Revisiting the controversy. *Diabetes Metabol* 29:579–585.

Wijesekara, I., Kim, S.K. 2010. Angiotensin-I-converting enzyme (ACE) inhibitors from marine resources: Prospects in the pharmaceutical industry. *Mar Drugs* 8:1080–1093.

Williams, R.J., Spencer, J.P., Rice-Evans, C. 2004. Flavonoids: Antioxidants or signaling molecules? *Free Radic Biol Med* 36:838–849.

Wollen, K.A. 2010. Alzheimer's disease: The pros and cons of pharmaceutical, nutritional, botanical, and stimulatory therapies, with a discussion of treatment strategies from the perspective of patients and practitioners. *Altern Med Rev* 15:223–244.

World Alzheimer Report. 2015. http://www.worldalzreport2015.org/ (accessed 29 April 2014).

World Health Organization. 2015. http://www.who.int/mediacentre/factsheets/fs404/en/ (accessed 7 June 2016).

Wu, A., Ying, Z., Gomez-Pinilla, F. 2006. Dietary curcumin counteracts the outcome of traumatic brain injury on oxidative stress, synaptic plasticity, and cognition. *Exp Neurol* 197:309–317.

Yao, Z., Drieu, K., Papadopoulos, V. 2001. The *Ginkgo biloba* extract EGb 761 rescues the PC12 neuronal cells from beta-amyloid-induced cell death by inhibiting the formation of beta-amyloid-derived diffusible neurotoxic ligands. *Brain Res* 889:181–190.

Youn, K., Jun, M. 2012. Inhibitory effects of key compounds isolated from *Corni fructus* on BACE1 activity. *Phytother Res* 26:1714–1718.

Youn, K., Jun, M. 2013. *In vitro* BACE1 inhibitory activity of geraniin and corilagin from *Geranium thunbergii*. *Planta Med* 79:1038–1042.

Zhu, Y.-G., Chen, X.-C., Chen, Z.-Z. et al. 2004. Curcumin protects mitochondria from oxidative damage and attenuates apoptosis in cortical neurons. *Acta Pharmacol Sin* 25:1606–1612.

22 Amyloid β Immunotherapy

Suhail Rasool and Charles Glabe

CONTENTS

22.1 INTRODUCTION

Alzheimer's disease (AD) is a progressive neurodegenerative disorder of the brain that leads to dementia [1]. According to the amyloid hypothesis, abnormal aggregation of amyloid-β (Aβ) in the brain triggers tau aggregation, microglial activation, synaptic dysfunction, and neuronal loss, ultimately resulting in cognitive decline [2]. Amyloid plaques that consist of Aβ peptides and intracellular neurofibrillary tangles (NFTs) are the two pathological hallmarks of AD. NFTs consist of tau and hyperphosphorylated tau [3]. Proteolytic cleavage of amyloid precursor protein (APP) leads to formation of Aβ, which is composed of 40–42 amino acids and Aβ42 is the major peptide plays an important role in the pathogenesis of AD [4]. There are some other factors associated with development of AD like mutation in APP genes that encodes presenlin 1 and 2 [5]. Different preclinical studies have emerged over the past decade using different transgenic mouse models of AD. Currently, there are more or less 13 different therapeutic modalities in clinical trials of humans under different phases worldwide [6]. In some of these clinical trials, autoimmune complications have been reported, like menigoencephalitis and vasogenic edema [7–9]. Active and passive Aβ immunotherapies are being investigated in great detail during these various clinical trials with an estimated enrollment of more than 10,000 patients. On the basis of results from preclinical and clinical studies, it is believed that active or passive immunization using Aβ or Aβ antibodies is not necessary to produce a protective immune response that specifically targets toxic conformation of Aβ. To facilitate the development of a vaccine that not only recues plaques and NFTs but also will avoid autoimmune inflammatory side effects, the use of different therapeutic modalities like nonhuman random antigens may be considered [10].

22.2   RESEARCH AND CLINICAL STUDIES

Transgenic animals exhibiting well-established AD-like pathology have been variable while testing the effects of Aβ immunotherapy. Vaccination with Aβ1–42 prevented the formation of new plaques and presumably led to a reduction in established plaques in plaque bearing aged PDAPP mice. The first evidence that reported that active Aβ immunotherapy could reduce Aβ pathology *in vivo* was in 1999 by Schenk et al. [11]. With preaggregated, synthetic Aβ1–42, researchers vaccinated PDAPP mice (transgenic animals that exhibit amyloid plaque pathology). A reduction in the number of preestablished plaques in the brains of these animals and prevented plaque depositions were caused by the anti-Aβ antibodies generated following immunization [11,12]. The same vaccine was much less effective in clearing preestablished plaques in aged mice, although in young mice, Aβ1–42 immunization was effective. Active immunization with fibrillar Aβ1–42 did not reduce plaque number or levels of insoluble Aβ or insoluble tau in 3xTg-AD mice [13]. However, there was significant improvement in cognition and reduction in soluble Aβ levels, which suggests that soluble Aβ species particular oligomers might be directly linked to the behavioral impairment observed in the 3xTgAD model [13]. Beneficial effects on synaptic plasticity and neuronal function have been shown through passive Aβ immunization. Synaptic loss and gliosis in Tg2576 transgenic mice are protected by intracerebroventricular infusion of anti-Aβ antibodies [14,15]. An increase in the occurrence of microhemorrhages in the areas of cerebral amyloid angiopathy (CAA) were caused by systemic injections of some anti-AB antibodies, while lowering Aβ-plaque burden [10,11,16,17]. Early lowering of Aβ might have downstream effects on tau pathology because this treatment did not, however, have any effect on more advanced tau pathology, such as NFTs, in older 3xTg-AD mice [18]. When immunization occurs before aggregation of Aβ and tau or before amyloid accumulates in cerebral blood vessels, Aβ immunotherapy's overall efficacy of this treatment might be higher, which has been proven by preclinical studies. The efficacy of several first-generation Aβ immunotherapies has been investigated using transgenic different animals models have already been moved into the later phases of clinical trials [19].

Elan and Wyeth developed an Aβ vaccine for humans in the late 1990s. A synthetic form of the Aβ1–42 peptide and the surface-active saponin adjuvant QS-21 are the two components that made up this vaccine, AN1792. The data from the two AN1792 trials are the only results of major active AB immunization clinical trials that have been made widely available to date. The first AN1792 trial was a Phase I safety study conducted in 80 patients with mild to moderate AD and was initiated in December 1999. The four different treatment groups were AN1792, AN1792 without QS-21, QS-21 only, and placebo, to which individuals were randomly assigned. Patients in each group received four intramuscular vaccinations that were given over a subsequent six-month period. There was no notable adverse events reported in this Phase I trial [20]. A Phase IIa 15-month trial was initiated in October 2001 to evaluate the efficacy of AN1792 plus QS-21 in AD following the results of the Phase I study. The double-bind, placebo-controlled, multicenter study had a total of 372 patients with mild to moderate AD. Out of the 300 patients who received AN1792 (plus QS-21), 223 completed the study. Out of 72 patients who received the placebo, 53 completed the study in the other arm of the trial. This trial was halted as 6% of patients displayed adverse side effects like aseptic menin-goencephalitis and leukoencephalopathy [21,22]. Immunized patients displayed a tendency toward slower cognitive decline than controls in the Phase I study designed to test safety of the AN1792 vaccine. The primary end points included safety, toler-ability, and point efficacy measures—multiple cognitive measures, volumetric MRI, and CSF levels of Aβ1–42, and phosphorylated tau in the Phase IIa trial. Phase IIa trial improvements were reported in some measures of cognitive performance in six patients with high antibody titers in a single-center analysis of a subgroup of 30 patients [23].

In 3 out of the 18 individuals who developed autoinflammatory complication like meningoencephalitis during the AN1792 trial, various neuropathological stud-ies were conducted. With numerous T cells and macrophages infiltrating the white matter and perivascular spaces in these brains, the study reported the presence of an unusual form of meningoencephalitis and leukoencephalopathy [24,25]. While other hallmarks of an AD brain, including CAA and NFTs, were identified in the CNS, amyloid plaques were sparse or absent throughout areas of neocortex in these patients (suggesting a favorable clearance of Aβ). As compared to unimmunized patients with AD, there was marked increase in cerebrovascular Aβ1–42 and Aβ1–40 deposition in the brain of nine AD patients during four months to five years after receiving AN1792 vaccine [26]. Therefore, to limit the Aβ-directed T-cell involve-ment, new approaches have been investigated. Active vaccines that use short Aβ peptide fragments conjugated to larger protein carriers and passive immunization approaches using monoclonal antibodies were two notable new approaches [27]. In Phase III clinical trials for treatment of mild/moderate AD, Bapineuzumab, which is an N-terminal-directed anti-Aβ monoclonal antibody (humanized anti-Aβ antibody 3D6) [28,29], failed to show overall clinical improvement or disease-modifying out-comes [30].

The second generation of Aβ vaccines currently in clinical trials has been designed to avoid stimulating adverse immune responses following the halting of dosing in the Phase II AN1792 trial for safety reasons [15]. Phase II trial of AN7192 was put on hold briefly due to the adverse side effects in skin; however, none of the adverse effects observed in the trial have been observed in the new study [31]. During the past 10 years, new discoveries and tools in imaging as well as in biomarkers have immensely increased the probability of identifying vari-ous risk factors that occurs in Aβ immunotherapy. Preventing downstream effects, such as dysfunction, neuronal damage, and cognitive impairment, can possibly be accomplished by being able to prevent aggregation of neurotoxic forms of Aβ immunization if given early. Improvements in cognition and reductions in plaques in APP/PS1 mice can be accomplished by a vaccination against a ran-dom sequence peptide encoded by read through of a stop codon in the nABri mRNA [25,32]. The ABri random peptide produced antibodies that recognize aggregated Aβ, reduced plaque deposition, and improved cognition, which are similar to the results as reported here, in mice immunized with this peptide. The immune response to this antigen is broader and includes antibodies that react with NFTs and plaques [32].

An outcome that can be achieved through genetic immunization and that has been shown to have a long-term effect on AB clearance is established by producing a safe vaccination against AD, which requires shifting the immune response to a Th2-type response [26,27,33,34]. The T-cell epitope has been mapped within Aβ15–42, while the B-cell epitope is located within the AB1–15 region [35]. The opportunity to use specific Aβ fragments that do not include potentially harmful T-cell epitopes is done through the segregation of T- and B-cell epitopes within the Aβ molecule [36,37]. Robust hormonal immune responses without triggering meningoencephalitis or enhancing CAA when mutated or presented as multiple copies have been reported to be stimulated by a gene vaccine encoding alternative immunogens encompassing the N-terminal epitope of Aβ but lacking the C terminal T-cell reactive sites, such as Aβ1–15, Aβ4–10, and Aβ1–10 [38–40]. Aβ4–10 is the dominant peptide that anti-Aβ42 antibodies specifically recognize with a high affinity, Aβ3–6 (EFRH) affects the solubility and disaggregation of Aβ fibrils, and its affinity for anti-Aβ antibodies is significantly decreased without the third amino acid, as shown by previous studies [41]. Tandem repeats of a small self-peptide are constructed to permit self-tolerance and increase molecular weight as well as reduce degradation of the peptide to overcome the hurdle of the low immunogenicity [42].

The issue of whether the different conformers are differentially associated with pathogenesis is raised by the increasing evidence that Aβ oligomers and fibrils are conformationaly and structurally diverse [43–46]. A single antibody may not be able to target all of the pathologically significant forms of Aβ because monoclonal antibodies recognize these conformers in a mutually exclusive fashion [47,48]. The polyclonal response to active immunization may be more beneficial in its ability to target more different comformers of Aβ than a single monoclonal because it is so broad [48].

22.3 CURRENT TRIALS

After the termination of the AN1792 clinical study, Banineuzumab was the first antibody to be tested in clinical trials [30]. This antibody binds to the N-terminal region of Aβ by a humanized IgGI (mAb) [49]. Mild to moderate AD patients revealed modest improvements related to the stabilization of Aβ burden in the analysis of the Phase II clinical trials. Reversible edema, which is considered an amyloid-related imaging abnormality (ARIA-E), was suffered by some individuals treated with the antibody [50,51]. The finalization of the clinical trials was caused by this event and the lack of clear benefits during Phase III. Based on the adverse side effects, like reduction of Ab in cerebral vessels, which is associated with the increase in permeability of blood–brain barrier and microhemorrhages, Bapineuzumab has been halted [52]. However, AAB-003 is currently in two Phase I trials, which is a humanized version of 3D6 (i.e., Bapineuzumab).

Ponezumab has a much stronger binding to Aβ40 than other monomers, oligomers, or fibrils and is a humanized IgG2a mAb against the C-terminal epitope of Aβ. Induced by the reduction of the peptide in plasma, Ponezumab diminishes the amyloid burden through an outflow of AB from the hippocampus [53]. At the moment, Ponezumab is being tested for the treatment of CAA, and results of the clinical trials evidenced no significant improvement in cognitive impairment of patients with mild to moderate AD [54,55].

The humanized version of the m266 IgG1 mAb that binds the central region of Aβ and has more affinity to monomers than to soluble and toxic species in patients with mild AD is Solanezumab [56]. The results of the Phase III clinical trials did not demonstrate significant improvements in individuals treated with the antibody at first [57]. Less cognitive and functional deterioration in AD patients was revealed in a complementary data analysis [58]. ARIA-E has been observed in 16 individuals enrolled in double-blind trials and their ongoing open-label extension trial, but the antibody has been tolerated well [59]. The magnitude of the benefits is at the same level of the inhibitors of acetylcholinesterase as well. Compared with those who received the placebo, the study showed that solanzumab was not able to slow down cognitive decline in patients with AD. This failure of the antibody could be explained by the possibility that it could be trapped in the blood and does not reach therapeutic concentrations in the brain [60].

The first fully human mAb designed to bind with subnanomolar affinity to a conformational epitope on Aβ fibrils was Gantenerumab [61]. In individuals with prodromal to moderate AD, it encompasses both N-terminal and central amino acids of Aβ binding to monomers, oligomers, and fibrils [62]. The antibody avoids plaque formation by reducing the amyloid load and by activating the microglia [62]. Some patients treated with high dosages developed transient ARIA; however, during Phase I clinical trials, the MRI was safe and well tolerated [63]. The post hoc analysis showed a slight benefit in patients with fast progression, but the Phase II studies indicated no efficacy in the enrolled cohort. A study to evaluate the effect of the antibody on safety, pharmacokinetics, cognition, and functioning in individuals with prodromal AD is included in the Phase III clinical trials in course. A Phase II/III study to determine whether the antibody improves the cognitive outcome of participants with dominantly inherited AD and a trial to test the efficacy and safety of gantenerumab in patients with mild AD are underway.

Created from healthy aged individuals, Aducanumab is a human IgG1 mAb developed from a B-cell library [64]. The antibody interacts with the fibrils and the Aβ N-terminal region binding oligomers of subjects with prodromal to mild AD [64,65]. Improvement of cognitive decline was shown in the Phase Ib clinical trial but caused ARIA in patients with high-dose treatment. Small sample sizes, the use of sequential dose-escalation design, and not being powered by exploratory clinical endpoints were some of the limitations of this study. The positive effects of cognition were less clear, but the trial did prove that the drug was safe and effective in amyloid clearance [64]. To evaluate the efficacy of aducanumab in slowing cognitive and functional impairment in participants based on interim data analysis and the promising results, it was decided to start two Phase III studies. In 150 centers in North America, Europe, Australia, and Asia, the trials will run until 2022. The amyloid hypothesis is out to the test so the expectations for results of these trials are great.

A humanized antibody detected against the mid-region of Aβ that uses an IgG4 isotyoe to reduce risk of microglial overactivation is known as Crenezumab or MABT. Even though it has less affinity for the first, it recognizes Aβ monomers, oligomers, and fibrils [66,67]. The Alzheimer's Prevention Initiative is recruiting

300 Colombian individuals, 200 harboring the E280A PS1 mutation and 100 non carriers currently [61]. The study's purpose is to evaluate the safety and efficacy of the antibody in a preclinical phase of AD.

Directed against APP bearing the E22G mutation in Aβ (Arctic mutation), BAN-2401 is a humanized mAb [68]. The antibody has the ability to recognize a specific conformation in AB protofibrils [68]. No serious adverse events were observed, which approved Phase I clinical trials with that the antibody was safe [68]. The clinical efficacy of BAN-2401 on mild cognitive impairment and mild AD is a current Phase II study enrolling participants.

22.4 CONCLUSION

The cross-reactivity and the inflammatory alterations observed in some patients, along with the efficiency of mAb to cross the blood–brain barrier regarding passive immunization, have to be improved. Immunization for decreasing amyloid pathology and improving cognitive function effectiveness does not depend on the titer. Compared to Aβ fibrils and oligomer antigens, the random peptide oligomer antigens give rise to relatively low antibody titers, but yet are equally effective. Nonhuman amyloid oligomer reduces the neuropathology in Tg2576 transgenic mice, as studied by Rasool et al. in 2012 [10]. Independent of the specific amino acid sequence, it is now concluded that the critical epitope is a pathology-specific confirmation of the peptide backbone. Vaccination against generic amyloid oligomer epitopes is capable of attenuating cognitive impairment and producing protective immune response is demonstrated through these results. A therapeutic strategy for developing an effective vaccine that also circumvents autoimflammatory immune complications is suggested to use a vaccination against a nonhuman amyloid oligomer epitope.

REFERENCES

1. Boche D, Zotova E, Weller RO et al. 2008. Consequence of Abeta immunization on the vasculature of human Alzheimer's disease brain. *Brain* 131:3299–3310.
2. Hardy J, Selkoe DJ. 2002. The amyloid hypothesis of Alzheimer's disease: Progress and problems on the road to therapeutics. *Science* 297:353–356.
3. Terry R, Hansen L, Masliah E. 1994. Structural basis of the cognitive alterations in Alzheimer disease. In *Alzheimer disease* (eds. Terry R, Katzman R), pp. 179–196. Raven, New York.
4. Wolfe MS. 2006. Shutting down Alzheimer's. *Sci. Am.* 294:72–79.
5. De Strooper B, Saftig P, Craessaerts K, Vanderstichele H, Guhde G, Annaert W, Von Figura K, Van Leuven F. 1998. Deficiency of presenilin-1 inhibits the normal cleavage of amyloid precursor protein. *Nature* 391:387–390.
6. ClinicalTrials.gov [online], http://www.clinicaltrials.gov (2009).
7. Pride M, Seubert P, Grundman M, Hagen M, Eldridge J, Black RS. 2008. Progress in the active immunotherapeutic approach to Alzheimer's disease: Clinical investigations into AN1792-associated meningoencephalitis. *Neurodegener. Dis.* 5(3–4):194–196.
8. Wisniewski T. 2005. Practice point commentary on "Clinical effects of Aβ immunization (AN 1792) in patients with AD in an interrupted trial." *Nat. Clin. Pract. Neurol.* 1:84–85.

9. Boche D, Nicoll JA. 2008. The role of the immune system in clearance of Aβ from the brain. *Brain Pathol.* 18(2):267–278.

10. Rasool S, Albay R, Martinez-Coria H et al. 2012. Vaccination with a non-human random sequence amyloid oligomer mimic results in improved cognitive function and reduced plaque deposition and micro hemorrhage in Tg2576 mice. *Mol. Neurodegener.* 7:37.

11. Schenk D, Barbour R, Dunn W et al. 1999. Immunization with amyloid-beta attenuates Alzheimer-disease-like pathology in the PDAPP mouse. *Nature.* 400(6740):173–177.

12. Das P, Murphy M, Younkin L et al. 2001. Reduced effectiveness of Aβ1–42 immunization in APP transgenic mice with significant amyloid deposition. *Neurobiol. Aging* 22:721–727.

13. Oddo S, Vasilevko V, Caccamo A et al. 2006. Reduction of soluble Aβ and tau, but not soluble Aβ alone, ameliorates cognitive decline in transgenic mice with plaques and tangles. *J. Biol. Chem.* 281:39413–39423.

14. Chauhan NB, Siegel GJ. 2002. Reversal of amyloid β toxicity in Alzheimer's disease model Tg2576 by intraventricular antiamyloid β antibody. *J. Neurosci. Res.* 69:10–23.

15. Chauhan NB, Siegel GJ. 2003. Intracerebroventricular passive immunization with anti-Aβ antibody in Tg2576. *J. Neurosci. Res.* 74:142–147.

16. Racke MM, Boone LI, Hepburn DL et al. 2005. Exacerbation of cerebral amyloid angiopathy-associated microhemorrhage in amyloid precursor protein transgenic mice by immunotherapy is dependent on antibody recognition of deposited forms of amyloid β. *J. Neurosci.* 25:629–636.

17. Wilcock DM, Rojiani A, Rosenthal A et al. 2004. Passive immunotherapy against Aβ in aged APP-transgenic mice reverses cognitive deficits and depletes parenchymal amyloid deposits in spite of increased vascular amyloid and microhemorrhage. *J. Neuroinflamm.* 1:24.

18. Oddo S, Billings L, Kesslak JP et al. 2004. Aβ immunotherapy leads to clearance of early, but not late, hyperphosphorylated tau aggregates via the proteasome. *Neuron* 43:321–332.

19. Lemere CA, Masliah E. 2010. Can Alzheimer disease be prevented by amyloid-beta immunotherapy? *Nat. Rev. Neurol.* 6:108–119. doi:10.1038/nrneurol.2009.219.

20. Bayer AJ, Bullock R, Jones RW et al. 2005. Evaluation of the safety and immunogenicity of synthetic Aβ42 (AN1792) in patients with AD. *Neurology* 64:94–101.

21. Orgogozo JM, Gilman S, Dartigues JF et al. 2003. Subacute meningoencephalitis in a subset of patients with AD after Abeta42 immunization. *Neurology* 61(1):46–54.

22. Gilman S, Koller M, Black RS et al. 2005. Clinical effects of Abeta immunization (AN1792) in patients with AD in an interrupted trial. *Neurology* 64(9):1553–1562.

23. Monsonego A, Zota V, Karni A et al. 2003. Increased T cell reactivity to amyloid β protein in older humans and patients with Alzheimer disease. *J. Clin. Invest.* 112:415–422.

24. Ferrer I, Boada Rovira M, Sánchez Guerra ML et al. 2004. Neuropathology and pathogenesis of encephalitis following amyloid-β immunization in Alzheimer's disease. *Brain Pathol.* 14:11–20.

25. Nicoll JA, Barton E, Boche D et al. 2006. Abeta species removal after abeta42 immunization. *J. Neuropathol. Exp. Neurol.* 65(11):1040–1048.

26. Boche D, Zotova E, Weller RO et al. 2008. Consequence of Aβ immunization on the vasculature of human Alzheimer's disease brain. *Brain* 131:3299–3310.

27. Schenk D. 2002. Opinion: Amyloid-beta immunotherapy for Alzheimer's disease: The end of the beginning. *Nat. Rev. Neurosci.* 3(10):824–828. doi:10.1038/nrn938.

28. Black RS, Sperling RA, Safirstein B et al. 2010. A single ascending dose study of bapineuzumab in patients with Alzheimer disease. *Alzheimer Dis. Assoc. Disord.* 24:198–203.

29. Panza F, Frisardi V, Imbimbo BP et al. 2011. Anti-beta-amyloid immunotherapy for Alzheimer's disease: Focus on bapineuzumab. *Curr. Alzheimer Res.* 8:808–817.

30. Farlow MR, Brosch JR. 2013. Immunotherapy for Alzheimer's disease. *Neurol. Clin.* 31(3):869–878.

31. Kounnas MZ, Danks AM, Cheng S et al. 2010. Modulation of gamma-secretase reduces beta-amyloid deposition in a transgenic mouse model of Alzheimer's disease. *Neuron* 67(5):769–780.

32. Goni F, Prelli F, Ji Y et al. 2010. Immunomodulation targeting abnormal protein conformation reduces pathology in a mouse model of Alzheimer's disease. *PLoS One* 5(10):e13391.

33. Qu B, Rosenberg RN, Li L et al. 2004. Gene vaccination to bias the immune response to amyloid-beta peptide as therapy for Alzheimer disease. *Arch. Neurol.* 61:1859–1864.

34. Qu BX, Xiang Q, Li L et al. 2007. Abeta42 gene vaccine prevents Abeta42 deposition in brain of double transgenic mice. *J. Neurol. Sci.* 260:204–213.

35. Agadjanyan MG, Ghochikyan A, Petrushina I et al. 2005. Prototype Alzheimer's disease vaccine using the immunodominant B cell epitope from beta-amyloid and promiscuous T cell epitope pan HLA DR-binding peptide. *J. Immunol.* 174:1580–1586.

36. Ghochikyan A, Mkrtichyan M, Petrushina I et al. 2006. Prototype Alzheimer's disease epitope vaccine induced strong Th2-type anti-Ab antibody response with Alum to Quil A adjuvant switch. *Vaccine* 24:2275–2282.

37. Monsonego A, Maron R, Zota V et al. 2001. Immune hyporesponsiveness to amyloid beta-peptide in amyloid precursor protein transgenic mice: Implications for the pathogenesis and treatment of Alzheimer's disease. *Proc. Natl. Acad. Sci. U S A* 98:10273–10278.

38. Seabrook TJ, Thomas K, Jiang L et al. 2007. Dendrimeric abeta1–15 is an effective immunogen in wildtype and APP-tg mice. *Neurobiol. Aging* 28:813–823.

39. McLaurin J, Cecal R, Kierstead ME et al. 2002. Therapeutically effective antibodies against amyloid-beta peptide target amyloid beta residues 4–10 and inhibit cytotoxicity and fibrillogenesis. *Nat. Med.* 8:1263–1269.

40. Moretto N, Bolchi A, Rivetti C et al. 2007. Conformation sensitive antibodies against Alzheimer amyloid-beta by immunization with a thioredoxin-constrained B-cell epitope peptide. *J. Biol. Chem.* 282:11436–11445.

41. Frenkel D, Kariv N, Solomon B. 2001. Generation of autoantibodies towards Alzheimer's disease vaccination. *Vaccine* 19:2615–2619.

42. Zou J, Yao Z, Zhang G et al. 2008. Vaccination of Alzheimer's model mice with adenovirus vector containing quadrivalent foldable Ab1–15 reduces Ab burden and burden and behavioral impairment without Ab-specific T cell response. *J. Neurol. Sci.* 272:87–98.

43. Kayed R, Head E, Thompson JL et al. 2003. Common structure of soluble amyloid oligomers implies common mechanism of pathogenesis. *Science* 300(5618):486–489. PubMed PMID: 12702875.

44. Petkova AT, Leapman RD, Guo Z et al. 2005. Self-propagating, molecular-level polymorphism in Alzheimer's beta-amyloid fibrils. *Science* 307(5707):262–265. PubMed PMID: 15653506.

45. Kodali R, Williams AD, Chemuru S et al. 2010. Abeta (1–40) forms five distinct amyloid structures whose beta-sheet contents and fibril stabilities are correlated. *J. Mol. Biol.* 401(3):503–517. PubMed PMID: 20600131.

46. Glabe CG. 2008. Structural classification of toxic amyloid oligomers. *J. Biol. Chem.* 283(44):29639–29643. PubMed PMID: 18723507.

47. Necula M, Kayed R, Milton S et al. 2007. Small molecule inhibitors of aggregation indicate that amyloid beta oligomerization and fibrillization pathways are independent and distinct. *J. Biol. Chem.* 282(14):10311–10324.

48. Kayed R, Canto I, Breydo L et al. 2010. Conformation dependent monoclonal antibodies distinguish different replicating strains or conformers of prefibrillar Abeta oligomers. *Mol. Neurodegener.* 5:57. PubMed PMID: 21144050.

49. Johnson-Wood K, Lee M, Motter R et al. 1997. Amyloid precursor protein processing and A beta42 deposition in a transgenic mouse model of Alzheimer disease. *Proc. Natl. Acad. Sci. U S A* 94(4):1550–1555.

50. Salloway S, Sperling R, Gilman S et al. 2009. A phase 2 multiple ascending dose trial of bapineuzumab in mild to moderate Alzheimer disease. *Neurology* 73(24):2061–2070.

51. Rinne JO, Brooks DJ, Rossor MN et al. 2010. 11C-PiB PET assessment of change in fibrillar amyloid beta load in patients with Alzheimer's disease treated with bapineuzumab a phase 2, double-blind, placebo-controlled, ascending-dose study. *Lancet Neurol.* 9(4):363–372.

52. Sperling R, Salloway S, Brooks DJ et al. 2012. Amyloid-related imaging abnormalities in patients with Alzheimer's disease treated with bapineuzumab: A retrospective analysis. *Lancet Neurol.* 11(3):241–249.

53. La Porte SL, Bollini SS, Lanz TA et al. 2012. Structural basis of C-terminal β-amyloid peptide binding by the antibody ponezumab for the treatment of Alzheimer's disease. *J. Mol. Biol.* 421(4–5):525–536.

54. Landen JW, Zhao Q, Cohen S et al. 2013. Safety and pharmacology of a single intravenous dose of ponezumab in subjects with mild-to-moderate Alzheimer disease a phase I, randomized, placebo-controlled, double-blind, dose escalation study. *Clin. Neuropharmacol.* 36(1):14–23.

55. Miyoshi I, Fujimoto Y, Yamada M et al. 2013. Safety and pharmacokinetics of PF-04360365 following a single-dose intravenous infusion in Japanese subjects with mild-to-moderate Alzheimer's disease a multicenter, randomized, double-blind, placebo-controlled, dose-escalation study. *Int. J. Clin. Pharmacol. Ther.* 51(12):911–923.

56. DeMattos RB, Bales KR, Cummins DJ et al. 2001. Peripheral anti-A beta antibody alters CNS and plasma A beta clearance and decreases brain A beta burden in a mouse model of Alzheimer's disease. *Proc. Natl. Acad. Sci. U S A* 98(15):8850–8855.

57. Doody RS, Thomas RG, Farlow M et al. 2014. Phase 3 trials of solanezumab for mild-to-moderate Alzheimer's disease. *N. Engl. J. Med.* 370(4):311–321.

58. Siemers ER, Sundell KL, Carlson C et al. 2016. Phase 3 solanezumab trials secondary outcomes in mild Alzheimer's disease patients. *Alzheimers Dement.* 12(2):110–120.

59. Carlson C, Siemers E, Hake A et al. 2016. Amyloid-related imaging abnormalities from trials of solanezumab for Alzheimer's disease. *Alzheimers Dement. (Amst).* 2:75–85.

60. Bohrmann B, Baumann K, Benz J et al. 2012. Gantenerumab a novel human anti-Aβ antibody demonstrates sustained cerebral amyloid-β binding and elicits cell-mediated removal of human amyloid-β. *J. Alzheimers Dis.* 28(1):49–69.

61. Tucker S, Möller C, Tegerstedt K et al. 2015. The murine version of BAN2401 (mAb158) selectively reduces amyloid-β protofibrils in brain and cerebrospinal fluid of tgArcSwe mice. *J. Alzheimers Dis.* 43(2):575–588.

62. Ostrowitzki S, Deptula D, Thurfjell L et al. 2012. Mechanism of amyloid removal in patients with Alzheimer disease treated with gantenerumab. *Arch. Neurol.* 69(2):198–207.

63. Sevigny J, Chiao P, Bussière T et al. 2016. The antibody aducanumab reduces Aβ plaques in Alzheimer's disease. *Nature* 537(7618):50–56.

64. Kastanenka KV, Bussiere T, Shakerdge N et al. 2016. Immunotherapy with aducanumab restores calcium homeostasis in Tg2576 mice. *J. Neurosci.* 36(50):12549–12558.

65. Muhs A, Hickman DT, Pihlgren M et al. 2007. Liposomal vaccines with conformation specific amyloid peptide antigens define immune response and efficacy in APP transgenic mice. *Proc. Natl. Acad. Sci. U S A* 104(23):9810–9815.

66. Adolfsson O, Pihlgren M, Toni N et al. 2012. An effector-reduced anti-β-amyloid (Aβ) antibody with unique aβ binding properties promotes neuroprotection and glial engulfment of Aβ. *J. Neurosci.* 32(28):9677–9689.

67. Ayutyanont N, Langbaum JB, Hendrix SB et al. 2014. The Alzheimer's prevention initiative composite cognitive test score sample size estimates for the evaluation of preclinical Alzheimer's disease treatments in presenilin 1 E280A mutation carriers. *J. Clin. Psychiatry* 75(6):652–660.

68. Logovinsky V, Satlin A, Lai R et al. 2016. Safety and tolerability of BAN2401—A clinical study in Alzheimer's disease with a protofibril selective Aβ antibody. *Alzheimers Res. Ther.* 8(1):14.

Section IV

Autism

23 Nutrition and Dance Movement Psychotherapy as Positive and Effective Interventions for Autism in Cyprus

Antonios C. Raftis and Silia Rafti

CONTENTS

23.1 A HOLISTIC APPROACH TOWARD AUTISM IN CYPRUS, THEN AND NOW

During the last two decades, autism spectrum disorder (ASD) has increased drastically among the population of Cyprus. This has caused an alarm among healthcare professionals. A lot more emphasis is given now to ASD by more specialists who are trying to deal with ASD with different kinds of treatments. Neuropsychiatrists, physicians, biochemists, psychologists, psychotherapists, and professionals in special education and, very recently, nutritionists are trying to reach different kinds of medical and alternative therapies. The questions to be asked are as follows: Why are there now so many specialists who are trying to cure or improve the health of the child with autism? Why are there now so many home care centers and why is the government giving so much emphasis on this disorder? Why are there now so many nonprofit organizations that are trying voluntarily to help these children?

The answer to all these questions is very simple but also very worrying. It is because in the last two decades, ASD has increased to the point that they are very common and visible in our society. We described a business term to this phenomenon as "supply and demand." In other words, this tremendous increase in ASD has led to an increase in different kinds of medical and alternative treatments. This tremendous increase in ASD has brought the need for a different set of questions to be answered. Why has ASD increased in the last several years so much? What are the causes for this to happen? Why did we not have so many cases of autism before and why do we have so many cases now? Is ASD inherited, genetic, or environmental or is it caused by pollution or food? Think of the lifestyle, nutrition, and living conditions of people two to three decades ago, the natural unprocessed foods consumed without chemicals, the unrefined products, and the unpolluted environment and many others.

How many people were suffering before from food allergies and how many are suffering now? Undoubtedly, food allergies have increased among the population in general and not only in the case of ASD. This is happening because in the modern world, food manufacturers are interfering with natural food and experimenting with chemicals, processes, food preservatives, refined food products, trans-fat, sugar, salt, milk and milk products, lactose intolerance, gluten protein allergy, and others. Another important factor that is responsible for the increasing effect on ASD and for its effect on the increase observed is caesarean deliveries and the relevant vaccinations. According to universal statistics, 83% of cases of children with autism were born through caesarean rather than with normal delivery.

According to the population report, Cyprus has the third largest percentage of caesarean deliveries in the world. Caesarean deliveries were performed very seldom during the 1960–1980s. Unfortunately, nowadays, caesarean deliveries are very popular. Many women choose not to give birth naturally, while in many cases, gynecologists encourage them to have caesarean deliveries! At the same time, ASD has started to spread in this new "modern" Cyprus in very rapidly increasing rates.

It is not yet proven that caesarean deliveries or vaccinations given before have an impact on the increase in ASD cases. The question that arises and needs to be investigated is whether it is really a coincidence that during the 1960–1980s, caesarean deliveries were very rare and were performed overall only for medical emergencies. At the same time, in those days, ASD was also very rare.

As mentioned before, the cases of autism were very rare and not much attention was given by the government or any other organization as there were no special institutions or home care centers available for children with ASD. Children with autism overall were not accepted as normal by parents and the society and were therefore considered a stigma. Children were criticized for their behavior, sometimes with people expressing pity against the child or the parents and other times by discriminating them. Before, as there were no special places to educate the child, the only place where children with autism were able to be educated were the special homes for Down syndrome cases. Unfortunately, most of the doctors, special educators, psychiatrists, psychologists, nutritionists, families of the children, and governmental bodies were considering autism as a genetic disorder with no chance for improvement. The therapists sometimes suggested medical treatment such as tranquilizers in order to calm and suppress hyperactive and deviant behavior of autistic child. This treatment was not for curing or improving the physical/psychological and mental stage of the child with autism.

As mentioned earlier, most specialists, families, and the society, in general, believed in the past that ASD was a rare and incurable disease and not too much attention was given for curing. The only treatment was drugs for anxiety and stress or antidepressants or epilepsy drugs. The surrounding society and the families believed that a mental-brain disorder existed, and many times, they did not have the patience and the support to cope with the hyperactive and imbalanced behavior and then ended up punishing the child. As a result of this aggressive behavior of the parents and the environment, the children were becoming worse and the "disease" became harder to treat. In these cases, it might not be that all cases were truly incurable or difficult to manage but because the approach and behavior of society, parents, and families usually made it impossible to deal with. The question that needs to be answered by the experts, family, and the society of those days is whether ASD was an incurable disorder or whether their ignorance and wrongdoings were creating further problems. Our opinion is that the answer rests on that people were ignorant and misinformed to a great extent. Sometimes, the approach toward a child with autism (similarly in cases such as epilepsy, schizophrenia or manic depression) was that the child had a kind of "curse" and its soul was "taken by the devil!" These parents and families would often take the "mentally ill, possessed" child to the priest for a special prayer in order to get rid of the "devil" and the "curse" from the soul and set the child free.

In addition, during those times, no attention was given to nutritional therapy. There was total ignorance in the areas of nutritional deficiency and nutritional needs. Little or no attention was given on gut problems, the gastrointestinal (GI) digestive track and everything related to nutrition and physical discomfort of the child. There was also not enough emphasis given on food allergy, environmental pollution, hygiene, food deficiency in carbohydrates, proteins, fats, vitamins, and minerals. Supplements were prescribed rarely and not based on actual blood or urine tests. A nonmedical, nonscientific approach was used, with over-the-counter supplements given with a "guessing" approach to prescriptions. All this information gives a brief description of the approach toward ASD in Cyprus around the period 1960–1990.

23.2   ARE MORE CHILDREN BEING DETECTED NOW WITH ASD OR ARE MORE CHILDREN BEING AFFECTED?

Another question to be answered is whether today more children are being affected or whether there are more children being detected with autism. The most possible answer is that both stances are under scientific investigations, but the definite outcome is the obvious increase in autism cases worldwide. All over the world, scientists are looking for the genetic base of autism. Dr. Michael Stone is a very well-known physician in Oregon who has done a very extensive clinical study on autism (Bland, 2014). According to him, "once you know one child with autism you know only one child with autism." What he was trying to prove is that autism appears in so many different ways and with such different severities that there is not one certain gene or one special reason to be the cause of autism. Dr. Michael Stone believes that the origin of the autism disorder usually varies from child to child. He concluded his final results by saying that a combination of genetics integrating with environmental factors increases the possibilities for the cause of autism. He believes that, in some cases, by changing the environment and the diet of the child and also by other alternative therapies, they can alter the genetic expression of autism (Bland, 2014).

Although the etiology of autism continues to be unclear, it has been observed that disorders of the GI and immune system frequently appear in children with autism. The effect of a child's diet and how certain foods and special nutritional plans on ASD promises to improve behavior, helps children to improve communication and relieve GI discomfort that usually accompanies ASD. This has led doctors, nutritionists, and other specialists to focus on nutrition and the role it may have on the alleviation of the symptoms. Healthcare specialists, parents, and the society in general must realize that autism is not a brain issue alone but it is a multiproblem that has to be treated with nutrition and a more holistic approach.

23.3   PHYSICAL AND BEHAVIORAL EFFECTS OF AUTISM

Many children with autism are facing various physical problems such as GI, nutrition, and other problems. The most common GI symptoms include chronic diarrhea, abdominal discomfort and bloating, gastroesophageal reflux diseases, excessive gas, constipation, food regurgitation, food sensitivities, esophagitis, parasite overloads, poor immune function, seizures, enuresis, and a leaky gut syndrome. Children with autism are also at risk for many other nutritional problems such as nutrient deficiencies, food allergies, food intolerance, feeding problems, and poor appetite. They are also more sensitive in their emotions such as anxiety, frustration, anger, and sensitivity in touch and sound (Bland, 2014). Specialists who deal with cases of autism should look first at the physiological conditions that might cause any discomfort or any other symptoms before they look into the psychological and behavioral causes of the problem and the kind of therapy treatment they should give.

23.3.1   NUTRITIONAL INTERVENTIONS AND SUPPLEMENTS

Different studies have been performed throughout the years that look at the many factors, such as medical, nutrition, and behavioral, that affect children with autism and how they contribute to the children's overall status. However, healthcare professionals need to give more emphasis on nutrition and should become more familiar with the evidence that exists on the different nutritional approaches in order to be able to manage the symptoms of autism better. Due to the unknown etiology of autism and the different conditions that exist in each case, there is strong evidence that nutrition may play an important role in managing the symptoms in the cases of children with autism.

Over the years, many investigations and studies have focused on nutritional interventions and on the addition of supplements to the diets of children with autism. These include supplementation with a variety of vitamins and minerals, including vitamin B6, magnesium, vitamin C, vitamin D, vitamin B12, dietary fatty acids (omega-3, cod liver oil), melatonin, folic acid, probiotics, L-carnitine, iron, zinc, and copper. The results on the effectiveness of a number of these supplements are not very clear and remain very limited. Even though the evidence to date is limited on the effectiveness of these supplements, the survey shows that healthcare professionals, in cooperation with the parents, may perceive many of these biomedical interventions to be effective.

23.3.2   IS IT POSSIBLE FOR A WELL-DESIGNED NUTRITIONAL PLAN TO HELP THE CHILD WITH AUTISM?

Even though what we know about autism is not completely clear, new scientific approaches answer this question with a strong "yes." What we know is that in the autism spectrum, many parts of the brain are involved, although it is not quite clear how. Recent studies show strong evidence that autism is not only an 'inner mind' case but may be caused by different biochemical disorders that happen to the body and influence ASD. According to specialists and researchers, a very important factor that can have a positive effect on ASD and balance some of the disorders is nutrition (Walsh, 2014). How can this really happen? In the cases of children with autism, if a healthy and balanced nutritional plan is followed, it can be beneficial. It can relieve the patient from allergy and GI problems and at the same time fulfill the needs and the deficiency of vitamins, minerals, healthy essential fatty acids, amino acids, and other nutritional supplements. In many cases, the deficiency and the imbalance of food nutrients may be the direct or indirect cause of many neurological disorders.

23.3.3 INTOLERANCE AND OTHER FOOD ALTERNATIVES

Another important issue that we must look into is the effect of allergy-producing foods on children. Among the foods for which evidence exists that they might lead to food allergies are gluten-containing grains and cow's milk proteins. In many cases, parents start giving a child these kinds of foods at a very early age, before the age of three. During this stage, the immune system has not completed developed tolerance toward gluten and cow's milk proteins. Of course, this does not mean that all cases of ASD are exposed to this kind of foods. Remember that autism varies from case to case. However, the evidence suggests that in some cases, a child's immune system could change because of the exposure to foods to which they are intolerant, that exposure leads to altered gene expression in the brain, which may cause autism.

In addition, Dr. S. Jill James, a well-known pediatrician and researcher from Arkansas, and her team also tried to analyze the connection between specific genes and environmental factors associated with autism (Bland, 2014). Dr. S. Jill James' team has identified specific gene characteristics that reduce the child's ability to properly metabolize folic acid which is so critical for brain function. They concluded that a child carrying this specific impaired gene has allergies towards cow's milk protein, gluten and also other food substances; this can alter the brain function development. It is very important to remove these kind of foods that cause alteration to the child's nervous system and one should add the foods to the diet that are essential for their development. Through the studies of Dr. S. Jill James' team, they have reached the conclusion that providing supplements like vitamins B12 and B6 and folic acid has a positive effect on autism cases' improvement. This approach is a very different strategy rather than relying on medical treatment in order to manage the symptoms of a child with autism.

23.3.4 NUTRITIONAL MECHANISM

There are some pathogenic organisms in the body that are called bacteria and yeast that can affect the energy levels, the GI and immune system, and the focus of the mind. When there are overdeveloped bacteria and yeast in the body, then its toxicity enters the brain through the blood, and as a result, different kinds of toxins can cause symptoms such as blurred thinking and dizziness in behavior (Delvinioti, 2011). In addition to this, when the biochemistry of methylation does not work effectively, the possibility arises for increased stress behavior, depression, and sleeping disorders (Walsh, 2014).

Methylation is a very important biochemical procedure in the body that adjusts almost all the body functions. If, for example, serotonin, which is produced by the brain, goes through methylation, then there is a great risk to have a deficiency of it, which can lead to depression. In other words, if the process of detoxification of the body does not happen correctly, then toxins accumulate in the brain and react like drugs, causing irritation, aggressiveness, and cell damage. At the same time, when there is poor digestion, nutrients are not absorbed correctly. As a result, this might lead to deficiencies that could lead to poor functioning of the cells, including brain cells. This is why people with autism tend to have more sensitive peptic systems and therefore need to improve their digestive system and to minimize colon inflammation. These are very important steps in order to improve behavior, allergies, GI disorders, and others.

The nutritional components that have a strong, negative impact and play an important role on the behavior of children with autism are for the most part the following four.

1. Gluten (a protein that is present in wheat)
2. Casein (a protein that is present in cow milk and other dairy products)
3. Soy
4. Sugar

On the other side, there are some foods that can cure the colon and help ASD, such as the following:

1. Foods like fish oil and fish that include omega-3 (Ω-3)
2. Foods that have been through fermentation such as yogurt, kefir (preferably from goat milk), or pickles
3. Probiotics: these are foods that favor the development of beneficial bacteria in the colon. Probiotics are found in sauerkraut, microalgae, miso soup, pickles, tempeh, kimchi, kombucha tea, kefir, and yogurt.

These therapies would be beneficial because they could reduce brain inflammation, which in turn could reduce irritability and enhance development of speech, cognition, and socialization. All this would help protein digestion, which could promote development of healthy brain cells (Nierengarten, 2014; Udell, 2016).

Following this thought from earlier, in order to improve digestion in children with autism, there are two main nutrition plans that could be followed. Plan A is a gluten-free, casein-free (GFCF) diet and Plan B is a probiotic diet. Gluten (protein found in wheat) and casein (protein found in dairy products) are two proteins that are known as GI irritants and are both common food allergens. From different studies carried out around the world, it is shown that these two proteins can cause problems for individuals with autism. This occurs by triggering an immune response that has an impact neurologically or by causing GI inflammation and irritation that leads to an increase in ASD symptoms and behavioral problems.

A research study was conducted by the Pennsylvania State University in order to investigate if children who followed a GFCF diet would notice a reduction in ASD symptoms (Vacon, 2014). The research outcomes demonstrated that the children who completely eliminated both gluten and casein from their diets for more than six months had the greatest benefits. They experienced an improvement in ASD symptoms such as greater eye contact, attention span, engagement, and social responsiveness and more independent behaviors. The foods and ingredients that contain gluten should be consumed with caution: wheat, kamut, spelt, barley, rye, and oatmeal (unless it is gluten free). These ingredients can be found in breads, pastas, any packaged foods, or dressings. These ingredients can sometimes be hidden in many other kinds of foods. Foods that are gluten free are brown rice, quinoa, amaranth, millet, beans and lentils, gluten-free bread products, sweet potatoes, starchy vegetables (squash, pumpkin, potato, beets and root vegetables), and gluten-free pasta.

23.3.4.1   Plan A: Go for GFCF

Foods that contain casein should be consumed with caution: yogurt, cheese, milk (especially cow milk), and all dairy products, including dairy ice cream and sour cream. Special care should be given to processed foods with milk or other dairy-based ingredients. Casein-free foods are coconut yogurt, almond, hemp or other nut milks, and coconut ice cream (in moderation due to high sugar content). Small amounts of organic butter are acceptable because of the positive effect on healing the digestive system. Eliminating gluten and casein from the diet does not mean that the person with autism will not enjoy the other food items. On the contrary, with more nutritional knowledge and more food options available nowadays, the person with autism has many options for gluten- and casein-free recipes to eat and enjoy.

23.3.4.2   Plan B: Go for Probiotic Foods

Another approach is a probiotic-rich diet. This would help anyone to improve gut health in general and may have significant benefits for children with autism. A recent study was published in the *Journal of Probiotics and Health* in which 25 children received a six-month supply of a probiotic supplement with 10 billion active cultures. Parents and caregivers were given a survey to complete for 21 days prior to starting the probiotics regarding GI problems as well as a score of ASD symptoms using the Autism Treatment Evaluation Checklist (ATEC). After 21 days of supplementing with the probiotic, 48% of participants saw a reduction in diarrhea severity, 52% saw a reduction in constipation, and 88% saw a reduction in overall ATEC scores (Vacon, 2014).

Considering this study, it seems that the outcomes are promising for all special-ists (nutritionists, physicians, psychologists, chemists, psychiatrists, naturopaths, and other healthcare professionals) who deal with ASD problems. These results encour-age researchers to study further the benefits of probiotics in the diet in order to explore more in-depth the benefits for children with autism. One can start just by simply add-ing probiotic-rich, nondairy fermented foods in the daily menu. Taking supplements with probiotics under the supervision of a healthcare practitioner can result in great improvements. Probiotic foods to be enjoyed are tempeh, miso, coconut kefir, yogurt, kimchi, sauerkraut, kombucha (instead of soda), and anything pickled (but definitely vinegar-free). By adding one or two of these foods in the daily menu, one can gain great benefits from the healing powers of probiotics (Vacon, 2014).

To summarize, it is concluded that the two plans (A and B) are excellent ways to start making improvements to the health of ASD cases in their daily care. However, it is very important to remember that before starting any new health plan, it is important to speak to a physician or health practitioner in order to ensure that the new food plan is safe. It is also important to remember that each individual is naturally and genetically different and will always have specific health and nutritional needs. Other additional nutritional recommendations to avoid are cow milk, sugar, fried and ready-made fast foods, seafood, microwave and fast cooking, food chemicals, food preservatives, artificial ingredients for taste and color, and others. Furthermore, foods such as sweets, cookies, fruit juices, dry fruits, and toxins that come from food preparation material/utensils such as aluminum or plastics for cooking, grapes, prunes, red meat, cheese, bread, pasta, wine vinegar, seed oil, and sunflower oil should also be consumed in moderation. The use of materials/utensils made of aluminum and plastics produce toxins and should therefore be avoided in cooking and food preparation.

Nutrition good for children with autism are goat milk; coconut milk or almond milk as a source of milk; herbs and antioxidants; 1–2 tsp of honey; clear filter water; slow cooking; nonprocessed, more natural, and biological foods that are easier to digest and absorb (contain fewer toxins that need to be eliminated); and olive oil. Diet rich in antioxidants; B-carotene: carrots, apricots, zucchini, cabbage, spinach, broccoli, liver, and egg; vitamin C: marrows, green leafy vegetables, broccoli, and citrus fruits (orange, lemons etc.); vitamin E: olive oil, sunflower seeds, almonds, green leafy vegetables, and hazelnuts; vitamin B6: walnuts, sunflower seeds, lentils, and beans; vitamin B12: liver, eggs, lamb, and beef; folic acid: beans, green leafy vegetables, and banana; and omega-3 fatty acids: fish oils, fatty small fish, and walnut are also good to improve autism symptoms.

23.3.5   MAIN FACTORS THAT CAN BE THE CAUSE OF ASD AND IMPORTANCE OF ANTIOXIDANT THERAPY

Dr. Walsh believes that the three main factors that can be the cause of the development of autism in a child are

1. Undermethylation
2. Oxidative stress
3. Epigenetics

In general, it seems that autism is a gene-programming disorder that develops in undermethylated children who experience environmental interaction that might produce overwhelming oxidative stress. Having all these in mind, Dr. Walsh suggests antioxidant therapies (Walsh, 2014). An antioxidant therapy is very important for an individual with autism in order to reduce cognitive deterioration as the person ages. This can be achieved under routine medical supervision with inexpensive supplementation. Reports and statistics often make reference to the improvement in constipation, diarrhea, self-harming tendencies, and intoxicated feelings just by cutting out food products, as mentioned earlier, that contain soy, gluten, and casein from their daily nutrition program. GFCF diets are more beneficial and therapeutic when a new balanced health nutrition program is tailor-made to the needs of the person with autism.

According to Dr. William J. Walsh, when children with autism undergo urine and blood tests, they often exhibit the following results:

- Zinc deficiency
- Copper overload
- B6 deficiency
- Elevated toxic metals

23.3.6 FINDINGS FROM RESEARCH CARRIED OUT ON MOST POPULAR DIETS FOR AUTISM IN CRETE

1. GFCF diet: It improves conditions such as diarrhea, constipation, and self-injuries
2. Diet Feingold: Avoids preservatives and adding food chemicals to foods: improves the general behavior, attention span, and GI disorders
3. Ketogenic diet: Better known as triglycerides diet with more fat, low in protein, and restriction in carbohydrates. The results of this diet show improvement in social communication, improvement in speech, more cooperation, more willingness to learn, and less hyperactivity. The effects of this diet were the results of a study that took place in Crete from the pediatric clinic of the University of Crete and at a hospital in Thessaloniki.
4. Special carbohydrate diet: This diet allows consumption of meat, fish, eggs, vegetables, and olive oil, while avoiding starch, legumes, wheat, pasta, and bread. The results of these food changes have positive effects on the improvement of the damages of the colon internal walls.

According to research, after using the previous food interactions, 60%–65% of individuals with autism were witnessed to show improvement; 20% had not shown any improvement but their condition was stable. There were no cases reported in which any of the autism incidents worsened with these special diets (Mbalasides, 2015).

23.4 VACCINATION: TRUE OR MYTH

Some reports made reference to autism that occurred after a vaccination. The majority of medical doctors do not believe this theory, but a genetic or acquired hypersensitivity to certain vaccines is something that needs more investigation. Dr. Andrew Wakefield, an academic pediatric gastroenterologist, and his associates published a paper in 1998 that identified a problem with the digestive system of children as strongly associated with autism. In this paper, it presumes that the alteration in the immune system of the children with autism was associated with vaccinations for mumps, measles, and rubella, the MMR vaccine (Bland, 2014). The inquiry panel that investigated Dr. Wakefield's results concluded that his data supporting this association had been falsified. It seems that this debate on the effects of vaccinations and the association to autism needs further research. Dr. Wakefield's results should be given more emphasis, and the possible connection that exists between vaccination, intestinal immune activation, and autism should be further investigated (Bland, 2014).

23.5 CONCLUDING REMARKS

To summarize in a few words the role of nutrition in ASD, it is very important to say that improvement of digestion and the GI system has a positive outcome on the improvement of the total picture of the person with autism. Clinical nutritionists and psychologists should always encourage the patient with autism to improve his/her health with nutrition and wellness strategies. One of the most fundamental strategies that are true for everybody is to improve one's digestion. Good digestion is very important to the health of an individual. Digestion is simply the process of eating food, breaking it down into base nutrients, and assimilating the nutrients into the body. Different researchers have discovered that there is a link between compromised digestion and ASD. Children with autism have higher rates of GI disturbances as well as symptoms of food allergy. It is well known that there is a major link between cognitive function and digestive health, making sense that poor digestion can worsen ASD symptoms and behavioral problems.

Although the etiology of autism remains unknown and many factors are involved in this complication, scientists must continue to perform more studies and more research for this disorder. In recent years, ASD has increased drastically in the modern world. Eating habits and food taste have changed overall throughout the years. The modern industrialized world has made the environment unhealthier and more polluted. Families are living every day in a stressful and sedentary lifestyle that has an impact on children's behavior and food habits as well. The psychological stage of each individual and the family in general is another aspect that has to be considered.

Since the last decade, autism among children has shown an enormous increase worldwide. It seems that it is the fastest growing disorder in children. According to research reports, more children will be diagnosed with autism in the incoming years compared to cancer, diabetes, Down syndrome, and AIDS combined. As of today, there is no known definite cause for autism and not one absolute course of action in the management and treatment of ASD. Each case is different and should be treated accordingly with caution.

Considering all factors analyzed, it is recommended that the complexity of the symptoms of the autism spectrum that include both physiological difficulties and environmental factors should be approached by a multidisciplinary team. This team should include physicians, pediatricians, biochemists, neuropsychiatrists, nutritionists, psychologists, movement psychotherapists, behavioral/developmental specialists, as well as other healthcare professionals who can help with particular challenges, such as speech therapist, physiotherapists, and other alternative therapy specialists.

23.6 WORKING THERAPEUTICALLY THROUGH DANCE MOVEMENT PSYCHOTHERAPY FOR AUTISM

Based on research, autism is a lifelong disability of normal development. This condition causes the isolation of the child with autism from the world around him/her. The cause of this isolation is due to the various communication problems faced by the child with autism, which also affect his relation to other people. "Living in their own world" is the phrase that characterizes people with autism. The term "autism" was given to this category of people to describe how they withdraw from everyday social life and go into their own self. According to Hardman et al. (1984), the word "autism" was taken from the Greek *autos*, meaning "self," to indicate the extreme sense of isolation and detachment from the world around them that characterizes these individuals. Levy (1988) described children with autism to be living in their own world, in which they shun human contact, are unable to relate meaningfully to others or to the environment, and frequently engage in idiosyncratic movement patterns. "Autism is a developmental disability that affects children and adults in a variety of ways and in varying degrees" (Wall, 2004).

There are various characteristics of each child with autism that have to be taken into consideration. Researchers and authors use the term "autism spectrum disorder" to refer to autism. The word "spectrum" is used to show that, while all people with autism share certain difficulties, the individual condition of each one will affect him or her in different ways. Not everyone with autism (ASD) has all the characteristics of autism. Individuals vary in the particular issues that they are struggling with, and none of the characteristics of autism are unique, nor is the level of the seriousness of each characteristic which may change over time. Although there are signs of autism in babies, a diagnosis of an ASD in the first years of life is considered inappropriate.

23.7   WING'S TRIAD OF IMPAIRMENTS IN AUTISM AND HOW IT IS USED IN DANCE MOVEMENT PSYCHOTHERAPY

Although autism is a spectrum condition and there are differences between individuals who suffer from autism, there are three main categories that are the common impairments of children with autism (Jordan and Powell, 1995). The three features provide the diagnostic criteria for autism and they are called Wing's Triad of Impairments in Autism.

The first category of common impairments is the social impairment in children's development. People are born in this world with a nervous system, which has definite features that enables people to learn and read socially significant information from very limited input. For children with autism, this system may not have developed so well or may not have been set up properly, which causes them to have many difficulties in their social behavior. With their difficulty to understand social information, children with autism might show inappropriate behavior toward other people without being aware of doing so. They might also misinterpret reactions of people toward them. For example, pleasure to them might be a scary feeling. A child with autism might choose not to communicate with someone when feeling uncomfortable because he or she does not know how to ask for something. Children with autism seek attention but do not know how to deal with it (Jordan and Powell, 1995). "Communication through movement helps a child to be more aware of him or herself and more able to interact with others" (Levy, 1995).

The second category of common impairments is that of language and communication. According to Hardman et al. (1984), children with autism exhibit an impairment of delayed function of language. "Approximately half of the children with autism do not develop speech—and those who do—often engage in strange language and speaking behavior such as echolalia." When too much speech is directed to them, they might start to repeat back things that are being told in order to hold words in their memory while they try to process them for meaning (Clements and Zarkowsa, 2000). They have difficulty in understanding anything that is said to them. Some of the children with autism face hearing difficulties, which interfere with the ability to understand and get the meaning of language. According to Jordan and Powell (1995), some of the children with autism have difficulties in understanding and using facial expressions, expressive gestures, and body postures. Levy (1995) elaborated on the importance of nonverbal communication, because the child with autism usually has not developed communicative speech but has a unique movement "language." Therefore, nonverbal communication is an effective means of contact.

The third category in the Triad of Impairments of autism is the impairments of the development of behavior and imagination of the child. There is lack of imaginative play. According to Wall (2004), autistic children tend to play in a rigid and repetitive way. They have difficulties and anxieties in coping with changes of their routines. Children with autism tend to play alone or watch others play (Phillips and Beavan, 2007). Body image is one of the basic concepts in human growth and development and one that seems to be lacking in children with autism. The more defined one's body image, the better one is able to differentiate oneself from the environment and from others. This differentiation is necessary for the formation of relationships. Therefore, movement and the body image are two of the Dance Movement Psychotherapist's major concerns when addressing the needs of children with autism.

23.8 AIMS AND TECHNIQUES USED IN DANCE MOVEMENT PSYCHOTHERAPY

As shown earlier in Wing's Triad of Impairments, children with autism clearly have problems in their development in the areas of social skills, language and communication, and behavior and imagination. The following paragraphs will demonstrate how movement can be used as a healing intervention and how children with autism could be aided to live a better life. According to Rogers (2000), movement is a human need. Even breathing is a kind of movement. There is a big connection between movement and emotions. By observing a person's body language and movement, one can identify how this person feels. Movement can affect how we feel and how it can affect our way of moving. A reciprocal relationship exists between movement and emotions. Levy (1995), also states that dance movement psychotherapy is an ideal intervention for working with autistic children being a universal means of communication through any kind of movement.

23.8.1 AIMS OF THE MOVEMENT PSYCHOTHERAPIST WHEN DEALING WITH ASD

23.8.1.1 Development of Body Image

Levy (1995) states that one of the most fundamental concepts in human growth and development is body image, which appears to be lacking in children with autism.

Without a body image, a symbol of one's own body, the psychic structures necessary for symbolic representation of other things, cannot be formed, since they depend on previous symbolization. Great emphasis is given to the relationship between movement and body image. Movement can help people to know their body better. We learn to know our body when we move it. Levy (1995) states that "movement experiences can lead from a change in body image to a change in the psyche." When you define your body image, it is better to differentiate yourself from the environment and from others. This differentiation is very important for the formation of relationships. Levy (1995) concludes that movement and the body image are two of the dance/movement therapist's major concerns when addressing the needs of children with autism.

23.8.1.2   Development of Social Skills

One of the most important guiding principles in a dance therapy session with children with autism is the combination of both physical and relational aspects of work (Levy, 1995). The primary emphasis in dance therapy whatever physical skills a child with autism has is on developing better relationships with human beings. Children with ASD's ability to develop friendships is generally limited, as is their capacity to understand other people's emotional expression (National Autistic Society, 2016). According to Levy (1995), by broadening or expanding a child's movement repertoire, we provide him/her with a wider range of skills to use in understanding and coping with the environment. He also claims that it is very important for a child to understand his/her own body and the skills that his/her body has in movement in order to cope with the external demands of the environment. Unless there is a sense of oneself as a separate entity, differentiated from others, one cannot effectively, or affectively, relate beyond oneself. The primary goals of dance/movement therapy are to assist the child to be functioning at the sensorimotor level, to build a relationship, and to work on formatting a normal body image.

23.8.1.3   Development of Communication Skills

According to Jordan and Powell (1995), one of the defining features of autism is the difficulty in communication. Through various techniques, the dance movement psychotherapist aims to create some dance patterns that provide support and acceptance to the child. This leads to a positive outcome by building trust and a relationship between the therapist and the child (Levy, 1988). Often, in children with autism, there is a lack of desire to establish relationships and friendships (Wall, 2004). Having this in mind, building a relationship between the movement therapist and the child is a very positive achievement for the therapist.

23.8.1.4   Developing Imagination and Thinking Behavior

According to Jordan and Powell (1995), in children with autism, there is a rigidity of thought and behavior and impoverished imagination. Play intervention activities can help those children in areas such as rapport and relationships, imitation, gaining attention, turn-taking, enjoyment, and structure impact on children's behavior, with a reduction in challenging behavior.

23.8.2 Techniques, Approaches, and Methods Used in Dance Movement Psychotherapy

As mentioned previously, autism is a spectrum condition and there are differences between individuals who suffer from autism. Therefore, according to the specific condition and needs of the individual child with autism, dance/movement has a variety of approaches, techniques, and methods to choose from, some of which are discussed here.

23.8.2.1 Mirroring

This is a technique through which the therapist reflects back on the movements of the child with autism without copying. This can help the therapist to understand a child's experience on a body level. The benefit is not only for the therapist to learn more about the child but also for the child to feel accepted as she/he is (Levy, 1995). Such acceptance often causes a child to shift his/her focus from inner stimuli to stimuli in the environment, which then leads to increased connectivity and paves the way for reciprocal interactions. Mirroring is very effective, but it has to take place only when the child is moving with the sense of boundaries and safety and when his/her movement offers the possibility for a positive change (Levy, 1995). Through techniques such as reflecting, sharing, and mirroring the child's movements, the movement therapy creates a dance that is reassuring in its familiarity and implicit acceptance of the child (Levy, 1988).

23.8.2.2 Group Therapy

When working in a group, the therapist observes the mood, tone, and energy level of the group in order to change activities or movement intervention (Levy, 1995). The dance/movement therapist aims to provide an accepting atmosphere while working on building a relationship with each child with autism. This is not an easy process and is often slow, according to each child's individual abilities. According to Levy (1995), experiences such as tactile stimulation, identification of body parts and boundaries, and visual-kinesthetic awareness development, which is strengthened by having children alternate between moving and observing others move, could make children with autism begin to develop self-confidence and body awareness. Helen Payne (1990) states in her book *Creative Movement and Dance in Group Work* that through dance, one can release tension and be aided in self-expression and integration. "The creative act of moving alone, or with others, can enable an integration of mind, body and spirit" (Payne, 1990). Movement is at the core of our development and has a profound influence on the learning of speech, socially acceptable behavior and cognitive skills.

23.8.2.3 Sensorimotor Activities

"Sensory integration" is a process that can be used in movement therapy that provides additional opportunities to explore further the area of body-image formation. It includes full body movements, and the goal is to improve the way the brain processes and organizes sensations so that all the parts of the nervous system work together to activate a person to interact with the environment effectively and experience appropriate satisfaction. According to Levy (1995), the use of both integrative therapy and movement therapy can provide the child with autism with integrative, supportive, and effective treatment. Sensorimotor activities contribute to the building of a relationship as well as to development of body image and self-esteem for self-confidence. "The development of body image is enhanced through isolation and identification of body parts. Naming body parts as they are touched and moved provides a cognitive link to the physical actions" (Levy, 1995).

23.8.2.4 Using Music Together with Movement

Music is a very important element especially when working in a group. Music can promote a mood of calmness at the end of the session or whenever the therapist finds appropriate. It is found that moving with other people in the same music and rhythm usually helps build relationships. Children with autism are often described as being socially inept. When a child communicates through movement, it helps him/her to be more aware of himself/herself and more able to collaborate with others (Levy, 1995). The movement therapy for the special child deals fundamentally with sensory motor and perceptual motor development and integration. It helps the child build the body image and developing the self-concept. According to Levy (1995), singing can help children with autism in their language development, communication, and expression. On the same subject, Jordan and Powell (1995) claim that singing provides a more readily comprehended means of communication that may help understanding of communication itself and of the accompanying spoken language.

23.8.2.5 Using Props

Props can be used to promote the focus attention of the child. They can also be used as a bridge between the therapist and the child, to help the child feel more comfortable and without feeling threatened. Props can also create more synchronous and creative movements. Props have broader and more far-reaching positive effects because they involve the social, emotional, cognitive, as well as physical realms. Some examples of props are puppets, different kinds of balls (Nerf, balance, and therapy balls), large mats, hoops, several large geometric shapes, and pieces of fabric of varying sizes. Props like a large therapy ball can be used by the children to lie on their stomachs or their backs while the therapist gently places his/her back to support them, which makes them feel safe. "The touch is important in helping to define body boundaries and in establishing a relationship with each child" (Levy, 1995).

23.8.2.6 Dance and Movement Psychotherapy Using Play as an Intervention for Children with Autism

Movement therapy uses play as an intervention. According to Phillips and Beavan (2007), play in all of the definitions includes enjoyment as the key defining feature. Playing is equal to "having a good time." Research suggests that play is not just about having fun. There are early signs of play behavior in most children in the first months of their life and follow a development. In children with autism, the pattern of development in play is impaired or absent. Therefore, it is important to assist these children in developing patterns of play. Children with autism often find it difficult to engage in play, as play requires a degree of flexibility and creativity in their behavior and thinking. Children with autism tend to engage in more repetitive play, for example, following obsessive interests, with an emphasis on sensory stimulation (Phillips and Beavan, 2007). In free play, children with autism tend to watch others playing or they play alone, avoiding any social interactions. According to Philips and Beavan (2007), the benefits of play, which may be missed if a child with an ASD is unable to independently engage in a typical play behavior, result in enhanced cognitive, social, and emotional skills.

The child with autism needs to be taught to play and to extend his/her play routines. With increased practice, it is hoped that an effective play therapy approach would lead to independent and imaginative play by the child. Using play as a therapy can help the child with autism to support oneself in imitation, turn-taking, social interaction, and cognitive skills (Wall, 2004).

23.9 A CASE STUDY WHERE MOVEMENT THERAPY WAS USED AS A POSITIVE INTERVENTION FOR AUTISM

In the following paragraphs, a case study is described as one-to-one interaction with an individual patient. In this study, some incidents and experiences with patients will also be described and how dance movement psychotherapy has helped in the progress of the client. The client was a four-year-old boy who came to attend sessions at a private clinic in Cyprus and was diagnosed with autism from the age of three. For confidentiality purposes, his name will be referred to as patient no. 1. All rules of the institute review board were followed with this patient. Patient no. 1 had 12 sessions in total, once a week for three months. In the first assessment of patient no. 1, it was observed that he was not making eye contact with anyone and he was not using verbal communication or facial expressions. He came to the therapy clinic with no hesitation and sat on the floor without even looking at the new space he was in. There were various props on the floor like balls, scarves, balance ball, and small toys. One important aspect of dance movement psychotherapy is to "start where the patients are at." As there is such a wide spectrum, there isn't any one way for building connection with children with autism. This is why the "entry point," in order to build relationship, has to start from where the person is sitting. With this approach, one can start connecting with the world of the patient with autism, and with this way, one can start a form of communication in a less threatening way. Additionally, with patient no. 1, a nondirective approach was applied to see the way he was relating to and participating in interactions with new people. Also, with this approach, he could follow and use his movement preferences as a dialogue for getting to know others.

Communicating through the body and movement is the only universal language. It is universal because everybody communicates nonverbally whether one has ASD or not. It is found that patient no. 1 was mostly nonverbal; his attention was drawn toward objects and toys rather than on social interaction. He started his session by moving his body side by side and warming up with gentle music. He often became overwhelmed, agitated, or anxious with loud noises. Being doctors, we encouraged him to explore his body by moving it and at the same time naming the body parts while he was touching or moving. He was listening and naming the body parts while he was touching, moving them and waiting for others to name them.

In this way, patient no. 1 was helped during sessions to develop his body image and a sense of self to form a relationship with others. As a patient with ASD, he has started to develop a relationship with himself and he is more able to develop social interactions and realize the presence of people around him. It is important to note that during this clinical study, the main intention is to first understand the person with autism, to join in with him or her, and then to help modify his or her communication in a way that repetitive restrictive behaviors can become channeled, the nervous system could settle, and social engagement can begin. This is the main aim and starting point for dance movement psychotherapy.

During the 10th session, his movement developed into twisting his body and facing toward and away from others. This movement metaphor of wanting but at the same time rejecting the social communication and interaction was really meaningful and powerful for this patient therapy journey. Sometimes, when a hand was offered while twisting, patient no. 1 touched the offered hand, which then developed into play for him to twist away and then touching offered hands while looking at others. Later, he then verbally said "hello" while touching an offered hand. This was a powerful moment for everybody, which showed that meeting and following his language first as an initial communication led to a verbal expression. All of these developed from the previous spontaneous play and movement expression, which led into social interaction. Moving together gave him the sense of connections with others.

The journey of patient no. 1 through the dance movement psychotherapy has been really meaningful. His capacity for expression both verbally and nonverbally has increased. He is able to recognize, understand, and respond to other people. The therapy has helped him reduce social isolation and has also helped him build confidence and the capacity to participate in social activities in school, something that his teachers were surprised to observe. Patient no. 1 has still a unique way to communicate. However, he is now open to people to come into his world and understand and follow his way of communication. Dance movement psychotherapy is used both as a contact bridge and as a means for communication between the practitioner and the person with autism. More sessions would continue on an individual basis with patient no. 1, with the aim to help him continue his journey of developing self-awareness, learning, and understanding himself and at the same time becoming less isolated and withdrawn from people around him. Patient no. 1 is expected to create a better ability to communicate with caring people and they are able to become more understanding for his feelings and needs. The present study also helps patient no. 1's parents to better learn how to connect, join, and understand their child through the use of nonverbal language. This is a means of support in order for them to create an empathic, unconditional, and satisfying relationship with their child.

Dance movement psychotherapy is an example of an effective intervention working with people with autism. Through this study, it was shown that the impairments of children with autism described in Wing's Triad, which are social, language and communication, and behavior and imagination impairments, can be addressed by dance movement psychotherapy interventions. Dance Movement Psychotherapy can act as a bridge of communication and understanding and can also create a safe environment with the right tools for further development of communication, social interaction and self-awareness. It can also help in the general understanding of the behaviors of people with autism. It is very important to remember that movement is a universal language and therefore it is especially important when it comes to working with people with autism.

REFERENCES

Bland, J. 2014. *The Disease Delusion: Conquering Causes of Chronic Illness for a Healthier, Longer and Happier Life.* New York: Harper Collins Publisher.

Clements, J. Zarkowsa, E. 2000. *Behavioral Concerns & Autistic Spectrum Disorders.* London: Jessica Kingsley Publisher.

Delvinioti, E. 2011. Διατροφή και αυτισμός (Nutrition and Autism). Article http://www .mednutrition.gr/portal/ygeia/alles-pathiseis/4624-diatrofi-aftismos

Hardman, M., Drew, C., Egan, W., Wolf, B.A. 1984. *Human Exceptionality: Society, School, and Family, Fourth Edition.* Boston, MA: Allyn & Bacon.

Jordan, R., Powell, S. 1995. *Understanding and Teaching Children with Autism.* England: Wiley.

Levy, F. 1988. *Dance Movement Therapy: A Healing Art: American Alliance for Health, Physical Education, Recreation and Dance.* Reston, VA: National Dance.

Levy, F. 1995. *Dance and Other Expressive Art Therapies: When Words Are Not Enough.* New York: Routledge.

Mbalasides, P. 2015. Αυτισμός και Δίαιτα του αυτιστικού παιδιού (Autism and Diet of autistic children). Article. http://keadd.gr.άρθρα/αυτισμος-και-διατροφη-αυτιστικων -παιδιων/64html.

National Autistic Society. 2016. Early years and autism. http://www.autism.org.uk/about /family-life/grandparents/grandchildren.aspx

Nierengarten, M.B. 2014. *Contemporary Pediatrics.* Managing autism symptoms through nutrition. April 1, 2014. http://contemporarypediatrics.modernmedicine.com/contemporary -pediatrics/content/tags/autism/managing-autism-symptoms-through-nutrition?page=full

Payne, H. 1990. *Creative Movement and Dance in Group work.* London: Speechmark Publishing Ltd.

Phillips, N., Beavan, L. 2007. *Teaching Play to Children with Autism.* London: SAGE Publications.

Rogers, N. 2000. *The Creative Connection Expressive Arts as Healing.* Hereford, UK: PCCS Books.

Udell, B. 2010. Diet and Autism. WebMD Interview February 1, 2010 http://www.webmd .com/diet/diet-and-autism

Vacon, A. 2014. Top Nutrition Strategies for Autism. *Joyous Health*, July 24. http://www .joyoushealth.com/blog/2014/07/24/top-nutrition-strategies-for-autism

Wall, K. 2004. *Autism and Early Years Practice.* 2nd edition. London: Sage Publications Ltd.

Walsh, W. 2014. *Nutrient Power: Heal Your Biochemistry and Heal Your Brain.* New York: Skyhorse Publishing, Inc.

Index

Page numbers followed by f and t indicate figures and tables, respectively.